THE CELL NUCLEUS

Volume III

LIST OF CONTRIBUTORS

L. BOLUND	J. L. MANDEL
HARRIS BUSCH	ROSEMONDE MANDEVILLE
P. CHAMBON	ANTHONY R. MEANS
YONG C. CHOI	M. MEILHAC
R. K. CRAIG	ROSS N. NAZAR
G. P. GEORGIEV	MARK O. J. OLSON
F. GISSINGER	BERT W. O'MALLEY
URSULA I. HEINE	N. R. RINGERTZ
C. KEDINGER	TAE SUK RO-CHOI
H. M. KEIR	ANDRÉ ROYAL
LEROY KUEHL	RENÉ SIMARD
YVES LANGELIER	BERNARD S. STRAUSS
NICOLE MAESTRACCI	ANDRZEJ VORBRODT

EDITORIAL ADVISORY BOARD

THE CELL NUCLEUS

Volume III

EDITED BY

HARRIS BUSCH

Department of Pharmacology
Baylor College of Medicine
Texas Medical Center
Houston, Texas

ACADEMIC PRESS New York and London 1974

A Subsidiary of Harcourt Brace Jovanovich, Publishers

ACADEMIC PRESS, INC.
111 Fifth Avenue, New York, New York 10003

United Kingdom Edition published by
ACADEMIC PRESS, INC. (LONDON) LTD.
24/28 Oval Road, London NW1

Library of Congress Cataloging in Publication Data
Main entry under title:

The Cell nucleus.

Includes bibliographies.
1. Cell nuclei. I. Busch, Harris, ed.
[DNLM: 1. Cell nucleus. QH595 B977c]
QH595.C44 574.8'732 73-18944
ISBN 0-12-147603-0 (v. 3)

PRINTED IN THE UNITED STATES OF AMERICA

Contents

PART II NUCLEAR PROTEINS

List of Contributors

Numbers in parentheses indicate the pages on which the authors' contributions begin.

L. BOLUND (417), Institute for Medical Cell Research and Genetics, Medical Nobel Institute, Karolinska Institutet, Stockholm, Sweden

HARRIS BUSCH (109, 151, 211), Department of Pharmacology, Baylor College of Medicine, Texas Medical Center, Houston, Texas

P. CHAMBON (269), Institut de Chimie Biologique, Faculté de Médecine de Strasbourg, Strasbourg, France

YONG, C. CHOI (109), Department of Pharmacology, Baylor College of Medicine, Texas Medical Center, Houston, Texas

R. K. CRAIG (35), Department of Biochemistry, Marischal College, University of Aberdeen, Aberdeen, Scotland

G. P. GEORGIEV (67), Institute of Molecular Biology, Academy of Sciences of USSR, Moscow, USSR

F. GISSINGER (269), Institut de Chimie Biologique, Faculté de Médecine de Strasbourg, Strasbourg, France

URSULA I. HEINE (489), Virus Studies Section, Viral Oncology Area, Division of Cancer Cause and Prevention, National Cancer Institute, National Institutes of Health, Bethesda, Maryland

C. KEDINGER (269), Institut de Chimie Biologique, Faculté de Médecine de Strasbourg, Strasbourg, France

H. M. KEIR (35), Department of Biochemistry, Marischal College, University of Aberdeen, Aberdeen, Scotland

LEROY KUEHL (345), Department of Biological Chemistry, College of Medicine, University of Utah, Salt Lake City, Utah

YVES LANGELIER (447), Department de Biologie Cellulaire, Centre Hospitalier Universitaire, Université de Sherbrooke, Sherbrooke, Quebec, Canada

NICOLE MAESTRACCI (447), Department de Biologie Cellulaire, Centre Hospitalier Universitaire, Université de Sherbrooke, Sherbrooke, Quebec, Canada

J. L. MANDEL (269), Institut de Chimie Biologique, Faculté de Médecine de Strasbourg, Strasbourg, France

ROSEMONDE MANDEVILLE (447), Department de Biologie Cellulaire, Centre Hospitalier Universitaire, Université de Sherbrooke, Sherbrooke, Quebec, Canada

ANTHONY R. MEANS (379), Department of Cell Biology, Baylor College of Medicine, Texas Medical Center, Houston, Texas

M. MEILHAC (269), Institut de Chimie Biologique, Faculté de Médecine, de Strasbourg, Strasbourg, France

ROSS N. NAZAR (109), Department of Pharmacology, Baylor College of Medicine, Texas Medical Center, Houston, Texas

MARK O. J. OLSON (211), Department of Pharmacology, Baylor College of Medicine, Texas Medical Center, Houston, Texas

BERT W. O'MALLEY (379), Department of Cell Biology, Baylor College of Medicine, Texas Medical Center, Houston, Texas

N. R. RINGERTZ (417), Institute for Medical Cell Research and Genetics, Medical Nobel Institute, Karolinska Institutet, Stockholm, Sweden

TAE SUK RO-CHOI (151), Department of Pharmacology, Baylor College of Medicine, Texas Medical Center, Houston, Texas

ANDRÉ ROYAL (417), Department de Biologie Cellulaire, Centre Hospitalier Universitaire, Université de Sherbrooke, Sherbrooke, Quebec, Canada

RENÉ SIMARD (447), Department de Biologie Cellulaire, Centre Hospitalier Universitaire, Université de Sherbrooke, Sherbrooke, Quebec, Canada

BERNARD S. STRAUSS (3), Department of Microbiology, The University of Chicago, Chicago, Illinois

ANDRZEJ VORBRODT (309), Department of Tumor Biology, Institute of Oncology, Gliwice, Poland

Preface

Although the cell nucleus is such an integral part of cell function, it has not been the subject of an extensive review in recent years. There have been important monographs on this subject including a conference on "The Cell Nucleus" chaired by J. S. Mitchell (Academic Press, 1960) and a conference on "The Nucleus of the Cancer Cell" (H. Busch, ed., Academic Press, 1963). When the monograph on "The Nucleolus" (H. Busch and K. Smetana, Academic Press, 1970) was undertaken it became apparent that there were so many contributory fields to nuclear and nucleolar function that a thorough review of the subject would be worthwhile. This three-volume treatise is designed to provide such a work.

It should be clear to researchers and students of the cell nucleus that there is such an enormous gap between our present information and the complete understanding of nuclear composition and function that this work represents only a small portion of the knowledge still to be developed in this field. It was simply not possible to cover the complete literature. Interested readers are urged to consult primary sources or special reviews.

I am indebted to my many colleagues around the world who have contributed to the actual writing of this work and particularly to the editorial advisors for their many suggestions that have brought this work to fruition. In addition, we are grateful for the aid provided for much of the basic research from the National Institutes of Health, the National Cancer Institute, the American Cancer Society, the National Science Foundation, and the Welch Foundation.

The cell nucleus is so important to the basic understanding of biological and medical problems that it holds a fascination for one and all. It is hoped that this treatise will provide a useful guide for research and study of this very exciting area of human endeavor.

Harris Busch

Contents of Other Volumes

Volume I

Introduction

Why a cell nucleus? Although there is little doubt that the presence of a cell nucleus has permitted a great extension of the numbers of combinations and permutations of cellular phenotypes in both single and multicellular species, the origin of the cell nucleus is so ancient a part of evolution that it is uncertain what circumstances induced its origin and its development. What appear to be primitive "nuclear structures" have been found in both yeast and bacterial cells, but further evolutionary development produced a much more complex and functional structure in higher organisms. At present, it is not clear whether in its current state of development in the most specialized animal species, the cell has yet achieved its total potential for functionality. Some advantages of the cell nucleus may relate to a variety of characteristics of eukaryotic cells; for example, many eukaryotic cells are extremely long-lived and specifically differentiated, particularly cells of the central nervous system and the endocrine glands.

It seems reasonably clear that the development of a cell nucleus carried with it significant new chemical and physical properties of cells. Included among these are such obvious features as the nuclear envelope (nuclear membrane, or the bileaflet nuclear shell). This structure has three interfaces: one with the cytoplasm, another with the internal nuclear structure, and the third the space between the two layers of the nuclear envelope. This nuclear envelope not only serves as a geographic marker between the nuclear and cytoplasmic boundaries but in addition contains pores that give it more of a "Swiss cheese" or "Wiffle ball" appearance than a solid membrane between two heterogeneous masses. Through these pores migrate not only nuclear products that are "gene readouts" on their way to the cytoplasm but also the "cytonucleoproteins" and other elements that may serve as communication mechanisms between the cytoplasm and the nucleus. The role of hormone protein receptors in nuclear function is an exciting current chapter in mechanisms of gene activation.

By scanning microscopy, the cell nucleus resembles a ball studded with small bodies since it is covered with ribosomes and probably with polysomes. Although it is not certain that synthesis of the nuclear proteins occurs on the outer nuclear surface, it seems likely that they are formed either there or close by, and rapidly penetrate the nuclear mass.

Nuclear constituents. The presence of nuclear DNA which is almost all of the genetic complement of the cell is the key characteristic of the nucleus, but there are other structures that are specialized nuclear constituents. Among these are the histones, whose evolutionary origin seems to be very close to that of the nuclear envelope itself. Although the histones are now extremely well defined in terms of structure and number, their functions are shrouded in almost as much mystery as ever. There are so many histone molecules per nucleus (10^8) and they are so few in types that their role has been currently relegated to that of either structural support for DNA or as a general gene repressor system which can be activated by combination of the histones with "acidic nuclear proteins" or nonhistone nuclear proteins. In any event, their presence in association with DNA is sufficiently universal in nucleated cells and even in the chromosomes that the rule is "where there is DNA, there are histones."

The nucleus contains defined structural elements which seem to increase in number as technical advances increase in electron microscopy. The largest of these structures and the most universal is the "nucleolus" which contains an intense concentration of RNA and is now known to produce most of the total RNA of the cell, especially the rRNA species which are the backbones of the ribosome. Its role in the production of other types of RNA, such as mRNA, remains to be defined. The ultrastructure of the nucleolus varies markedly in various cell types but its responsiveness to the variations in cell function is both ordered and harmonious with the other events and requirements of the cell.

Among the other structural elements of the nucleus are the *interchromatin dense granules* that are probably parts of the processing elements of the nucleus; *perichromatin dense granules,* dark RNP particles surrounded by a light halo (by usual electron microscopic studies); *intranuclear rodlets;* and other structures, of which the juxtanucleolar channel system is one of the most intriguing. The functional roles of such "nucleus-specific" bodies are not defined, and manifest the very great requirement for research for understanding of the nuclear "government" of the cell.

The nucleus produces polysomes for export but retains for itself certain RNA molecules. Of these, the low molecular weight nuclear RNA species (LMWN RNA) are now being analyzed chemically, and the nucleotide sequences for three are defined. One of these, the U3 RNA, is "nucleolus specific." Others appear to be limited to the chromatin, and may exist juxtaposed to proteins in small RNP particles.

The nuclear proteins are composed of the histones, already noted above, many enzymes including the polymerases for RNA synthesis, structural proteins for ribosomal precursor elements, and other specialized processing elements of the nucleolus and nuclear nonhistone proteins (NHP) some of which may be "gene derepressor" proteins. Although the "gene derepressors" are clearly of enormous interest and objects of intensive research interest at present, it is only recently with the development of two-dimensional gel systems that the overall number of nuclear proteins has been approximated as several hundred. It is not yet clear which of these serve specific regulatory functions. It remains to be seen whether in individual chromosomes one or more of these nonhistone proteins (NHP) is specifically present. At present, methods for chromosome isolation seem to be improving to the point where it may be possible to ascertain whether any proteins have a special chromosome localization.

One of the more amazing aspects of the cell nucleus is the variety of changes that occurs during cell division. Not the least remarkable is the disappearance of the nuclear envelope. In metaphase there is the precise and equal separation of chromosomes of the daughter cells. It must be remembered, however, that there are other events accompanying metaphase that are of great importance and that all of the cellular components are distributed to the daughter cells approximately equally. Aspects of the formation of spindles and other nuclear elements are dealt with as specific topics in these volumes.

Although a definitive answer as to "why a cell nucleus" requires some consideration of its components, one may ask whether the functions subserved within this structure could not as well be served in a "nucleus-free" system? One may ask many other questions. Does the nuclear envelope protect delicate nuclear structures from enzymatic attack? Does the nuclear segregation provide for multilog specialization of function? Does the segregation of specific reactions for gene control and gene readout provide improved concentration of reactants and increased efficiency of these reactions? Does the nuclear envelope provide for penetrance of specific cellular elements into the nucleus? At present one can only speculate on these questions.

History of the cell nucleus. Improvements in light microscopy in the early nineteenth century permitted Robert Brown to discover the cell nucleus in 1830. The finding of "one nucleus per cell" led to the cell theory of Schleiden and Schwann in 1838. This concept provided a base for many developments including the understanding of Virchow that cells are all derived from pre-existing cells (*omnis cellula e cellula*) by extraordinary complex molecular events. The biological and clinical sequelae to the development of this concept have been truly astonishing in the last century and a half.

Definition of the nuclear contents emerged from development of staining methods and the improvements for isolation and analysis of nuclear products. After Miescher found DNA, the Dische stain established that DNA was largely localized to the nucleus in mammalian cells. By the use of appropriate staining techniques it was also found that the nucleus contained a nucleolus and, further, that the nucleolus contained vacuoles and nucleolini. With the Unna and other RNA stains, Brachet showed that RNA was concentrated in the nucleolus and cytoplasm. Development of microscopic spectrophotometry enabled Caspersson to show that the nucleolus is an island of RNA in a nuclear sea of DNA and histones.

Readily visualized by specific staining procedures, the chromosomes were observed in metaphase. Initially observed in 1873 by Butschlii, Flemming, Schneider, and others, they were named "chromosomes" by von Waldeyer–Hartz in 1888. Their separation into daughter cells was visible support for the concepts of Mendelian segregation. Chromosomal aberrations in special diseases and alterations in membranes and type of chromosomes in cell hybridization are topics of intensive current studies. Almost all of the elegant light microscopic studies on nucleoli that were beautifully reviewed by Montgomery were subjected to the criticism that staining procedures produce many artifacts. It remained for the development of light and electron microscopy to confirm and extend many features of the nucleus including the fascinating characteristics of the nucleolus and nuclear envelope. Not only were the characteristics of these structures defined by Bernhard, Swift, Smetana, and others but, in addition, important new structures were found that included nucleolar vacuoles, granular and fibrillar elements, perichromatin granules, interchromatinic granules, a variety of cytoplasmic invaginations, rodlets, and intranuclear tubular structures.

The preoccupation of biochemists with nuclear structures began in earnest after the finding of DNA by Miescher and the very rapid evolution of information of protamines and histones by Kossel, Lilienfeld, Mirsky, and others. After the Stedmans suggested that gene control might be exerted by nuclear proteins, an extensive series of investigations on nuclear proteins developed that continue with increasing excitement at present.

"The Cell Nucleus" is designed to mark the state of our understanding in the mid 1970's at a time when an enormous number of new and exciting developments are occurring in morphological, biochemical, and biological comprehension of nuclear function. While the nucleus is generally regarded as the "governor" of the cell, information is still accumulating on what it governs, how it governs, and the input that produces specific responses. Although our understanding is incomplete, the great enthusiasm in the field is well supported by its many accomplishments. The Tables of Contents of these volumes show the breadth of our current concepts and information.

Harris Busch

PART I

Nucleic Acids

1

Nuclear DNA

Bernard S. Strauss

Soothsayer: "In nature's infinite book of secrecy
A little I can read"

Antony and Cleopatra (I, 1)

At least four properties distinguish the DNA of eukaryotic organisms from that of prokaryotes: (*a*) the amount of DNA in eukaryotic cells is much greater than that found in bacteria and the increased amounts are greater than the increased complexity of the organisms seems to require; (*b*) the DNA of eukaryotic organisms is organized into separate chromosomes, each with a distinctive morphology and containing protein and RNA as well as DNA; (*c*) within each chromosome the DNA occurs in two forms, heterochromatin and euchromatin, characterized by different intensity of staining; and (*d*) whereas the DNA in prokaryotic organisms replicates from only a single origin, there are numerous, independent, replicating units within each chromosome and these replicating units, or replicons, have a characteristic time at which they start replication.

I. The Organization of Bacterial and Viral DNA

Eukaryotic DNA is best understood when compared with the organization of DNA in simpler organisms. For the most part, DNA exists in prokaryotic cells as a molecule of unique sequence with a circular structure or, if linear, with terminal repetitions. Viral DNA may exist as molecules of from 1.6×10^6 to 250×10^6 molecular weight. The molecules may be single-stranded circles (ϕX174), double-stranded circles (PB-1), or linear molecules with terminal repetitions (phage T7) and circularly permuted ends (phage T4). Phage DNA may contain single-stranded ends complementary to each other as in mature phage λ (MacHattie and Thomas, 1970). These special features of DNA structure are almost certainly related to the mode of virus replication. Circles and permuted ends may be devices to permit the replication of a complete gene set, since, as Watson (1972) has pointed out, it would be impossible to complete replication of a linear structure at the 3′ ends without redundancy. Bacteriophages also contain modified bases; certain viruses substitute hydroxymethylcytosine for cytosine, and uracil or hydroxymethyluracil for thymine, but these changes do not alter the codons and presumably function only to permit the enzymatic distinction of viral and host DNA.

Both bacterial and viral DNA are tightly coiled (Table I) but the nature of the coiling process and the question of whether the molecule is coiled in specific and reproducible ways is unanswered. If it were possible to produce filled virus heads *in vitro* this question could be answered. Although the morphogenesis of large phages can be studied *in vitro*, such experimental systems start with filled heads. None of the studies on bacteriophage T4 morphogenesis has resulted in the *de novo* formation of viral heads filled with DNA although several small RNA viruses have been completely reconstructed *in vitro* (Hohn and Hohn, 1970).*

In the smallest viruses all of the nucleic acid functions. An RNA virus of 1.2×10^6 molecular weight contains information for about three structural genes.† These may be identified as the RNA polymerase, the main

* Kaiser and Masuda have just (1973) reported the assembly of bacteriophage λ heads from λ DNA and head proteins.

† This value may be calculated as follows: Small viral RNA's have molecular weights of about 1.15×10^6 or about $1.15 \times 10^6/3.35 \times 10^2$ molecular weight units/

TABLE I
Coiling of DNA in Organisms and Organelles

Organism or organelle	Dimensions (μm)	Amount of DNA (MW units)	Length of DNA (μm)	Ratio: length of DNA/ length of organism or organelle
T2 bacteriophage	0.065×0.095 (head)	1.6×10^8	61.5	650
Escherichia coli	2.5×0.5	2.5×10^9	1350	675
Human metaphase chromosome				
Longest[a]	6.8 ± 1.4			
Average		6.4×10^{10}	34,000	5000
Shortest	1.36 ± 0.31			

[a] Data from Puck (1972).

structural protein of the capsid, and an organizing structural protein. Given these three gene products, a complete virus particle can be constructed. In more complex organisms, not all genes function continually. In complex viruses a program ensures the orderly transcription of genes and the separation in time between "early" and "late" messages (Colendar, 1970). Most bacterial DNA is not transcribed much of the time. Genes controlling the production of sporulation products, for example, do not function during exponential growth. McCarthy and Bolton (1964) claim that all bacterial DNA can be transcribed under certain conditions and is therefore informational. They demonstrated that the RNA extracted from *Escherichia coli* can be hybridized to half the possible DNA sequences. Since RNA is copied from only one of the DNA strands, essentially all of the *E. coli* genome produces RNA although the abundance of the different RNA species varies; less than 1% of the DNA sites are responsible for 20% of the message. More recently Kennell (1968) showed that 10% of *E. coli* DNA is complementary to more than 99.85% of the RNA made. The methods available make it impossible to tell whether the remaining sites are occasionally transcribed, but it is evident that some DNA may not be transcribed. It is not known whether all nontranscribed DNA serves a regulatory function, e.g., as an operator or promotor region.

---

nucleotide = 3.4×10^3 nucleotides. Assuming the average protein to have a subunit molecular weight of 45,000, there are about 333 amino acids or 1000 nucleotides in a gene. Each viral genome therefore has enough information to code for 3 or 4 different products.

Protists show a peculiar variation in the bases used to form codons. The relative molar proportion of guanine + cytosine (G + C) varies in the prokaryotic organisms from about 25% in some *Mycoplasma* to over 75% in *Micrococcus luteus* (Normore and Brown, 1970). Notwithstanding this great variation in the composition of the codons, the variation in protein amino acid composition throughout the prokaryotes is relatively slight (Sueoka, 1964), testifying to the degeneracy of the genetic code. The eukaryotic protists have as wide a variation as do the bacteria. *Tetrahymena* has a G + C composition of 25% whereas *Chlamydomonas angulosa* has a DNA with 68% G + C (Mandel, 1970). No such variation occurs in the higher eukaryotes (Shapiro, 1970) but this relative uniformity may merely reflect the difference in the amount of DNA in the two types of organisms. The higher eukaryotes may contain a variety of very heterogeneous molecules which give rise to an identical overall average composition. Furthermore, much of the DNA in multicellular organisms may be noninformational (see below). Therefore the average base composition in eukaryotes is not related to the composition of the codons.

Replication in prokaryotic organisms occurs from a single starting point and proceeds simultaneously along both strands of the Watson-Crick double helix but in two directions (Schnöss and Inman, 1970) (Fig. 1). This has been satisfactorily demonstrated in *E.coli* (Bird *et al,* 1972; Prescott and Kuempel, 1972) and has been deduced from the electron micrographs of replicating virus molecules (i.e., bacteriophage T7) in which "bubbles" are seen to extend for increasing distances, indicating a process which starts in the interior of a linear molecule and works out-

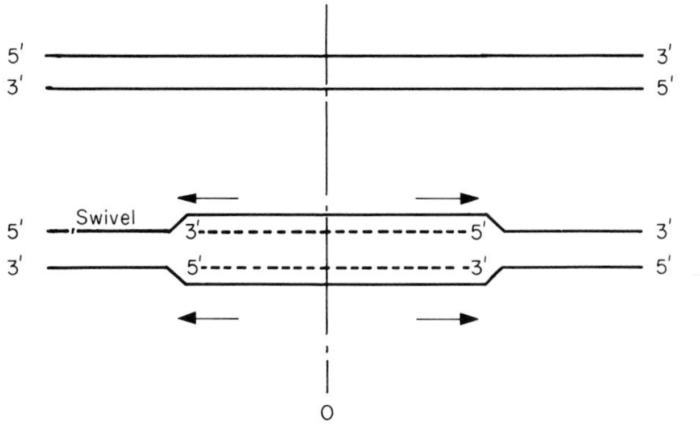

Fig. 1 Scheme of DNA replication showing replication in both directions and along both strands. The "swivel" required for untangling the strands is arbitrarily placed and indicated as a break. Arrows indicate the macroscopic direction of DNA synthesis. 0, origin of replication. The polarity of the chains is indicated by the designation as 3′ (OH) or 5′ (OH).

ward (Dressler *et al.*, 1972). There is generally only one replicating unit in the bacteria. Even when a second replicating unit exists (i.e., an RTF or F factor) some control mechanism generally keeps the number of copies of the plasmid replicons correlated with the number of copies of the chromosomal gene (Kasamatsu and Rownd, 1970; Clowes, 1972).

No enzyme is known to add nucleotides to the 5′ end of a polynucleotide chain and all replication appears to occur by reaction of 5′ nucleotide triphosphates with free 3′ OH groups on a primer strand (Goulian, 1971). Since replication occurs simultaneously along both strands, elongation along one of the strands is accomplished by synthesis "backwards" in the 5′ to 3′ direction for lengths of about 1000 nucleotides (Fig. 2). These short pieces are then joined together by DNA ligase (Okazaki *et al.*, 1968).

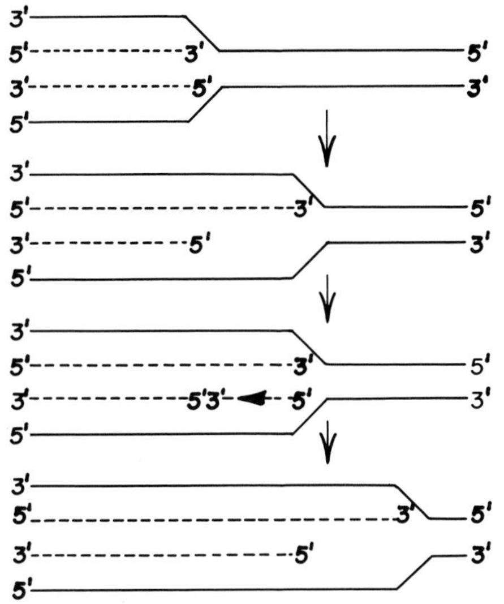

Fig. 2 A model for discontinuous synthesis of DNA. In this particular model chain elongation always occurs in the 5′ to 3′ direction and does not occur simultaneously on both strands at the growing point.

DNA polymerases can not initiate new strands but require a primer with a free 3′ OH group for chain elongation. RNA polymerases initiate polynucleotide chains and there is evidence that DNA is synthesized using an RNA chain of about 80 to 100 ribonucleotides as a primer (Sugino *et al.*, 1972; Brutlag *et al.*, 1971). The ribonucleotide segment is then excised and replaced with deoxynucleotides by an excision-type reaction. There are at least three different DNA polymerases in bacteria (Gass and Cozzarelli, 1973). DNA polymerase III lengthens the DNA chains. DNA polymerase I is most likely a repair enzyme but has a role in replication, particularly of some small plasmids and episomes.

II. The Nuclear DNA of Eukaryotic Organisms

A. The Amount of DNA in Cells

Theoretical considerations make it seem possible that a free-living organism could be constructed with less than 1000 functions; some of the smallest bacteria, the mycoplasmas, have only enough DNA to code for about this number of genes (Manloff and Morowitz, 1972). *Escherichia coli* has enough DNA to code for 4000 gene products (Table II) and there is a general increase in the amount of DNA per organism with increasing complexity (Rees and Jones, 1972). However, the progression is not uniform and the increases in the amount of DNA per organism seem out of proportion to the increase in complexity. Man, for example, has enough DNA in the haploid complement to code for approximately 2,000,000 functions. While one might argue that the human must be at least 500 times more complex than a colon bacillus, the distribution of the amount of DNA in related organisms is erratic. For example, the newt *Triturus* has enough DNA to code for 45,000,000 gene functions whereas the closely related frog, *Rana pipiens,* has only one sixth as much DNA (Table II).

Comparisons of organisms can also be made on the basis of the average amount of DNA per chromosome since related organisms can have different chromosome numbers. The smallest yeast chromosomes have amounts of DNA comparable to some of the larger DNA viruses (Table II). Some of the variation in the amount of DNA per cell in higher plants can be accounted for as a result of speciation due to polyploidy, the duplication of whole sets of chromosomes. However, the variation of chromosome number alone can not account for the variation in amounts of DNA since *Rana* with 13 chromosomes has less DNA than *Triturus* with 11. It does seem that there are real and unaccountable differences in the amounts of DNA in closely related organisms and that some "simpler" organisms may have more DNA than some complicated ones.

This variation in the amount of DNA is the basis for two related hypotheses. The first suggests that many of the genes may be present in repeated copies (not to be confused with the short, repeated sequences discussed below). The second hypothesis suggests that not all of the DNA carries genetic information, i.e., either codes for structural genes or is

TABLE II
DNA Content of Cells[a,b]

Organism	DNA per cell (pg)	DNA per chromosome set (pg)	Molecular weight of haploid DNA	Nucleotide pairs per haploid set	Length of total DNA (μm)	Chromosome number (haploid)	Average length per chromosome (μm)	Nucleotide pairs per chromosome	S Period	Required replication rate: μm/min/chromosome[c]
Escherichia coli	0.00422[d]	0.00422	2.5×10^9	3.8×10^6	1.35×10^3	1	1.35×10^3	3.8×10^6	41 min	33
Bacillus subtilis	0.0051	0.0051	3.1×10^9	4.7×10^6	1.63×10^3	1	1.63×10^3	4.7×10^6	55 min	30
Saccharomyces cerevisiae	0.0245	0.0245	1.48×10^{10}	2.26×10^7	7.8×10^3	17	4.6×10^2	1.3×10^6	30 min[e]	15
Drosophila melanogaster:	0.241	0.12	7.2×10^{10}	1.1×10^8	3.8×10^4	4	9.6×10^3	2.8×10^7	8–10 min[f]	960
Salivary gland	284		1.7×10^{14}	2.6×10^{11}	9.1×10^7	4(2000)	2.3×10^7	6.5×10^{10}	12–24 hr(16)[g]	24,000
Rana pipiens	15.6	7.8	4.7×10^{12}	7.2×10^9	2.5×10^6	13	1.9×10^5	5.5×10^8	48 hr[h]	66
Triturus viridescens	98	49	2.95×10^{13}	4.5×10^{10}	1.6×10^7	11	1.4×10^6	4.1×10^9		
Amphiuma	168	84	5.1×10^{13}	7.7×10^{10}	2.7×10^7					
Mouse (*Mus musculus*)	5.31	2.66	1.6×10^{12}	2.5×10^9	8.5×10^5	20	4.2×10^4	1.2×10^8	6 hr[i]	117
Rat (*Rattus norvegicus*)	6.05	3.03	1.8×10^{12}	2.8×10^9	9.6×10^5	21	4.6×10^4	1.3×10^8	7 hr[j]	110
Chinese hamster (*Cricetulus griseus*)	5.4	2.7	1.5×10^{12}	2.5×10^9	8.4×10^5	11	7.7×10^4	2.3×10^8	6.7 hr	192
Homo sapiens Lymphocyte	5.06	2.53	1.5×10^{12}	2.3×10^9	8.0×10^5	23	3.5×10^4	1×10^8	9.6 hr	61
Sperm	2.44	2.44	1.47×10^{12}	2.3×10^9	8.0×10^5					

[a] General references: amount of DNA per organism, Sober (1970); S period; Cleaver (1967); chromosome number; Altman and Dittmer (1964); Hsu and Benirschke (1968).

[b] The calculations are based on the following: average G + C content (7 eukaryotes); 41.4 mole % (Mandel, 1970); average molecular weight; 654/nucleotide pair; distance between bases; 3.4 Å. Therefore 1 μm = 1.92×10^6 molecular weight units; 1 μm = 2.94×10^3 base pairs.

[c] Calculated by dividing the average length per chromosome in μm by the S period in minutes. Assuming a replication rate of 1 μm/min/replicon (Huberman and Riggs, 1968), this equals the minimum number of replicons per eukaryotic chromosome assuming replication throughout S.

[d] Data from Cooper and Helmstetter (1968). Autoradiographic measurement (Cairns, 1963b) gives a length of 1100 μm.

[e] Average value for S in synthetic medium (M. Esposito, personal communication). Petes and Fangman (1972) give slightly different values for DNA content 8.4 to 12.0 $\times 10^9$ molecular weight units per haploid cell or about 4.9 to 7.1 $\times 10^8$ molecular weight units (3.1×10^2 μm) per chromosome.

[f] Oocytes.

[g] The calculation of replication rate per chromotid is based on 4 salivary chromosomes each resulting from the fusion of homologs and with a total of 2000 chromatids. An S value of about 16 hr is assumed.

[h] Epithelium at 30°C.

[i] Fibroblasts.

[j] Regenerating liver.

involved in regulation. According to this hypothesis much of the DNA of *Triturus* and of man has no function. Although we have a complex nervous system and manufacture antibodies it can be argued that these properties do not require millions of genes. Both hypotheses may be correct and some eukaryotic DNA is certainly not transcribed, but it is not possible to state with certainty that there is a class of "junk" DNA (Comings, 1972a).

B. The Length of DNA Molecules

The length or size of DNA molecules can be studied by centrifugation, by electron microscopy, and by autoradiography. However, none of these methods is completely satisfactory for eukaryotic DNA because of the large size of the molecule. If the DNA of a human chromosome were a single molecule, that molecule would most likely be sheared in the process of centrifugation. The average chromosome would yield a DNA filament of over 3 cm, wider than most of the centrifuge tubes used in such experiments. Centrifugation of a completely extended molecule would result in the formation of a fibrous network or aggregate and the molecules that adsorbed to the side of the tube would be subject to large shear forces. Certainly the equations derived for the study of DNA molecules would not apply to such a conglomerate (Lehmann and Ormerod, 1970; McBurney *et al.*, 1972; Ormerod and Lehmann, 1971).

Alkaline sucrose-gradient velocity sedimentation depends on the denaturation of DNA. This technique (McGrath and Williams, 1966) has been used to study DNA repair in eukaryotic cells by determination of the gradual elongation of molecular fragments. Such studies are particularly useful for exposing discontinuities in the DNA structure since even a single-strand DNA break, a single apurinic site, or a single ribonucleotide in the DNA causes a large change in sedimentation velocity because of the alkali lability of the bond. However, the number of cells lysed on the gradient, the time of incubation in alkali, and the composition of the lysing solution affect the sedimentation of DNA from untreated cells, making it difficult to interpret such experiments. Repair experiments have been most successfully performed by a technique which introduces a standard number of breaks into the DNA either immediately before lysis by X-ray treatment (Lehmann, 1972) or by digestion with alkali. Digestion of DNA in 0.25 N NaOH + 0.1 M EDTA at room temperature for 18 hr results in a stepwise degradation until a limit-digest sedimenting at about 165 S is obtained (Lett *et al.*, 1970). These results have prompted Lett and his co-workers to suppose that the DNA is composed of units separated by alkali-labile sites. DNA is degraded on exposure to alkali.

The unresolved question is whether alkali treatment exposes a naturally occurring discontinuity in the DNA or produces the discontinuity by a random chemical event. The problem is complicated because large numbers of alkali-labile sites are continually produced in DNA molecules. As Lindahl and Nyberg (1972) have pointed out, the spontaneous rate of depurination of human DNA at physiological temperature and pH is sufficient to produce about 10,000 apurinic sites in a human cell in 24 hr. In fact, there must be an active repair process since the DNA is not as alkali-sensitive as would be expected with 10^4 alkali-labile apurinic sites randomly scattered throughout the DNA. Some apurinic sites do remain which might account for the immediate *in vitro* alkali lability. The spontaneous production of apurinic sites indicates that the chromosome is not a continuous deoxyribonucleotide strand. It is not clear whether the breaks produced on incubation with alkali occur at rigidly predetermined sites or whether some are due to depurination. If RNA is involved in eukaryotic DNA synthesis, the continued presence of remnants of the ribonucleotide template might provide additional sites for attack by alkalai.

Thus, centrifugation methods cannot define whether the DNA in the chromosome is an uninterrupted double helix. Strands of DNA can be obtained which are as long as the DNA packed in a chromosome (Sasaki and Norman, 1966) but the autoradiographic evidence showing strands of over 2 cm in length does not eliminate the possible occurrence of single-strand interruptions in the DNA. Electron microscopy is still of only limited use because of the length of the molecule to be studied. The yeast chromosome is smaller (Table II) and therefore these objections need not apply to it. DNA of molecular weight 6.2×10^8 has been observed in sucrose gradients, indicating that in yeast there is a single DNA duplex per chromosome (Petes and Fangman, 1972).

C. *Repetitious DNA*

1. HETEROCHROMATIN

Cytologists observe DNA by staining it with particular dyes. One of the most important cytological distinctions of DNA into heterochromatin and euchromatin is based on its staining characteristics (see Volume I, Chapter 12). Dense staining heterochromatin can be found in both the interphase nucleus and in mitotic chromosomes. The distinction of heterochromatin and euchromatin has both functional and cytological significance (Rudkin, 1965a; Brown, 1966) as, for example, in the phenomenon of variegated position effect in which the function of a particular gene depends on its location in the chromosome next to heterochromatin or euchromatin (Baker, 1968).

Recent studies on chromosome staining with quinacrine dyes have made it possible to identify bands of fluorescent material characteristic of individual chromosomes (Caspersson *et al.*, 1970). The bands are heterochromatin (although not all heterochromatic regions need give fluorescence) and are due to the reaction of the quinacrine dye with AT-rich regions of the DNA (Weisblum and de Haseth, 1972). A similar banding pattern is observed with Giemsa staining (Fig. 3) after preliminary denaturation of the DNA *in situ* (Drets and Shaw, 1971; Schnedl, 1971, 1972).

Heterochromatin stains more deeply partly because of its base composition but also because the regions of heterochromatic staining contain DNA that is more tightly coiled (Brown, 1966). There is evidence that some of the chromosomal coiling may be related to actual changes in the structure of the constituent molecule, that is, to the change from the β-form in which DNA is generally found in solution and in crystals to the α-form in which the bases are tilted somewhat from the perpendicular to the axis of the phosphodiester chain (Comings, 1972a; Shih and Fasman, 1970).

2. REANNEALING EXPERIMENTS

Native DNA can be denatured by heating or by treatment with alkali or acid. When fully denatured the DNA strands separate. Denatured DNA can be renatured by long-term incubation. The renaturation or annealing process occurs most rapidly at a temperature about 25°C below the T_m for denaturation and is a second-order reaction (Wetmur and Davidson, 1968) dependent on the concentration of the denatured DNA.

Several experimental methods for the determination of the degree of renaturation are available. Since the speed of renaturation is affected by the size of the DNA fragments, DNA is sheared to a uniform size of about 400 nucleotides (1.3×10^5 molecular weight units) before denaturation and annealing. The earliest method for studying renaturation was based on the decrease in hyperchromicity at 260 nm on annealing. More recent methods are based on the ability of hydroxyapatite to distinguish double-stranded DNA fragments from purely single-stranded molecules (Bernardi, 1971) so that the precent remaining unannealed can now be operationally defined by the proportion of DNA which is eluted from hydroxyapatite at phosphate concentrations of 0.14 M (Britten and Kohne, 1968).

These methods do not require absolute base complementarity since

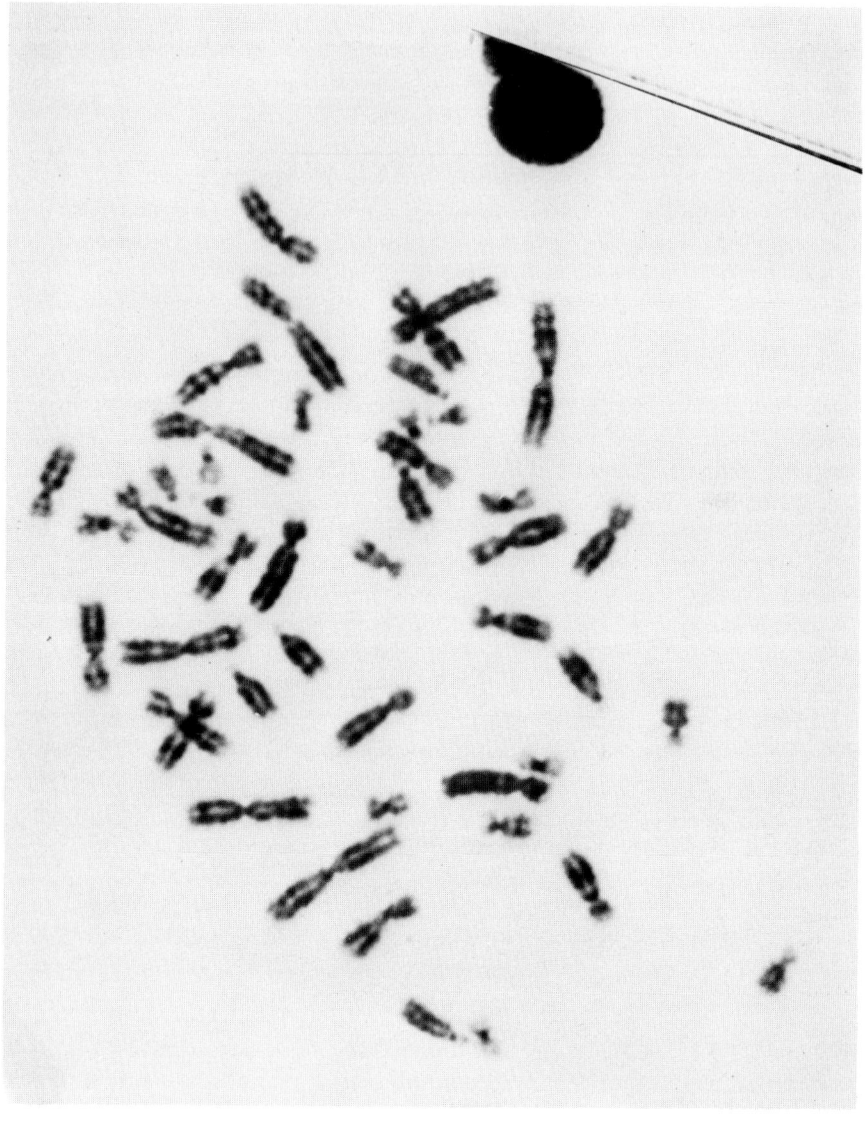

Fig. 3 Giemsa-stained, phytohemagglutinin-stimulated, human peripheral blood lymphocytes. (Preparation and photograph courtesy of Dr. Janet Rowley.)

renaturation requires only that the complementary bases of two strands be in register for long stretches. The length of the stretches required to make a stable helix is not precisely determined (see Thomas, 1966) but it appears that between 12 and 25 nucleotide pairs in register give a double helix. Noncomplementary bases which constitute less than 1% of the total do not affect the stability of the double helix and a structure of proportionally less stability results when noncomplementary bases make up 1–20% of the total (McCarthy and Church, 1970).

Renaturation is a second-order reaction which fits the relation

$$C/C_0 = 1/1 + k_2(C_0t)$$

where C_0 is the concentration of single-stranded DNA and C is the concentration after annealing for a time (Bostock, 1971). At half renaturation:

$$1/2 = 1/1 + k_2(C_0t);\ k_2C_0t = 1$$

The second-order rate constant, k_2, is found to be inversely proportional to the complexity of the genome (i.e., to the amount of DNA with unique sequences) and the value of C_0t at half renaturation is directly proportional to the complexity of the genome. Determinations of the degree of renaturation as a function of C_0t have provided a great deal of information about the structure of eukaryotic DNA (Fig. 4) (Siu *et al.*, 1972).

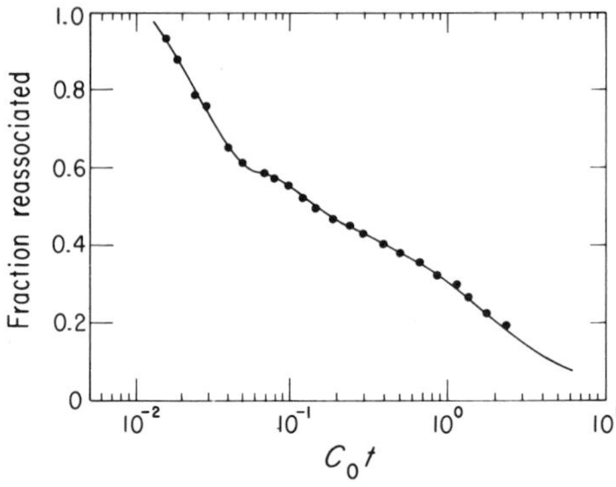

Fig. 4 Reassociation of *Polytoma obtusum* β-satellite DNA plotted as a function of C_0t. Renaturation in 1 × SSC at 60°. Satellite density = 1.683 gm/cc (Siu, Chiang, and Swift, unpublished data).

Denatured DNA has a greater absorbancy than does native DNA and an alternate way of determining genetic complexity and the rate constant is by a plot of the reciprocal of the hyperchromicity factor as a function of the time of annealing

$$1/A - A_\infty = 2.04 \times 10^{-4}k_2t + 1/0.36A_\infty$$

where A represents the absorbancy and A_∞ the absorbancy of native DNA (Wetmur and Davidson, 1968). Since this is a linear plot, the value of k_2 can be easily determined (Fig. 5).

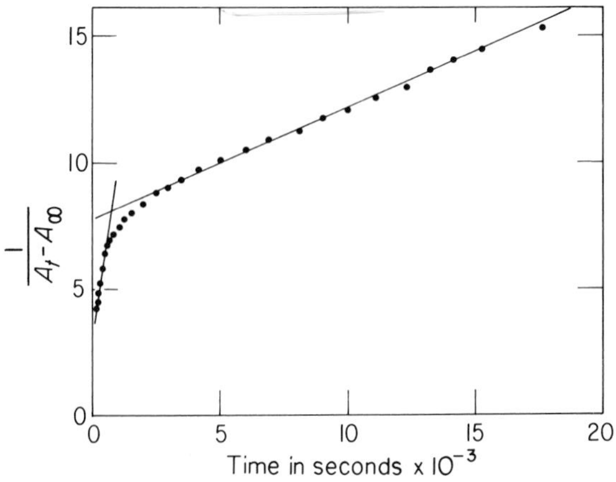

Fig. 5 Reassociation of *Polytoma obtusum* β-satellite DNA plotted by the method of Wetmur and Davidson (1968). The data are the same as used for the C_0t plot of Fig. 4. A = 0.641; $C_0 = 32.05/340 \times 10^3 = 0.942 \times 10^{-4}$ mole (Siu, Chiang, and Swift, unpublished data).

The kinetic complexity, measured as described above and defined as equal to the amount of DNA with unique sequences, is often compared with the analytical complexity, i.e., the haploid amount of DNA as determined by direct analysis. Although the size of the genome measured analytically equals the complexity calculated from the results of reannealing experiments for the DNA of bacteria and viruses (Wetmur and Davidson, 1968), this equality is not obtained with the DNA of higher organisms. In such organisms the amount of DNA determined analytically is greater than the genetic complexity measured kinetically. Therefore a portion of the DNA occurs in repeated sequences since, if certain fragments in the denatured DNA solution occur repeatedly, the chance increases that two complementary fragments meet. The reannealing rate of the repetitious sequences is therefore greater, resulting in a lower value for the kinetic complexity than for the unique sequences, for which only a single complement is possible.

Repeated sequences are stretches of nucleotides which form stable double helices on denaturation and reannealing and which recur in the DNA of an organism. These sequences must be long enough so that they recur necessarily and not by chance. If a nucleus contains about 10^9 base pairs and has only 4 base pairs to choose from, certain sequences will recur by chance if the order of bases is statistically random, as would be deduced from the random nature of the amino acid sequences in protein (Ycas, 1969). The repeated sequences must therefore be longer than about 25 nucleotides.* Repeated sequences vary in length from as few as 150 to well over 10^5 nucleotide pairs. The number of copies of each sequence may vary from about 50 to well over 10^6 and there is an inverse relationship between the number of copies and their complexity (Table III). The types of repeated sequences have been classified as follows

TABLE III
Repeated Sequences in DNA[a]

Species	Nuclear DNA (%) in repeated sequences	No. of copies	Complexity (nucleotide pairs)
Calf	38	66,000	1.7×10^4
	5	1×10^6	1.5×10^2
Sea urchin	20	50	3×10^6
	10	1,200	6×10^4
	3	14,000	1.6×10^3
Human	3	300	4×10^5
	15	40,000	10^4
	10	300,000	10^3
Mouse	25	10^3–10^4	—
	10	10^6(satellite)	3×10^2
Green monkey	20	1.5 ± 10^6	4.5×10^2
Guinea pig	5.5	2.2×10^3	8×10^4
	2.5	1.6×10^5	5×10^2
	2.5	5×10^5	1.5×10^2
Ilyanossa	12	2×10^1	1.7×10^7
	15	10^3	4.5×10^5

[a] Taken and adapted from Britten and Davidson (1971). These are the best estimates available in early 1971 and may be in error. In several cases the major intermediate frequency repetitive DNA is not listed.

(Walker, 1971; Bostock, 1971; Flamm, 1972): (1) Rapidly reassociating fragments repeated 10^5 times or more; (2) intermediate fragments, with C_0t values of 0.001–1 and repeated 10^2–10^5 times; and (3) unique, slow renaturing fragments, not repeated or repeated once. Rapidly reassociating DNA renatures at C_0t values of below 0.001 moles × sec/liter and unique DNA reassociates at a C_0t of about 100. The number of repeating fragments and their size is obtained in all cases from a quantitative analysis of the reannealing curves. The repetitions in the class of inter-

* It can be shown that a sequence of from 12 to 25 bases has a probability of almost one of recurring by chance in a sequence of 10^9 bases, whereas the probability of a sequence of 40 recurring by chance is practically zero.

mediate fragments need not be exact; instead there may be slight variations within each sequence which do not interfere with the formation of a stable, double-stranded helix. These intermediate sequences occur dispersed throughout the genome.

3. SATELLITE DNA

Some repetitive fragments that are present in large numbers can be distinguished from the bulk of the DNA by their characteristic density in equilibrium CsCl gradients. Depending on their base composition, such fragments centrifuge at a position somewhat removed from the bulk of the DNA and form satellite DNA (Fig. 6). In some cases the density difference is great enough so that a second band distinct from the bulk of the DNA is formed. Some organisms may produce more than one satellite band (Table III; Walker, 1971) which become evident only after special methods such as centrifugation in a heavy metal-containing Cs_2SO_4 gradient such as Ag^+–$CsSO_4$ or Hg^{2+}–$CsSO_4$ (Corneo *et al.*, 1970). Satellite DNA cannot be observed in all organisms; in fact, most discrete satellites have only been seen in, and isolated from, rodent tissues. However, organisms without satellite bands of grossly different density can have multiple repeated sequences whose average density does not lead to separation in CsCl.

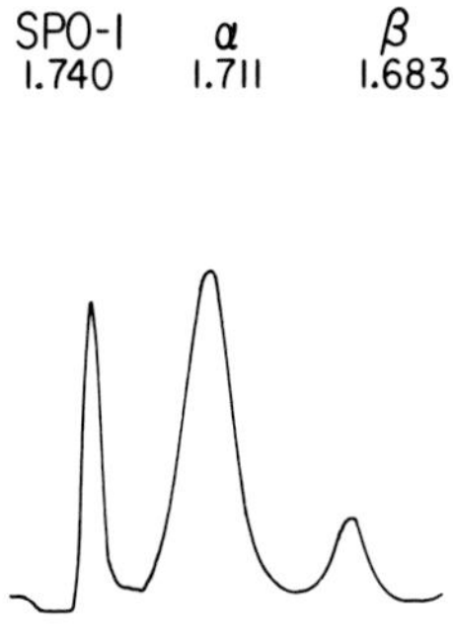

Fig. 6 Analytical CsCl density gradient centrifugation of *Polytoma obtusum* showing the β-satellite used in the experiments plotted in Fig. 4 and Fig. 5. SPO-1 bacteriophage DNA is used as a density marker. Densities are indicated on the figure; 6 μg native DNA was centrifuged (Siu, Chiang, and Swift, unpublished data).

This phenomenon and the observation of satellite DNA bands in preparations of bacterial DNA are difficult to distinguish. Bacteria sometimes contain independently replicating plasmids such as the F (fertility) factors, and the RTF (drug transfer) factors which can be transferred from organism to organism (Clowes, 1972). The F or RTF factors form a satellite band which consists of distinct, often circular DNA molecules and does not result from the shear degradation of DNA as in most eukaryotic satellites. Mitochondrial DNA also may form a satellite band which represents discrete, often circular molecules. Since shearing forms artificially induced fragments, the separation of chromosomal DNA into main and satellite bands is only a chemical coincidence.

Some satellite DNA is composed of relatively simple repeated sequences of bases. Southern (1970) has found that the α-satellite of the guinea pig consists of a basic sequence, CCCTAA, repeated 50 to 100 times with some substitution. The ratio of 50% G + C in this satellite is sufficiently different from the 42% G + C average of the main band to result in a 0.008 unit difference in density of CsCl centrifugation. Mouse satellite can be separated into heavy and light strands in CsCl after denaturation indicating a differentiation into purine-rich and pyrimidine-rich strands, but this is not true of all satellites. The origin and evolution of the satellites remains a matter for speculation (Britten and Davidson, 1971; Walker, 1971; Comings, 1972a).

4. CYTOLOGICAL LOCALIZATION OF REPEATED SEQUENCES

About 35% of the human genome occurs in repeated sequences (Saunders and Shirakawa, 1972). The question is: Where are they located? If some of the repetitions occur together in blocs, then it should be possible to visualize them by *in situ* hybridization with radioactive complementary RNA (cRNA) (Pardue and Gall, 1970; John *et al.*, 1969; Gall and Pardue, 1971). Radioactive DNA, or RNA complementary to DNA sequences (cRNA) is either isolated or prepared enzymologically. For example, the satellite DNA can be isolated and used as a template for the production of cRNA in a polymerase-catalyzed reaction. The ability of polymerase to transcribe such sequences *in vitro* does not relate to *in vivo* transcription. Cells with chromosomes in metaphase are squashed and treated with alkali to denature the DNA. The alkali is quickly neutralized and the section is incubated with radioactive, denatured DNA or RNA which hybridizes at the homologous chromosomal regions. Autoradiography results in an exposed emulsion in which silver grains cover the regions of homology. The method requires regions of repeated sequences and so far has been restricted to the localization of DNA complementary to ribosomal RNA, 5 S RNA, and histone messenger RNA. Price *et al.* (1972) have claimed autoradiographic localization of the genes involved in hemoglobin synthesis (but see Bishop and Jones, 1972).

The data on the localization of satellite DNA may be summarized as follows:

1. Satellite DNA is located at the constitutive heterochromatic regions around the centromere (Eckhart and Gall, 1971; Jones and Robertson, 1970; Arrighi *et al.*, 1970), but these regions do not consist exclusively of repetitious satellite sequences (Eckhart and Gall, 1971). The difficulty in preparing repetitive DNA sequences without contamination by unique sequences and the large amount of hybridization of centromeric heterochromatin with RNA complementary to main band DNA show that repetitive sequences alternate with unique sequences.
2. Satellite DNA is not only found at the centromere but also elsewhere in the chromosome. Eckhart and Gall (1971) report hybridization at the telomere regions of specific chromosomes of *Rhynchosciara* and Hennig *et al.* (1970) showed that satellite DNA can also occur throughout the chromosome.
3. There are AT- and GC-rich satellites as well as repetitive fragments with the same density as main band DNA (Comings and Mattoccia, 1972).
4. The hybridization techniques show satellite sequences to be species specific (Hennig *et al.*, 1970; Flamm *et al.*, 1969a).

It is unlikely that the highly repeated sequences code for protein because the base sequence would code for an unknown amino acid sequence. For example, guinea pig satellite is made up of fragments of about 150 nucleotide pairs with the repeated sequence (Southern, 1970):

L strand   5′-CCCTAA—
H strand   3′-GGGATT—

These fragments would code for 50 amino acids to give a repetitious peptide of 8000 molecular weight units. Flamm *et al.* (1969b) were unable to detect RNA complementary to mouse satellite DNA and concluded therefore that the satellite sequences are not transcribed.

III. Genetic Organization of Eukaryotic DNA

A. *Ribosomal DNA*

Proteins are synthesized on ribosomes and organisms have mechanisms to provide the necessary amount of ribosomal RNA. These mechanisms involve repetition of the genes coding for ribosomal RNA. The bacterium *Bacillus subtilis*, for example, has 6–8 such genes (Smith *et al.*, 1968). Eukaryotes may contain hundreds of such genes located at the nucleolar organizer region of the chromosome (Brown and Weber, 1968); the toad *Xenopus* has about 450 per haploid set. In addition special mechanisms provide the oocytes of amphibia and other animals with an extra supply of DNA coding for ribosomal genes (rDNA).

The nucleolus of oocytes is generated by "looping off" differentially replicated rDNA cistrons. It is not known whether the formation of the circular molecules containing hundreds of ribosomal cistrons (Miller and Beatty, 1969a) occurs by an actual excision, but at least the first extra replica is copied from the chromosomes rather than being present as an episomal element (Brown and Blackler, 1972). There is some evidence that an RNA–rDNA complex is an intermediate in the formation of rDNA copies, perhaps by operation of the reverse transcriptase (Brown and Tocchini-Valentini, 1972; Mahdavi and Crippa, 1972). Referred to as amplification, this phenomenon occurs at the pachytene stage of oogenesis (Gall, 1968). There is not a 1:1 correspondence between the number of ribosomal cistrons in the chromosome and the number looped off to form the nucleolus (Miller and Beatty, 1969b).

Amplification has been demonstrated in a variety of organisms (Gall, 1969). In amphibian oocytes 1500 nucleoli are spread around the periphery of the nucleus (Brown and Dawid, 1968). Some higher vertebrates and insects contain a single large nucleolus. Gene amplification appears to be a variant of gene replication and could involve the same sort of excision process as that which follows the induction of a lysogenic virus in bacteria.

The discovery of rDNA indicates that at least some genes recur in many copies. Furthermore, the detailed study of this DNA and of its RNA products by Dawid *et al.* (1970), by Miller and Beatty (1969a), and others shows that the ribosomal RNA genes have a special structure in which nonfunctional spacer, corresponding to about two-thirds of the base pairs required to code for ribosomal RNA, separates each gene in the ribosome.

In vertebrates, there may be anywhere from 100 to 600 rDNA cistrons per haploid genome. Although there are some reports of sequence heterogeneity in ribosomal RNA molecules (Kurland, 1972), the ribosomal RNA in an organism is identical for the most part.* In *Xenopus,* there are thousands of genes for the 5 S RNA that is also found in ribosomes but very few different base sequences. Kidney cells produce only one major sequence, ovaries produce an additional three (Ford and Southern, 1973). The limited number of sequences is unexpected because one might have assumed some evolutionary divergence. There is either some strong selection process which imposes uniformity or some mechanism makes all the DNA copies conform to a master pattern.

* There is good evidence for the heterogeneity of bacterial ribosomes themselves, based on the stoichiometry of their protein composition.

B. The Chromomere Hypothesis

Many insects have large chromosomes in the salivary glands which when stained show a distinct pattern. Such salivary gland chromosomes are polytene, and result from numerous replications in which the daughter chromosomes remain laterally aligned, possibly because the heterochromatic centromere regions fail to replicate (Rudkin, 1965a). Chromomeres in the individual chromosomal strands (the unineme), in which there is an aggregation of DNA, make the bands visible. The lateral apposition of the chromomeres in polytene chromosomes produces the specific and dramatic banding pattern. Occasionally, puffs emanate from some particular bands. Different bands may puff depending on the functional state of the organism. The puffs in sciarid flies are either DNA or RNA and transform that region from heterochromatin to euchromatin as a result of the onset of gene function (Pavan and da Cunha, 1969).

Each chromomere is the site of a single genetic unit. For example, all of the mutations of genes located in a restricted region of the X chromosome of *Drosophila* can be classified within a limited number of cistrons corresponding to particular chromomeric bands (Judd *et al.*, 1972). One expects therefore that each chromomere should contain just enough DNA for a single structural gene. However, each band contains too much DNA. Rudkin (1965b) estimates that the amount of DNA per band per single chromatid varies from a low of 5000 nucleotide pairs to an average of about 60,000 nucleotide pairs, that is, about 60 times the amount of DNA required to code for the average protein. When related organisms differ in the amount of DNA these differences can be traced to the relative amount of DNA in homologous chromomeres (see Thomas, 1971). Furthermore, the finding that the radiation-induced mutation rates of widely disparate species are the same *per unit of nuclear DNA* can be interpreted to mean that the size of a gene or complementation group is proportional to the total DNA content of the haploid genome (Abrahamson *et al.*, 1973).

Studies on the lampbrush chromosomes of amphibia give similar results (Callan, 1967). When treated with deoxyribonuclease the large lampbrush chromosomes behave as if their structural backbone were a single Watson-Crick double helix. A DNA loop is a part of each chromomere. As in *Drosophila*, loops and the chromomeres seem to contain identical and duplicated regions of DNA.

These discoveries can be explained by the hypothesis that structural genes in eukaryotic organisms occur in many copies in tandem repeats. Although this accounts for the cytochemical findings, it does not explain the phenomenon of point mutation, since one would not expect to find mutations other than large deletions in a gene made up of a series of tandem repeats. To resolve this paradox, Callan (1967) and Whitehouse (1967) postulated that any change in the first, or "master" gene was automatically transferred to the others by a process they called "rectification." According to this concept, changes in any gene other than the master would not be transmitted because of the rectification process. The identity of most ribosomal RNA molecules and the absence of any evolutionary divergence of the rDNA genes within a species are presumptive evidence of the efficacy of the rectification (Brown *et al.*, 1972). The different amount of DNA in the nuclei of related organisms with a similar chromosome number (i.e., *Rana pipiens,* 13 chromosomes, 15.6 pg DNA/cell and *Triturus viridescens,* 11 chromosomes, 98 pg DNA/cell) is accounted for by tandem duplication.

If this is correct, large numbers of repeated sequences should be obtained from sheared DNA. Treatment of these repeated sequences with exonuclease III or λ-exonuclease followed by denaturation and reannealing should give circles of DNA as a result of helix formation between different fragments of identical sequence. Such circles are obtained from salmon, trout, *Necturus,* and calf thymus (Thomas *et al.*, 1970) and from *Drosophila* (Thomas, 1971). The repetitious sequences which are responsible for the rings obtained from *Drosophila* DNA are thought to be clustered into short regions of about 5 μm and to be about equal in number to the number of salivary bands or chromomeres (Lee and Thomas, 1973). However, these results seem to be at odds with the finding that most of the haploid DNA sequences of *Drosophila* anneal as though they represent unique copies (Laird, 1971) although Lee and Thomas (1973) argue that the two types of data are not mutually exclusive.

According to the master gene concept, most of the DNA in the chromosome (chromomere) is structural and codes for protein. It has been shown, for example, that more than 12% of the nonrepeated sequences in the newborn mouse are transcribed (Gelderman *et al.*, 1971), indicating that the functional complexity is at least 4×10^8 nucleotide pairs or approximately 400,000 genes. An alternative model has been proposed by Crick (1971) based on the suggestions of Britten and Davidson (1969). Crick assumes that the structural genes are to be found in the interband regions and that the DNA in the chromomeric bands serves a regulatory function. Britten and Davidson (1969) suppose that most of the repetitious DNA in eukaryotic nuclei is involved in the regulation of gene activity. They argue that the number of functions in mammalian cells, for example, enzymatic activities, need not be too distinct from that required by the bacteria, and that the major difference lies in the more rigorous requirement for control to permit the precise expression of patterns of development. The presence of numerous reiterative sequences, spread throughout the genome, and occurring next to unique DNA, provides a mechanism for the common control of unlinked genes. Crick adds the suggestion that the coding sequences of the DNA are in the interbands. At present the two hypotheses appear to have equal merit. Another, but by no means alternative, suggestion offered by Walker (1971) proposes that a portion of the repetitive DNA plays a role in chromosomal "housekeeping" functions, for example, in folding and pairing.

IV. Replication of Eukaryotic DNA

A. *Multiple Replicating Sites*

The rate of DNA chain elongation can be measured by autoradiography (Cairns, 1966) and also by the time required for DNA molecules to become hybrid when incubated in bromodeoxyuridine (BUdR)-containing medium (Taylor, 1968). Both methods require that the intracellular pool of thymidine derivatives be low, or that corrections be made for this pool, and measurements made by both methods are in fair agreement. *Escherichia coli* DNA is about 1100 μm long (Cairns, 1963a) and the time taken for a complete round of replication as measured from the DNA content of synchronized cells is about 41 min (Cooper and Helmstetter, 1968), indicating a rate of synthesis of about 27 μm/min. Direct autoradiographic measurement gives a chain growth rate of about 30 μm/min/growing point (Cairns, 1963b), indicating that there need be only one growing point per chromosome even though recent experiments (see above) show that in some strains and under some conditions replication proceeds in two directions. Measurements with mammalian cells give very different rates. Huberman and Riggs (1968) have measured the rate of growth of the DNA in Chinese hamster cells in culture as 0.5–1.2 μm/min/growing point, in agreement with the results of Taylor (1968), who came to a figure of 1–2 μm/min as a result of BUdR density transfer experiments, and of Weintraub (1972a), who calculated a chain elongation rate of 70 base pairs/sec/piece of 40×10^6.

The amount of DNA in the haploid complement of the Chinese hamster is about 3.2 pg, equivalent to 9.4×10^5 μm in length (see Table II). Since there are 11 chromosome pairs, the average chromosome contains about 8.5×10^4 μm of DNA. At the measured rate of 1.2 μm/min it would take 1183 hr to replicate the DNA in a chromosome. However, measurements of the length of the S period in CHO Chinese hamster cells (Puck *et al.*, 1964) indicate that the S period is only 4.1 hr long. Therefore, the only way in which the DNA can be replicated is to have many separate growing points per chromosome. A minimum estimate of the number necessary is given as 1183 hr/4.1 hr = 288 growing points. Huberman and Riggs (1968) have determined that there are two growing points per replicon (see below) and therefore there must be a minimum of 144 replicons, each $8.5 \times 10^4/144 = 590$ μm long. However, the size of the actual replicating sections is only 30 μm long (*loc. cit.*). This means that there are more replicons than the minimal estimate and that each replicon takes only $30/1.2 \times 2$ growing points = 12.5 min to complete synthesis. Therefore, each replicon is active for only $12.5/4.1 \times = 0.05$ of the S period. A replicon length of 30 μm corresponds to a molecular weight of about 6×10^7 or about 90,000 base pairs and is much smaller than the bacterial replicon, for example, the *E. coli* chromosome of about 3.2×10^6 base pairs. Not only is this replicating unit smaller, but the rate of synthesis at each replicating point is slower, 27 μm/min for *E. coli*, 1.2 μm/min for the hamster cell. These calculations assume a constant replication rate and replicon size which is an oversimplification, since the length of the replicon may differ in different cells of the same organism (Callan, 1972).

B. Bidirectional Replication

Autoradiographs of cells incubated with [^{3}H]thymidine show a pattern of dense label trailing off at both ends. This pattern is best explained by supposing that replication proceeds in two directions from a single point of origin (Huberman and Riggs, 1968). As in bacteria, both chains are synthesized at about the same time in each direction. Therefore, the elongation of chains terminating with 5′ groups also occurs in eukaryotic organisms by the synthesis of Okazaki pieces (see Goulian, 1971 for references). Huberman and Riggs' (1968) conclusions have been confirmed by Callan (1972) and by a method based on the radiation sensitivity of BUdR (Weintraub, 1972b). If replication proceeds outward in two directions, and if the middle is radiation sensitive as a result of the incorporation by synchronized cells of BUdR, the molecular weight of the synthesized DNA will be halved by radiation. If, however, replication proceeds from one direction, irradiation of DNA synthesized according to the same protocol will have little effect on molecular size. The data obtained by this method indicate that replication occurs in two directions.

A diagram showing the minimal features of DNA replication in eukaryotes (Fig. 7) has the following characteristics: (*a*) portions of the DNA are duplicated while others are still in the original double-stranded state; (*b*) replication in two directions starting from the interior of a DNA molecule involves some break in the DNA to provide a swivel for the

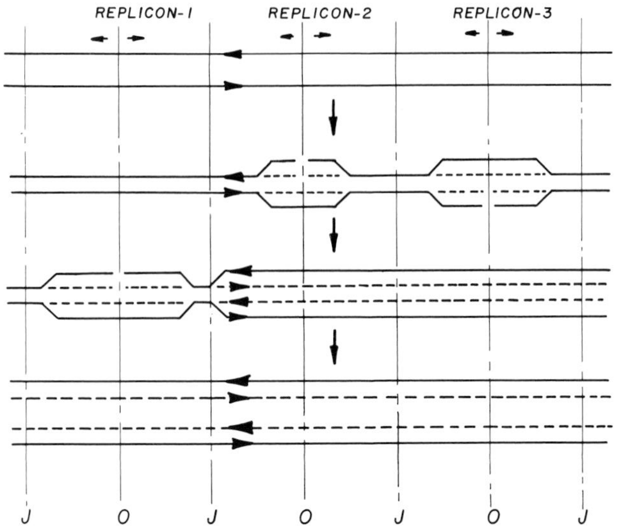

Fig. 7 Scheme of replication in eukaryotic chromosomes. Arrows indicate the direction ←→ or polarity—>— of DNA synthesis; o, origin of replicon; j, juncture of two replicons. The swivel in this diagram is arbitrarily placed at the origin.

unwinding and rewinding that is a part of replication; and (*c*) if there are fixed termination points for each replicon [some of the autoradiograms have the distribution of grains expected on this hypothesis (Huberman and Riggs, 1968)], then there must be "stationary growing points" at the junction of an already completed replicon and one not yet started or in the process of growth. Such growing points have regions of single-stranded DNA (Scudiero and Strauss, in press).

Single-stranded regions have been observed in both nuclear and mitochondrial DNA. Vinograd and his co-workers (Robberson *et al.*, 1972) have observed that the DNA of the mitochondria replicates first along one of the two strands. The resulting replicative intermediate forms what they (*loc. cit.*) have called a "D" loop made up of a double-stranded bar with a single-stranded loop. Single-stranded regions have also been observed in replicating nuclear DNA. Painter and Schaeffer (1969) found small single-stranded pieces of less than 2000 nucleotides in replicating HeLa cells. Similar pieces were observed by Scudiero and Strauss (1974), who found an accumulation of single-stranded DNA of about 1300 nucleotides in cells treated with methyl methanesulfonate. Although replication occurs simultaneously along both strands in a macrosense, it may be out of phase and discontinuous for regions of about 1 μm as in bacteriophage T7 replication (Dressler *et al.*, 1972). If there is a delay in the formation of Okazaki pieces synthesized in the "backward" direction, short, single-stranded pieces would result.

C. Late Replicating DNA

The base composition of newly synthesized DNA can be determined by comparing the density of the newly synthesized DNA to that of the bulk DNA in synchronized cells (Bostock *et al.*, 1972; Comings, 1972a, b; Tobia *et al.*, 1970). For the most part, it appears that regions rich in G + C content may replicate early in S whereas the A + T-rich regions replicate later. These observations reflect the late replication of relatively AT-rich satellite DNA associated with heterochromatin. However, not all satellites are AT-rich (Comings and Mattoccia, 1972) and much of the repetitious DNA is arranged interspersed between unique sequences.

The time of replication may be determined by two factors: (1) late replication of heterochromatin unrelated to base composition (Lima-de-Faria and Jaworska, 1968; Comings, 1972b); for example, as pointed out by Comings (1972a), the randomly inactivated X chromosome (Lyon, 1968) has the same base composition as the active X, yet one replicates much later than the other; (2) the base composition of those regions at which replication occurs first. Although G + C-rich regions are generally supposed to replicate first (Bostock *et al.*, 1972; Tobia *et al.*, 1970), both Comings (1972b) and Taylor *et al.* (1970) have suggested that replication may start at A + T-rich initiator sites and Comings (1972b) has evidence that the very earliest regions to replicate are A + T rich. Comings points out (*loc. cit.*) that quail heterochromatin is both late replicating and G + C rich and he supposes therefore that the late replication of A + T-rich regions is due to the composition of the heterochromatin in the particular species. At present we cannot describe the detailed program for DNA replication in any eukaryote.

D. The Control of DNA Replication

DNA synthesis in prokaryotes seems to be controlled by an initiator protein which, by combination with DNA at some replicator site, initiates DNA synthesis at a membrane (Jacob *et al.*, 1963). In eukaryotes DNA synthesis has been reported to initiate on the nuclear membrane (Comings and Kakefuda, 1968; Milner, 1969; Friedman and Mueller, 1969) but Huberman *et al.* (1973) show that it is heterochromatin which condenses on the nuclear membrane while euchromatin replication occurs throughout the nucleus.

The eukaryotic cell divides according to a rigorously programed pattern. DNA synthesis (S) is preceded by a phase (G_1) of synthetic activity and is followed by a pause (G_2) before mitosis (M), which is closely linked to cell division (see Puck, 1972). The stages G_1–M are reflected in the composition of the cell surface (Fox *et al.*, 1971). In many cells and tissues, reactions at the surface provide the signals for division. Tumor viruses, when integrated into the cell, may upset the control mechanism, change the surface, and keep the cells dividing (Eckhart *et al.*, 1971).

We do not know what causes eukaryotic cells to enter the S period but both the initiation and maintenance of DNA synthesis requires protein synthesis and a cytoplasmic factor. Cell fusion experiments indicate that the signal for DNA synthesis is cytoplasmic since the fusion of dividing and nondividing cells results in DNA synthesis in the previously inactive nucleus (Harris, 1970).

Inhibitors of protein synthesis can inhibit both the initiation and continuation of DNA synthesis (Mueller, 1969; Rusch, 1969; Weiss, 1969; Terasima and Yasukawa, 1966), unlike the prokaryotic organisms in which once started, a "round" of DNA synthesis continues until completion. Protein synthesis might be required during the S period for at least two reasons. First, since all replicons are not turned on simultaneously, continued synthesis of initiator proteins is necessary to maintain the temporal order of DNA synthesis (Rusch, 1969). Second, histone must be synthesized to neutralize the charge on the DNA. Histones are (mostly) made in the cytoplasm (Robbins and Borun, 1967; Borun *et al.*, 1967) and there is coupling between the amount of DNA and that of histone synthesized through a short-lived intermediate which may not be message (Gallwitz and Mueller, 1969; Mueller, 1969; and see Sagopal and Bonner, 1969). Proteins other than histones probably control the synthesis of replicons which are individually controlled as shown by the differential replication in rDNA amplification and in the formation of polytene chromosomes (Watson, 1971).

E. Repair Reactions

DNA synthesis involves many of the same enzymes utilized for repair reactions of damaged DNA (see Goulian, 1971). Repair systems are present in eukaryotes and appear to be very similar to their counterparts in bacteria (Altmann, 1972). The system of excision repair occurs also in animal cells (Regan *et al.*, 1971) and a human mutation prevents cells from excising ultraviolet-induced damage. Individuals carrying this mutation in homozygous form suffer from the serious disease, xeroderma pigmentosum (Cleaver, 1969). A UV-endonuclease has been reported in mammalian cells (Brent, 1972) and has many of the properties of the bacterial enzyme. Repair synthesis can be demonstrated in eukaryotic cells after treatment with ultraviolet light, X-rays (Painter and Young, 1972), or alkylating agents (Coyle *et al.*, 1971; Ayad *et al.*, 1969). The number of bases replaced by repair synthesis averages about 14–25 after UV-irradiation (Regan *et al.*, 1971) as compared to 3–4 after X-irradiation of human cells (Painter and Young, 1972). The major difference between the repair systems of bacteria and of the eukaryotes seems to be a limitation in the number of pyrimidine dimers which can be excised (Regan *et al.*, 1968) and the failure of eukaryotic cells to degrade DNA to acid-soluble fragments (Painter and Young, 1972).

The studies to date ignore the question of whether repair occurs equally in different regions of the chromosome or whether A + T-rich or heterochromatic regions are differentially affected. Chromosome aberrations do occur more frequently in heterochromatic regions following treatment with ionizing radiation (Natarajan and Ahnström, 1970) and alkylating agents (Natarajan and Upadhya, 1964; Rao and Natarajan 1967). Flamm *et al.* (1969b) have shown that the replication of mouse satellite DNA is selectively inhibited by bifunctional but not by monofunctional alkylating agents.

Although the UV-survival curves of certain lines of mouse or hamster cells indicate biologically significant recovery processes, UV-induced pyrimidine dimers are not removed (Regan *et al.*, 1968).* These cells probably recover as a result of postreplication repair (Rupp *et al.*, 1971). Postreplication repair in eukaryotic cells may not involve recombination (Lehman, 1972). Xeroderma pigmentosum cells which are unable to excise damage nonetheless tolerate a limited number of pyrimidine dimers. DNA synthesis in such cells requires a "bypass" mechanism (Buhl *et al.*, 1972) which may be "error prone" (Witkin and George, 1973). Such error prone mechanisms are associated with mutation and perhaps, by extension, with carcinogenesis.

In bacteria, an unrepaired break or scission is lethal and stops DNA synthesis (see Strauss, 1968) whereas in eukaryotic organisms replication of DNA with single-strand breaks leads to the formation of chromatid fragments (Evans and Scott, 1969). Chromosome fragments may replicate and be lost on subsequent division, which parallels the observation that the DNA formed by damaged cells does not replicate (Myers and Strauss, 1971; Coyle *et al.*, 1971). This provides additional evidence for the reproductive independence of the sections of the chromosome, notwithstanding their final unification into a single molecule.

* Repair synthesis can be detected in such cells, especially when DNA containing bromuracil is irradiated.

V. Conclusions and Summary

Ten years ago as a result of the major advances in molecular biology, particularly the elucidation of the genetic code (Crick, 1966) and the confirmation of the colinearity of gene, nucleotide sequence, and protein structure (Yanofsky *et al.*, 1966), it seemed evident that most DNA coded for gene products and that biological functions could be deduced from the sequence of bases in DNA. This concept continues to hold for the microorganisms. However, even ten years ago it was clear that the eukaryotic nucleus contained much more DNA than necessary to code for the required number of functions (Strauss, 1964). Today the problem has been further complicated by the observations that eukaryotic DNA contains numerous small, repeated sequences, that certain genes may be repeated and amplified, and that extensive sequences of the DNA may not be transcribed but may serve as "spacer" regions.

Ten years ago it was thought that DNA synthesis was the result of the *in vivo* operation of DNA polymerase I. Today we are less certain that we understand DNA replication *in vivo* in spite of the isolation of several DNA polymerases and a variety of proteins associated with replication. Were it not for the very large difference in replication rates, there would seem to be no reason to distinguish between eukaryotic and prokaryotic DNA synthesis at the replication fork. We are now aware of the presence of numerous replicons in the DNA of the eukaryote and of the difference in the timing of the replication of euchromatic and heterochromatic regions, but we do not know how the different replicons are independently controlled nor how the replication pattern is determined.

I consider that there are two fundamental problems which remain unsolved. The first is the nature of the three dimensional structure of the DNA in the interphase nucleus and in the metaphase chromosome. The second and perhaps most interesting problem is still the apparent excess of DNA in the eukaryotic nucleus. We know that DNA is the genetic material of the prokaryotes and that most of it has a function. All the evidence with eukaryotic systems *in vitro* convinces us that DNA is genetic material for eukaryotes as well. However, notwithstanding our anthropomorphic conviction that we are, if not infinitely, at least much more complex than the bacteria, it seems possible that much of the DNA in vertebrates and in higher plants has no informational role. But if that is so, why has this DNA persisted throughout evolutionary development?

REFERENCES

Abrahamson, S., Bender, M., Conger, A., and Wolff, S. (1973). *Nature* **245,** 460.

Altmann, H. (ed.) (1972). "DNA-Repair Mechanisms." Schattauer Verlag, Stuttgart.

Altman, P., and Dittmer, D. (1964). "Biological Data Book." Fed. Amer. Soc. Exp. Biol., Washington D.C.

Arrighi, F., Hsu, T., Saunders, P., and Saunders, G. (1970). *Chromosoma* **32,** 224.

Ayad, S., Fox, M., and Fox, B. (1969). *Mutat. Res.* **8,** 639.

Baker, W. (1968). *Advan. Genet.* **14,** 133.

Bernardi, G. (1971). *In* "Methods in Enzymology" (L. Grossman and K. Moldave, eds.), Vol. 21, pp. 95–139, Academic Press, New York.

Bird, R., Louarn, J., Martuscelli, J., and Caro, L. (1972). *J. Mol. Biol.* **70,** 549.

Bishop, J., and Jones, K. (1972). *Nature (London) New Biol.* **240,** 149.

Borun, T., Scharff, M., and Robbins, E. (1967). *Proc. Nat. Acad. Sci. U.S.* **58,** 1977.

Bostock, C. (1971). *Advan. Cell Biol.* **2,** 153.

Bostock, E., Prescott, D., and Hatch, F. (1972). *Exp. Cell Res.* **74,** 487.

Brent, T. (1972). *Nature (London) New Biol.* **239,** 172.

Britten, R., and Davidson, E. (1969). *Science* **165,** 349.

Britten, R., and Davidson, E. (1971). *Quart. Rev. Biol.* **46,** 111.

Britten, R., and Kohne, D. (1968). *Science* **161,** 529.

Brown, D., and Blackler, A. (1972). *J. Mol. Biol.* **63,** 75.

Brown, D., and Dawid, I. (1968). *Science* **160,** 272.

Brown, D., and Weber, C. (1968). *J. Mol. Biol.* **34,** 661.

Brown, D., Wensink, P., and Jordan, E. (1972). *J. Mol. Biol.* **63,** 57.

Brown, R., and Tocchini-Valentini, G. (1972). *Proc. Nat. Acad. Sci. U.S.* **69,** 1746.

Brown, S. (1966). *Science* **151,** 417.

Brutlag, D., Schekman, R., and Kornberg, A. (1971). *Proc. Nat. Acad. Sci. U.S.* **68,** 2826.

Buhl, S., Stillman, R., Setlow, R., and Regan, J. (1972). *Biophys. J.* **12,** 1183.

Cairns, J. (1963a). *Cold Spring Harbor Symp. Quant. Biol.* **28,** 43.

Cairns, J. (1963b). *J. Mol. Biol.* **6,** 208.

Cairns, J. (1966). *J. Mol. Biol.* **15,** 372.

Callan, H. (1967). *J. Cell Sci.* **2,** 1.

Callan, H. (1972). *Proc. Roy. Soc. London B* **181,** 19.

Caspersson, T., Zech, L., Johansson, C., and Modest, E. (1970). *Chromosoma* **30,** 215.

Cleaver, J. (1967). "Thymidine Metabolism and Cell Kinetics." North-Holland Publ., Amsterdam.

Cleaver, J. (1969). *Proc. Nat. Acad. Sci. U.S.* **63,** 428.

Clowes, R. (1972). *Bacteriol. Rev.* **36,** 361.

Colendar, R. (1970). *Ann. Rev. Microbiol.* **24,** 241.

Comings, D. (1972a). *Advan. Hum. Genet.* **3,** 237.

Comings, D. (1972b). *Exp. Cell Res.* **71,** 106.

Comings, D., and Kakefuda, T. (1968). *J. Mol. Biol.* **33,** 225.

Comings, D., and Mattoccia, E. (1972). *Exp. Cell Res.* **71,** 113.

Cooper, S., and Helmstetter, C. (1968). *J. Mol. Biol.* **31,** 519.

Corneo, G., Ginelli, E., and Polli, E. (1970). *J. Mol. Biol.* **48,** 319.

Coyle, M., McMahon, M., and Strauss, B. (1971). *Mutat. Res.* **12,** 427.

Crick, F. (1966). *Cold Spring Harbor Symp. Quant. Biol.* **31,** 3.

Crick, F. (1971). *Nature (London)* **234,** 25.

Dawid, I., Brown, D., and Reeder, R. (1970). *J. Mol. Biol.* **51,** 341.

Dressler, D., Wolfson, J., and Magazin, M. (1972). *Proc. Nat. Acad. Sci. U.S.* **69,** 998.

Drets, M., and Shaw, M. (1971). *Proc. Nat. Acad. Sci. U.S.* **68**, 2073.
Eckhardt, R., and Gall, J. (1971). *Chromosoma* **32**, 407.
Eckhart, W., Dulbecco, R., and Burger, M. (1971). *Proc. Nat. Acad. Sci. U.S.* **68**, 283.
Evans, H., and Scott, D. (1969). *Proc. Roy. Soc. London B* **173**, 491.
Flamm, W. (1972). *Int. Rev. Cytol.* **32**, 1.
Flamm, W., Bernheim, N., and Spalding, J. (1969a). *Biochim. Biophys. Acta* **195**, 273.
Flamm, W., Walker, P., and McCallum, M. (1969b). *J. Mol. Biol.* **40**, 423.
Ford, P., and Southern, E. (1973). *Nature* (*London*) *New Biol.* **241**, 7.
Fox, T., Sheppard, J., and Burger, M. (1971). *Proc. Nat. Acad. Sci. U.S.* **68**, 244.
Friedman, D., and Mueller, G. (1969). *Biochim. Biophys. Acta* **174**, 253.
Gall, J. (1968). *Proc. Nat. Acad. Sci. U.S.* **60**, 553.
Gall, J. (1969). *Genetics Suppl.* **61,1**, 121.
Gall, J., and Pardue, M. (1971). *Methods Enzymol.* **21**, 470.
Gallwitz, D., and Mueller, G. (1969). *J. Biol. Chem.* **244**, 5947.
Gass, K., and Cozzarelli, N. *Methods Enzymol.* (1973) **29**, 27.
Gelderman, A., Rake, A., and Britten, R. (1971). *Proc. Nat. Acad. Sci. U.S.* **68**, 172.
Goulian, M. (1971). *Ann. Rev. Biochem.* **40**, 855.
Harris, H. (1970). "Cell Fusion." Harvard Univ. Press, Cambridge, Massachusetts.
Hennig, W., Hennig, I., and Stein, H. (1970). *Chromosoma* **32**, 31.
Hohn, T., and Hohn, B. (1970). *Advan. Virus Res.* **16**, 43.
Hsu, T., and Benirschke, K. (1968). "An Atlas of Mammalian Chromosomes." Springer-Verlag, Berlin and New York.
Huberman, J., and Riggs, A. (1968). *J. Mol. Biol.* **32**, 327.
Huberman, J., Tsai, A., and Deich, R. (1973). *Nature* (*London*) **241**, 32.
Jacob, F., Brenner, S., and Cuzin, R. (1963). *Cold Spring Harbor Symp. Quant. Biol.* **28**, 329.
John, H., Birnstiel, M., and Jones, K. (1969). *Nature* (*London*) **223**, 582.
Jones, K., and Robertson, F. (1970). *Chromosoma* **31**, 331.
Judd, B., Shen, M., and Kaufman, T. (1972). *Genetics* **71**, 139.
Kaiser, D., and Masuda, T. (1973). *Proc. Nat. Acad. Sci. U.S.* **70**, 260.
Kasamatsu, H., and Rownd, R. (1970). *J. Mol. Biol.* **51**, 473.
Kennell, D. (1968). *J. Mol. Biol.* **34**, 85.
Kurland, C. (1972). *Ann. Rev. Biochem.* **41**, 377.
Laird, C. (1971). *Chromosoma* **32**, 378.
Lee, C., and Thomas, C. (1973). *J. Mol. Biol.* **77**, 25.
Lehmann, A. (1972). *J. Mol. Biol.* **66**, 319.
Lehmann, A., and Ormerod, M. (1970). *Biochim. Biophys. Acta* **217**, 268.
Lett, J., Klucis, E., and Sun, C. (1970). *Biophys. J.* **10**, 277.
Lima-de-Faria, A., and Jaworska, H. (1968). *Nature* (*London*) **217**, 138.
Lindahl, T., and Nyberg, B. (1972). *Biochemistry* **2**, 3610.
Lyon, M. (1968). *Ann. Rev. Genet.* **2**, 31.
MacHattie, L., and Thomas, C. A. Jr. (1970). *In* "Handbook of Biochemistry" (H. Sober, ed.), pp. H3–8. Chem. Rubber, Cleveland, Ohio.
Mahdavi, V., and Crippa, M. (1972). *Proc. Nat. Acad. Sci. U.S.* **69**, 1749.
Mandel, M. (1970). *In* "Handbook of Biochemistry" (H. Sober, ed.), pp. H75–H79. Chem. Rubber, Cleveland, Ohio.
Manloff, J., and Morowitz, H. (1972). *Bacteriol. Rev.* **36**, 263.
McBurney, M., Graham, F., and Whitmore, G. (1972). *Biophys. J.* **12**, 369.
McCarthy, B., and Bolton, E. (1964). *J. Mol. Biol.* **8**, 184.

McCarthy, B., and Church, R. (1970). *Ann. Rev. Biochem.* **39,** 131.
McGrath, R., and Williams, R. (1966). *Nature (London)* **212,** 534.
Miller, O., and Beatty, B. (1969a). *Science* **164,** 955.
Miller, O., and Beatty, B. (1969b). *Genetics Suppl.* **61:1,** 133.
Milner, G. (1969). *J. Cell Sci.* **4,** 569.
Mueller, G. (1969). *Fed. Proc.* **28,** 1780.
Myers, T., and Strauss, B. (1971). *Nature (London) New Biol.* **230,** 143.
Natarajan, A., and Ahnstrom, G. (1970). *Chromosoma* **30,** 250.
Natarajan, A., and Upadhya, M. (1964). *Chromosoma* **15,** 156.
Normore, W., and Brown, J. (1970). *In* "Handbook of Biochemistry" (H. Sober, ed.), pp. H24–H74. Chem. Rubber, Cleveland, Ohio.
Okazaki, R., Okazaki, T., Sakabe, K., Sugimoto, K., Kainuma, R., Sugino, A., and Iwatsuki, N. (1968). *Cold Spring Harbor Symp. Quant. Biol.* **33,** 129.
Ormerod, M., and Lehmann, A. (1971). *Biochim. Biophys. Acta* **247,** 369.
Painter, R., and Schaeffer, A. (1969). *Nature (London)* **221,** 1215.
Painter, R., and Young, B. (1972). *Mutat. Res.* **14,** 225.
Pardue, M., and Gall, J. (1970). *Science* **168,** 1356.
Pavan, C., and da Cunha, A. (1969). *Ann. Rev. Genet.* **3,** 425.
Petes, T., and Fangman, W. (1972). *Proc. Nat. Acad. Sci. U.S.* **69,** 1188.
Prescott, D., and Kuempel, P. (1972). *Proc. Nat. Acad. Sci. U.S.* **69,** 2842.
Price, P., Conover, J., and Hirschhorn, K. (1972). *Nature (London)* **237,** 340.
Puck, T. (1972). "The Mammalian Cell as a Microorganism." Holden-Day, San Francisco, California.
Puck, T., Saunders, P., and Petersen, D. (1964). *Biophys. J.* **4,** 441.
Rao, R., and Natarajan, A. (1967). *Genetics* **57,** 821.
Rees, H., and Jones, R. (1972). *Int. Rev. Cytol.* **32,** 53.
Regan, J., Trosko, J., and Carrier, W. (1968). *Biophys. J.* **8,** 319.
Regan, J., Setlow, R., and Ley, R. (1971). *Proc. Nat. Acad. Sci. U.S.* **68,** 708.
Robberson, D., Kasamatsu, H., and Vinograd, J. (1972). *Proc. Nat. Acad. Sci. U.S.* **69,** 737.
Robbins, E., and Borun, T. (1967). *Proc. Nat. Acad. Sci. U.S.* **57,** 409.
Rudkin, G. (1965a). *Genet. Today* **2,** 359.
Rudkin, G. (1965b). *Genetics* **52,** 665.
Rupp, W., Wilde, C., Reno, D., and Howard-Flanders, P. (1971). *J. Mol. Biol.* **61,** 25.
Rusch, H. (1969). *Fed. Proc.* **28,** 1761.
Sagopal, A., and Bonner, J. (1969). *Biochim. Biophys. Acta* **186,** 349.
Sasaki, M., and Norman, A. (1966). *Exp. Cell Res.* **44,** 642.
Saunders, G., and Shirakawa, S. (1972). *J. Mol. Biol.* **63,** 323.
Schnedl, W. (1971). *Nature (London) New Biol.* **233,** 93.
Schnedl, W. (1972). *Chromosoma* **38,** 319.
Schnoss, M., and Inman, R. (1970). *J. Mol. Biol.* **51,** 61.
Scudiero, D., and Strauss, B. (1974). *J. Mol. Biol.* **82,** in press.
Shapiro, H. (1970). *In* "Handbook of Biochemistry" (H. Sober, ed.), pp. H80–H116. Chem. Rubber, Cleveland, Ohio.
Shih, T., and Fasman, G. (1970). *J. Mol. Biol.* **52,** 125.
Siu, C., Chiang, K., and Swift, H. (1972). *J. Cell. Biol.* **55,** 241a.
Smith, I., Dubnau, D., Morell, P., and Marmur, J. (1968). *J. Mol. Biol.* **33,** 123.
Sober, H. (ed.). (1970). "Handbook of Biochemistry," 2d ed. Chem. Rubber, Cleveland, Ohio.
Southern, E. (1970). *Nature (London)* **227,** 794.

Strauss, B. (1964). *Progr. Med. Genet.* **3,** 1.
Strauss, B. (1968). *Current Topics Microbiol. Immunol.* **44,** 1.
Sueoka, N. (1964). *In* "The Bacteria" (I. Gunsalus and R. Stanier, eds.). Vol. 5, p. 419. Academic Press, New York.
Sugino, A., Hirose, S., and Okazaki, R. (1972). *Proc. Nat. Acad. Sci. U.S.* **69,** 1863.
Taylor, J. (1968). *J. Mol. Biol.* **31,** 579.
Taylor, J., Mego, W., and Evenson, D. (1970). *In* "The Neurosciences, A Second Study Program" (F. Schmitt, ed.), pp. 998–1013. Rockefeller Univ. Press, New York.
Terasima, T., and Yasukawa, M. (1966). *Exp. Cell Res.* **44,** 669.
Thomas, C. (1966). *Prog. Nucl. Acid Res. Mol. Biol.* **5,** 315.
Thomas, C. (1971). *Ann. Rev. Genet.* **5,** 237.
Thomas, C., Hamkalo, B., Misra, D., and Lee, C. (1970). *J. Mol. Biol.* **51,** 621.
Tobia, A., Schildkraut, C., and Maio, J. (1970). *J. Mol. Biol.* **54,** 499.
Walker, P. (1971). *Progr. Biophys. Mol. Biol.* **23,** 147.
Watson, J. D. (1971). *Advan. Cell Biol.* **2,** 1.
Watson, J. (1972). *Nature (London) New Biol.* **239,** 197.
Weintraub, H. (1972a). *J. Mol. Biol.* **66,** 31.
Weintraub, H. (1972b). *Nature (London) New Biol.* **236,** 195.
Weisblum, B., and de Haseth, P. (1972). *Proc. Nat. Acad. Sci. U.S.* **69,** 629.
Weiss, G. B. (1969). *J. Cell Physiol.* **73,** 85.
Wetmur, J., and Davidson, N. (1968). *J. Mol. Biol.* **31,** 349.
Whitehouse, H. (1967). *J. Cell Sci.* **2,** 9.
Witkin, E., and George, D. (1973). *Genetics Supplement* **73,** 91.
Yanofsky, C., Ito, J., and Horn, V. (1966). *Cold Spring Harbor Symp. Quant. Biol.* **31,** 151.
Ycas, M. (1969). "The Biological Code." Amer. Elsevier, New York.

2

Nuclear DNA Polymerases

R. K. Craig and H. M. Keir

I. Introduction

Some eight years ago, a general assessment was made of the occurrence and properties of a soluble, DNA-dependent DNA polymerase (DNA nucleotidyltransferase, E.C. 2.7.7.7; hereafter to be termed DNA polymerase) activity readily extractable from a wide variety of mammalian tissues and cell culture lines (Keir, 1965). In this assessment a clear distinction was drawn between two types of soluble enzymatic activities, both capable of polymerizing deoxyribonucleotides.

One of these, the terminal DNA polymerase activity, has now been described in great detail (Yoneda and Bollum, 1965; Chang and Bollum, 1971a). It displays a requirement for a single-stranded polydeoxyribonucleotide or oligonucleotide primer, to which it is capable of sequentially adding deoxyribonucleotidyl residues of a single species at the 3′-hydroxy terminus. This activity has recently been shown to be probably peculiar to thymus tissue (Chang, 1971) where it is present in a soluble form in both nucleus and cytoplasm. Wang (1968a) solubilized a similar enzyme activity from the nonhistone chromatin proteins from calf thymus and has suggested that it is different from the soluble one. The terminal DNA polymerase will not be discussed further in this chapter.

In contrast, the replicative DNA polymerase has an absolute requirement *in vitro* for a polydeoxyribonucleotide template, catalyzing the sequential polymerization of deoxyribonucleotide monomers in the 5′ ⟶ 3′ direction, in a complementary manner dictated by the base sequence of the template.

The survey of early observations (Keir, 1965) on the soluble, replicative DNA polymerase suggested the existence of one such enzyme in each cell type, although some information, notably that derived from experiments on intracellular distribution, implied that the enzyme might exist in two functionally different forms.

Accordingly a hypothesis was presented for the mechanism of action of the replicative DNA polymerase inside the cell, such that the enzyme adopted one molecular conformation while engaged in the replication of DNA in the nucleus, and subsequently a second, different conformation when not so occupied, the enzyme then being distributed between nucleus and cytoplasm. Although this hypothesis has been neither proved nor disproved, it has served a useful purpose in the continuing investigation of DNA polymerases. Since 1965, further data on replicative DNA polymerases from mammalian cells have appeared, and the work has been extended to other eukaryotic cells including those from avian, amphibian, molluscan, echinoderm, plant, algal, fungal, and protozoan cells.

An important step forward was taken by Patel *et al.* (1967), who partially purified a replicative DNA polymerase after solubilization of the activity from rat liver chromosomal protein. This partially purified enzyme, however, although displaying characteristics somewhat different from those of the soluble enzyme activity, was not shown at this stage to be a physically distinct enzyme.

The use of improved subcellular fractionation and enzyme purification techniques has more recently led to the identification of physically distinct DNA polymerase activities within a single cell or tissue type. Bellair (1968), using Sephadex G-200 chromatography, separated two distinct DNA polymerase activities from regenerating rat liver, while Meyer and Simpson (1968) demonstrated that mitochondria contain a DNA polymerase distinct from the chromosome-bound enzyme. More recently, Baril *et al.* (1971) and Poulson *et al.* (1973), on the basis of pH optima, divalent cation requirement, and affinity for DEAE-cellulose, have presented evidence for the existence of three distinct DNA polymerase activities within a single tissue type. These activities correspond to chromosomal, cytoplasmic, and mitochondrial DNA polymerases. Chloroplasts contain a distinct DNA polymerase also (Tewari and Wildman, 1967; Keller *et al.*, 1973; McLennan and Keir, 1973).

Apart from the mitochondrial and chloroplast enzymes, it can now be stated in summary that eukaryotic cells contain at least two replicative DNA polymerases, one of which, the chromosome-bound enzyme, is solubilized only after extraction with buffer solutions of high ionic strength, and which possesses a sedimentation coefficient of 3.3 S (Chang and Bollum, 1971b), while the other probably corresponding to the original, soluble DNA polymerase, is readily extractable from cells in buffer solutions of low ionic strength, and has a sedimentation coefficient covering the range 6–8 S (Chang and Bollum, 1971b).

Each polymerase is primarily associated with a specific subcellular fraction and the current status of cellular polymerases can be summarized broadly thus:

1. The cell nucleus contains predominantly the 3.3 S species, tightly bound to chromatin, while the 6–8 S species is apparently less abundant and is readily extractable at low ionic strength.

2. The soluble cell sap contains predominantly the 6–8 S species and only small amounts of the 3.3 S species.

Through the remainder of this chapter, we propose to describe these two enzyme activities in the context of the cell nucleus, referring to them as the 3.3 S and 6–8 S polymerase activities, and comparing relevant observations from several laboratories on the nuclear enzymes from a variety of tissues. However, although we shall consider the 6–8 S enzyme as a single species in the context of this chapter, it does in fact show heterogeneity, and can be separated into specific fractions sedimenting at about 6 and 8 S (Chang *et al.*, 1973; Holmes *et al.*, 1973; Momparler *et al.*, 1973). Moreover, recent evidence points to the existence of a further polymerase activity derived from the 6–8 S species, sedimenting at about 5 S. Such heterogeneity has been reported in mouse L cells (Chang *et al.*, 1973), calf thymus (Chang *et al.*, 1973; Holmes *et al.*, 1973), and rat spleen (Holmes *et al.*, 1973).

II. DNA Polymerases of *Escherichia coli*

In order to provide a comprehensive comparative background to what will be presented later on the function of nuclear DNA polymerases, we have deemed it appropriate to refer very briefly to the DNA polymerases found in a prokaryotic system. The organism most fully described in this respect is *Escherichia coli.* These bacterial cells are known to contain three distinct DNA polymerases, termed I, II, and III. Their properties and likely roles in DNA repair and replication have recently been reviewed by Otto (1973) and Smith (1973). The present position can be summarized as follows:

1. DNA polymerase I has a molecular weight of 109,000 and there are about 400 molecules of it per cell. *In vitro* it catalyzes the polymerization of deoxyribonucleotides at only 1% of the rate estimated for DNA replication *in vivo.* The enzyme has a prominent role in DNA excision repair, but seems also to be involved (to a limited extent) in DNA replication.

2. DNA polymerase II has a molecular weight of about 120,000 and there are about 20 molecules of it per cell. It can synthesize polydeoxyribonucleotide only at about 0.3% of the *in vivo* rate. Its function in the cell is not known at the present time.

3. DNA polymerase III has a molecular weight of 140,000 and there are about 10 molecules of it per cell. It can synthesize polydeoxyribonucleotide at about 15% of the *in vivo* rate. Therefore, if all 10 molecules in each cell are involved in replication, DNA polymerase III has the capacity to make DNA *in vivo* at the required rate. Studies with the enzyme purified from wild-type *E. coli* cells and from certain mutant cells defective in DNA replication clearly show that DNA polymerase III is essential for replication.

4. All three DNA polymerases synthesize polydeoxyribonucleotide in the $5' \longrightarrow 3'$ direction and all of them have associated nuclease activities. They have similar template requirements although none of them has yet been shown to be capable of completely replicating double-helical DNA semiconservatively.

The point emphasized in this section is that *E. coli* cells contain at least three DNA polymerases and that two of them appear to have roles in different aspects of DNA biosynthesis, namely, repair and replication.

III. DNA Polymerases from Nuclei

A. Background

Early work on DNA polymerases (see Keir, 1965) indicated that the mammalian cell enzyme was located primarily in the cytoplasmic fraction after subcellular fractionation was conducted in isotonic or hypotonic aqueous media, usually at low ionic strength. However, there was always clear evidence of the nuclear or particulate fractions retaining some polymerase activity even after extensive washing procedures, an observation that reflected to some extent the divalent cation composition of the medium (Main and Cole, 1964). Bazill and Philpot (1963) found that washing of calf thymus nuclei with buffer solutions at low ionic strength failed to remove the residual polymerase activity from the nuclei.

Isolation of nuclei from regenerating rat liver, calf thymus, and rabbit thymus in nonaqueous solvent systems (Keir *et al.*, 1962; Behki and Schneider, 1963; Smith and Keir, 1963), followed by extraction of the nuclei with aqueous buffers at low ionic strength, revealed higher amounts of DNA polymerase in the nuclei than had been observed with nuclei prepared in aqueous media. Moreover, the nuclei usually contained more polymerase than the corresponding nonaqueous cytoplasm. Inclusion of KCl at 0.15 *M* in the aqueous extraction buffer promoted release of yet more DNA polymerase from the nonaqueous nuclei, but did not increase the yield of the enzyme when the nonaqueous cytoplasmic fraction was likewise extracted. It was concluded that both nucleus and cytoplasm normally contain DNA polymerase activity. At the same time, it was recognized that chromatin-bound proteins were minimally soluble at 0.15 *M* KCl and therefore that the extracted nuclear enzyme was probably not chromatin-bound. Meaningful comparison of the properties of the nuclear and cytoplasmic enzymes was difficult at that time largely because of contaminating deoxyribonuclease activities. Nevertheless these early experiments have a strong bearing on the current status of work with nuclear DNA polymerases. The problems have always been to reconcile the apparent cytoplasmic location of some of the cell polymerase with the knowledge that replication of DNA is a nuclear event, and to identify nuclear enzymatic systems that have the ability to carry out replication.

B. *Synthesis of DNA by Isolated Nuclei*

More recently, Friedman and Mueller (1968) and Kidwell and Mueller (1969) established an *in vitro* nuclear system for the replication of DNA using nuclei isolated from HeLa cell cultures synchronized with respect to the growth cycle. The nuclei, which were isolated in buffer solutions at low ionic strength, displayed the ability to synthesize DNA using as precursors dATP, dCTP, dGTP, and dTTP, thereby providing evidence for the participation of nuclear polymerase(s) in DNA synthesis. At the same time it was demonstrated that the cytoplasmic DNA polymerase activity apparently bore no temporal relationship to DNA synthesis. The nuclear system used endogenous DNA template, was stimulated by ATP, and displayed a requirement for a heat-labile cytoplasmic factor (cf. Gurdon, 1967; Harris, 1967; Thompson and McCarthy, 1968) distinct from the cytoplasmic DNA polymerase (see also Kidwell, 1972; Hershey *et al.*, 1973). This type of system, although very complex and poorly defined at the molecular level, indicates that meaningful DNA synthesis can proceed *in vitro*. Friedman (1970) went on to show that the nuclei contained 25–30% of the total deoxyribonucleotide polymerizing activity of the cell, and that it could not be extracted by extensive washing of the nuclei.

Subsequently a number of reports on DNA synthesis in isolated nuclei appeared, confirming and extending the earlier observations. Among the isolated nuclear systems employing the four deoxyribonucleoside 5′-triphosphates and therefore implicating nuclear DNA polymerases in DNA replication are the following examples: rat thymus (Lagunoff, 1969; Burgoyne *et al.*, 1970a), rat liver (Burgoyne *et al.*, 1970b; Lynch *et al.*, 1970, 1972; Probst *et al.*, 1972; Grisham *et al.*, 1972), Ehrlich tumor cells (Teng *et al.*, 1970), adult and embryo tissues of *Xenopus laevis* (Arms, 1971), mouse 3T3 or 3T6 cells noninfected or infected by polyoma virus (Winnacker *et al.*, 1971), mouse L cells (Kidwell, 1972; Adams and Wood, 1973), rat brain (Shimada and Terayama, 1972), *Rana pipiens* embryos (Klose and Flickinger, 1972), human KB cells noninfected or infected by adenovirus type 5 (van der Vliet and Sussenbach, 1972) and HeLa cells (Kumar and Friedman, 1972; Hershey *et al.*, 1973). These examples serve to fortify the view that isolated nuclei represent an *in vitro* system for DNA replication that simulates the *in vivo* situation. It seems reasonable to conclude therefore, that some or all of the enzymatic machinery, notably the polymerizing activity, is located in the cell nucleus.

C. *A Distinct Species of DNA Polymerase Associated with the Cell Nucleus*

In the meantime, two major advances were made. First, a replicative DNA polymerase was partially purified by solubilization at high ionic strength of the nonhistone chromosomal proteins from rat liver (Patel *et al.*, 1967; Howk and Wang, 1969), from calf thymus (Wang, 1967), and from rat Walker 256 carcinosarcoma (Wang, 1968b). Second, it became clear that two separable types of DNA polymerase could be distinguished in fetal, adult, and regenerating rat liver and in rat hepatomas (Bellair, 1968; Iwamura *et al.*, 1968; Ove *et al.*, 1969). Concurrently, Meyer and Simpson (1968), who were studying mitochondrial DNA synthesis in rat liver, not only detected a distinct mitochondrial DNA polymerase but also showed that purified nuclei contained a DNA polymerase; the latter was extracted from nuclei prepared in a medium of low ionic strength by treating them with NaCl at 1 *M*.

The residual nuclear activity extractable at high ionic strength from these and other cells and tissues is attributable to the chromatin-bound enzyme first described by Patel *et al.* (1967). The enzyme has since been shown to be physically distinct from the soluble cytoplasmic enzyme. Chang and Bollum (1971b), using rabbit bone marrow and spleen tissue, prepared particulate ("nuclear") and soluble ("cytoplasmic") fractions; the nuclear polymerase activity was solubilized by treatment of the nuclear fraction with buffer containing NaCl at 1 *M*. The polymerases of the soluble and nuclear fractions were shown to be separable by centrifugation in sucrose density gradients, the former sedimenting at 6–8 S and the latter at 3.3 S. Separation of these two major polymerase activities was also achieved by gel filtration on Sephadex G-100 and by column chromatography on phosphocellulose. The 3.3 S nuclear activity was shown to be present also in rat liver, in calf fetal liver, lung, kidney, and spleen, and in all lymphocyte tissue culture lines examined. Chang and Bollum (1971b) also confirmed the observation of Howk and Wang (1969) that the nuclear DNA polymerase extracted at high ionic strength has an alkaline pH optimum (7.6–8.6, depending on the template used).

Haines *et al.* (1971) showed that rat liver has distinct nuclear and cytoplasmic DNA polymerases. The isolated nuclei were extracted with buffer solution containing NaCl at 1 *M* in a manner similar to that described by Patel *et al.* (1967) and Meyer and Simpson (1968). Chromatography of the nuclear extract on Sepharose 6B clearly showed the presence of a single nuclear polymerase of relatively low molecular weight (about 65,000), although the cytoplasm also contained some of this species together with two other polymerase activities, the major one of which had a molecular weight of about 400,000. Haines *et al.* (1972) purified the nuclear enzyme extensively and, using polyacrylamide gel electrophoresis and column chromatography on Sephadex G-100, provided evidence that it has a molecular weight of about 60,000 and consists of two polypeptide chains each of molecular weight about 30,000.

Chang and Bollum (1972a), continuing their work on the 3.3 S and 6–8 S DNA polymerase activities in rabbit bone marrow, prepared cytoplasm and purified nuclei from that tissue. The nuclei were prepared by the standard method of Blobel and Potter (1966), were purified by treatment with the nonionic detergent Triton X-100 at 0.5% in a buffer containing 0.25 *M* sucrose and 5 m*M* $MgCl_2$, and were finally extracted with buffer at high ionic strength (0.2 *M* potassium phosphate buffer, pH 7.5). The cytoplasmic and nuclear fractions were centrifuged at high speed to give supernatant fractions for analysis of DNA polymerase content. The ensuing sucrose gradient analysis revealed only the 3.3 S polymerase in the nuclei; its activity was much greater at pH 8.6 than at pH 7.0. The cytoplasmic fraction contained predominantly the 6–8 S heterogeneous polymerase species; its activity was greater at pH 7.0 than at 8.6. The cytoplasm contained also a small amount of the 3.3 S species. Column chromatography of the nuclear 3.3 S enzyme on phosphocellulose gave a single peak eluting at a phosphate concentration of 0.3 *M*. The molecular weight of the 3.3 S polymerase was estimated from sucrose gradient analysis and gel filtration on Sephadex G-100 to be 40,000–50,000.

Recently it has been reported (Chang, 1973) that the 3.3 S enzyme isolated from calf thymus chromatin and purified to apparent homogeneity consists of a single polypeptide chain with a molecular weight of 44,000 as assessed by polyacrylamide-SDS gel electrophoresis.

Heterogeneity of cell DNA polymerases was found also by Baril *et al.* (1971) in an extensive study on subcellular fractions prepared from rat liver. Notably, nuclei were prepared and purified by the method of Blobel and Potter (1967) as modified by Whittle *et al.* (1968); the purification included detergent treatment with Triton at 1%. The nuclei were then extracted at high ionic strength in order to solubilize the nuclear DNA polymerase, and the extract fractionated by column chromatography on DEAE-cellulose and phosphocellulose; each elution profile showed only one peak of DNA polymerase, located in the column wash in the case of DEAE-cellulose, and eluted with 0.5 *M* potassium phosphate buffer pH 7.2 in the case of phosphocellulose. This enzyme corresponds to the nuclear 3.3 S activity with an alkaline pH optimum described by Chang and Bollum (1971b, 1972a) and to the low-molecular-weight species described by Haines *et al.* (1971, 1972). Baril *et al.* (1971) found that the same enzyme species was associated with ribosomes, and that bulk of the remaining cytoplasmic polymerase corresponding to the 6–8 S species described by others was associated with smooth membranes. Chiu and Sung (1972a) also solubilized at high ionic strength a low-molecular-weight DNA polymerase from a chromatin–membrane complex prepared from rat liver nuclei. The enzyme had a sedimentation coefficient of about 3–4 S and was separable by column chromatography on DEAE-cellulose from the cytoplasmic polymerase which had a sedimentation coefficient of about 9 S.

Rat brain also has been shown to contain two DNA polymerases (Chiu and Sung, 1971), which are different in properties and which have different sedimentation coefficients on sucrose density gradient analysis, one sedimenting at about 9 S and the other at 3–4 S (Chiu and Sung, 1972c). The low-molecular-weight species is solubilized from purified brain nuclei with 0.2 *M* phosphate buffer, pH 7.4 (Chiu and Sung, 1972b, d), and has an alkaline pH optimum; this contrasts with the soluble 9 S species which has a neutral pH optimum.

Definitive evidence for a primarily nuclear 3.3 S DNA polymerase and a primarily cytoplasmic 6–8 S enzyme species has also been obtained using preparations from regenerating rat liver (Chang and Bollum, 1972b), mouse L cells, and a variety of other mammalian systems, including antigen-stimulated rabbit spleen, phytohemagglutinin-stimulated human lymphocytes, dimethylbenzanthracene-induced leukemic rat spleen and liver, phenylhydrazine-induced erythropoietic mouse spleen, bovine spleen, and calf thymus (Chang and Bollum, 1972c; Chang *et al.*, 1973).

Lindsay *et al.* (1970) inferred the existence of multiple forms of DNA polymerase in mouse L cells, and in particular solubilized a polymerase activity from the cell nuclei (prepared without treatment by detergent) either by extraction at high ionic strength (1.5 *M* KCl) or by ultrasonication. Further work with the same system (Adams *et al.*, 1973) involved fractionation of nuclear and cytoplasmic polymerase preparations on columns of Sephadex G-200 and DEAE-cellulose. The observations suggested the presence of up to three polymerases in the nucleus; the polymerases were estimated to have molecular weights of 35,000, 70,000, and 140,000 and were bound within the nuclei. A high-molecular-weight polymerase from the cytoplasm was estimated to have a molecular weight of 250,000.

Nuclei prepared from rat intestinal mucosa at low ionic strength in the absence of detergent, and subsequently extracted at low ionic strength with vigorous homogenizing, apparently yield an extract containing three DNA polymerase activities as assessed by column chromatography on DEAE-cellulose, Sephadex G-150 and G-200, and by sucrose density gradient analysis (Leung and Zbarsky, 1970). The molecular weights of the three polymerase fractions were estimated to be 25,000, 180,000, and

300,000. Further work involving subcellular fractionation of the same tissue (Poulson *et al.*, 1973) gave evidence of only three cellular DNA polymerases. On the basis of their pH optima, Mg^{2+} ion requirement and affinity for DEAE-cellulose, these polymerases were judged to correspond respectively to the 6–8 S activity (present in mucosal soluble nuclear, soluble cytoplasmic, and ribosomal fractions), to the 3.3 S activity (present in the mucosal nuclear extract prepared at high ionic strength), and to the mitochondrial enzyme (present also in mucosal smooth membranes). The behavior of the mucosal enzymes on DEAE-cellulose and their intracellular distribution are somewhat at variance with the work of others, notably Baril *et al.* (1971), who conducted similar experiments with rat liver. The disparities presumably reflect tissue variations and differences in technique. Poulson *et al.* (1973) suggested, on the basis of electron microscopic examination of their nuclei and mitochondria, that damage to the outer membrane of these organelles accounts for the presence of two polymerases of nuclear origin and one of mitochondrial origin in certain cytoplasmic fractions.

Human cell nuclei contain DNA polymerases extractable by buffer solutions of high ionic strength. Weissbach *et al.* (1971) showed that HeLa cells and WI–38 cells (a human lung diploid cell line) contain two separable polymerases. It seems likely from the subcellular fractionation techniques used and from the behavior of the enzymes on DEAE-cellulose, phosphocellulose, and Sephadex G-200 that one of the nuclear activities corresponds to the low-molecular-weight nuclear species (3.3 S) while the other is similar or identical to the high-molecular-weight cytoplasmic polymerase (6–8 S). Likewise, human KB cells display heterogeneity of DNA polymerase activity (Sedwick *et al.*, 1972) in possessing two nuclear polymerases and one cytoplasmic enzyme. Evidence was adduced that one of the nuclear enzymes was similar to, but distinct from, the cytoplasmic DNA polymerase; the other nuclear polymerase was clearly different, had a molecular weight of about 38,000, and corresponded in other respects to the 3.3 S enzyme isolated from other cells and tissues.

D. Multiple Molecular Forms of DNA Polymerase in Other Systems

1. MAMMALIAN SYSTEMS

Among the many other related experimental observations which include one or more of the characteristics that emerged from the definitive work described above for nuclear DNA polymerase, or which show heterogeneity of DNA polymerases are the following: guinea pig adrenocortical nuclei (Masui and Garren, 1970), Walker 256 tumor cells and nuclei (Ballal *et al.*, 1970; Furlong and Gresham, 1971), a variety of rat and mouse tissues (Wallace *et al.*, 1971), Ehrlich ascites tumor cells (Harm and Hilz, 1971), rat brain and liver (Murthy and Bharucha, 1971, 1972), a nuclear membrane–chromatin fraction from rat ascites hepatoma cells (Tsuruo *et al.*, 1972a, b), human blood lymphocytes (Smith and Gallo, 1972), bovine adrenal nuclei (Long and Garren, 1972), rat and guinea pig liver nuclei (Deumling and Franke, 1972), differentiating embryonic muscle cells (Wicha and Stockdale, 1972), mouse testes (Hecht and Davidson, 1973), and BALB/3T3 cells (Ross *et al.*, 1971).

2. NONMAMMALIAN EUKARYOTIC SYSTEMS

Reports on the existence of multiple molecular forms of DNA polymerase in eukaryotic cells are not restricted to mammalian systems. For example, at least two species of polymerase have been described for the protozoon *Tetrahymena pyriformis* (Westergaard, 1970; Crerar and Pearlman, 1971; Westergaard and Lindberg, 1972), the yeast *Saccharomyces cerevisiae* (Wintersberger and Wintersberger, 1970a, b; Helfman, 1973), the slime mold *Physarum polycephalum* (Schiebel and Bamberg, 1973), the unicellular green alga *Euglena gracilis* (Keller *et al.*, 1973; McLennan and Keir, 1973), *Xenopus laevis* oocytes (Grippo and Lo Scavo, 1972), and chick embryo (Stavrianopoulos *et al.*, 1971, 1972a, b).

DNA polymerase activity has also been described in other nonmammalian cells, although without any clear indication of heterogeneity; these include yeast (Eckstein *et al.*, 1967), sea urchin embryos (Loeb, 1969), maize seedlings (Stout and Arens, 1970), nuclei from *Tradescantia* pollen grains (Wever and Takats, 1970; Takats and Wever, 1971), *Lilium* (Hecht, 1971; Hecht and Stern, 1971; Howell and Hecht, 1971), unicellular green algae (Schönherr and Keir, 1971), brain tissue of *Octopus vulgaris* (Libonati *et al.*, 1972), avian erythroid cells (Williams, 1972), and crown gull tumor tissue culture of tobacco (Srivastava, 1973).

E. General Unifying Comments on the Intracellular Location of DNA Polymerases

1. THE 3.3 S DNA POLYMERASE

It appears that whatever techniques of cell disruption, subcellular fractionation, and enzyme extraction are applied to a cell or tissue system, some 3.3 S DNA polymerase is invariably found to be associated with the cell nucleus. It is firmly bound to chromatin and may be solubilized by extraction of chromatin by buffer solutions of high ionic strength. In certain cases, some of the nuclear enzyme may be released by vigorous homogenization (Leung and Zbarsky, 1970) in buffers of low ionic strength.

The 3.3 S enzyme may be found also in a soluble form in cytoplasmic fractions (Haines *et al.*, 1971; Chang and Bollum, 1971b, 1972a, b; Chang *et al.*, 1973), and it has been reported (Chiu and Sung, 1972c, d) that the cytoplasmic content of it is highest in growing tissue. Chang and Bollum (1972a) have suggested that the ratio of 3.3 S polymerase in the cytoplasmic fraction to that in the nuclear fraction depends on the tissue studied and that the ratio may be partially related to the ratio of cytoplasmic to nuclear volume. Furthermore, Chang and Bollum (1972b) observed changes in the intracellular distribution of the 3.3 S polymerase in the course of the regeneration of liver that follows partial hepatectomy in the rat. These changes may be summarized (see also Section IV): (*a*) About equal amounts of the 3.3 S and 6–8 S polymerase were present in the livers of control animals; (*b*) about 10% of the 3.3 S enzyme was in the nuclear fraction and about 90% in the cytoplasmic fraction; (*c*) at 24 and 48 hr after partial hepatectomy the activity of the cytoplasmic 3.3 S polymerase rose by approximately 19 and 33%, respectively, over the control values (as assessed by sucrose gradient analysis); (*d*) at 24 and 48 hr after partial hepatectomy the activity of the nuclear 3.3 S polymerase had similar rises, by about 29 and 28%, respectively. It is quite likely that the low-molecular-weight species of DNA polymerase purified from soluble cell extracts, made at low ionic strength, from normal and regenerating rat liver (e.g., Berger *et al.*, 1971) represents the soluble 3.3 S polymerase of the cytoplasm.

In contrast to these observations, Weissbach *et al.* (1971) and Sedwick *et al.* (1972) failed to find any activity corresponding to the 3.3 S species in cytoplasmic fractions.

2. THE 6–8 S DNA POLYMERASE

There is considerable doubt concerning the true intracellular distribution of the heterogeneous 6–8 S polymerase. Subcellular fractionation conducted in nonaqueous solvents to give nuclei and cytoplasm (see Keir, 1965) suggests that there are substantial amounts of the enzyme in both fractions. However, the recent results of subcellular fractionation in aqueous media tend rather to support the view that the 6–8 S species has primarily, if not exclusively, a cytoplasmic location.

Weissbach *et al.* (1971), Sedwick *et al.* (1972), Chiu and Sung (1972a, d) and Poulson *et al.* (1973) claim that while the bulk of the enzyme corresponding to the 6–8 S species is in the cytoplasm, there is clearly a similar or identical species in the nucleus. On the other hand, the observations of Chang and Bollum (1972a, b), Chang *et al.* (1973), Haines *et al.* (1971, 1972), and Baril *et al.* (1971) indicate quite clearly that there is no 6–8 S activity in the nucleus. The controversy, which remains unresolved, no doubt stems from properties peculiar to different tissues and cell types and from variations in the techniques practiced in different laboratories.

3. INTERPRETATION OF THE EXPERIMENTAL DATA

The interpretation of results of experiments designed to determine the intracellular distribution of the 3.3 S and notably the 6–8 S DNA polymerase activities appears to be a complex function of the following parameters: (*a*) the *in vivo* state of the tissue or cell population with respect to DNA replication, (*b*) the method of cell disruption, (*c*) the method of purification of the nuclei, and (*d*) the ionic strength of the extraction buffer.

With regard to the 6–8 S enzyme, it is conceivable that this species is nuclear in origin and that its apparent predominance in isolated cytoplasmic fractions is attributable solely to an artifactual situation created by leakage from nuclei during cell disruption in aqueous media.

If the enzyme has a predominantly or exclusively cytoplasmic location, the experimental observations might be explained in several ways. First, since the enzyme is presumably synthesized on cytoplasmic polyribosomes, it is possible that it is stored in the cytoplasm until called upon for temporary translocation to the nucleus when it is required for participation in a DNA synthetic event. According to this view, there might be varying amounts of the 6–8 S enzyme associated with nuclei, depending upon the overall physiological state of the tissue or cell population at the time of the experiment. Second, if the enzyme is truly cytoplasmic, detection in isolated nuclear preparations might be attributable to contamination of the nuclei by whole cells. Third, conditions of cell disruption might dictate the final distribution of the 6–8 S enzyme in the subcellular fractions. For example, disruption in hypotonic media (Weissbach *et al.*, 1971; Sedwick *et al.*, 1972) might promote an influx of the polymerase into the nucleoplasm from the cytoplasm. Fourth, the 6–8 S enzyme might normally be associated with cytoplasmic structures such as ribosomes and membranes (cf. Baril *et al.*, 1971; Poulson *et al.*, 1973) and might therefore be a normal component of the outer nuclear membrane. Removal of the outer nuclear membrane during purification of nuclei with the aid of a detergent would then give nuclei lacking 6–8 S enzyme activity (Chang and Bollum, 1971b, 1972a). On the other hand, failure to remove this outer membrane would yield nuclear preparations displaying 6–8 S polymerase activity (Chang and Bollum, 1971b).

It is quite impossible at the present time to arrive at a definite conclusion concerning disparities in the observations reported from different laboratories. However, it is apparent that action should be taken to conduct an exhaustive survey on a single tissue or cell type with a view to defining experimental parameters that might influence the redistribution of DNA polymerases between the two major cell compartments during cell disruption.

IV. Evidence for the Involvement of the 6–8 S DNA Polymerase Activity in Nuclear Events

Circumstantial evidence has implicated the 6–8 S polymerase species in DNA replication as distinct from DNA repair synthesis. Much of this evidence is based upon observations that demonstrate, on the one hand, the presence of substantial amounts of the enzyme in proliferating tissues and cells and, on the other, the absence of it from tissues and cells not engaged in DNA replication.

The work of Iwamura *et al.* (1968) and Ono and Umehara (1968) on adult, fetal, and regenerating rat liver and rat hepatomas demonstrated the existence in these tissues of high- and low-molecular-weight DNA polymerases that were separable by gel filtration on columns of Sephadex G-100. The activities of the high-molecular-weight enzyme in each tissue correlated closely with the growth rate of the tissue, values being high in rapidly growing hepatomas, in fetal and regenerating liver, but low in slowly growing hepatomas and adult liver. Similar observations were made by Stockdale (1970), who investigated the soluble DNA polymerase of embryonic muscle tissue. The activity of the enzyme was high in proliferating cells, declining rapidly as the cells began to differentiate with consequential loss of the ability to replicate DNA. This soluble polymerizing activity has since been shown to be attributable to a 6–8 S polymerase species (Wicha and Stockdale, 1972).

An even closer correlation between the activity of the 6–8 S polymerase and the *in vivo* rate of DNA replication has been shown during the developmental stages of the rat brain. In this system, DNA synthetic activity rises to a peak approximately 6 days after birth, decreasing rapidly thereafter (Sung, 1969). Differential extraction, using low and high ionic strength buffers of brain tissues at different stages of development, followed by separation of the 6–8 S and 3.3 S activities by sucrose density gradient centrifugation, clearly revealed a close correlation between the maximum rate of DNA synthesis *in vivo* and the maximum levels of activity of the 6–8 S enzyme. Conversely, when the *in vivo* rate of DNA synthesis was at a minimum (as in the brain of the adult rat) the 6–8 S species accounted for only 4% of the total DNA polymerase activity; throughout this maturation process the activity of the 3.3 S polymerase remained relatively constant (Chiu and Sung, 1972b, c, d).

Similar experiments using regenerating rat liver (Chang and Bollum, 1972b) indicated that DNA synthesis was correlated with increased activity of the 6–8 S cytoplasmic DNA polymerase. Moreover, experiments with stationary phase mouse L cells, stimulated to proliferate by dilution into fresh growth medium, showed that the activity of the 6–8 S cytoplasmic polymerase declined substantially when the cells were in the stationary phase and subsequently increased five- to twelve-fold, in parallel with DNA synthesis, measured *in vivo* by thymidine incorporation, as the cells proceeded from stationary phase into the logarithmic phase of growth (Chang *et al.*, 1973). In contrast there were only relatively minor changes in the levels of activity of the 3.3 S polymerase (in both nuclei and cytoplasm) during the cycle of events, both in regenerating rat liver and in mouse L cells. Other observations implying a similar temporal relationship between DNA replication and the level of activity of the 6–8 S polymerase have been reported by Chang *et al.* (1973) in systems as diverse as antigen-stimulated rabbit spleen, phytohemagglutinin-stimulated human lymphocytes, dimethylbenzanthracene-induced leukemic rat spleen and liver, and phenylhydrazine-induced erythropoietic mouse spleen.

Although there undoubtedly exists a direct correlation between the activity level of the 6–8 S enzyme and the *in vivo* rate of DNA replication, there is unfortunately little evidence to suggest association of the enzyme with the replication machinery of the cell nucleus. Moreover, there is but scant evidence for a possible redistribution of total cell DNA polymerase during DNA replication. For example, cell cultures synchronized with respect to the growth cycle by the use of metabolic inhibitors showed an increase in the total DNA polymerase activity associated with the nuclear fractions as the cells entered S phase; this was accompanied by a corresponding decrease of polymerase in the cytoplasmic fraction. After completion of S phase, a similar migration of enzyme back to the cytoplasm from the nucleus occurred, reestablishing the original intracellular distribution (Littlefield *et al.*, 1963). Similar observations were described by Gold and Helleiner (1964) and Lindsay *et al.* (1970). Increases of activity of total DNA polymerase in crude nuclear fractions during S phase (but not the concomitant decrease in the total cytoplasmic activity) have been reported (Friedman, 1970; Madreiter *et al.*, 1971).

Unfortunately, at no time has the polymerizing enzyme(s) been directly identified with either the 6–8 S or the 3.3 S type activity.

Intracellular migration of a DNA polymerase from the cytoplasm to the nucleus has been demonstrated in early developing sea urchin embryos (Loeb and Fansler, 1970); the migration occurred over a series of cell divisions at the end of which up to 95% of the total DNA polymerase activity became associated with the nucleus. Over the entire period, the total DNA polymerase activity of the system remained constant. An extension of this work (Fansler and Loeb, 1973) revealed a reversible association of the DNA polymerase with the nucleus, the enzyme content of the nucleus being highest at the beginning of DNA synthesis and subsequently declining as replication approached completion. The polymerase was then released back into the cytoplasm and subsequently reassociated with the chromosomes during the latter stages of the ensuing mitosis. The translocation was repeated with each cell cycle. It is difficult to relate this echinoderm DNA polymerase activity to that of mammalian systems. There is evidence for the existence in sea urchin nuclei of two DNA polymerases, one of which has been purified and characterized (Loeb, 1969; Loeb and Fansler, 1970). However, it has yet to be demonstrated by the criteria of Chang and Bollum (1972a, b) that the DNA polymerase initially monitored in the cytoplasm is the same activity as is found later in the nucleus.

The release of enzyme from the nucleus at the end of S phase in the sea urchin embryo system has several parallels in mammalian systems. Craig *et al.* (1973), using cultures of baby hamster kidney cells (BHK-21/C13) stimulated to enter the growth cycle in synchrony, demonstrated that although total nuclear DNA polymerase activity passed through a maximum in the S phase, the total cytoplasmic enzyme increased in activity as the S phase progressed, reaching a maximum only at the termination of the S phase. The cytoplasmic polymerase activity was shown to be due to the 6–8 S species (Craig and Keir, 1973), one possible interpretation of the observations being that the 6–8 S polymerase was released into the cytoplasm as a consequence of replication. A similar concept has been proposed by Chang *et al.* (1973) while earlier work on subcultured mouse L cells (Lindsay *et al.*, 1970) also showed that total cytoplasmic DNA polymerase activity rose to a maximum toward the end of S phase. This cytoplasmic activity is now known to be attributable to the high-molecular-weight DNA polymerase (Adams *et al.*, 1973).

The implications of the apparent release of DNA polymerase from the nucleus into the cytoplasm as a result of DNA replication will be considered in Section VII.

V. A Comparison of the Properties of the 3.3 S and 6–8 S DNA Polymerases

A. Perspectives

Chang and Bollum (1972a) have emphasized that characterization of DNA polymerases, on the basis of template preferences and other reaction properties, in crude extracts or in preparations at a low level of purification, is highly questionable. This is the consequence of the presence in such enzyme preparations of certain contaminating activities, notably deoxyribonucleases, which are capable of altering the state of DNA templates used in the polymerization reaction. Bearing in mind this note of caution, we decided to compare and contrast only those polymerases that have been purified to a degree sufficient, in our estimation, to allow a meaningful assessment of the *in vitro* activities of the enzymes concerned. It is our opinion that this approach will avoid problems of interpretation of artifactual results that might arise from modification or masking the true polymerase reaction mechanism as a consequence of the presence of such contaminating factors. It will also spare the reader the daunting prospect of absorbing a lengthy catalogue of enzyme parameters derived from experimental observations made on a wide variety of tissues and cells.

B. pH Optima

In agreement with the early work on the chromatin-bound enzyme of rat liver (Howk and Wang, 1969), the 3.3 S DNA polymerases in general display a pH optimum in the alkaline range, 7.7 (Berger *et al.*, 1971) to 9.2 (Sedwick *et al.*, 1972). In contrast, the 6–8 S polymerase activities have a pH optimum nearer to neutrality, covering the range 6.8 (Chiu and Sung, 1972a, b) to 8.0 (Baril *et al.*, 1971). An exception to this appears to be the 6–8 S type activity purified by Sedwick *et al.* (1972) from KB cells; it has a pH optimum of 9.2, identical to that of the 3.3 S DNA polymerase activity isolated from the same cell line. However, these values must not be regarded as absolute. Chang and Bollum (1972a) clearly demonstrated that the figures quoted are dependent upon the template used. For example, the 3.3 S DNA polymerase from rabbit bone marrow, which is apparently free from nuclease, is optimally active at pH 8.6 when utilizing as a template activated DNA or alternatively the homopolymer, poly(dC), provided with an oligonucleotide primer as an initiator. However, the pH optimum is 7.6 when the template utilized is poly(dA) with an oligonucleotide initiator. Weissbach *et al.* (1971) also showed that the apparent pH optimum of the 6–8 S polymerase from HeLa cells varied, depending on the ratio of enzyme to template. When assayed at a low template concentration, the pH optimum was between 6.5 and 7.0, but the same enzyme assayed at high template concentration displayed a pH optimum in the range 8.5–9.0.

It is obvious from these reports that definition of the pH optimum of a DNA polymerase reaction is not a simple matter but is heavily dependent upon reaction conditions, especially with respect to template.

C. *Monovalent and Divalent Metal Cation Requirements*

1. MONOVALENT CATIONS

The effects of monovalent cations on the 3.3 S and 6–8 S DNA polymerase activities are complex. At high ionic strength (for example, 150–200 m*M* KCl), both the 3.3 S polymerase (Howk and Wang, 1969; Weissbach *et al.*, 1971; Chiu and Sung, 1972b) and the 6–8 S activity (Weissbach *et al.*, 1971; Sedwick *et al.*, 1972; Chiu and Sung, 1972b) show marked inhibition. The 6–8 S enzyme in particular is unstable under such conditions (Haines *et al.*, 1971), while the activity of the 3.3 S enzyme, although largely inhibited at high ionic strength, shows optimum stimulation by monovalent cations at a concentration of about 50 m*M* NaCl or KCl (Haines *et al.*, 1972; Chiu and Sung, 1972a, b). Berger *et al.* (1971) and Sedwick *et al.* (1972) made interesting observations with respect to increased stability of the 3.3 S polymerase in 0.2 *M* KCl. Berger *et al.* (1971), for example, demonstrated that although the initial reaction rate of the enzyme using an activated DNA template in the presence of 0.2 *M* KCl is slower than in its absence, the final level of deoxyribonucleotide incorporation achieved in the presence of 0.2 *M* KCl is 3 times that obtained in its absence over a time course of 12 hr. Hence, the 3.3 S polymerase appears to be a more stable activity in its partially purified form at high salt concentrations than is the 6–8 S polymerase.

2. DIVALENT CATIONS

All DNA polymerases which have been isolated and characterized require the presence of a divalent cation for activity.

When activated DNA is used as a template, Mg^{2+} ions are preferred, the 3.3 S polymerase displaying optima within the range 10–20 m*M*. Mn^{2+} ions may to some extent replace Mg^{2+} ions although they are less effective, enzyme activity ranging from 4% (Berger *et al.*, 1971) to 33% (Chang and Bollum, 1972a) of that obtained using Mg^{2+} ions. However, this situation is not generally applicable. For example, Stavrianopoulos *et al.* (1972a) purified from chick embryo a low-molecular-weight DNA polymerase that shows virtually absolute dependence on Mn^{2+} ions, Mg^{2+} ions being only 3% as effective. Further, the 3.3 S polymerase from rabbit bone marrow displays a distinct preference for Mn^{2+} ions over Mg^{2+} ions when the activated DNA in the assay is replaced as template by the initiated homopolymers $dA_n \cdot dT_{12}$ and $dC_n \cdot dG_{5\text{-}12}$ (Chang and Bollum, 1972a).

The 6–8 S polymerases show similar divalent cation requirements, Mg^{2+} ion optima covering the range 3 m*M* (Chiu and Sung, 1971; Tsuruo *et al.*, 1972a, b) to 10 m*M* (Sedwick *et al.*, 1972). Again, Mn^{2+} ions can to varying degrees replace Mg^{2+} ions (Baril *et al.*, 1971; Chiu and Sung, 1971).

The wide range of divalent cation optima exhibited by the various polymerase preparations is in some measure due to the disparate assay conditions used in different laboratories. The higher the concentration of template and/or deoxyribonucleoside triphosphate precursors in the assay, the higher is the optimum concentration of divalent cation. This situation can be ascribed to some extent to the chelating action of the triphosphates and to the binding of divalent cation to the template, especially at the higher pH values.

D. Requirement for Deoxyribonucleoside Triphosphates

Replicative DNA polymerases require the presence of all four deoxyribonucleoside 5′-triphosphates for maximum activity with DNA templates. Nevertheless, relatively pure preparations of both the 3.3 S and 6–8 S polymerase species can show considerable activity on DNA templates when the full complement of triphosphates is lacking. In the past, this situation has been interpreted as an indication of contamination of the replicative polymerases by terminal DNA polymerase (terminal deoxynucleotidyltransferase) activity. Such activity appears in general to be higher for the 3.3 S polymerase than for the 6–8 S species (Baril *et al.*, 1971; Weissbach *et al.*, 1971; Chiu and Sung, 1972d; Sedwick *et al.*, 1972; Tsuruo *et al.*, 1972a, b; Wicha and Stockdale, 1972). However, the substantial increase in activity observed in the presence of all four deoxyribonucleoside triphosphates, the lack of incorporation of ribonucleoside triphosphate, and the requirement for base-pairing of the incorporated deoxyribonucleoside triphosphates with the template (Baril *et al.*, 1971; Chiu and Sung, 1972d; Sedwick *et al.*, 1972) are responses not compatible with the activities reported for the terminal DNA polymerase (Krakow *et al.*, 1961; Bollum *et al.*, 1964; Gottesman and Canellakis, 1966; Kato *et al.*, 1967; Wang, 1968; Chang, 1971; Felix, 1972; Roychoudhury, 1972).

Chang and Bollum (1972a) have shown that although the rabbit bone marrow 3.3 S DNA polymerase gave up to 36% of its maximal activity on an activated DNA template when provided with only one deoxyribonucleoside triphosphate, the same enzyme, when assayed with a poly(dA) initiator and dGTP (Chang, 1971) displayed no detectable activity and hence is not a terminal DNA polymerase. Similar experiments using oligodeoxyribonucleotide initiators such as $(pT)_3$, $(pT)_5$, and $(pT)_9$, or the polyribonucleotide initiator poly(A), or the polydeoxyribonucleotide initiator poly[d(A-T)], with a single species of deoxyribonucleoside triphosphate, have been used to demonstrate the absence of terminal DNA polymerase activity from both the 6–8 S and the 3.3 S polymerases (Weissbach *et al.*, 1971; Chiu and Sung, 1972d; Sedwick *et al.*, 1972; Tsuruo *et al.*, 1972a, b; Schlabach *et al.*, 1971). Sedwick *et al.* (1972) measured the number of available 3′-hydroxy termini in their DNA primer templates using DNA polymerase I from *E. coli*, a single deoxyribonucleoside triphosphate, and limiting amounts of DNA primer template (Adler *et al.*, 1958). This allowed them to show that the amount of a single species of nucleotide incorporated by the 3.3 S and 6–8 S polymerases of KB cells is always substantially less than the number of initiation sites available for single nucleotide addition, strongly suggesting that this incorporation is not due to a contaminating or associated terminal DNA polymerase activity.

Evidence in a similar vein has been presented by Chang and Bollum (1972a), who showed that the relaxed requirement for all four deoxyribonucleoside triphosphates is due at least in part to the nature of the DNA template. If DNA treated with deoxyribonuclease I is used as template ("nicked" DNA), less than 1% of the DNA can be replicated; hence the nucleotide incorporation observed in the presence of one to three species of triphosphate is probably due to the replication of many short regions of DNA template, each being satisfied by a monomer provision of less than the full complement of four. However, if the nicked DNA is further treated with exonuclease III from *E. coli* (Richardson and Kornberg, 1964; Richardson *et al.*, 1964) to extend the nicks into single-strand gaps, consequently exposing greater lengths of the template for replicative polymerization, then the requirement for all four deoxyribonucleoside triphosphates becomes more stringent. Thus, if the incorporation of a single species of triphosphate is contrasted with that of all four over a time course, the relative activity in the absence of three triphosphates is low (Chang and Bollum, 1972a; Haines *et al.*, 1972; Sedwick *et al.*, 1972). Hence the stringency of the triphosphate requirement of the 3.3 S and 6–8 S DNA polymerases appears to be a function of the state of the DNA template used.

E. Associated or Contaminating Enzyme Activities

DNA polymerases have been extensively purified and characterized from a variety of prokaryotic sources (see Section II), investigations revealing the presence in the polymerases of associated nuclease activities. All three *E. coli* DNA polymerases possess an exonuclease activity which hydrolyzes single-stranded DNA in the 3′–5′ direction, and DNA polymerase I has additionally a 5′–3′-exonuclease activity which requires a double-stranded DNA substrate (Kornberg, 1969; Gefter *et al.*, 1972; Kornberg and Gefter, 1972; Otto, 1973; Smith, 1973). Moreover, associated nuclease activities have been found in bacteriophage-induced DNA polymerases (see Smith, 1973).

In addition to associated nuclease activities, a nucleoside diphosphokinase activity is known to be present in highly purified preparations of DNA polymerases I and II from *E. coli* and *Micrococcus luteus* (Miller and Wells, 1971). Therefore it seems appropriate to examine the current status of associated enzyme activities in eukaryotic DNA polymerases.

Sedwick *et al.* (1972) have reported the presence of nucleoside diphosphokinase activity in their highly purified 6–8 S polymerase from KB cells and also in the 3.3 S species. In addition, Haines *et al.* (1972) described a similar activity in the 3.3 S polymerase isolated from rat liver nuclei. Although none of the above polymerases has been conclusively shown to be purified to homogeneity, there does appear to be a distinct possibility that nucleoside diphosphokinase may prove eventually to be a constituent catalytic activity of all DNA polymerases.

The situation with regard to the presence or absence of associated nuclease activity is less clear. Evidence generally favors the total absence of associated exo- or endonuclease from relatively pure preparations of the 3.3 S polymerase (Berger *et al.*, 1971; Chang and Bollum, 1972a; Haines *et al.*, 1972; Stavrianopoulos *et al.*, 1972a). The absence of associated nuclease activities from the 6–8 S DNA polymerase isolated from calf thymus was established 10 years ago (Bollum, 1963a, b), while the apparent absence of such activities has been reported also for sea urchin embryo DNA polymerase (Loeb, 1969) and rat liver 6–8 S DNA polymerase (Baril *et al.*, 1971). On the other hand, Sedwick *et al.* (1972) claim the retention of low levels of activity of exonuclease in the KB cell 6–8 S polymerase, while Momparler *et al.* (1973), in demonstrating heterogeneity within the calf thymus 6–8 S polymerase species, also show the presence of some endonuclease activity associated with each of the fractions obtained by chromatography on DEAE-cellulose, and limited exonuclease activity associated with one of them.

Evidence for association of an exonuclease activity with a highly purified yeast DNA polymerase has been presented by Helfman (1973). However, the polymerase has yet to be purified to homogeneity; unless this is achieved without loss of the nuclease activity, there will remain only circumstantial evidence to suggest the association of exonuclease and/or endonuclease with yeast DNA polymerase.

At the present time, therefore, in the light of the positive evidence available, we are inclined to favor the view that eukaryotic and certainly mammalian DNA polymerases contain no associated nuclease activity.

F. Effects of Inhibitors

1. THE 3.3 S DNA POLYMERASE

The activity of this enzyme from rat liver (Baril *et al.*, 1971; Berger *et al.*, 1971; Haines *et al.*, 1971), from human cell lines (Weissbach *et al.*, 1971; Sedwick *et al.*, 1972), and from rat ascites hepatoma (Tsuruo *et al.*, 1972b), is inhibited somewhat by *p*-mercuribenzoate (*p*-chloromercuribenzoate, *p*-chlormercuribenzoate) at concentrations of the latter about 100 μM and above. The degree of inhibition varies somewhat according to the tissue under investigation and according to the precise experimental circumstances, but *p*-mercuribenzoate at 25–30 μM exerts no inhibition at all (Weissbach *et al.*, 1971; Sedwick *et al.*, 1972), while at concentrations of 300 μM and beyond, the extent of inhibition ranges from 16–20% (Baril *et al.*, 1971; Tsuruo *et al.*, 1972b) to 94% (Berger *et al.*, 1971). Haines *et al.*, (1971) observed that the inhibition of the 3.3 S polymerase activity by *p*-mercuribenzoate is reduced by half if the enzyme is first bound to DNA before exposure to the mercurial.

N-Ethylmaleimide at about 1.3 m*M*–2.5 m*M* inhibits the activity of the 3.3 S enzyme by 30–62% (Chiu and Sung, 1972a, d; Adams *et al.*, 1973), depending on the cell system used. At a concentration of 0.5 m*M*, *N*-ethylmaleimide does not inhibit the 3.3 S polymerase of human blood lymphocytes (Smith and Gallo, 1972). Other compounds also exert inhibitory effects on the 3.3 S polymerase; they include KCN at 4 m*M* (20% inhibition; Berger *et al.*, 1971) and ethidium bromide at 10 μM (10% inhibition; Berger *et al.*, 1971). However, Sedwick *et al.* (1972) claim no inhibition by ethidium bromide at concentrations up to 20 μM.

2. THE 6–8 S DNA POLYMERASE

This polymerase species is much more susceptible to inhibition by the thiol-active reagents, *p*-mercuribenzoate and *N*-ethylmaleimide, than is the 3.3 S enzyme. For example, the 6–8 S enzyme is inhibited to the extent of 80–100% by *p*-mercuribenzoate at concentrations of 10–40 μM

(Baril *et al.*, 1971; Haines *et al.*, 1971; Weissbach *et al.*, 1971; Sedwick *et al.*, 1972; Tsuruo *et al.*, 1972a, b). Unlike the situation with the 3.3 S enzyme, no protection against the inhibitory effect of *p*-mercuribenzoate was afforded by prior incubation of the 6–8 S polymerase with DNA (Haines *et al.*, 1971).

N-Ethylmaleimide is also strongly inhibitory with the 6–8 S polymerase relative to the 3.3 S species (Chiu and Sung, 1972a, b, d; Smith and Gallo, 1972; Adams *et al.*, 1973; Momparler *et al.*, 1973). Ethidium bromide also exerts a more potent inhibitory effect on the activity of the 6–8 S enzyme than it does on that of the 3.3 S enzyme (Sedwick *et al.*, 1972).

VI. Primer Requirements and Template Specificity

A. *Fidelity of Replication*

One of the major problems involved in the understanding of the mechanism of DNA replication concerns the molecular process responsible for the initiation of synthesis of new polynucleotide chains. Such a statement requires that the DNA polymerases involved in replication must recognize the DNA primer, presumably at a preformed initiation site, and must also be capable of catalyzing faithful, complementary synthesis of daughter strands using the parental strands as templates.

Sedwick *et al.* (1972) have demonstrated the ability of the 3.3 S and 6–8 S KB cell DNA polymerases to produce a faithful, complementary copy of the template using the alternating copolymer poly[d(A-T)]. The enzymes catalyze incorporation of dTTP and dATP into a polydeoxyribonucleotide product; no such incorporation occurs when dCTP and dGTP are provided as the monomeric substrates. Similar evidence has been presented by Momparler *et al.* (1973) using the 6–8 S activity from calf thymus, and by Baril *et al.*, (1972) using both the 3.3 S and 6–8 S activities of rat liver. Furthermore, Chang and Bollum (1972a) have demonstrated that the rabbit bone marrow 3.3 S DNA polymerase utilizes the template information for replication, the margin of error being less than 1%, using all four deoxyribonucleoside triphosphates as substrates and initiated homopolymers as templates.

B. *Nature of the Initiative 3′-Hydroxy Terminus*

All known replicative DNA polymerase activities require a DNA template in order to incorporate deoxyribonucleoside triphosphates into polymer chains. However, their ability to use such templates depends in turn on the availability of 3′-hydroxy termini. Hence, maximum incorporation catalyzed by highly purified 3.3 S and 6–8 S DNA polymerases has been obtained, not on native or denatured DNA primer templates, which contain relatively few 3′-hydroxy termini, but on the so-called "activated" DNA, a native DNA primer template, which has been nicked by limited digestion with pancreatic deoxyribonuclease I, producing a large number of 3′-hydroxy termini (Berger *et al.*, 1971; Weissbach *et al.*, 1971; Baril *et al.*, 1971; Chang and Bollum, 1972a; Haines *et al.*, 1972; Sedwick *et al.*, 1972; Tsuruo *et al.*, 1972a, b; Momparler *et al.*, 1973). A similar degree of digestion with micrococcal nuclease, which produces 3′-phosphoryl termini, results in the abolition of template activity, demonstrating the absolute requirement of DNA polymerases for 3′-hydroxy termini (Bollum 1963a, b; Schlabach *et al.*, 1971; Sedwick *et al.*, 1972; Tsuruo *et al.*, 1972a, b).

Sedwick *et al.* (1972) have also noted that, in contrast to the *E. coli* polymerases, there is no stimulation of activity if a "gapped" DNA primer template (a nicked DNA primer template digested with exonuclease III; this treatment converting nicks to gaps with relatively little increase in 3′-hydroxy termini) is used to measure 3.3 S and 6–8 S activity. In fact, the decisive reduction of activity of the 3.3 S enzyme on such a primer template has lead them to speculate whether this reflects the inability of such enzymes to traverse long gaps. A similar lack of stimulation has been reported for the 3.3 S activity purified to a similar degree from rat liver (Haines *et al.*, 1972).

The nature of the initiator required to prime both the 3.3 S and the 6–8 S DNA polymerizing activities has been extensively studied by Bollum (1963a, b). Using the extensively purified 6–8 S enzyme from calf thymus gland, which is known to be free from extraneous degradative enzyme activities (Bollum, 1963a, b; Yoneda and Bollum, 1965), Chang *et al.* (1972) demonstrated that synthesis of polydeoxyadenylate and polydeoxythymidylate as catalyzed by the 6–8 S enzyme could proceed only in the presence of a complementary oligodeoxyribonucleotide, such an initiating oligodeoxyribonucleotide being incorporated into the product chain. Similar evidence has also been presented by de Recondo *et al.* (1973). Thus the primary role of the oligodeoxyribonucleotide is to provide a 3′-hydroxy terminus from which point replication may proceed. A similar requirement for an oligodeoxyribonucleotide initiator to provide a 3′-hydroxy terminus has also been shown by the 3.3 S DNA polymerase from rabbit bone marrow (Chang and Bollum, 1972a).

Although the activity of both these 3.3 S and the 6–8 enzymes on homopolymer templates can be initiated using oligodeoxyribonucleotides, a major divergence in activity occurs if an oligoribonucleotide is used to initiate a polydeoxyribonucleotide template. Using 3.3 S and 6–8 S enzyme preparations known to be free from ribonuclease H—a ribonuclease capable of excising RNA from an RNA–DNA hybrid (Hausen and Stein, 1970)—Chang and Bollum (1972c), demonstrated that only the 6–8 S activities were capable of using oligoribonucleotides to initiate polydeoxyribonucleotide templates. A similar observation has been made by de Recondo *et al.* (1973), who demonstrated the ability of a purified high-molecular-weight 6–8 S type DNA polymerase from regenerating rat liver to utilize the polyribonucleotide strand of a synthetic RNA–DNA hybrid as an initiator for the synthesis of the complementary polydeoxyribonucleotide strand. Sedwick *et al.* (1972) also report preliminary experiments which demonstrate the limited activity, not only of the 6–8 S, but also the 3.3 S polymerase from KB cells utilizing a poly(A) primer to initiate polymerization on the template poly[$d(T)_{5000}$], with dATP serving as the monomeric precursor. Limited incorporation on a dT·rA primer template using dATP as substrate was observed by Schlabach *et al.* (1971) using the HeLa cell 3.3 S polymerase.

These observations obviously have far-reaching implications with regard to the *in vivo* initiation mechanism of DNA replication in eukaryotes. It is now well established that the synthesis of RNA is a prerequisite for DNA synthesis in prokaryotes. The association of RNA with nascent DNA has been demonstrated both *in vivo* (Sugino *et al.*, 1972) and *in vitro* (Sugino and Okazaki, 1973), while there is considerable evidence for the direct involvement of RNA in the initiation of DNA replication in *E. coli* (Lark, 1972), in the synthesis *in vitro* of the replicative form DNA of bacteriophage ϕX174 (Schekman *et al.*, 1972), in the conversion of bacteriophage M13 DNA to its replicative form *in vitro* (Wickner *et al.*, 1972), and in the replication of minicircular DNA in *E. coli* (Messing *et al.*, 1972).

Direct evidence linking RNA with the initiation of DNA synthesis in eukaryotic systems remains somewhat circumstantial. Preliminary evidence implicating the association of RNA to nascent DNA in Ehrlich ascites tumor cells *in vivo* has been presented by Sato *et al.* (1972). Magnusson *et al.* (1973), using *in vitro* systems, have isolated RNA-linked, short, nascent strands of DNA from nuclei of polyoma-infected cells. Keller (1972) has found that *in vitro* DNA synthesis on single-strand circular DNA can be initiated by RNA primers, the RNA chains made *in vitro* by *E. coli* RNA polymerase being covalently extended by the 6–8 S-type enzymes from KB cells. *In vitro* evidence for a coupling of replication to transcription has also been presented by Stavrianopoulos *et al.* (1972b).

Thus, there appears to be a distinct possibility that the initiation of new chains *in vivo* may proceed via the limited transcription of the parental DNA strands (presumably at the replication fork) by an RNA polymerase producing an oligoribonucleotide initiator. Such speculation directly implicates the 6–8 S enzyme activity with DNA replication, as chain propagation by the 3.3 S enzyme using an oligoribonucleotide as an initiator appears to be limited. It should perhaps be emphasized that at present no enzymatic mechanisms known for the *in vivo* generation of oligodeoxyribonucleotides.

C. Utilization of RNA–DNA Hybrid Templates

So far in this section we have described only the DNA-dependent DNA polymerizing activities of both the 6–8 S and the 3.3 S enzymes. However, there is a considerable amount of conflicting evidence, implicating the ability of the 3.3 S activity to utilize a polyribonucleotide template, if provided with a complementary oligodeoxyribonucleotide as an initiator. However, no such activity can be associated with the 6–8 S DNA polymerizing activity isolated from HeLa cells (Fridlender *et al.*, 1972), KB cells (Sedwick *et al.*, 1972), regenerating rat liver (de Recondo *et al.*, 1973), or calf thymus (Chang and Bollum, 1972a; Momparler *et al.*, 1973).

Evidence for the association of an RNA-dependent DNA polymerizing activity with a highly purified 3.3 S DNA polymerase from rat liver has been presented by Haines *et al.* (1972), who demonstrated a threefold increase in the incorporation of dTTP into poly(A)·poly(dT) over that of activated calf thymus DNA. Similar activities associated with 3.3 S type DNA polymerizing activities have been reported in chicken embryo (Stavrianopoulos *et al.*, 1972a), rabbit bone marrow (Chang and Bollum, 1972a), normal human blood lymphocytes (Smith and Gallo, 1972), and BALB/3T3 cells (Ross *et al.*, 1971).

The RNA-dependent DNA polymerizing activity, however, cannot be identified with purified 3.3 S DNA polymerizing activities from HeLa cells (Fridlender *et al.*, 1972) or KB cells (Sedwick *et al.*, 1972), neither enzyme showing the ability to incorporate dTTP into primer templates such as poly(A)·oligo$(dT)_{12}$. Moreover, it is of interest to note that the RNA-dependent DNA polymerizing activities of both HeLa cells (Fridlender *et al.*, 1972) and KB cells (Sedwick *et al.*, 1972) appear to reside in a different enzyme. The existence of distinct RNA-dependent and DNA-dependent DNA polymerases has also been reported in rat liver (Ward *et al.*, 1972) and chick embryo (Maia *et al.*, 1971).

Whether Haines *et al.* (1972) are in fact monitoring a true activity of the rat liver 3.3 S DNA polymerase or a contaminant must remain speculation, since although both activities copurify (Wickremasinghe *et al.*, 1973), the 3.3 S DNA polymerase has yet to be purified to homogeneity.

VII. Interrelation between 3.3 S and 6–8 S DNA Polymerase Activities

Chang and Bollum (1972d) have shown, in experiments utilizing rabbit antibody prepared against the calf thymus 6–8 S DNA polymerase, that the activities of both the 3.3 S and the 6–8 S DNA polymerases from a variety of animal tissues and cell lines were inhibited by the antibody preparation. This demonstrates the presence of common antigenic determinants not only in the two major polymerase species of a single cell type, but also in the corresponding enzymes from distinct mammalian sources. The activity of *E. coli* DNA polymerase I and the terminal deoxynucleotidyltransferase remained unaffected by the antibody preparation.

The presence of common antigenic determinants is indicative of polypeptide sequences or subunits common to both the 3.3 S and 6–8 S DNA polymerases from all mammalian sources. Recent work involving the fractionation of the heterogeneous 6–8 S (see Section I) has led to the identification of a 5 S subunit common to both the 6 S and the 8 S polymerases isolated from both calf thymus and mouse L929 cells (Chang *et al.*, 1973). Moreover, work on the 3.3 S and 6–8 S DNA polymerizing activities isolated from mouse testis (Hecht, 1972; Hecht and Davidson, 1973) suggests that an active subunit can be dissociated from the 6–8 S DNA polymerase by treatment with $(NH_4)_2SO_4$, and that the resulting subunit appears identical with the 3.3 S DNA polymerase. This dissociation is apparently reversible. Hecht and Davidson (1973), on the basis of this dissociation phenomenon, propose that the 3.3 S activity shuttles between nucleus and cytoplasm associating with other molecules in the cytoplasm to produce the 6–8 S species, the relative distribution between 3.3 S and 6–8 S species depending on the physiological state of the cell.

In the light of an earlier hypothesis (Keir, 1965) and the work of Chang and Bollum (1972d), Chang *et al.* (1973), and Hecht and Davidson (1973), it seems quite feasible that an as yet unspecified number of subunits participate in association–dissociation phenomena, the nature of which is related to the physiological requirement of the cell, resulting in the passage of an appropriately modified form of a basic DNA polymerizing subunit to a specific nuclear site, these promoting polymerization reactions for nuclear events as DNA replication, DNA repair, and gene amplification.

The nature and function of such subunits remains open to speculation. The replication of chromosomal DNA is a complex event not just restricted to the polymerization of deoxyribonucleotides. Experimental evidence has implicated cytoplasmic factors in the replication process (see Section III,B); hence it would not be unreasonable to suppose that these factors might well be involved, not only in such mechanisms as the unwinding of the DNA at the replication fork, but also in the recognition of initiator sites on different replicons, hence ensuring that the particular pattern and temporal sequence of chromosomal DNA replication are maintained (Kidwell, 1972).

Momparler *et al.* (1973) have already drawn analogies between the apparent situation with respect to eukaryote DNA polymerases and the case that is known to exist with the *E. coli* RNA polymerase. In the latter situation, the functional enzyme can be separated into two components, the core enzyme and the σ-factor (Burgess *et al.*, 1969). A similar situation could not only explain the known heterogeneity of eukaryotic DNA polymerases, but also the apparent accumulation in the cytoplasm of the heterogeneous 6–8 S species during the later stages of S phase. Each such 6–8 S complex would presumably be specific to a particular part of the genome, becoming redundant once this has been replicated, and consequently ejected into the cytoplasm.

The apparent absence of such a complex from the cell nucleus in a large number of preparations (see Section III,C) could well be due to some aspect of the isolation procedure (see Section III,E). A particular point worth emphasizing is the known instability of the 6–8 S enzyme in conditions of high ionic strength and the extraction procedures used in the isolation of nuclear enzymes.

REFERENCES

Adams, R. L. P., and Wood, W. (1973). *Biochem. Soc. Trans. 537th Meeting, Canterbury.*

Adams, R. L. P., Henderson, M. A. L., Wood, W., and Lindsay, J. G. (1973). *Biochem. J.* **131,** 237–246.

Adler, J., Lehman, I. R., Bessman, M. J., Simms, E. S., and Kornberg, A. (1958). *Proc. Nat. Acad. Sci. U.S.* **44,** 641–647.

Arms, K. (1971). *Develop. Biol.* **26,** 497–502.

Ballal, N. R., Collins, M. S., Halpern, R.. M., and Smith, R. A. (1970). *Biochem. Biophys. Res. Commun.* **40,** 1201–1208.

Baril, E. F., Brown, O. E., Jenkins, M. D., and Laszlo, J. (1971). *Biochemistry* **10,** 1981–1992.

Bazill, G. W., and Philpot, J. St.L. (1963). *Biochim. Biophys. Acta* **76,** 223–233.

Behki, R. M., and Schneider, W. C. (1963). *Biochim. Biophys. Acta* **68,** 34–44.

Bellair, J. T. (1968). *Biochim. Biophys. Acta* **161**, 119–124.
Berger, H., Huang, R. C. C., and Irvin, J. L. (1971). *J. Biol. Chem.* **246**, 7275–7283.
Blobel, G., and Potter, V. R. (1966). *Science* **154**, 1662–1665.
Blobel, G., and Potter, V. R. (1967). *J. Mol. Biol.* **26**, 279–292.
Bollum, F. J. (1963a). *Progr. Nucl. Acid Res.* **1**, 1–26.
Bollum, F. J. (1963b). *Cold Spring Harbor Symp. Quant. Biol.* **28**, 21–26.
Bollum, F. J., Groeniger, E., and Yoneda, M. (1964). *Proc. Nat. Acad. Sci. U.S.* **51**, 853–859.
Burgess, R. R., Travers, A. A., Dunn, J. J., and Bautz, E. K. F. (1969). *Nature (London)* **221**, 43–46.
Burgoyne, L. A., Waqar, M. A., and Atkinson, M. R. (1970a). *Biochem. Biophys. Res. Commun.* **39**, 254–259.
Burgoyne, L. A., Waqar, M. A., and Atkinson, M. R. (1970b) *Biochem. Biophys. Res. Commun.* **39**, 918–922.
Chang, L. M. S. (1971). *Biochem. Biophys. Res. Commun.* **44**, 124–131.
Chang, L. M. S. (1973). *J. Biol. Chem.* **248**, 3789–3793.
Chang, L. M. S., and Bollum, F. J. (1971a). *J. Biol. Chem.* **246**, 909–916.
Chang, L. M. S., and Bollum, F. J. (1971b). *J. Biol. Chem.* **246**, 5835–5837.
Chang, L. M. S., and Bollum, F. J. (1972a). *Biochemistry* **11**, 1264–1272.
Chang, L. M. S., and Bollum, F. J. (1972b). *J. Biol. Chem.* **247**, 7948–7950.
Chang, L. M. S., and Bollum, F. J. (1972c). *Biochem. Biophys. Res. Commun.* **46**, 1354–1360.
Chang, L. M. S., and Bollum, F. J. (1972d). *Science* **175**, 1116–1117.
Chang, L. M. S., Cassani, G. R., and Bollum, F. J. (1972). *J. Biol. Chem.* **247**, 7718–7723.
Chang, L. M. S., Brown, M., and Bollum, F. J. (1973). *J. Mol. Biol.* **74**, 1–8.
Chiu, J.-F., and Sung, S. C. (1971). *Biochim. Biophys. Acta* **246**, 44–50.
Chiu, J.-F., and Sung, S. C. (1972a). *Biochem. Biophys. Res. Commun.* **46**, 1830–1835.
Chiu, J.-F., and Sung, S. C. (1972b). *Biochim. Biophys. Acta* **262**, 397–400.
Chiu, J.-F., and Sung, S. C. (1972c). *Biochim. Biophys. Acta* **269**, 364–369.
Chiu, J.-F., and Sung, S. C. (1972d). *Nature (London)* **239**, 176–178.
Craig, R. K., and Keir, H. M. (1973). Manuscript in preparation.
Craig, R. K., Costello, P. A., and Keir, H. M. (1973). Manuscript in preparation.
Crerar, M., and Pearlman, R. E. (1971). *FEBS Lett.* **18**, 231–237.
Deumling, B., and Franke, W. W. (1972). *Hoppe-Seyler's Z. Physiol. Chem.* **353**, 287–297.
Eckstein, H., Paduch, V., and Hilz, H. (1967). *Eur. J. Biochem.* **3**, 224–231.
Fansler, B., and Loeb, L. A. (1973). *Exp. Cell Res.* **75**, 433–441.
Felix, G. (1972). *Biochem. Biophys. Res. Commun.* **46**, 2141–2147.
Fridlender, B., Fry, M., Bolden, A., and Weissbach, A. (1972). *Proc. Nat. Acad. Sci. U.S.* **69**, 452–455.
Friedman, D. L. (1970). *Biochem. Biophys. Res. Commun.* **39**, 100–109.
Friedman, D. L., and Mueller, G. C. (1968). *Biochim. Biophys. Acta* **161**, 455–468.
Furlong, N. B., and Gresham, C. (1971). *Texas Rep. Biol. Med.* **29**, 75–82.
Gefter, M. L., Molineux, I. J., Kornberg, T., and Khorana, H. G. (1972). *J. Biol. Chem.* **247**, 3321–3326.
Gold, M., and Helleiner, C. W. (1964). *Biochim. Biophys. Acta* **80**, 193–203.
Gottesman, M. E., and Canellakis, E. S. (1966). *J. Biol. Chem.* **241**, 4339–4352.
Grippo, P., and Lo Scavo, A. (1972). *Biochem. Biophys. Res. Commun.* **48**, 280–285.

Grisham, J. W., Kaufman, D. G., and Stenstrom, M. L. (1972). *Biochem. Biophys. Res. Commun.* **49**, 420–427.
Gurdon, J. B. (1967). *Proc. Nat. Acad. Sci. U.S.* **58**, 545–552.
Haines, M. E., Holmes, A. M., and Johnston, I. R. (1971). *FEBS Lett.* **17**, 63–67.
Haines, M. E., Wickremasinghe, R. G., and Johnston, I. R. (1972). *Eur. J. Biochem.* **31**, 119–129.
Harm, K., and Hilz, H. (1971). *Hoppe-Seyler's Z. Physiol. Chem.* **352**, 1469–1479.
Harris, H. (1967). *J. Cell Sci.* **2**, 23–32.
Hausen, P., and Stein, H. (1970). *Eur. J. Biochem.* **14**, 278–283.
Hecht, N. B. (1971). *Exp. Cell Res.* **70**, 248–250.
Hecht, N. B. (1972). *J. Cell. Biol.* **55**, 109a.
Hecht, N. B., and Davidson, D. (1973). *Biochem. Biophys. Res. Commun.* **51**, 299–305.
Hecht, N. B., and Stern, H. (1971). *Exp. Cell Res.* **69**, 1–10.
Helfman, W. B. (1973). *Eur. J. Biochem.* **32**, 42–50.
Hershey, H. V., Stieber, J. F., and Mueller, G. C. (1973). *Eur. J. Biochem.* **34**, 383–394.
Holmes, A. M., Hesslewood, I. P., and Johnston, I. R. (1973). *Biochem. Soc. Trans. 540th Meeting, Oxford.*
Howell, S. H., and Hecht, N. B. (1971). *Biochim. Biophys. Acta* **240**, 343–352.
Howk, R., and Wang, T. Y. (1969). *Arch. Biochem. Biophys.* **133**, 238–246.
Iwamura, Y., Ono, T., and Morris, H. P. (1968). *Cancer Res.* **28**, 2466–2476.
Kato, K., Goncalves, J. M., Houts, G. E., and Bollum, F. J. (1967). *J. Biol. Chem.* **242**, 2780–2789.
Keir, H. M. (1965). *Progr. Nucl. Acid Res. Mol. Biol.* **4**, 81–128.
Keir, H. M., Smellie, R. M. S., and Siebert, G. (1962). *Nature (London)* **196**, 752–754.
Keller, S. J., Biedenbach, S. A., and Meyer, R. R. (1973). *Biochem. Biophys. Res. Commun.* **50**, 620–628.
Keller, W. (1972). *Proc. Nat. Acad. Sci. U.S.* **69**, 1560–1564.
Kidwell, W. R. (1972). *Biochim. Biophys. Acta* **269**, 51–61.
Kidwell, W. R., and Mueller, G. C. (1969). *Biochem. Biophys. Res. Commun.* **36**, 756–763.
Klose, J., and Flickinger, R. A. (1972). *Develop. Biol.* **29**, 214–219.
Kornberg, A. (1969). *Science* **163**, 1410–1418.
Kornberg, T., and Gefter, M. L. (1972). *J. Biol. Chem.* **247**, 5369–5375.
Krakow, J. S., Kammen, H. O., and Canellakis, E. S. (1961). *Biochim. Biophys. Acta* **53**, 52–64.
Kumar, K. V., and Friedman, D. L. (1972). *Nature (London)* **239**, 74–76.
Lagunoff, D. (1969). *Exp. Cell Res.* **55**, 53–56.
Lark, K. G. (1972). *J. Mol. Biol.* **64**, 47–60.
Leung, F. Y. T., and Zbarsky, S. H. (1970). *Can. J. Biochem.* **48**, 529–536.
Libonati, M., Liguori, G., and Guiditta, A. (1972). *J. Neurochem.* **19**, 1959–1965.
Lindsay, J. G., Berryman, S., and Adams, R. L. P. (1970). *Biochem. J.* **119**, 849–860.
Littlefield, J. W., McGovern, A. P., and Margeson, K. B. (1963). *Proc. Nat. Acad. Sci. U.S.* **49**, 102–107.
Loeb, L. A. (1969). *J. Biol. Chem.* **244**, 1672–1681.
Loeb, L. A., and Fansler, B. (1970). *Biochim. Biophys. Acta* **217**, 50–55.
Long, G. L., and Garren, L. D. (1972). *Biochem. Biophys. Res. Commun.* **46**, 1228–1235.
Lynch, W. E., Brown, R. F., Umeda, T., Langreth, S. G., and Lieberman, I. (1970). *J. Biol. Chem.* **245**, 3911–3916.

Lynch, W. E., Umeda, T., Uyeda, M., and Lieberman, I. (1972). *Biochim. Biophys. Acta* **287**, 28–37.
Madreiter, H., Kaden, P., and Mittermayer, C. (1971). *Eur. J. Biochem.* **18**, 369–375.
Magnusson, G., Pigiet, V., Winnacker, E. L., Abrams, A., and Reichard, P. (1973). *Proc. Nat. Acad. Sci. U.S.* **70**, 412–415.
Maia, J. C. C., Rougeon, F., and Chapeville, F. (1971). *FEBS Lett.* **18**, 130–134.
Main, R. K., and Cole, L. J. (1964). *Nature (London)* **203**, 646–648.
Masui, H., and Garren, L. D. (1970). *J. Biol. Chem.* **245**, 2627–2632.
McLennan, A. G., and Keir, H. M. (1973). *Biochem. Soc. Trans. 538th Meeting, Birmingham.*
Messing, J., Staudenbauer, W. L., and Hofschneider, P. H. (1972). *Nature (London)* **238**, 202–203.
Meyer, R. R., and Simpson, M. V. (1968). *Proc. Nat. Acad. Sci. U.S.* **61**, 130–137.
Miller, L. K., and Wells, R. D. (1971). *Proc. Nat. Acad. Sci. U.S.* **68**, 2298–2302.
Momparler, R. L., Rossi, M., and Labitan, A. (1973). *J. Biol. Chem.* **248**, 285–293.
Murthy, M. R. V., and Bharucha, A. D. (1971). *Can. J. Biochem.* **49**, 1285–1291.
Murthy, M. R. V., and Bharucha, A. D. (1972). *Can. J. Biochem.* **50**, 186–189.
Ono, T., and Umehara, Y. (1968). *Gann Monogr.* **6**, 97–107.
Otto, B. (1973). *Biochem. Soc. Trans. 537th Meeting, Canterbury.*
Ove, P., Brown, O. E., and Laszlo, J. (1969). *Cancer Res.* **29**, 1562–1576.
Patel, G., Howk, R., and Wang, T. Y. (1967). *Nature (London)* **215**, 1488–1489.
Poulson, R., Krasny, J., and Zbarsky, S. H. (1973). *Biochim. Biophys. Acta* **299**, 533–544.
Probst, G. S., Bikoff, E., Keller, S. J., and Meyer, R. R. (1972). *Biochim. Biophys. Acta* **281**, 216–227.
de Recondo, A.–M., Lepesant, J.–A., Fichot, O., Grasset, L., Rossignol, J.–M., and Cazillis, M. (1973). *J. Biol. Chem.* **248**, 131–137.
Richardson, C. C., and Kornberg, A. (1964). *J. Biol. Chem.* **239**, 242–250.
Richardson, C. C., Lehman, I. R., and Kornberg, A. (1964). *J. Biol. Chem.* **239**, 251–258.
Ross, J., Scolnick, E. M., Todaro, G. J., and Aaronson, S. A. (1971). *Nature (London)* **231**, 163–167.
Roychoudhury, R. (1972). *J. Biol. Chem.* **247**, 3910–3917.
Sato, S., Ariake, S., and Saito, M. (1972). *Biochem. Biophys. Res. Commun.* **49**, 827–834.
Schekman, R., Wickner, W., Westergaard, O., Brutlag, D., Geider, K., Bertsch, L. L., and Kornberg, A. (1972). *Proc. Nat. Acad. Sci. U.S.* **69**, 2691–2695.
Schiebel, W., and Bamberg, U. (1973). *Biochem. Soc. Trans. 537th Meeting, Canterbury.*
Schlabach, A., Fridlender, B., Bolden, A., and Weissbach, A. (1971). *Biochem. Biophys. Res. Commun.* **40**, 879–888.
Schönherr, O. T., and Keir, H. M. (1972) *Biochem. J.* **129**, 285–290.
Sedwick, W. D., Wang, T. S.-F., and Korn, D. (1972). *J. Biol. Chem.* **247**, 5026–5033.
Shimada, H., and Terayama, H. (1972). *Biochim. Biophys. Acta* **287**, 415–426.
Smith, D. W. (1973). *Progr. Biophys. Mol. Biol.* **26**, 321–408.
Smith, R. G., and Gallo, R. C. (1972). *Proc. Nat. Acad. Sci. U.S.* **69**, 2879–2884.
Smith, M. J., and Keir, H. M. (1963). *Biochim. Biophys. Acta* **68**, 578–588.
Srivastava, B. I. S. (1973). *Biochim. Biophys. Acta* **299**, 17–23.
Stavrianopoulos, J. G., Karkas, J. D., and Chargaff, E. (1971). *Proc. Nat. Acad. Sci. U.S.* **68**, 2207–2211.

Stavrianopoulos, J. G., Karkas, J. D., and Chargaff, E. (1972a). *Proc. Nat. Acad. Sci. U.S.* **69**, 1781–1785.

Stavrianopoulos, J. G., Karkas, J. D., and Chargaff, E. (1972b). *Proc. Nat. Acad. Sci. U.S.* **69**, 2609–2613.

Stockdale, F. E. (1970). *Develop. Biol.* **21**, 462–474.

Stout, E. R., and Arens, M. Q. (1970). *Biochim. Biophys. Acta* **213**, 90–100.

Sugino, A., and Okazaki, R. (1973). *Proc. Nat. Acad. Sci. U.S.* **70**, 88–92.

Sugino, A., Hirose, S., and Okazaki, R. (1972). *Proc. Nat. Acad. Sci. U.S.* **69**, 1863–1867.

Sung, S. C. (1969). *Can. J. Biochem.* **47**, 47–50.

Takats, S. T., and Wever, G. H. (1971). *Exp. Cell Res.* **69**, 25–28.

Teng, C., Bloch, D. P., and Roychoudhury, R. (1970). *Biochim. Biophys. Acta* **224**, 232–245.

Tewari, K. K., and Wildman, S. G. (1967). *Proc. Nat. Acad. Sci. U.S.* **58**, 689–696.

Thompson, L. R., and McCarthy, B. J. (1968). *Biochem. Biophys. Res. Commun.* **30**, 166–172.

Tsuruo, T., Satoh, H., and Ukita, T. (1972a). *Biochem. Biophys. Res. Commun.* **48**, 769–775.

Tsuruo, T., Tomita, Y., Satoh, H., and Ukita, T. (1972b). *Biochem. Biophys. Res. Commun.* **48**, 776–782.

van der Vliet, P. C., and Sussenbach, J. S. (1972). *Eur. J. Biochem.* **30**, 584–592.

Wallace, P. G., Hewish, D. R., Venning, M. M., and Burgoyne, L. A. (1971). *Biochem. J.* **125**, 47–54.

Wang, T. Y. (1967). *Arch. Biochem. Biophys.* **122**, 629–634.

Wang, T. Y. (1968a). *Arch. Biochem. Biophys.* **127**, 235–240.

Wang, T. Y. (1968b). *Proc. Soc. Exp. Biol. Med.* **129**, 469–472.

Ward, D. C., Humphryes, K. C., and Weinstein, I. B. (1972). *Nature (London)* **237**, 499–503.

Weissbach, A., Schlabach, A., Fridlender, B., and Bolden, A. (1971). *Nature (London)* **231**, 167–170.

Westergaard, O. (1970). *Biochim. Biophys. Acta* **213**, 36–44.

Westergaard, O., and Lindberg, B. (1972). *Eur. J. Biochem.* **28**, 422–431.

Wever, G. H., and Takats, S. T. (1970). *Biochim. Biophys. Acta* **199**, 8–17.

Whittle, E. D., Bushnell, D., and Potter, V. R. (1968). *Biochim. Biophys. Acta* **161**, 41–50.

Wicha, M., and Stockdale, F. E. (1972). *Biochem. Biophys. Res. Commun.* **48**, 1079–1087.

Wickner, W., Brutlag, D., Schekman, R., and Kornberg, A. (1972). *Proc. Nat. Acad. Sci. U.S.* **69**, 965–969.

Wickremasinghe, R. G., Holmes, A. M., and Johnston, I. R. (1973). *Biochem. Soc. Trans. 537th Meeting, Canterbury.*

Williams, A. F. (1972). *J. Cell. Sci.* **11**, 785–798.

Winnacker, E. L., Magnusson, G., and Reichard, P. (1971). *Biochem. Biophys. Res. Commun.* **44**, 952–957.

Wintersberger, U., and Wintersberger, E. (1970a). *Eur. J. Biochem.* **13**, 11–19.

Wintersberger, U., and Wintersberger, E. (1970b). *Eur. J. Biochem.* **13**, 20–27.

Yoneda, M., and Bollum, F. J. (1965). *J. Biol. Chem.* **240**, 3385–3391.

3

Precursor of mRNA (Pre-mRNA) and Ribonucleoprotein Particles Containing Pre-mRNA

G. P. Georgiev

I. Precursor of Messenger RNA*

A. Introduction

The nuclei of eukaryotes contain significant amounts of RNA which is characterized by a very rapid labeling. In 1961 this RNA was sepa-

* Abbreviations used: dRNA, RNA with DNA-like base composition (synonyms: heterogeneous RNA, HnRNA; messengerlike RNA, mlRNA; see pre-mRNA); pre-mRNA, precursor of mRNA (the same as dRNA); mRNA, messenger RNA; ps-mRNA, pseudo-mRNA.

rated into two classes; ribosomal-like RNA with a base composition rich in G + C (rRNA) and RNA with DNA-like base composition (dRNA) (Georgiev, 1961; Georgiev and Mantieva, 1962a).

The method of hot phenol fractionation has been described for the separation of nuclear rRNA and dRNA (Georgiev and Mantieva, 1962b). It consists of successive extractions of the cells with a 0.14 *M* NaCl–phenol, pH 6, mixture at 4°, 40°, 55°, 63–65°, and 85°C. At 4°C the bulk of cytoplasmic RNA is extracted. The 40°C extraction releases the nucleolar RNA, containing rRNA precursor (pre-rRNA), into the aqueous layer; the 55°C treatment results in the extraction of a mixed RNA fraction. The extractions at 65° and 85°C release a dRNA population with about 90% purity (Table I). Some minor modifications for obtaining very high molecular-weight nuclear rRNA have been summarized recently (Georgiev *et al.,* 1972).

Since 1962, numerous papers have appeared providing evidence for the existence of nuclear dRNA in all eukaryotic cells (Sibatani *et al.,* 1962; Scherrer *et al.,* 1963; Brawerman, 1963; see review by Georgiev, 1967). Brawerman (1963) fractionated nuclear RNA by phenol treatment at increasing pH. Although this method is less efficient than the phenol-temperature method of dRNA purification from rRNA precursor, it clearly demonstrated the presence of nuclear dRNA. This technique is also useful in isolating dRNA from cytoplasmic homogenates. In most studies, the total cellular or nuclear RNA was fractionated by zonal sedimentation. Because the bulk of newly formed nuclear dRNA has a very high molecular weight, it may be separated from cytoplasmic RNA's and pre-rRNA (Scherrer *et al.,* 1963; Scherrer and Marcaud, 1965; Penman *et al.,* 1966; Warner *et al.,* 1966). Low-molecular-weight nuclear dRNA is lost in these experiments.

Another approach was developed by Penman *et al.* (1966), who separated nuclear homogenates into nucleolar and nucleoplasmic fractions by DNase treatment at high ionic strength followed by differential centrifugation. The RNA isolated from the "nucleoplasmic fraction" was enriched in dRNA. This technique can only be used for tissues which lack nuclear RNase as degradation of dRNA takes place during the DNase incubation.

Nuclear dRNA may be also obtained from the purified nuclear particles containing dRNA (see Section I,B,2); however, it is very difficult to avoid partial dRNA degradation.

Various authors have used different terms for the designation of nuclear dRNA. In the first papers it was called "nuclear AU-rich RNA" or "dRNA" (Georgiev and Mantieva, 1962a, b); later, the term "heterogeneous," "heterodisperse" RNA or HnRNA was widely used (Warner

TABLE I
Base Composition of RNA Fractions Isolated by Hot Phenol Fractionation Cells[a] from Ehrlich Carcinoma

RNA fraction	Newly synthesized RNA (mole %)					Total RNA (mole %)				
	A	U	G	C	G+C/A+U	A	U	G	C	G+C/A+U
0°C Whole	—	—	—	—	—	19.3	19.6	31.8	29.2	1.57
40°C Whole	14.5	21.7	30.6	33.2	1.77	17.9	21.8	30.7	29.6	1.52
55°C Whole	17.6	25.2	28.2	29.0	1.33	21.7	23.7	28.0	26.6	1.20
63°C Whole	28.7	26.6	21.0	23.6	0.81	29.1	24.6	23.2	23.1	0.86
85°C Whole	28.4	28.1	21.5	21.8	0.77	27.0	28.7	24.3	20.0	0.80
63°C <18 S fraction	34.1	23.9	19.8	22.2	0.73	—	—	—	—	—
18 S	31.3	23.8	19.3	25.6	0.81	28.6	24.4	23.3	23.7	0.89
23 S	27.7	25.7	21.8	24.8	0.87	—	—	—	—	—
28 S	28.0	26.3	22.2	23.5	0.83	—	—	—	—	—
32 S	27.4	27.1	22.1	23.4	0.83	—	—	—	—	—
45 S	26.3	28.6	21.8	23.3	0.80	—	—	—	—	—
>45 S	26.2	27.9	21.6	24.3	0.85	—	—	—	—	—
85°C 18 S	26.1	26.4	22.3	25.2	0.90	—	—	—	—	—
28–45 S	26.4	27.9	21.5	24.2	0.84	—	—	—	—	—
45 S	26.4	29.1	21.3	23.2	0.80	—	—	—	—	—
>45 S	28.0	28.0	21.0	23.0	0.78	—	—	—	—	—

[a] From Markov and Arion (1973).

et al., 1966). Recently the term "messenger-like" or mlRNA was proposed (Scherrer and Marcaud, 1968). All of these terms are quite arbitrary and operational. The role of this RNA as a messenger-RNA precursor has been confirmed; therefore, one may use the functional term "precursor of mRNA" or pre-messenger RNA (pre-mRNA) in a manner analogous to the use of such terms as pre-rRNA or pre-tRNA to describe rRNA and tRNA recursors. In this paper dRNA or pre–mRNA are used to describe this nuclear RNA.

B. Properties of Nuclear dRNA

1. DNA-LIKE BASE COMPOSITION

Nuclear dRNA is characterized by a low G + C content, equal to about 40–45% of the total nucleotides. In this respect it resembles total cellular DNA and is very different from high G + C rRNA or pre-rRNA. The base composition of nuclear dRNA is not necessarily symmetrical since in some cases nuclear dRNA has a very high U content as reflected by a U/A ratio of more than 1. In other cases, especially in low molecular-weight nuclear dRNA and in cytoplasmic dRNA, the A content is higher and the U/A ratio is less than 1 (Georgiev and Mantieva, 1962b; Georgiev *et al.*, 1963; Brawerman, 1963; Samarina *et al.*, 1965; Scherrer and Marcaud, 1965, 1968; Warner *et al.*, 1966).

Usually the base composition of dRNA is measured on the basis of the distribution of ^{32}P among 2′, 3′-nucleotides obtained from alkaline hydrolysis of RNA. Such experiments yield information about the base composition of RNA synthesized during the labeling period. In the case of hot phenol fractionation where the direct measurement of nucleotide optical density was possible, the base composition of corresponding RNA's (65° and 85°C fractions) was also found to be DNA-like (Georgiev and Mantieva, 1962b; Samarina *et al.*, 1965; Markov and Arion, 1973) (Table I).

2. HIGH MOLECULAR WEIGHT AND MOLECULAR HETEROGENEITY OF dRNA

It has been demonstrated that the nuclear dRNA is heterogeneous in respect to molecular weight. The peak of distribution of preformed dRNA was observed in the ~18 S region of sucrose gradients, whereas the newly formed, rapidly labeled nuclear dRNA possessed much higher sedimentation coefficients and correspondingly higher molecular weights (Georgiev and Lerman, 1964; Samarina *et al.*, 1965). A major peak of labeled dRNA is usually localized in the 30–40 S region with smaller amounts sedimenting much faster (Yoshikawa-Fukada, 1964, 1965; Samarina *et al.*, 1965a; Scherrer and Marcaud, 1965; Warner *et al.*, 1966; Gazaryan *et al.*, 1966; Attardi *et al.*, 1966).

In all cases some labeled material may be found in the very heavy zones, 70–100 S, which correspond to molecular weights of between 10–20 × 10^6. Thus the molecular weight range of rapidly labeled nuclear dRNA is from 2 to 20 × 10^6 (Fig. 1). After long labeling times the optical density and radioactivity curves are coincident (Georgiev *et al.*, 1972).

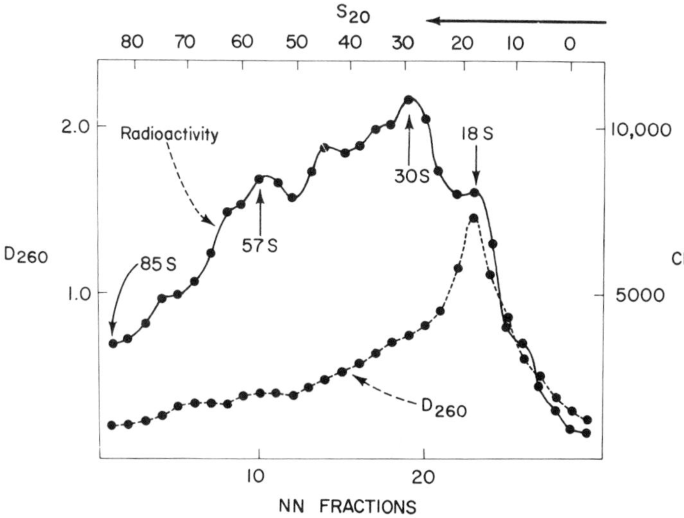

Fig. 1 Sedimentation graph of isolated nuclear dRNA from Ehrlich ascites carcinoma cells labeled with [^{14}C]orotic acid for 45 min. dRNA obtained by hot phenol fractionation in the temperature interval between 55° and 85° C. Ultracentrifugation in SW-25 rotor in 5–20% sucrose gradient in 0.5 × SSC, 0.005 *M* EDTA–Na, pH 7.5, for 12 hr at 21,000 rpm.

Some authors doubt the reality of these high values for molecular weights of dRNA estimated from sedimentation and electrophoretic experiments and suspect the possibility of macromolecular aggregation. However, many different treatments which destroy the weak bonds responsible for aggregation do not decrease the observed molecular weights (Warner *et al.*, 1966; Granboulan and Scherrer, 1969). The determination of 5′- and 3′-ends have demonstrated good agreement between the number of molecule ends and the molecular weights calculated on the basis of sedimentation coefficients (Georgiev *et al.*, 1972).

The very high molecular weight of nuclear dRNA is in contrast to the rather low molecular weight of cytoplasmic mRNA. Experiments using total cytoplasmic mRNA (Brawerman *et al.*, 1963; Penman *et al.*, 1963; Georgiev *et al.*, 1963), and individual partially purified mRNAs (Becker and Rich, 1966; Borun *et al.*, 1967; Williamson and Askonas, 1967) showed the size of mRNA to be only slightly higher than that calculated for a monocistronic messenger template of the corresponding protein.

It was suggested that giant nuclear dRNA is a precursor of cytoplasmic mRNA, which is cleaved into shorter chains during nuclear processing mechanisms (Samarina, 1964; Samarina *et al.*, 1965a).

3. RAPID SYNTHESIS AND DEGRADATION OF dRNA

After short pulses of radioactive precursors nuclear dRNA has the highest specific activity of the various RNA fractions. Only nucleolar pre-rRNA approaches a comparable specific activity. The rapid incorporation indicates a high rate of the dRNA synthesis inside the cell nucleus. The approximate calculations show that even in rapidly growing and dividing tissues the amount of precursor incorporated into total rRNA after a short pulse is several times lower than that incorporated into dRNA. Since dRNA content in the cell constitutes 2–3% of the total RNA compared to 70–80% for rRNA, degradation of a significant part of newly synthesized dRNA must take place (Harris, 1963).

Actinomycin D chase experiments have shown that the kinetics of dRNA decay are rather complex. Some dRNA fractions disappear very rapidly with half-lives of about 5–10 min or less while some dRNA molecules are degraded more slowly. The rapid decay of dRNA has also been demonstrated in chase experiments with cold precursors, although the continuing synthesis from the labeled intracellular precursor pool make the results harder to interpret (Gvozdev and Tikhonov, 1964; Gazaryan *et al.*, 1966; Warner *et al.*, 1966; Scherrer and Marcaud, 1968). The degradation of dRNA probably takes place inside the nucleus since very little labeled RNA appears in the cytoplasm in these chase experiments.

4. EFFICIENT HYBRIDIZATION OF dRNA WITH HOMOLOGOUS DNA

Nuclear dRNA is efficiently hybridized to DNA (Samarina *et al.*, 1965b). dRNA forms complementary hybrids with a much larger fraction of the genome than does rRNA. In saturation experiments at very high RNA/DNA ratios, 5–10% of the total DNA forms hybrids (Church and Brown, 1972; Brown and Church, 1972), whereas rRNA saturation occurs with only 0.03%. The hybridization of RNA with DNA is similar to DNA/DNA renaturation since the rate depends on the initial concentration of reacting components and on the time of reaction (Cot) (Britten and

Kohne, 1968). The concentration of a given DNA sequence is a function of its reiteration frequency in a genome. The copies of highly reiterative sequences hybridize much more rapidly than do the copies of unique sequences; very rarely will complementary DNA sequences or RNA transcript react with locus specificity (Ananieva *et al.*, 1968; Church and McCarthy, 1968). Most of the early hybridization experiments were developed at low or intermediate Cot values under conditions where only copies of reiterative sequences could hybridize. Later more careful experiments were carried out at different Cot values and with DNA fractionated as to different degrees of repetitiveness.

Nuclear dRNA was found to be transcribed from the intermediate fraction of reiterated DNA sequences and from the unique sequences. RNA molecules complementary to the most highly repetitive fraction of mouse DNA (the AT satellite) are absent from nuclear dRNA (Melli *et al.*, 1971; Flamm *et al.*, 1969).

5. TEMPLATE ACTIVITY OF dRNA

Many authors have demonstrated that nuclear dRNA may serve as an efficient template for the incorporation of amino acids into acid-precipitable polypeptides in cell-free systems (Brawerman *et al.*, 1964; Lang and Sekeris, 1964; Samarina *et al.*, 1965b). In most of these studies, the nature of the acid-insoluble product was not determined; therefore, the template activity may be the result of peculiarities of the secondary structure of dRNA. The incorporation of amino acids into specific proteins has been claimed (Zimmerman and Turba, 1965; Lang, 1965), but these results await confirmation. Recently Williamson and Dewienkiewicz (1972) demonstrated the synthesis of hemoglobin after the injection of erythroblast nuclear dRNA into the oocytes.

C. *Nature and Function of Nuclear dRNA*

In the first descriptions of nuclear dRNA it was suggested that some of this RNA was a precursor of cytoplasmic mRNA. Mechanisms for the cleavage of giant dRNA before the transfer of presumptive mRNA into the cytoplasm have been postulated (Samarina, 1964). RNA/DNA hybridization experiments favored this idea. The hypothesis was further tested by the use of specific markers, like virus-specific RNA or by the use of markers of the terminal ends of dRNA. The preparation of DNA antimessenger permitted unequivocal demonstration of the presence of mRNA in the giant nuclear dRNA.

1. RESULTS OF HYBRIDIZATION EXPERIMENTS

Nuclear dRNA is effectively hybridized to DNA while nonlabeled polysomal RNA added to the hybridization medium competes to a limited extent for the DNA-binding sites. These results suggest the transfer of some portion of the nuclear dRNA to the cytoplasm (Georgiev, 1966; Arion and Georgiev, 1967; Scherrer and Marcaud, 1967). Both the giant and low-molecular-weight dRNA's were examined. Polysomal RNA competed more efficiently with light dRNA, but the hybridization of giant dRNA was also inhibited, although to a less extent (Arion and Georgiev, 1967). Saturation experiments showed that the complexity of nuclear dRNA is much higher than that of polysomal RNA (Shearer and McCarthy, 1967). The competition of polysomal RNA against nuclear dRNA is not complete for any fraction. With giant dRNA the competition reached only 15–20%, while for light dRNA it is higher. Conversely, either giant or light nuclear dRNA completely inhibits hybridization of polysomal mRNA to DNA (Georgiev *et al.*, 1972). The conclusion from these experiments was that not all sequences of nuclear dRNA are transferred to the cytoplasm. The RNA sequences transferred to the cytoplasm and those RNA molecules degraded inside the nucleus are therefore transcripts of different DNA sequences. These RNA fractions were designated as $dRNA_1$ and $dRNA_2$, respectively. In the light fraction, the content of $dRNA_1$ is more prominent (Fig. 2).

These data are in agreement with results of experiments on cross competition between heavy (>35 S) and light nuclear dRNA's (10–20 S). The heavy dRNA competes completely with light dRNA, while the latter, even in excess, only partly inhibits the hybridization of the heavy dRNA (Arion, unpublished). Some hybridizable RNA is only present in the heavy dRNA and therefore the light dRNA appears more similar to the cytoplasmic mRNA. The following general scheme was drawn (Georgiev, 1966).

$$\text{Giant nuclear dRNA}\begin{cases} dRNA_1 \rightarrow \text{light nuclear dRNA} \rightarrow \text{cytoplasmic mRNA} \\ dRNA_2 \rightarrow \text{degradation} \end{cases}$$

The question arises as to whether the $dRNA_1$ and $dRNA_2$ are parts of the same nascent RNA chain or are they grouped into two different kinds of giant transcriptional molecules. In actinomycin D chase experiments with Ehrlich ascites carcinoma cells and rat liver, the decay of newly formed dRNA was associated with a corresponding decrease of average molecular weight of dRNA and by an increase in $dRNA_1$ content. The most probable explanation was that giant dRNA chains contain both kinds of sequences, $dRNA_2$ being destroyed during processing (Georgiev *et al.*, 1972). In the case of some other cells actinomycin D chase experiments failed to give clear-cut results (Scherrer *et al.*, 1970).

Another difficulty in the acceptance of the above mentioned scheme is that all hybridization results were obtained in conditions of rather low Cot values which only involve the reiterated base sequences and are subject to base-pair mismatching (McCarthy and Church, 1970). The contribution of the dRNA transcripts complementary to unique DNA sequences (which make up 70–80% of the dRNA) has not been examined.

The recent experiments made at high Cot values showed that most of unique sequences of nuclear dRNA from brain do not reach the cytoplasm (R. Church, personal communication). Ideally some specific indi-

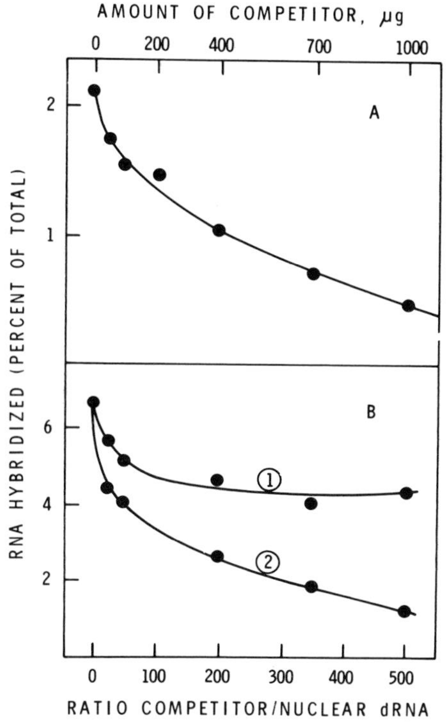

Fig. 2 The competitive hybridization of DNA and RNA fractions isolated from rat liver. (A) Hybridization of labeled polysomal RNA (obtained in conditions of low actinomycin concentration after 4 hr labeling) in the presence of nonlabeled giant nuclear dRNA (35–70 S). The amount of DNA in each tube was 1.5 mg; of polysomal RNA, 3 mg. (B) Hybridization of labeled heavy nuclear dRNA (35–70 S fraction of total nuclear dRNA) in the presence of polysomal RNA (1) or total cellular RNA (2). Each hybridization mixture contained 500 μg DNA, 10 μg nuclear dRNA, 0–5000 μg polysomal (or total) RNA, and *Escherichia coli* RNA, added in such amount as to make the final quantity of nonlabeled RNA in each tube equal to 5 mg. The final volume was 2 ml.

vidual dRNA in a purified form should be used in unique DNA sequence hybridization reactions to answer this question.

2. VIRUS-SPECIFIC SEQUENCES IN CELLS TRANSFORMED BY ONCOGENIC DNA-CONTAINING VIRUSES

In the last decade it has been shown that in transformed cells one to a few viral genomes are integrated with the host genome by covalent bonding to form one linear structure (Oda and Dulbecco, 1968; Westphal and Dulbecco, 1968). The viral DNA is transcribed and the mRNA formed is translated, giving rise to some virus-specific protein(s). Thus viral DNA may be considered to be a model of a specific gene in a mammalian genome. The advantage of the system is that the hybridization to purified viral DNA permits the facile detection of the virus-specific RNA sequences.

In SV40-transformed cells a significant portion of newly formed virus-specific RNA was found in giant RNA with a molecular weight higher than that of the RNA transcript from the whole viral genome. Most of the nuclear virus-specific RNA in the nucleus was found in the heavy zones of sucrose gradients (>35 S). These giant molecules also contain the host-specific sequences. On the other hand, cytoplasmic, virus-specific RNA was of rather low molecular weight (Tonegawa *et al.*, 1970; Lindberg and Darnell, 1970; Wall and Darnell, 1971).

Similar results were obtained in cells transformed by polyoma virus (Georgiev *et al.*, 1972b). Thus, in respect to the SV40 mRNA, the general scheme propcsed above seems to be correct. In the next step, the fate of RNA transcripts from a regular gene in mammalian cells is considered.

3. EXPERIMENTS WITH ANTIMESSENGER DNA

The discovery of reverse transcriptase (Temin and Mizutani, 1970; Baltimore, 1970), which transcribes mRNA into DNA (Verma *et al.*, 1972), allows the synthesis of specific antimessenger DNA populations.

Pure hemoglobin mRNA has been isolated and used as a template in the synthesis of "anti-mDNA" using poly(dT) as a primer. This DNA was then hybridized to giant nuclear RNA isolated from erythroblasts. About 3% of the giant nuclear RNA was complementary to hemoglobin antimessenger DNA. These data clearly indicate that the part of giant molecule is transferred to the cytoplasm and represents a true mRNA (Imaizumi *et al.*, 1972).

The authors used kinetic studies to determine the content of messenger sequences. A rough calculation has given about one mRNA sequence per mole of giant dRNA. Treatment with dimethylsulfoxide reduced the hybridizability of giant RNA to anti-mDNA but some of it survived the treatment. Recently some of these results were reproduced with anti-messenger RNA transcribed from hemoglobin mRNA with the aid of micrococcus RNA polymerase (Melli and Pemberton, 1972). Williamson and Dewienkiewicz (1972) observed some synthesis of rabbit hemoglobin after injection of giant RNA isolated from rabbit erythroblasts into the frog oocytes.

In summary, nascent giant nuclear dRNA is a high-molecular-weight precursor of mRNA and therefore may be designated as pre-mRNA. The sequences degraded in the course of nuclear processing may be referred to as pseudo-mRNA (ps-mRNA). Nuclear dRNA is a mixture of complete nascent pre-mRNA, partly processed pre-mRNA, some mature mRNA ready for transport to the cytoplasm, and some dRNA which is degraded in the nucleus (ps-mRNA).

D. *Structure of Pre-mRNA and Organization of Transcriptional Units in Eukaryotes*

To explain gene activity it is important to understand how the transcriptional unit is organized in eukaryotes. In prokaryotes this question has been solved mainly by genetic manipulation analysis. Unfortunately for eukaryotes, molecular genetics is not yet sufficiently developed to permit genetic analysis. The main information on the structure of transcriptional unit was obtained from the studies on giant nuclear RNA. It appears that before processing giant dRNA is a copy of the whole transcriptional unit. The results obtained allow construction of a general scheme for the possible organization of transcriptional units in eukaryotes.

1. Localization of Messenger RNA in Pre-mRNA

a. End Sequence Analysis. One of the main approaches to the question of the position of mRNA in giant pre-mRNA is the anlysis of the hybridization properties of 5′- and 3′-end sequences (Georgiev *et al.*, 1972). 5′-Ends in nuclear pre-mRNA are monophosphorylated or triphosphorylated. Only triphosphorylated 5′-ends can be considered as true markers of starting sequences in pre-mRNA, since the monophosphorylated end may be formed either by dephosphorylation of original ends or by endonuclease breaks that occur during processing.

After alkaline hydrolysis triphosphorylated 5′-end nucleotides are recovered as nucleoside tetraphosphates (pppNp). Thus, the identification of pppNp allows detection of the 5′-end of pre-mRNA.

3′-Ends in nuclear pre-mRNA are not phosphorylated and after alkaline hydrolysis may be isolated as free nucleosides. After $NaIO_4$ oxidation followed by NaB^3H_4 reduction, the 3′-ends of RNA are liberated by alkali as the [^{3}H]labeled nucleoside derivatives. Unfortunately, it is difficult to distinguish original 3′-ends and 3′-ends formed in the course of processing.

pppNp is present only in giant pre-mRNA hydrolysates since in fractions with sedimentation coefficients of less than 35 S only traces of pppNp are present, although the concentration of pNp is much higher. This indicates that only giant pre-RNA's are the primary products of transcription and that the transcriptional units in eukaryotes are presumably polycistronic since they are much larger than solitary structural genes. The shorter chains of nuclear pre-mRNA probably result from giant RNA cleavage during nuclear processing.

The starting sequences in the pre-mRNA (containing pppNp groups) are transcripts from reiterative DNA base sequences. They hybridize very effectively to DNA even at rather low Cot values. On the other hand, polysomal RNA does not compete with the sequences containing pppNp groups. One can conclude, therefore, that mRNA is not at the 5′-end of pre-mRNA (Georgiev *et al.*, 1972a) (Table II).

3′-End nucleosides are represented mainly by adenosine, which comprises 75% of all terminal nucleosides, while 20% consists of uridine. This is typical for all classes of pre-mRNA—giant, >35 S; intermediate, 20–30 S; and light, 10–20 S (Georgiev *et al.*, 1972).

When polyadenylic 3′-end sequences were analyzed, about 20% of either light or heavy pre-mRNA 3′-ends were represented by short polyadenylic sequences ($n = 4$–7) (Ryskov *et al.*, 1972). The purified hemoglobin mRNA also was found to contain 5 and 6 adenylic acid residues at the 3′-end (Barr and Lingrel, 1971). These data may be explained on the assumption that the 3′-end of the pre-mRNA is conserved during processing and transferred into the cytoplasm in support of a 3′-end location mRNA.

At least some part of 3′-end sequences in pre-mRNA are rapidly hybridized with DNA; therefore some 3′-ends are represented by transcripts of reiterated DNA base sequences. The hybridization of these sequences is efficiently inhibited by polysomal RNA. The total hybridization of giant pre-mRNA is only slightly inhibited by an excess of polysomal RNA while the competition with 3′-end sequences reached 60–70% (Georgiev et al., 1972) (Table II). Thus, 3′-end sequences are transferred to the cytoplasmic polysomes as part of mRNA molecules. Althtough the hybridizable 3′-ends may correspond to some short ancillary sequences in pre-mRNA and mRNA, such as terminators of transcription and/or translation, this is as yet uncertain.

TABLE II

Hybridization and Competition Properties of 5′- and 3′-Ends of Giant Pre-mRNA from Ehrlich Ascites Carcinoma Cells[a,b]

		Cpm of nonhybridized RNA	Hybridized RNA		RNA hybridized in the presence of polysomal RNA as competitor		
			Cpm	Hybridization (%)	Cpm	Hybridization (%)	Competition (%)
5′-end analysis	^{32}P in pppNp	370	200	35	230	40	0
	^{32}P in Np	1,380,000	110,000	7.4	95,000	6.4	−14
	pppNp/Np $\times 10^{-2}$	0.027	0.18	—	0.24	—	—
3′-end analysis	[^{3}H]end nucleoside	2200	494	18.3	155	5.7	−69
	[^{14}C]internal nucleotides	3940	212	5.2	194	4.7	−10
	$^{3}H/^{14}C$	0.56	2.3	—	1.25	—	—

[a] From Georgiev *et al.* (1972) and Georgiev (unpublished data).

[b] Figures of two typical experiments at intermediate Cot values.

The main conclusion is that mRNA is localized near the 3′-end of giant pre-mRNA, whereas the 5′-end does not contain mRNA sequences.

b. Polyadenylic Acid Studies. Another approach to the study of nuclear pre-mRNA appeared recently with the discovery of poly(A) sequences in mRNA and nuclear dRNA. Several years ago, Edmonds and her colleagues discovered the synthesis of poly(A) from ATP in the nuclei of mammalian cells (Edmonds and Caramela, 1969). Poly(A), or A-rich sequences, have been observed in nuclear and polysomal RNA (Edmonds and Caramela, 1969; Lim and Canellakis, 1970). Later poly(A) sequences of about 150–250 nucleotides long were found to be bound covalently to mRNA and to nuclear pre-mRNA. In cytoplasmic mRNA the content of poly(A) reaches 5–10% of the total RNA, which corresponds to one poly(A) per mRNA molecule (Edmonds *et al.*, 1971; Lee *et al.*, 1971; Darnell *et al.*, 1971).

Poly(A) was isolated from RNA with the aid of pancreatic and T_1 RNase treatment at rather high ionic strength (~0.3 *M* NaCl) followed by purification on gel electrophoresis and hybridization to filter with immobilized poly(T) or poly(U) or by trapping on Millipore filters. The same techniques may be applied to RNA before RNase digestion to separate RNA chains with or without poly(A) sequences. According to the latter test most mRNA chains examined so far contain poly(A) sequences. An exception has been noted for the histone mRNA (Adesnik and Darnell, 1972).

Poly(A) was also found in nuclear pre-mRNA although its content is much lower, about 0.5–1% (Edmonds *et al.*, 1971; Lee *et al.*, 1971; Darnell *et al.*, 1971). This gives additional support to a precursor role of nuclear dRNA and enables the localization of mRNA in giant pre-mRNA.

Newly synthesized pre-mRNA probably does not contain poly(A). For example, no poly(A)–poly(T) sequences were found in the adenovirus genome, which suggests that the latter cannot serve as a template for poly(A) production. However, the adenovirus mRNA in the cytoplasm does contain poly(A) of normal size (Philipson *et al.*, 1971). It seems probable that poly(A) is attached to pre-mRNA during nuclear processing, but the mechanism of attachment remains unclear. It may be either stepwise nontemplate addition of AMP residues at the 3′-end of pre-mRNA by the Edmonds' enzyme, or by discrete DNA template synthesis of poly(A) followed by its addition to pre-mRNA by a ligase reaction. In the latter case poly(A) could be added to any end of the mRNA chain.

The location of poly(A) in mRNA and pre-mRNA is probably at the 3′-end of the molecule. Poly(A) labeled with [^{3}H]-adenosine produced about 0.5% of free adenosine after alkaline hydrolysis (Mendecki *et al.*, 1972). Treatment of mRNA with nuclear exonuclease, which digests

RNA from the 3′-end and only if it has a free 3′-hydroxyl group, destroyed poly(A) more rapidly than mRNA itself (Malloy *et al.*, 1972).

The results of experiments with $NaIO_4$–NaB^3H_4 labeling of mRNA and nuclear dRNA suggests that the 3′-ends were located after short poly(A) sequences ($n = 4$–7) (Barr and Lingrel, 1971; Ryskov *et al.*, 1972b). The results may be explained if degradation of poly(A) by RNase occurred at the low ionic strength used in the experiments.

A small amount of pAp was isolated from alkaline hydrolyzates of poly(A) prepared from light nuclear pre-mRNA. This fact suggests the possibility of two 5′-end locations of some poly(A) in light dRNA (Ryskov *et al.*, 1972a).

The concentration of poly(A) in giant pre-mRNA is much lower than in light pre-mRNA. The difference in the case of Ehrlich ascites carcinoma cells is greater than may be calculated on the basis of molecular weight (Ryskov *et al.*, 1972c). The addition of poly(A) may be delayed and take place after some processing steps, in accordance with the idea that not all the pre-mRNAs contain poly(A).

The presence of poly(A) in some of pre-mRNA was used to localize specific mRNA in giant pre-mRNA. Heavy RNA from rat cells transformed by adenovirus was isolated and the sequences containing poly(A) were selected. The main part of adenovirus-specific sequences as well as a significant amount of host RNA was detected in RNA bound to poly(U). After cleavage of long RNA's to short chains by mild alkaline hydrolysis most of the host RNA was no longer bound to poly(U). However, most of the adenovirus-specific sequences still formed complexes with the poly(U) Sepharose column. These results suggest that adenovirus-specific sequences in the giant mixed RNA molecule are very close to the poly(A) sequences (Darnell, personal communication). If the latter are at the 3′-end, one can conclude that the virus-specific mRNA is at the 3′-end of pre-mRNA in agreement with conclusions of the previous section.

The addition of poly(A) is probably a necessary step in mRNA maturation. Before the addition of poly(A), most mRNA cannot be transported into the cytoplasm and used for translation. Cordycepin (3′-deoxyadenosine) at low concentrations very efficiently inhibits poly(A) synthesis without stringent inhibition of pre-mRNA synthesis (Penman *et al.*, 1970). The appearance of mRNA in the cytoplasm is strongly inhibited by cordycepin. Only traces of mRNA reached polysomes and these mRNA molecules contained some poly(A) sequences shorter than usual (Darnell, personal communication). It is possible that poly(A) is necessary for interaction with some specific protein involved in transport or translation of mRNA. Interestingly, "old" mRNA's possess shorter poly(A) sequences than nascent molecules (Brawerman, 1973). The results of poly(A) experiments are in good agreement with the 3′-end localization of mRNA in giant pre-mRNA.

2. NONINFORMATIVE SEQUENCES IN PRE-mRNA

The 5′-end sequences of pre-mRNA are not normally transferred into the cytoplasm, and are transcribed from reiterated DNA; therefore it seems very probable that corresponding sequences in DNA may have some special regulatory functions (Georgiev *et al.*, 1972). Many other RNA transcripts from reiterative base sequences are also of a non-informative nature since ps-mRNA contains more transcripts from DNA reiterative sequences than mRNA (Arion and Georgiev, 1967). The RNA sequences forming low Cot hybrids obtained with mRNA and pre-mRNA are different in respect to base composition and the extent of reaction.

Scherrer (1971) observed that mRNA and dRNA hybrids are rich in A. Darnell *et al.* (1971) reported that A-rich hybrids were only formed with mRNA and not with nuclear pre-mRNA. Besson *et al.* (1972) found that in the hybrids formed from nuclear RNA of different sizes that only light RNA was A rich. Depending on the Cot values obtained in the reaction and nuclease treatment, the A content can approach 50–60%. This value is similar to that found with mRNA hybrids. However, hybrids from giant pre-mRNA are quite different since they are relatively rich in A (~30%) but contain a significant excess of G (30–35%). The G-rich transcripts of reiterative sequences are probably very unstable and may disappear very early during nuclear processing.

Some interesting RNA sequences characteristic of pre-mRNA, the so-called RNA "hairpin structures," were observed recently (Ryskov *et al.*, 1972c) (Fig. 3). After RNase treatment at high ionic strength some RNase-resistant material remains undegraded. These RNase stable sequences contain poly(A) as well as some other diverse base sequences. Poly(A) can be removed by hybridization to poly(U) or by binding to Millipore filters so that the remaining sequences can be analyzed. Recently it was found that these sequences may be separated by gel filtration or hydroxyapatite fractionation into two fractions: a GC-rich (GC content 75%) component of low molecular weight; and an AT-rich (GC content 45%) of relatively higher molecular weight (Ryskov *et al.*, 1973). The base composition is symmetrical, indicating a double-stranded RNA structure that is easily reconstituted after melting the secondary RNA structure. However, if RNase treatment precedes melting the reconstitution does not take place, suggesting the probability of digestion of the loops of the hairpin structures (Jelinek and Darnell, 1972; Ryskov *et al.*, 1973).

These RNA hairpin structures are probably analogous to the "double-helical" nuclear RNA isolated previously from animal cells (Harel and Montagnier, 1971). It was found that after loop cleavage and subsequent melting, the hairpin single-strand RNA sequences hybridize very efficiently to DNA (Jelinek and Darnell, 1972; Ryskov *et al.*, 1973). Half-hybridization takes place at a DNA-driven Cot value equal to about 10 (with total DNA), indicating the complementarity of RNA hairpin sequences to reiterated DNA base sequences (intermediate fraction). The size of long RNA hairpins is about 100 base pairs or more. Short hairpin regions consist of about 10–20 base pairs.

Recently Church and Georgiev (1973) found about 2% of similar hairpinlike structures in denatured mouse DNA. The hybridization between RNA and DNA hairpin sequences was observed. Very rough calculation based on the content of hairpincoding structures of DNA and kinetics of hybridization or RNA renaturation show that the genome

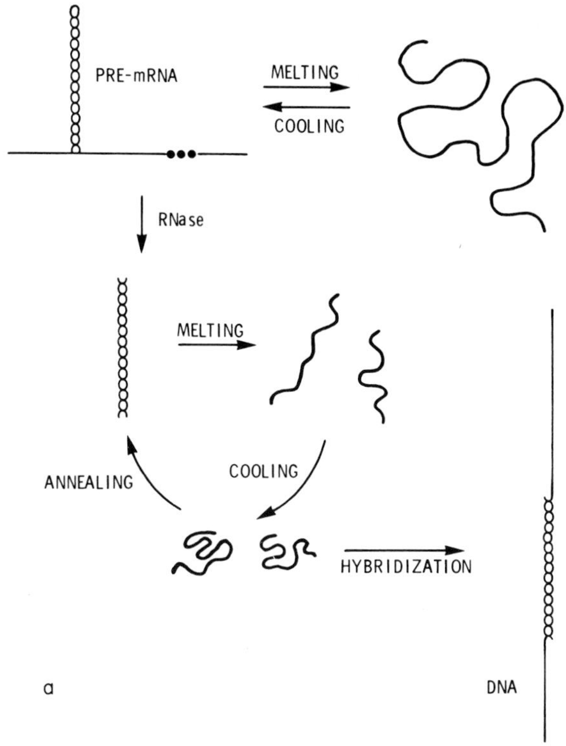

Fig. 3 Hairpinlike structures in giant pre-mRNA. (a) Scheme of the isolation and renaturation of hairpin regions. (b) Hybridization of hairpinlike structures to the excess of DNA at different Cot values. A, Long hairpins; B, short hairpins; C, total dRNA. (Ryskov *et al.*, 1973.)

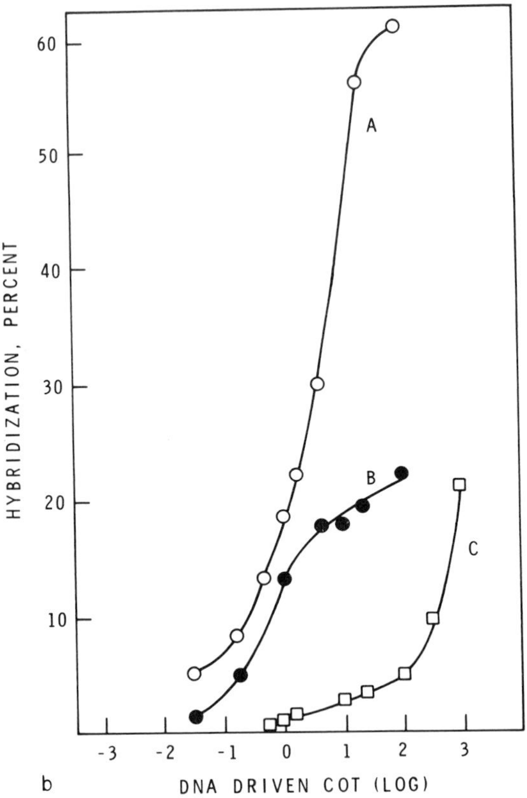

Fig. 3b (See preceding page for legend.)

contains about 500 families of sequences coding long hairpin regions of about 500 members each (Ryskov *et al.*, 1973).

RNA hairpins are predominantly found in the giant pre-mRNA and are absent from the light nuclear pre-mRNA as well as from cytoplasmic mRNA. In control experiments pre-rRNA was shown to be nearly free from long RNA hairpinlike structures (Ryskov *et al.*, 1972c). Therefore, RNA sequences present in hairpin structures are typical of ps-mRNA which is cleaved and degraded in the course of nuclear processing. The various experimental approaches described have all contributed to our understanding of the principles of the structural organization of pre-mRNA. The interpretation of the experimental data provides an insight into the organization of transcriptional units in eukaryotes.

3. STRUCTURE OF THE TRANSCRIPTIONAL UNIT

Several models have been suggested to describe the organization of the transcriptional unit, or the structure of the functional genetic element in eukaryotes (Scherrer *et al.*, 1968; Britten and Davidson, 1969; Georgiev, 1969). The model proposed (Georgiev, 1969) is in a good agreement with the data obtained up to this date (Fig. 4).

According to this model (Georgiev, 1969) the transcriptional unit consists of two parts: (*a*) an acceptor noninformative zone, adjacent to the promoter region and (*b*) a structural, informative zone at the end of transcriptional unit. The structural region may consist of one or more structural genes. The acceptor region does not contain any structural genetic information since it consists of a sequence (acceptor site) which is recognized by specific, mainly regulatory, proteins. The acceptor zone may be much longer than the structural zone.

While the whole of the transcriptional unit is transcribed, the structural region may only be transcribed after the acceptor region. The transcription of the acceptor region is under the control of regulatory proteins which interact specifically with acceptor sites. Some of acceptor sites in a transcriptional unit are different; therefore, one transcriptional unit may be under the control of many different proteins. Conversely, the same acceptor sequences may be present in many transcriptional units. Therefore, when the same acceptor sequence is located in a number of transcriptional units, all of these units are under the control of the same regulatory agent. This may explain the massive switching on or off of different genome patterns observed for different stages of differentiation.

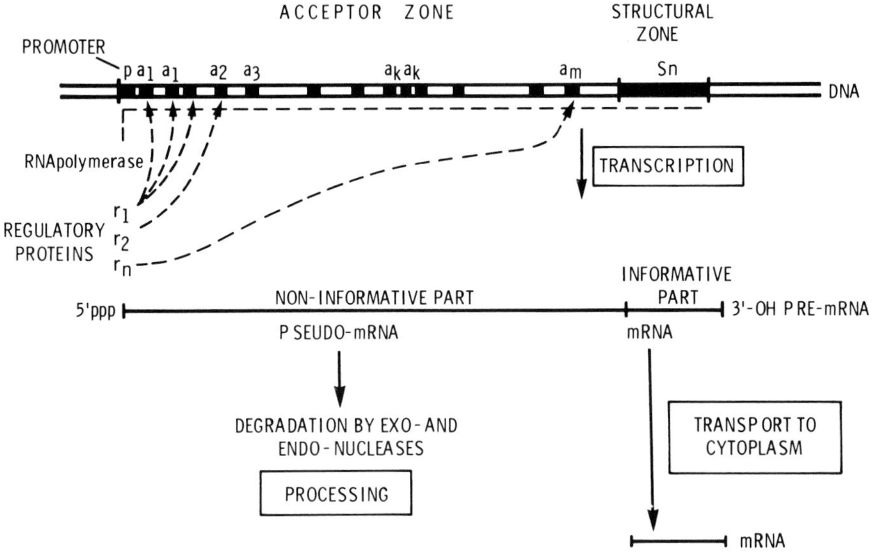

Fig. 4 Hypothetical model of the structure of transcriptional unit in eukaryotic cells. (Modified scheme of Georgiev, 1969.)

The primary product of transcription, the nascent pre-mRNA is a copy of the whole transcriptional unit. It consists of a noninformative part at the 5′-end, and a short informative part at the 3′-end of the nascent pre-mRNA. The noninformative part, or ps-mRNA, is degraded and the true presumptive mRNA, located at the 3′-end of the nascent precursor molecule, is transferred to the polysomes to be translated.

The evidence which has accumulated so far on the structure of pre-mRNA confirms some important postulates of this model, namely, the localization of mRNA at the 3′-end of the pre-mRNA and the degradation of the 5′-end sequences during nuclear processing.

One of the interesting questions, at present, is the identification of acceptor sequences and their subsequent isolation. The readily hybridizable RNA hairpin sequences may be transcripts of the DNA acceptor sequences or sites. Recently, operators in bacterial cells were found to contain short repetitive sequences which allow the formation of the two symmetrical, double-stranded, hairpin branches from the main chain. Such a secondary structure may efficiently interact with four subunit repressor protein molecules (Watson, personal communication). In this respect the studies of RNA hairpins in eukaryotic pre-mRNA and in the corresponding DNA acceptor regions may be a promising way of analyzing the structure and function of acceptor sites.

There are some genetic data which support the above hypothesis of the structure of transcriptional units and pre-mRNA. The haploid amount of DNA per band in *Drosophila* is very high, many times higher than the number of base pairs required for coding all known proteins. It varies from 5×10^6 to 70×10^6, comprising about 30×10^6 on the average (Beermann, 1972). On the other hand, the number of complementation groups, which may be analogous to transcriptional units, is correlated with the number of bands seen in some regions of polytene chromosomes (Kaufman *et al.*, 1968; Shannon *et al.*, 1969). It has also been found that the structural genes are nonrepetitive (Bishop *et al.*, 1972; Suzuki *et al.*, 1972). However, there are exceptions such as the genes for histones (Kedes and Birnstiel, 1971) and probably genes for saliva proteins in salivary glands of *Chironomus* (Daneholt *et al.*, 1969; Lambert *et al.*, 1972). Therefore one can conclude that the major portion of the DNA in a chromosome band does not correspond to the sequences coding for structural genes. A significant part of the chromosomal band may be deleted without the loss of the genetic activity. According to the Crick (1971) model of chromosome structure, structural genes are located in the chromosomal interband while the acceptor sites correspond to the chromosomal bands containing promoter and operator regions.

In summary, we now understand much more about the nature and structure of nuclear pre-mRNA than we did a few years ago. Studies of the transcriptional unit hold great promise for understanding of regulation of gene expression in eukaryotes.

II. Ribonucleoprotein Particles Containing Pre-mRNA

A. *Introduction*

In bacterial cells the growing RNA chain is combined with ribosomes such that translation starts before the termination of RNA synthesis. In eukaryotes, transcription and translation take place in different cellular compartments: RNA synthesis in the nucleus, and protein synthesis in the cytoplasm. Therefore, a new step in the pathway appears, namely, RNA transport. In analogy with bacteria it has been suggested that ribosomes or ribosomal subunits are engaged in the mRNA transport in eukaryotes. Evidence to the contrary appeared with the discovery of newly formed cytoplasmic RNA in embryonic cells that is not combined with ribosomes but with some other protein (Spirin *et al.*, 1964). At present three main classes of ribonucleoproteins containing nonribosomal dRNA have been described: (1) free cytoplasmic RNP or informosomes (Spirin *et al.*, 1964); (2) polysome-bound mRNP released from polysomes by EDTA treatment (Perry and Kelley, 1968; Henshaw, 1968); and (3) nuclear particles containing pre-mRNA (Samarina *et al*, 1965b).

Some authors refer to all of these particles as informosomes. (Spirin *et al.*, 1964) but this should be done only after proof is provided that the three kinds of particles represent the same cellular elements. Designation of only free cytoplasmic mRNP as informosomes seems preferable.

The following sections describe the properties of nuclear particles which contain pre-mRNA.

B. *Isolation of Nuclear Particles and Their Properties*

1. ISOLATION PROCEDURES

The main isolation procedure is very simple; an extraction of the isolated cell nuclei with 0.1 *M* NaCl–0.001 *M* Tris–0.001 $MgCl_2$ at pH 7.0 and then several times at pH 8.0 in the cold. In the case of rat liver practically all nuclear pre-mRNA is liberated during the second to fourth pH 8.0 extractions. Almost all particles are recovered by sucrose gradient ultracentrifugation as a homogeneous peak with a sedimentation coefficient of about 30–40 S. The main form of pre-mRNA in the nuclei of such tissues as rat liver or Ehrlich ascites carcinoma cells is in these 30–40 S particles (Samarina *et al.*, 1965b, 1966; Fig. 5).

The 30 S particles are the monomers of much bigger complexes that are organized like polysomes. These large complexes may be isolated from rat liver nuclei if RNase activity is inhibited with RNase inhibitor from rat liver supernatant. In the presence of inhibitor one can isolate a heterogeneous population of particles with sedimentation coefficients from 30 to 400 S, containing high-molecular-weight pre-mRNA. Mild RNase treatment of these complexes cleaves them to monomeric 30 S particles (Samarina *et al.*, 1968).

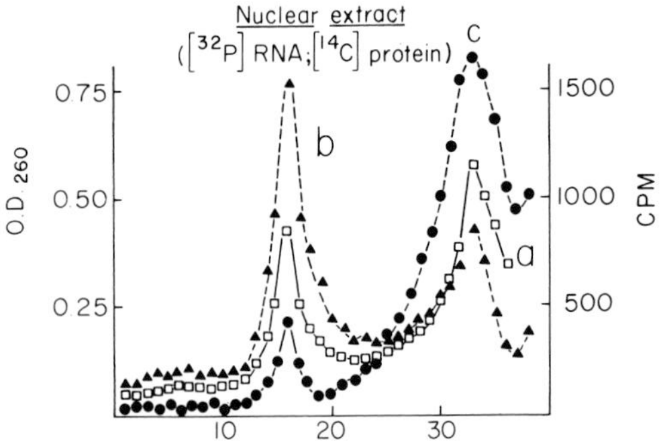

Fig. 5 Sedimentation diagram of rat liver nuclear extract containing pre-mRNP. a, optical density; b, [^{32}P]-labeled RNA; c, [^{14}C]-labeled protein. Ultracentrifugation in SW-25 rotor (Spinco L 40) in 15–30% sucrose gradient for 14 hr at 24,000 rpm. (Samarina *et al.*, 1965a.)

Nuclear particles could not be isolated in this way from some tissues. However, if the extraction is performed at elevated temperatures (25°–35°C) the RNP may be removed from the nuclei very easily, during a 5–10 min treatment. This behavior is typical of nuclei with very low RNase activity. Large complexes may not be extractable but after slight heating they are cleaved enzymatically to particles of lower size. Usually such extracts contain polyparticles as well as 30 S particles. This technique permits isolation of nuclear–particle complexes from many different tissues (Lukanidin *et al.*, 1972).

Some modifications have been described but they usually do not result in a higher yield or better quality of particles (Moule and Chauveau, 1966, 1968; Ishikawa *et al.*, 1969; Faiferman *et al.*, 1970).

2. PROPERTIES OF 30 S PARTICLES

Almost all nuclear pre-mRNA (80–90%) may be recovered in the form of 30 S particles. These particles consist of pre-mRNA complexed with protein. The RNA to protein ratio is between 1:4 and 1:5 according to direct measurement; however, the values were obtained with very small amounts of material. The buoyant density of the formaldehyde-fixed particles in CsCl density gradient is equal to 1.4 gm/cm^3. Particles containing 15–20% RNA would have such a density, although again an accurate measurement is impossible as the density of protein and RNA inside the particle is unknown. In sucrose and CsCl density gradients the particles are very homogeneous. Under electron microscopy the purified particles look like rather homogeneous globules about 200 Å in diameter. They have some internal ultrastructure but it is not well resolved by the electron microscope (Samarina *et al.*, 1965, 1966, 1967; Moule and Chauveau, 1966, 1968; Monneron and Moule, 1968) (Fig. 6).

RNA isolated from 30 S particles by the phenol–detergent method is degraded and has sedimentation coefficients about 4–6 S. Its base composition is DNA-like indicating the absence of significant amounts of ribosomal RNA. This RNA hybridizes efficiently with DNA and stimulates incorporation of amino acids in cell-free systems. The hybridization properties are identical to those of nuclear pre-mRNA isolated by other techniques such as the hot phenol fractionation. RNA isolated from particles competes completely with nuclear pre-mRNA and vice versa. These results suggest the coincidence of the RNA's isolated by two different techniques (Samarina *et al.*, 1965b, 1967a).

The protein component of 30 S particles has also been isolated in different ways. At first more or less denaturing procedures were used such that the particles were treated with RNase, precipitated with TCA, and the protein dissolved in a urea-containing solution (Samarina *et al.*, 1968). Another technique involved the elution of 30 S particles in 7 *M* urea from a DEAE–cellulose column. The RNA is retained but at least 90% of the protein passes through the column and is additionally purified and partly fractionated on a CM–cellulose column (Krichevskaya and Georgiev, 1969). Electrophoretic studies have been done with such protein preparations. In the first electrophoretic experiments using urea–Tris buffer as a solution, three main (A, B, and C) and some minor components were found (Samarina *et al.*, 1968). However, after reduction with mercaptoethanol, A, C, and minor components were converted to B. A + C purified by CM–cellulose chromatography were also converted to B by mercaptoethanol treatment. Thus only one component (B) is present in 30 S particles. A is probably a dimer of B, and C contains intermolecular S-S bonds making the chain more compact. Electrophoretic properties of this protein are quite different from those of ribosomal proteins or histones (Krichevskaya and Georgiev, 1969).

In SDS-polyacrylamide gel electrophoresis the molecular weight of the

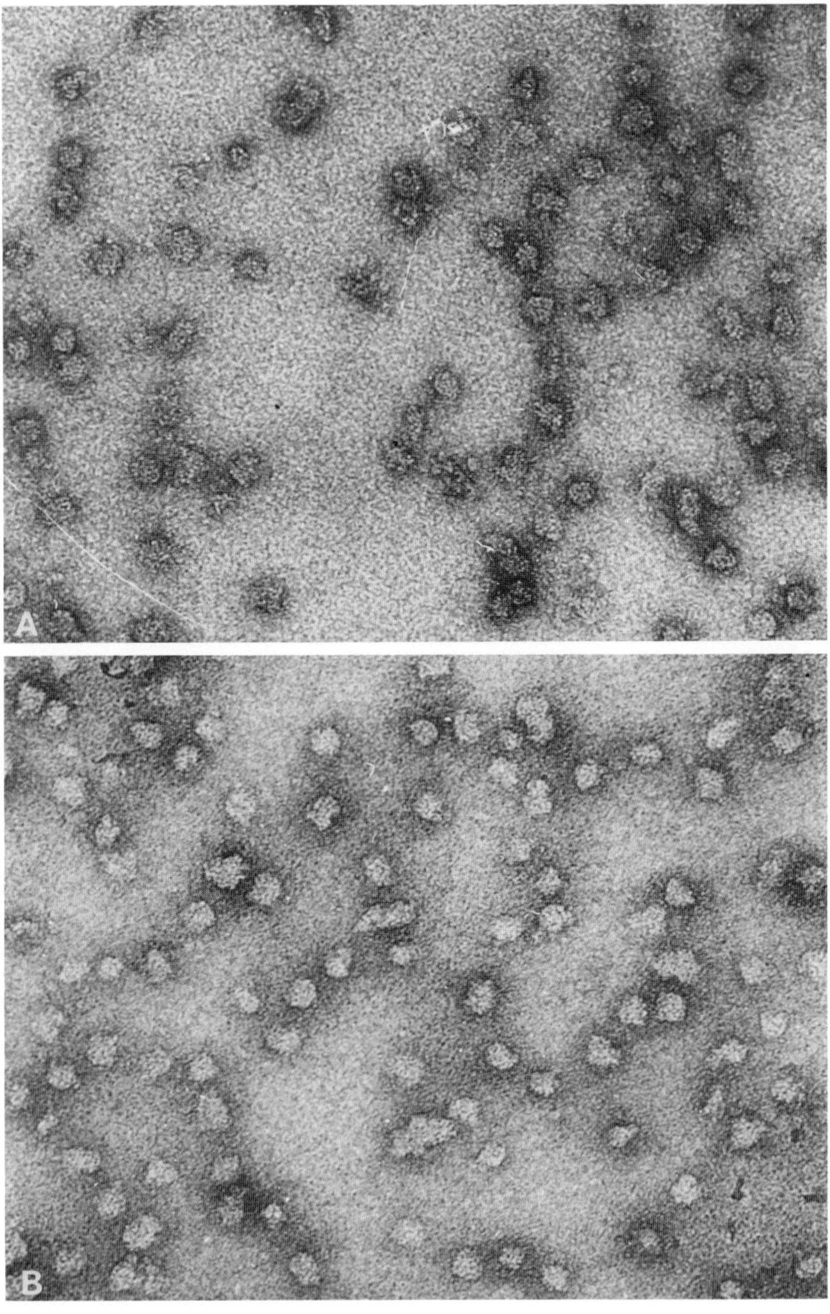

Fig. 6 Electron micrograph of 30 S particles containing dRNA and (A) free informofers and (B) negative contrast with uranyl acetate (× 187,000). (From Samarina *et al.*, 1967; Lukanidin *et al.*, 1972.)

major protein was found to be 40,000. This homogeneous protein with a molecular weight of 40,000 was called "informatin" (Krichevskaya and Georgiev, 1973).

Besides informatin some additional bands were found in SDS–polyacrylamide gel but the amount of these proteins was not greater than 10%. Some other authors found much more material in noninformatin bands of higher molecular weight (Faiferman *et al.*, 1971; Niessing and Sekeris, 1971). The reason for this discrepancy is not yet clear but it is possible that some denaturation procedures produce irreversible aggregation of informatin and lead to the formation of stable artificial complexes. The amino acid composition of informatin is typical of a more or less neutral protein (Krichevskaya and Georgiev, 1969; Saracin, 1969).

Recently, a mild isolation technique for informatin that excludes the use of denaturing agents has been described. The 30 S particles are concentrated by precipitation with ammonium sulfate, dissolved in a small volume of water, and then $CaCl_2$ is added to a 3 *M* concentration. The particles are dissociated and the RNA precipitates, whereas the protein remains soluble. The protein in solution is present in the form of subunits. $CaCl_2$ is removed by dialysis and the purified informatin may then be used for further studies (Krichevskaya, unpublished).

The main conclusion from the studies on 30 S particles was that nuclear pre-mRNA is combined with a very specific type of protein.

3. POLYPARTICLES

As was mentioned above, the use of RNase inhibitor allows one to isolate larger complexes sedimenting between 30 and 400 S (Samarina *et al.*, 1968). As a rule several discrete peaks are recovered: 30, 45, 57, and 65 S, as well as heterogeneous material with higher sedimentation coefficients (Fig. 7).

These complexes consist of the same components as 30 S particles, namely, pre-mRNA and informatin. No ribosomal RNA has been found and the main protein recovered in the heavy particles was informatin. The continuity of the polyparticles depends on the RNA. Very mild RNase treatment quantitatively converts polyparticles into monomers or 30 S particles, suggesting polysomelike structure. This structure has been confirmed by electron microscopic studies. In the 30 S region of sucrose gradients single particles occur; in the 45 S zone, dimers; in the 57 S, trimers; and in the 65–75 S, tetra- and pentamers. Near the bottom of such gradients large polyparticles consisting of up to 10–12 monomers may be observed. After mild RNase treatment, only single particles 200 Å in diameter are present (Samarina *et al.*, 1968) (Fig. 8).

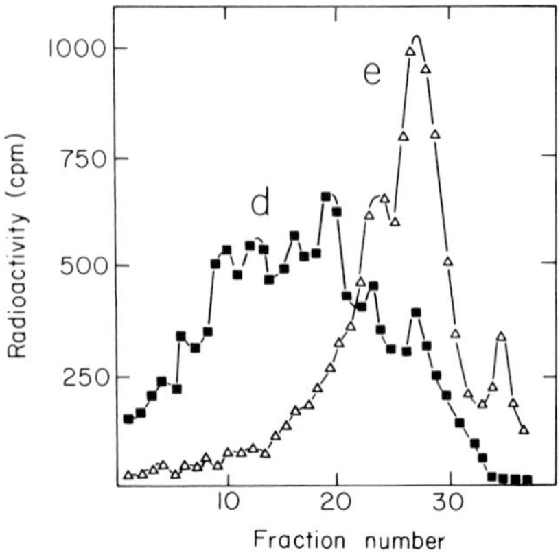

Fig. 7 Sedimentation diagram of rat liver nuclear pre-mRNP particles isolated in the presence of RNase inhibitor. Ultracentrifugation during 4.5 hr at 25,000 rpm. d, nontreated nuclear extract; e, nuclear extract treated by low RNase. (Samarina *et al.*, 1968.)

There is a good correlation between the size of the particles and the size of RNA isolated from them. Thus from the 30 S region of sucrose gradients, the RNA obtained is 9 S; RNA isolated from 45 S particles is 14 S, and from 57 S particles the RNA is 18 S. The calculations show that in all kinds of particles the molecular weight of RNA divided by the number of monomers in the complex is a constant value of about 200,000 daltons. Thus polyparticles have a very regular structure, in which each monomer contains exactly the same amount of RNA. This conclusion was confirmed by CsCl density gradient ultracentrifugation of fixed particles. The buoyant density that reflects the characteristic protein to RNA ratio was the same (1.4 gm/cm^3) for 30 S particles as for complexes up to 400 S as well as for 30 S particles obtained from giant structures by mild RNase treatment (Samarina *et al.*, 1968).

On the basis of these data the general scheme of a polysomelike structural organization of nuclear complexes containing pre-mRNA was postulated (Fig. 9). According to the model giant pre-mRNA is distributed on the surface of a number specific protein macroglobular particles. These protein particles were called "informofers." Each informofer is combined with a part of RNA chain with a molecular weight of 200,000 daltons. The size of the informofer itself has not been determined accurately but is probably about 1–2 $\times$ 10^6 daltons (Samarina *et al.*, 1968).

The localization of pre-mRNA on the surface of the informofer was confirmed by the following observations. First, RNA of nuclear particles is very sensitive to any kind of RNase. It is not protected by the protein moiety. Second, 30 S particles are able to combine specifically with some additional pre-mRNA added either to the nuclear extract or to the isolated 30 S zone of sucrose gradients. Third, this RNA artificially bound to 30 S particles has exactly the same kinetics of RNase degradation as the endogenous RNA of the particle (Samarina *et al.*, 1967; Lukanidin, unpublished). However, the most important proof is the ability to isolate RNA-free informofers; this is discussed in the next section.

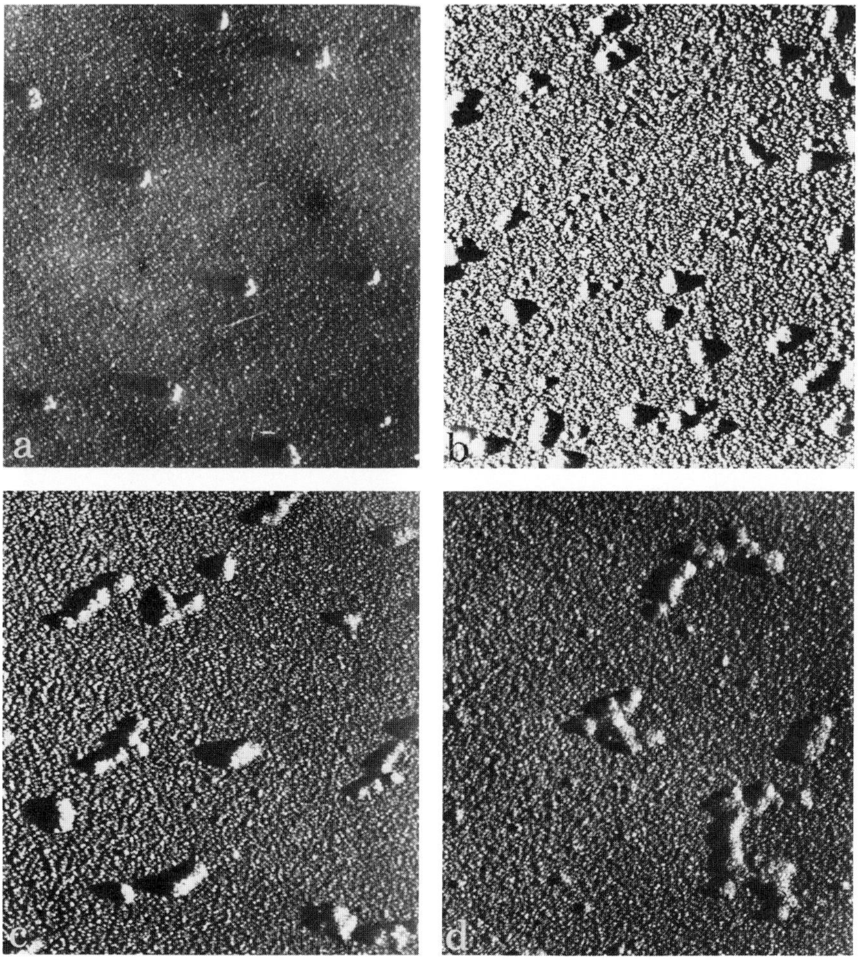

Fig. 8 Electron microscopy of particles obtained from different zones of sucrose gradient and fixed with 2% formaldehyde. Shadow-cast preparations (× 72,000). (a) preparation of 30 S particles obtained without inhibitor. (b) particles from 45 S zone; (c) from 70 S; and (d) from 90–100 S zone. Preparations b–d are obtained from the sample isolated in the presence of RNase inhibitor. (From Samarina *et al.*, 1968.)

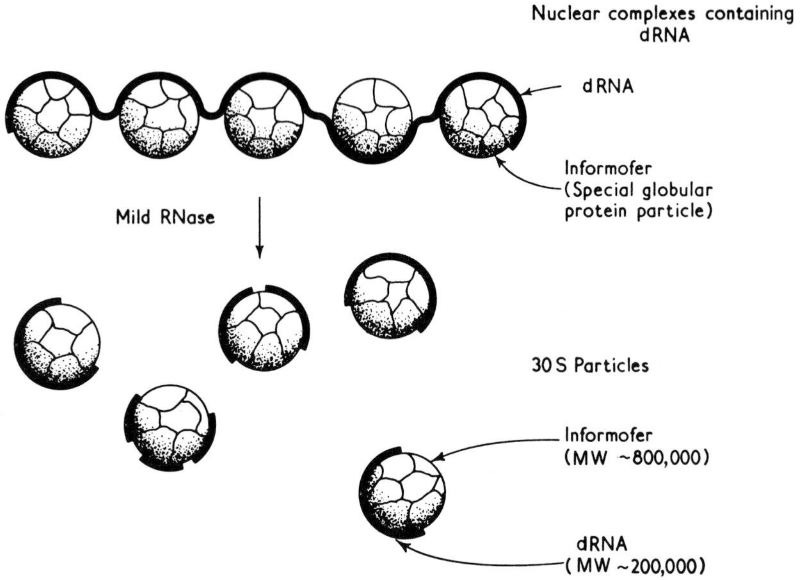

Fig. 9 The scheme of organization of nuclear dRNA-containing complexes. (From Samarina *et al.*, 1968.)

4. ISOLATION OF FREE INFORMOFERS AND THEIR PROPERTIES

The general scheme of the experiment is as follows: As the informatin is labeled very slowly *in vivo,* the isolated 30 S particle proteins are labeled chemically with ^{125}I. The particles are then treated with 2 *M* NaCl to dissociate protein and RNA, and ultracentrifuged in sucrose density gradients containing 2 *M* NaCl. Degraded RNA is recovered from the 4–10 S zone, whereas the protein is present in a sharp peak with a sedimentation coefficient of ~30 S (Fig. 10). Therefore a significant part of the informofers survived the isolation treatment and were not dissociated to subunits. This characterizes free informofers as having the same size as 30 S particles; in the electron microscope they could not be distinguished from the 200-Å diameter 30 S particles (Fig. 6). The buoyant density of informofers is equal to 1.34 gm/cm^3 corresponding to the density of pure protein. Informofers purified in CsCl density gradient do not contain any traces of pre-mRNA (Lukanidin *et al.*, 1972).

The important property of free informofers is their ability to react with free pre-mRNA with the reconstitution of 30 S particles or even polyparticles, when high-molecular-weight pre-mRNA is added. Reconstitution requires the removal of 2 *M* NaCl by dialysis. The 30 S particles or polyparticles formed have buoyant density of about 1.4 gm/cm^3 indicating the same protein to RNA ratio as in the original particle complexes (Lukanidin *et al.*, 1972).

The reaction seems to be specific to some extent for pre-mRNA. Binding of ribosomal RNA is less effective (Samarina *et al.*, 1967; McParland *et al.*, 1972).

These experiments substantiate the existence of informofers as protein entities and the primary localization of pre-mRNA on the surface of informofers. Otherwise it would be difficult to visualize the easy association and dissociation of long pre-mRNA chains with informofer without changes of its structure. The possibility of isolating free informofers and 30 S particles should allow the following analysis of their structural organization. For example, it is very important to understand how RNA is distributed on the surface of informofers. The length of RNA bound to one informofer is about 4000 Å compared with a 200-Å diameter of the informofer. One might expect the formation of regular helical RNA structures around the informofer.

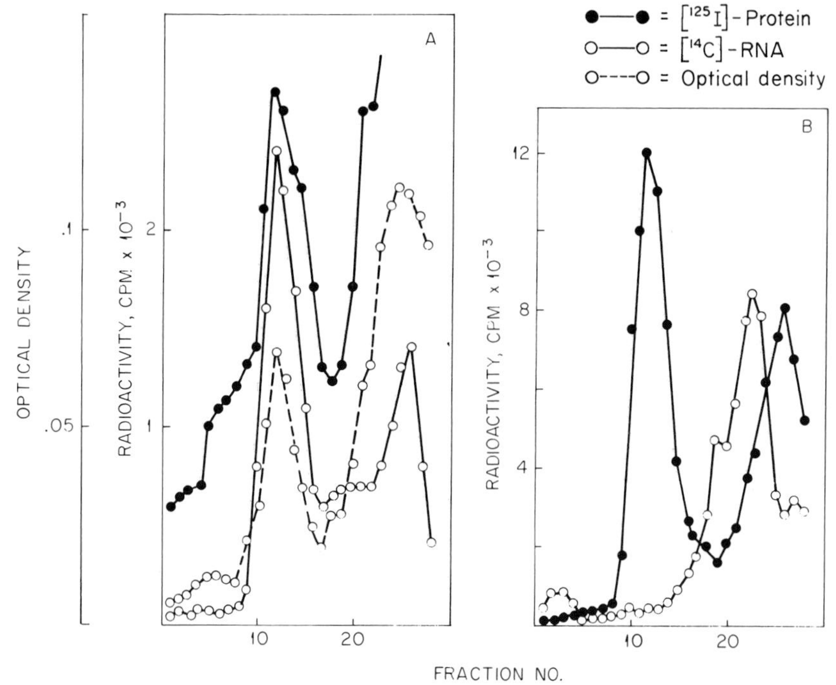

Fig. 10 Isolation of free informofers. (A) Control 30 S particles containing [^{14}C]-RNA and [^{125}I]-protein. (B) The same particles treated with 2 *M* NaCl. In all cases the material was ultracentrifuged during 14 hr in 15–30% sucrose gradient at 24,000 rpm. (Lukanidin *et al.*, 1972.)

C. *The Biological Role of Informofers in mRNA Transport*

To understand the role of informofers in mRNA transport one needs to have more information about the behavior of informofers under varying conditions as well as their general properties. A summary of some facts which may be used in the construction of a general hypothetical scheme of informofer participation in mRNA transfer from the nucleus to cytoplasm is presented below.

1. ALL KINDS OF NUCLEAR dRNA ARE BOUND TO INFORMOFERS

About 90% of nuclear pre-mRNA may be isolated in the form of complexes with informofers. However, pre-mRNA contains two types of sequences, mRNA and ps-mRNA, and the question is whether both of them are combined with informofers.

Pseudo-mRNA comprises the main part of nuclear pre-mRNA and it is evidently complexed with informofers. After short pulses, when the concentration of mRNA sequences is low, practically all pre-mRNA may be isolated in complexes with informofers. After long labeling or after actinomycin D chases, when a significant part of the ps-mRNA is degraded and the concentration of mRNA sequences is increased, practically all pre-mRNA is combined with informofers. This is shown by differences in the sizes of the complexes after short pulse labeling; the polyparticles isolated with RNase inhibitor have higher sedimentation coefficients compared to particles after long labeling or after actinomycin D chase. Thus, during the whole processing, pre-mRNA remain combined with informofers (Mantieva *et al.*, 1969).

Another approach is by competitive hybridization. RNA isolated from 30 S particles completely inhibits hybridization of nuclear pre-mRNA or polysomal RNA to DNA. On the other hand, polysomal RNA only partly competes with RNA isolated from 30 S particles, suggesting the presence of both mRNA and ps-mRNA complexes with informofers (Mantieva *et al.*, 1969, Drews, 1969).

The weak point of these experiments is that hybridization at low Cot values involves only those RNA transcripts complementary to reiterative DNA base sequences. Therefore, experiments with individual mRNA's are very important to the interpretations. One possible model is the use of cells infected with adenovirus, a DNA-containing virus replicating inside the cell nucleus. Adenovirus-specific RNA is formed in the nucleus and later all sequences reach the cytoplasm. Thus, adenovirus-specific RNA may be considered as a reasonable model of true mRNA (Velicer and Ginsberg, 1968; Parsons *et al.*, 1971). The adenovirus-infected cells are labeled and nuclear complexes, 30 S particles, and polyparticles are isolated, fixed with formaldehyde, and ultracentrifuged in CsCl density gradients. The peak with a buoyant density of 1.40 gm/cm^3 is collected, deproteinized with pronase and SDS–phenol, and hybridized to adenovirus DNA. The percent of RNA hybridized to viral DNA is the same in total nuclear pre-mRNA as in 30 S particles, and polyparticles purified by CsCl density gradient ultracentrifugation. In separate experiments, it was confirmed that the particles obtained from the infected cells contain informofers. Thus like other nuclear pre-mRNA's adenovirus-specific RNA is combined with informofers (Lukanidin *et al.*, 1972a). The general conclusion is that most pre-mRNA synthesized inside the cell nucleus immediately interacts with informofers and remains bound to them during the whole processing.

2. THE UNIVERSALITY OF INFORMOFERS

In all tissues studied up to this time (rat, mouse, and rabbit liver, Ehrlich carcinoma cells, calf thymus, bird erythroblasts, human culture cells, L cells, etc.) the general properties of ribonucleoproteins containing pre-mRNA are very similar, suggesting the binding of pre-mRNA to protein particles or informofers in all cases (Lukanidin *et al.*, 1972a). Recently, antibodies against rat liver 30 S particles have been prepared (Lukanidin *et al.*, 1972b). These antibodies interact specifically and efficiently with 30 S particles, shifting the peak in CsCl density gradients into a heavier zone. The buoyant density becomes lower, i.e., 1.34–1.35 gm/cm^3. No interaction is detectable when 30 S particles are mixed with nonimmune serum or in the addition of antibodies to purified polysomal mRNP. Antibodies against rat-liver 30 S particles interact with the same efficiency with human 30 S particles or polyparticles indicating a lack of species specificity (Lukanidin and Zalmanzon, unpublished).

Finally, the proteins isolated from all the above mentioned tissues are indistinguishable one from another in polyacrylamide gel electrophoresis either in urea or in SDS–urea (Lukanidin *et al.*, 1971, 1972a). Thus the binding of newly formed pre-mRNA to informofers is a general event at least in mammals and birds.

3. ELECTRON MICROSCOPIC OBSERVATIONS OF mRNA TRANSPORT

Structures which might be identified as mRNA containing ribonucleoproteins were described in giant nuclei of *Chironomus* salivary glands (Beermann and Bahr, 1954; Swift, 1959). These particles were about 300–400 Å in diameter, concentrated in Balbiani rings, scattered in the nuclear sap, and also observed in the pores of the cell nucleus. In the pores their shape was changed and they converted to fibrils. These structures were not found in the cytoplasm. The cytochemical properties of these particles suggested they were RNA-protein complexes, and autoradiography showed them to contain mRNA (or pre-mRNA) moving from the site of synthesis in the Balbiani ring to the cytoplasm.

In the nuclei of mammalian cells other kinds of ribonucleoproteins have been observed (Monneron and Bernhard, 1969; Fakan and Bernhard, 1971). A combination of electron microscopy and autoradiography have demonstrated that the rapidly labeled RNA is concentrated in the regions containing the "perichromatin fibrils" (Fig. 11). The latter are located on the periphery of the chromatin masses as irregular fibrils from 30 to 200 Å in diameter and granules 200 Å in diameter. Isolated 30 S particles treated in this way very often contain fibrillike structures in the electron microscope. The perichromatin fibrils may well correspond to the complexes of newly formed pre-mRNA and informofers. Besides perichromatin fibrils some other ribonucleoprotein structures were found: "perichromatin granules" similar to structures described in the Balbiani ring of *Chironomus* and "interchromatin granules" which are represented by 200 Å granules connected with thin fibrils. Interchromatin granules seem to be similar to isolated nuclear particles, except that they are labeled very slowly. It is possible that they contain more of the stable pre-mRNA which is enriched in mRNA but this is uncertain. In general, the identification of perichromatin granules and interchromatin granules awaits more definitive experiments.

In the nuclei of mammalian cells the flow of particles from the chromosomes to the "pores" of the nuclear membranes is less clear. Nevertheless, ribonucleoprotein particles have been observed near the pores. Inside the pores themselves the granular structures may be converted into thin fibrils which are evident, although no evidence has been found for them in the cytoplasm. Although electron microscopy does not allow one to describe the details of transport process it gives indications that the protein is associated with RNA from the moment of the formation of pre-mRNA to the transfer of mRNA to cytoplasm.

4. COMPARISON OF NUCLEAR PARTICLES WITH CYTOPLASMIC INFORMOSOMES AND POLYSOMAL mRNP

An important question is the relationship between the nuclear particles and cytoplasmic particles containing mRNA. The most direct way to answer this question is comparison of properties of the protein moiety.

Free mRNP particles or informosomes have mainly been studied in detail in embryonic tissues. In loach embryos informosomes give several dis-

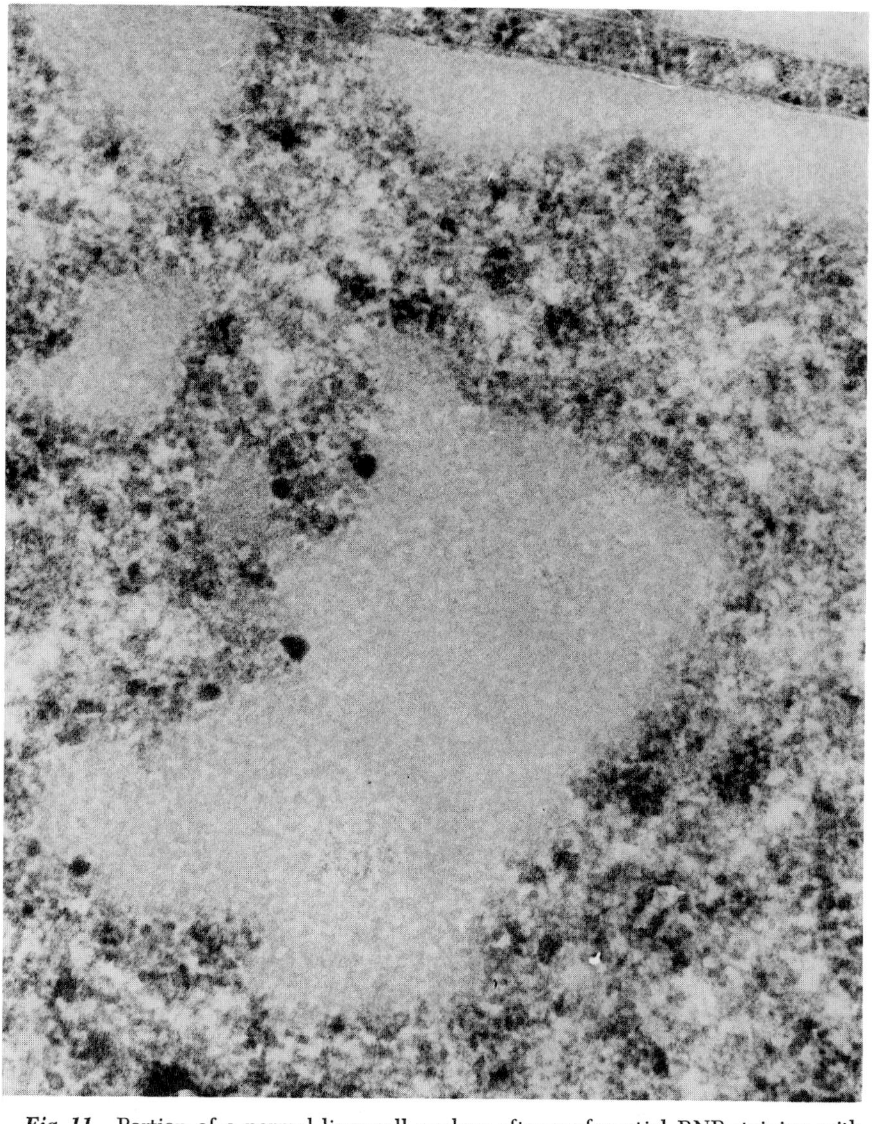

Fig. 11 Portion of a normal liver cell nucleus after preferential RNP staining with EDTA according to Bernhard. In the center and at the periphery of the nucleus is bleached, condensed chromatin, surrounded by many heavily stained perichromatin fibrils and perichromatin granules representing extranucleolar RNP (× 60,000). (From Bernhard, 1972.)

crete peaks in sucrose density gradients. The buoyant density of the main part of material in the heavy peaks (40 S and more) is about 1.4 gm/cm^3, similar to nuclear particles. The lighter informosomes have the lower buoyant density (1.35 gm/cm^3). Informosomes and nuclear particles have other common properties: RNase sensitivity; instability in the presence of deoxycholate and high salts; and a correlation between the size of the particles and RNA isolated from them (see Spirin, 1969; Georgiev and Samarina, 1971). There are also some differences, since informosomes have lower sedimentation coefficients than nuclear particles containing RNA of the same molecular weight. It is desirable to compare these two types of particles by analysis of their protein moieties. However, at present this direct experiment has not been feasible since pure informosomes were not available. A promising method of comparison is that of antibodies to nuclear 30 S particles.

In cells of adult organisms it is very difficult to obtain free informosomes. Careful homogenization to exclude leakage of material from the nucleus usually prevents the appearance of particles with buoyant density of 1.4 gm/cm^3 in cytoplasm. Only particles with higher density of about 1.45–1.48 gm/cm^3 may be observed. These particles correspond either to complexes of small ribosomal subunits with mRNA (or mRNP?) or to real mRNP similar to polysomal mRNP (Perry and Kelley, 1968; Huang and Baltimore, 1970; Ivanyi, 1971; Leytin *et al.*, 1972). Recently free mRNP was observed in reticulocytes (where nuclear leakage was excluded) but these particles were very similar to polysomal mRNP of reticulocytes (see below) and different from nuclear particles (Jacobs-Lorena *et al.*, 1972). The presence of particles with a density of 1.4 in the cytoplasm of early embryos and their absence in adult tissues may depend on the very high rate of mitosis in the former. It has been shown recently that during mitosis pre-mRNA is liberated into the cytoplasm and after completion of mitosis comes back to the nucleus (Abramova and Neifakh, 1973). It seems very probable that during mitosis pre-mRNA remains complexed with informofers. Embryonic informofers may be the nuclear particles which stay in cytoplasm during the mitosis. Of course, other possibilities cannot be excluded.

In polysomes mRNA is also combined with a specific protein since when polysomes are dissociated with EDTA and ultracentrifuged through sucrose gradients one can find 50 S and 30 S ribosomal subunits and heterogeneous RNP material, containing rapidly labeled RNA, presumably mRNA. In CsCl density gradients the latter is banded as a heterogeneous peak with a buoyant density of about 1.45–1.48 gm/cm^3, which corresponds to ribonucleoprotein with a lower protein to RNA ratio than nuclear particles (Perry and Kelly, 1968; Henshaw, 1968).

Polysomal mRNP of reticulocytes was studied in more detail as it may be easily separated from polysomes (Burny *et al.*, 1969) and the possibility of nuclear contamination is excluded. After EDTA treatment, polysomal mRNP moves in sucrose gradients as a homogeneous peak with a sedimentation coefficient of about 14 S. The particles contain 9 S mRNA and some protein with a particle buoyant density of 1.48 gm/cm^3. The particles have been purified, the protein isolated, and when compared to informatin found to be quite different (Lukanidin *et al.*, 1971; Morel *et al.*, 1971). Two main bands were found in polyacrylamide gels, neither of which coincide with informatin. Also, unlike informatin, mercaptoethanol treatment does not influence their electrophoretic pattern. Finally, proteins from mRNP exhibit different behavior on exchange resins than does informatin, suggesting that informatin is absent in polysomal mRNP. Therefore one can conclude that informofers do not reach cytoplasmic polysomes.

Olsnes (1971) found that polysomal mRNP from rat liver, like reticulocyte mRNP, contains several main proteins. However, several controls have shown that only one of them (with molecular weight 160,000) really associates with mRNA in polysomes. mRNP from rat liver polysomes has been mixed with antibodies against nuclear 30 S particles isolated from rat liver. No interaction was observed (Lukanidin *et al.*, 1972.)

5. IN VITRO FORMATION OF 30 S PARTICLES

It is possible to obtain the complexes of informofers and RNA synthesized in isolated cell nuclei. Rat liver nuclei were incubated with labeled nucleoside triphosphate and the product of the reaction was studied. Practically all labeled high-molecular weight RNA (S values higher than 4–6) isolated from these nuclei was found in 30 S particles with buoyant densities equal to 1.4 gm/cm^3. Some RNA, in contrast to the *in vivo* situation, remained bound to chromatin. Thus nearly all RNA liberated from chromatin was combined with informofers (Samarina *et al.*, 1973).

6. HYPOTHETICAL SCHEME OF THE PARTICIPATION OF INFORMOFERS IN mRNA TRANSPORT (INFORMOFER CYCLE)

On the basis of above mentioned data we proposed the following scheme of the first step of mRNA transport (Lukanidin *et al.*, 1972a) (Fig. 12).

a. First Step: The Binding of Pre-mRNA. Immediately after synthesis the pre-mRNA is combined with informofers, there are two possibilities: that it is combined with preexisting free informofers; or that pre-

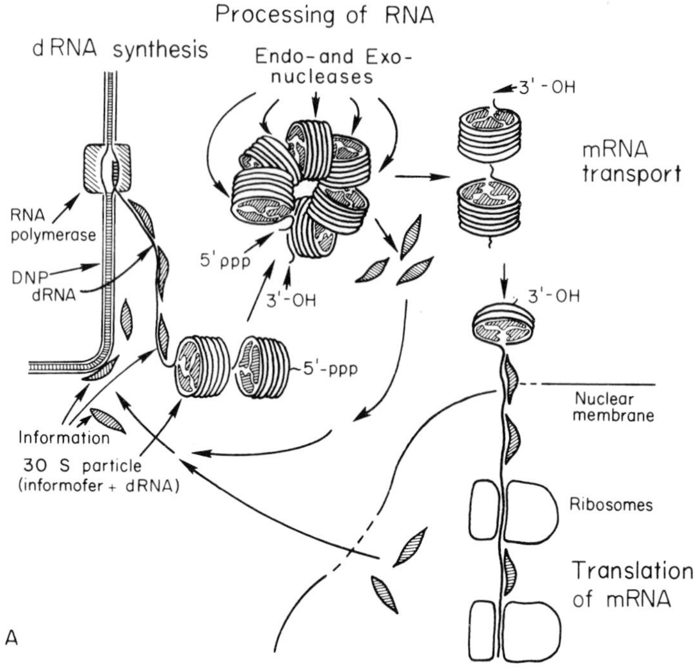

Fig. 12 Two hypothetical schemes of the informofer cycle. (A) The scheme with dissociation of informofers into informatin units and partial transfer of informatin to cytoplasm (Georgiev and Samarina, 1971). (B) The scheme with informofer survival.

mRNA interacts with informatin molecules and later the informatin collapses to form protein globules with the RNA distributed on the surface.

The significance of this first step is (1) the prevention of an interaction of pre-mRNA with the basic proteins of the chromatin and (2) the reduction of the linear size of RNA. It is known that RNA may easily combine with chromatin histones (Ilyin *et al.*, 1971). On the other hand, RNA bound to informofers neither interact with histones nor with ribosomes. The length of mammalian newly formed pre-mRNA is about 5–10 μm, or the same order of size as the diameter of the nucleus. However, after binding to informofers the linear size of this RNA decreases twentyfold (the diameter of 30 S particles is 200 Å whereas the length of RNA bound to one informofer is 4000 Å) and comprises only 250–500 nm.

b. Second Step: Processing of Pre-mRNA. After the separation of pre-mRNA from the chromatin template (or even before this moment) the processing begins. The pre-mRNA distributed on the surface of informofers is attacked by specific endo- and exonucleases. As a result ps-mRNA is degraded and some informofers (those bound to ps-mRNA) are lib-

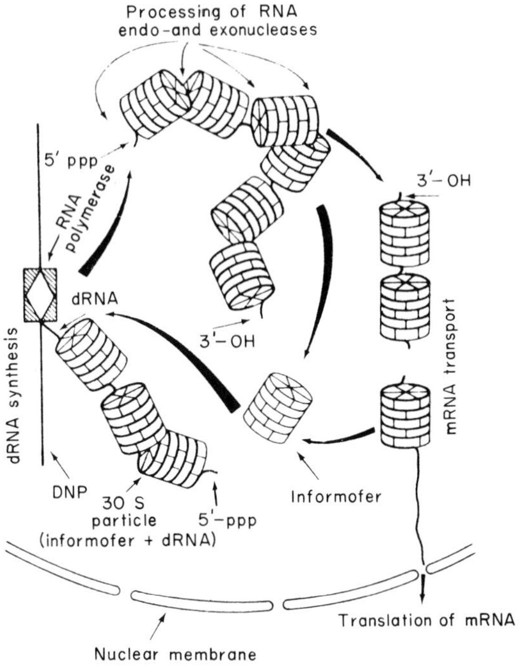

B

erated. The others remain combined with true mRNA. The significance of informofers at this step is that they provide a surface on which pre-mRNA is distributed in such a way that all its sites are available to processing enzymes allowing the latter to recognize the corresponding base sequences.

All pre-mRNA molecules at different steps of processing including newly formed giant pre-mRNA are combined with informofers. Some enzymes that presumably participate in processing were found to be associated with 30 S particles. These include a specific RNase which splits RNA at a limited number of points and a poly(A) synthetase (Niessing and Sekeris, 1970, 1972). The latter may be involved in the addition of poly(A) to mRNA sequences. Both enzymes are probably not the constituents of informofers themselves but associate with RNA of nuclear particles in the course of processing.

c. Third Step: The Injection. After the processing some mRNA survives and is transferred to the nuclear membrane. Some modification of this mRNA takes place by the addition of poly(A). Nothing is clearly defined about the mechanism of mRNA transfer to the nuclear membrane. It is possible that a simple diffusion of nuclear particles takes place but more complex and specific processes cannot be excluded.

In any case the complex of mRNA and informofers reaches the nuclear membrane, the mRNA crosses the membrane, whereas the informofers remain inside the nucleus. The transfer of mRNA from the nucleus to cytoplasm then proceeds. In general, nuclear particles may be compared with viruses and considered as "inverted viruses." Complex formation of functional RNA with protein particles consisting of identical subunits is characteristic of mRNA and viral RNA, but in contrast to the virus, pre-mRNA is localized on the surface of the informofer. The informofer protein does not protect pre-mRNA against nucleases but provides favorable conditions for this attack. Finally, like viruses only mRNA is released compared to virus transport out of the cell.

After the transfer of mRNA into the cytoplasm as well as after degradation of ps-mRNA two possibilities should be considered (see Section II, C,6a). Either the informofers survive this step and are incorporated into the pool of free informofers, or they are degraded to subunits and the pool consists of free informatin molecules which are assembled into informofers only after binding to new pre-mRNA. In any case, informofers (or informatin molecules) are involved in a new cycle and may be used for several cycles of mRNA transport.

The mechanism of transfer is quite obscure but membrane proteins or some ribosomal proteins may be involved. Data supporting the transfer of mRNA from informofers to ribosomes in cell-free systems have been reported (Ishikawa *et al.*, 1972). The reuse of informofers (or informatin) for mRNA transport can be followed from the low rate of informatin synthesis (Ishikawa *et al.*, 1970; Georgiev and Samarina, 1971). Only in rapidly dividing cells is it rather high.

The hypothesis presented is theoretical, but it is in good agreement with many facts about mRNA processing in animal cells. Further studies of nuclear particles containing pre-mRNA should clarify the mechanisms involved in the nuclear steps of mRNA processing and transport.

ACKNOWLEDGMENT

The author is indebted to Professor Robert B. Church for reading the paper and for his valuable advice.

REFERENCES

Abramova, N. B., and Neifakh, A. A. (1973). *Exp. Cell Res.* **77**, 136.
Adesnik, M., and Darnell, J. E. (1972). *J. Mol. Biol.* **67**, 397.
Ananieva, L. N., Kozlov, Yu. V., Ryskov, A. P., and Georgiev, G. P. (1968). *Mol. Biol. (USSR)* **2**, 736.
Arion, V. Ya., and Georgiev, G. P. (1967). *Proc. Acad. Sci. USSR* **172**, 716.

Attardi, G., Parnas, H., Huang, M.I.H., and Attardi, B. (1966). *J. Mol. Biol.* **20**, 145.
Baltimore, D. (1970). *Nature (London)* **226**, 1209.
Barr H., and Lingrel, J. B. (1971). *Nature (London) New Biol.* **233**, 41.
Becker, M., and Rich, A. (1966). *Nature (London)* **212**, 142.
Beermann, W. (1972). *In* "The Cell Nucleus. Morphology, Physiology and Biochemistry" (I. B. Zbarsky and G. P. Georgiev, eds.), p. 190, Nauka, Moscow.
Beermann, W., and Barr, G. F. (1954). *Exp. Cell Res.* **6**, 195.
Bernhard, W. (1972). *In* "The Cell Nucleus. Morphology, Physiology and Biochemistry, p. 15, Nauka, Moscow.
Besson, J., Farashyan, V. R., and Ryskov, A. P. (1972). *Cell Differentiation* **1**, 127.
Bishop, J. D., Pemberton, R., and Baglioni, C. (1972). *Nature (London) New Biol.* **11**, 351.
Borun, T. W., Scharff, M. D., and Robbins, E. (1967). *Proc. Nat. Acad. Sci. U.S.* **58**, 1977.
Brawerman, G., (1963). *Biochim. Biophys. Acta* **76**, 322.
Brawerman, G. (1973). *Mol. Biol. Repts.*, **1**, 7.
Brawerman, G., Biezunski, N., and Eisenstadt, J. (1963). *Biochim. Biophys. Acta* **49**, 240.
Brawerman, G., Gold, L., and Eisenstadt, J. (1964). *Proc. Nat. Acad. Sci. U.S.* **50**, 630.
Britten, R. J., and Kohne, D. E. (1968). *Science* **161**, 529.
Britten, R. J., and Davidson, E. H. (1969). *Science* **165**, 349.
Brown, I. R., and Church, R. B. (1972). *Develop. Biol.* **29**, 73.
Burny, A., Huez, G., Marbaix, G., and Chantrenne, H. (1969). *Biochim. Biophys. Acta* **190**, 228.
Church, R. B., and Brown, I. R. (1972). *In* "Results and Problems in Cell Differentiation" (H. Ursprung, ed.), Vol. 3, pp. 11–24. Springer-Verlag, Berlin and New York.
Church, R. B., and McCarthy, B. J. (1968). *Biochem. Genet.* **2**, 55.
Church, R. B., and Georgiev, G. P. (1973). *Mol. Biol. Repts.*, **1**, 21.
Crick, F. (1971). *Nature (London)* **234**, 25.
Daneholt, B., Edström, J. E., Egyhazi, E., Lambert, B., and Ringborg, U. (1969). *Chromosoma* **28**, 379, 399, 418.
Darnell, J. E., Wall, R., and Tushinsky, R. J. (1971). *Proc. Nat. Acad. Sci. U.S.* **68**, 1321.
Drews, J. (1969). *Eur. J. Biochem.* **9**, 263.
Edmonds, M., and Caramela, M. G. (1969). *J. Biol. Chem.* **244**, 1314.
Edmonds, M., Vaughan, M. H., and Nakazoto, H. (1971). *Proc. Nat. Acad. Sci. U.S.* **68**, 1336.
Faiferman, J., Hamilton, M. G., and Pogo, A. O. (1970). *Biochim. Biophys. Acta* **204**, 550.
Fakan, S., and Bernhard W. (1971). *Exp. Cell Res.* **67**, 129.
Flamm, W. G., Walker, P. M. B., and McCallum, M. (1969). *J. Mol. Biol.* **40**, 423.
Gazaryan, K. G., Schuppe, N. G., and Prokoshkin, B.D. (1966). *Biokhimiya* **31**, 108.
Georgiev, G. P. (1961). *Biokhimiya* **26**, 1095.
Georgiev, G. P. (1966). *In* "The Cell Nucleus; Metabolism and Radiosensitivity," p. 79. Taylor and Francis, London.
Georgiev, G. P. (1967). *Progr. Nucl. Acid Res. Mol. Biol.* **6**, 259.
Georgiev, G. P. (1969). *J. Theoret. Biol.* **25**, 473.
Georgiev, G. P., and Lerman, M. I. (1964). *Biochim. Biophys. Acta* **91**, 678.
Georgiev, G. P., and Mantieva, V. L. (1962a). *Vop. Med. Khim.* **8**, 92.

Georgiev, G. P., and Mantieva, V. L. (1962b). *Biokhimiya* **27**, 949.
Georgiev, G. P., and Samarina, O. P. (1971). *Advan. Cell Biol.* **2**, 47.
Georgiev, G. P., Samarina, O. P., Lerman, M. I., and Smirnov, M. N. (1963). *Nature (London)* **200**, 1291.
Georgiev, G. P., Ryskov, A. P., Coutelle, Ch., Mantieva, V. L., and Avakyan, E. R. (1972a). *Biochim. Biophys. Acta* **259**, 259.
Georgiev, G. P., Samarina, O. P., and Irlin, I. S. (1972b). *Proc. Acad. Sci. USSR*, **205**, 969.
Granboulan, N., and Scherrer, K. (1969). *Eur. J. Biochem.* **9**, 1.
Gvozdev, V. A., and Tikhonov, V. H. (1964) Biakhimiya **29**, 1083.
Harel, L., and Montagnier, L. (1971). *Nature (London) New Biol.* **229**, 406.
Harris, H. (1963). *Progr. Nucl. Acid Res. Mol. Biol.* **2**, 20.
Henshaw, E. C. (1968). *J. Mol. Biol.* **36**, 401.
Huang, A. S., and Baltimore, D. (1970). *J. Mol. Biol.* **47**, 275.
Ilyin, Yu. V., Varshovsky, A. Ya., Mickelsaar, U. N., and Georgiev, G. P. (1971). *Eur. J. Biochem.* **22**, 235.
Imaizumi, M. T., Diggelman, M., and Scherrer, K. (1972). *Proc. Nat. Acad. Sci. U.S.* **70**, 1122.
Ishikawa, K., Kurode, C., and Ogata, K. (1969). *Biochim. Biophys. Acta* **179**, 316.
Ishikawa, K., Kurode, C., and Ogata, K. (1970). *Biochim. Biophys. Acta* **213**, 505.
Ishikawa, K., Ueki, M., Nagai, K., and Ogata, K. (1972). *Biochim. Biophys. Acta* **259**, 138.
Ivaniy, J. (1971). *Biochim. Biophys. Acta* **238**, 303.
Jacobs-Lorena, M., and Baglioni, C. (1972). *Proc. Nat. Acad. Sci. U.S.* **69**, 1425.
Jelinek, W., and Darnell, J. E. (1962). *Proc. Nat. Acad. Sci. U.S.* **69**, 2537.
Kedes, L. H., and Birnstiel, M. L. (1971). *Nature (London)* **230**, 165.
Krichevskaya, A. A., and Georgiev, G. P. (1969). *Biochim. Biophys. Acta* **194**, 619.
Krichevskaya, A. A., and Georgiev, G. P. (1973). *Mol. Biol. (USSR)* **7**, 168.
Kwan, S. W., and Brawerman, G. (1972). *Proc. Nat. Acad. Sci. U.S.* **69**, 3247.
Lambert, B., Wieslander, L., Daneholt, B., Egyhazi, E., and Ringborg, U. (1972). *J. Cell Biol.* **53**, 407.
Lang, N. (1965). *J. Cell Comp. Physiol.* **66**, S.1,132.
Lang, N., and Sekeris, C. E. (1964). *Life Sci.* **3**, 161.
Lee, S. Y., Mendecki, T., and Brawerman, G. (1971). *Proc. Nat. Acad. Sci. U.S.* **68**, 1331.
Leytin, N. L., Podobed, O. V., and Lerman, M. I. (1972). *Biokhimiya* **37**, 65.
Lindberg, U., and Darnell, J. E. (1970). *Proc. Nat. Acad. Sci. U.S.* **65**, 1089.
Lukanidin, E. M., Georgiev, G. P., and Williamson, R. (1971). *FEBS Lett.* **19**, 152.
Lukanidin, E. M., Zalmanzon, E. S., Komaromi, L., and Georgiev, G. P. (1972a). *Nature (London) New Biol.* **238**, 193.
Lukanidin, E. M., Olsnes, S., and Phil, A. (1972b). *Nature (London) New Biol.* **240**, 90.
Malloy, G. R., Sporn, M. B., Kelley, D. E., and Perry, R. P. (1972). *Biochemistry* **11**, 3256.
Mantieva, V. L., Avakyan, E. R., and Georgiev, G. P. (1971). *Mol. Biol. (USSR)* **5**, 321.
Markov, G. G., and Arion, V. Ya. (1973). *Eur. J. Biochem.* **35**, 186.
McCarthy, B. J., and Church, R. B. (1970). *Ann. Rev. Biochem.* **39**, 131.
McParland, R., Crooke, S., and Busch, H. (1972). *Biochim. Biophys. Acta* **269**, 78.
Melli, M., and Pemberton, R. E. (1972). *Nature (London) New Biol.* **236**, 172.

Melli, M., Whitefield, C., Rao, K. V., Richardson, U., and Bishop, J. O. (1971). *Nature (London) New Biol.* **231**, 8.
Mendecki, J., Lee, S. Y., and Brawerman, G. (1972). *Biochemistry* **11**, 792.
Monneron, A., and Bernhard, W. (1969). *J. Ultrastruct. Res.* **27**, 266.
Monneron, A., and Moule, Y. (1968). *Exp. Cell Res.* **51**, 531.
Morel, C., Kayibanda, B., and Scherrer, K. (1971). *FEBS Lett.* **18**, 84.
Moule, Y., and Chauveau, J. (1966). *C. R. Acad. Sci.* Paris **D263**, 75.
Moule, Y., and Chauveau, J. (1968). *J. Mol. Biol.* **33**, 465.
Niessing, J., and Sekeris, C. E. (1970). *Biochim. Biophys. Acta* **209**, 484.
Niessing, J., and Sekeris, C. E. (1971). *FEBS Lett.* **19**, 39.
Niessing, J., and Sekeris, C. E. (1972). *FEBS Lett.* **22**, 83.
Oda, K., and Dulbecco, R. (1968). *Proc. Nat. Acad. Sci. U.S.* **60**, 525.
Olsnes, S. (1971). *Eur. J. Biochem.* **23**, 557.
Parsons, J. T., Gardner, J., and Green, M. (1971). *Proc. Nat. Acad. Sci. U.S.* **68**, 557.
Penman, S., Scherrer, K., Becker, I., and Darnell, J. E. (1963). *Proc. Nat. Acad. Sci. U.S.* **49**, 654.
Penman, S., Smith, I., and Holtzman, E. (1966). *Science* **154**, 786.
Penman, S., Fan, H., Perlman, S., Rosbash, M., Weinberg, R., and Zylber, Z. (1970). *Cold Spring Harbor Symp. Quant. Biol.* **35**, 561.
Perry, R. P., and Kelley, D. E. (1968). *J. Mol. Biol.* **35**, 37.
Philipson, L., Wall, R., Glickman, G., and Darnell, J. E. (1971). *Proc. Nat. Acad. Sci. U.S.* **68**, 2806.
Ryskov, A. P., Farashyan, V. R., and Georgiev, G. P. (1972a). *FEBS Lett.* **20**, 355.
Ryskov, A. P., Farashyan, V. R., and Georgiev, G. P. (1972b). *FEBS Lett.* **22**, 227.
Ryskov, A. P., Farashyan, V. R., and Georgiev, G. P. (1972c). *Biochim. Biophys. Acta* **262**, 568.
Ryskov, A. P., Saunders, G., Farashyan, V. R., and Georgiev, G. P. (1973). *Biochim. Biophys. Acta* **312**, 152.
Samarina, O. P. (1964). *Biochim. Biophys. Acta* **91**, 688.
Samarina, O. P., Asriyan, I. S., and Georgiev, G. P. (1965a). *Proc. Acad. Sci. USSR* **163**, 1510.
Samarina, O. P., Lerman, M. I., Timanyan, V. G., Ananieva, L. N., and Georgiev, G. P. (1965b). *Biokhimiya* **30**, 880.
Samarina, O. P., Krichevskaya, A. A., and Georgiev, G. P. (1966), *Nature (London)* **210**, 1319.
Samarina, O. P., Krichevskaya, A. A., Molnar, J., Bruskov, V. I., and Georgiev, G. P. (1967a). *Mol. Biol. (USSR)* **1**, 129.
Samarina, O. P., Molnar, J., Lukanidin, E. M., Bruskov, V. I., Krichevskaya, A. A., and Georgiev, G. P. (1967b). *J. Mol. Biol.* **27**, 187.
Samarina, O. P., Lukanidin, E. M., Molnar, J., and Georgiev, G. P. (1968). *J. Mol. Biol.* **33**, 251.
Samarina, O. P., Kholodenko, L. V., and Aitkhozhina, N. A. (1973). *Mol. Biol. (USSR)* **6**, 712.
Sarasin, A. (1969). *FEBS Lett.* **4**, 327.
Scherrer, K. (1971). *FEBS Lett.* **17**, 68.
Scherrer, K., and Marcaud, L. (1965). *Bull. Soc. Chim. Biol.* **47**, 1697.
Scherrer, K., and Marcaud, L. (1968). *J. Cell Physiol.* **72**, *Suppl.* **1**, 181.
Scherrer, K., Latham, H., and Darnell, J. E. (1963). *Proc. Nat. Acad. Sci. U.S.* **49**, 240.

Scherrer, K., Spohr, G., Granboulan, N., Morel, C., Grosclaude, J., and Chezzi, C. (1970). *Cold Spring Harbor Symp. Quant. Biol.* **35**, 539.
Shearer, R. W., and McCarthy, B. J. (1967). *Biochemistry* **6**, 283.
Sibatani, A., DeKloet, S. R., Allfrey, V. G., and Mirsky, A. E. (1962). *Proc. Nat. Acad. Sci. U.S.* **48**, 471.
Spirin, A. S., (1969). *Eur. J. Biochem.* **10**, 20.
Spirin, A. S., Belitsina, N. V., and Aitkhozhin, M. A. (1964). *J. Gen. Biol. (Moscow)* **24**, 321.
Suzuki, Y., Gage, L. P., and Brown, D. D. (1972). *J. Mol. Biol.* **70**, 637.
Swift, H. (1959). *Brookhaven Symp. Biol.* **12**, 134.
Temin, H. M., and Mizutani, S. (1970). *Nature (London)* **226**, 1211.
Tonegawa, S., Walter, G., Bernardini, A., and Dulbecco, R. (1970). *Cold Spring Harbor Symp. Quant. Biol.* **35**, 823.
Velicer, L. F., and Ginsberg, H. S. (1968). *Proc. Nat. Acad. Sci. U.S.* **61**, 1264.
Wall, R., and Darnell, J. E. (1971). *Nature (London)* **232**, 73.
Warner, J. R., Soeiro, R., Birnboim, H. C., Girard, M., and Darnell, J. E. (1966). *J. Mol. Biol.* **19**, 349.
Verma, I. M., Temple, G. F., Fan, H., and Baltimore, D. (1972). *Nature New Biol.* **235**, 163.
Westphal, H., and Dulbecco, R. (1968). *Proc. Nat. Acad. Sci. U.S.* **59**, 1158.
Williamson, A. R., and Askonas, B. A. (1967). *J. Mol. Biol.* **23**, 201.
Williamson, R., Drewienkiewicz, C. E., and Paul, F. (1973). *Nature New Biol.* **241**, 66.
Yoshikawa-Fukada, M., Fukada, T., and Kawada, Y. (1964). *Biochem. Biophys. Res. Commun.* **15**, 23.
Yoshikawa-Fukada, M., Fukada, T., and Kawada, Y. (1965). *Biochim. Biophys. Acta* **103**, 383.
Zimmerman, E., and Turba, F. (1964). *Biochem. Z.* **339**, 469.

4

Nuclear High-Molecular-Weight RNA

Yong C. Choi, Ross N. Nazar, and Harris Busch

I. Introduction*

Importance of Nuclear Function in Controls of RNA Metabolism

During the last decade when characterization of high-molecular-weight nuclear RNA (HMW nRNA) became feasible, great progress has been made in understanding of not only the structure and function of nuclear RNA but also of nuclear genes (see Smetana and Busch, Volume I, Chapter 2; Birnstiel, 1967; Georgiev, 1967, 1971; Perry, 1967; Darnell, 1968; Loening, 1968; Busch and Smetana, 1970; Attardi and Amaldi, 1970; Burdon, 1971; Maden, 1971; Darnell *et al.*, 1973). The cell nucleus contains most of the genes and consequently is the synthetic site of many RNA species, most of which are transported into the cytoplasm where genetic expression takes place in the form of protein synthesis. Figure 1 shows the sedimentation patterns of nuclear and nucleolar RNA (of normal rat liver preparations), obtained by sucrose density gradient centrifugation. High molecular weight nRNA constitutes a great part of nuclear and nucleolar RNA, i.e., 75% of whole nuclear RNA and 85% of whole nucleolar RNA on a weight basis. Arbitrarily, HMW RNA has a sedimentation rate of 20 S or greater, or a molecular weight greater than 10^6 daltons. Although the functional significance of all the HMW nRNA species is not fully defined, current knowledge allows their categorization into informational and noninformational RNA groups. The first group is heterogeneous nuclear RNA (HnRNA) or premessenger RNA (pre-mRNA) described in Chapter 3 of this volume. The designation "heterogeneous" refers to physical characteristics of polydispersity and characteristics of the nucleotide composition (such as a high A + U content) and consequently the term DNA-like or DRNA (or dRNA) has often been used. As

* Abbreviations used: HMW RNA, high-molecular-weight RNA; mRNA, messenger RNA; nRNA, nucleolar RNA; pre-rRNA, pre-ribosomal RNA or ribosomal precursor RNA; rRNA, ribosomal RNA; ψ, pseudouridine; m, methyl moiety: (on the left it represents base methylation, on the right it represents ribose methylation). Location of methyl group and multipliers are indicated by superscripts and subscripts, respectively. Symbols for modified nucleosides are used according to the nomenclature recommended by the IUPAC-IUB Commission on Biochemical Nomenclature.

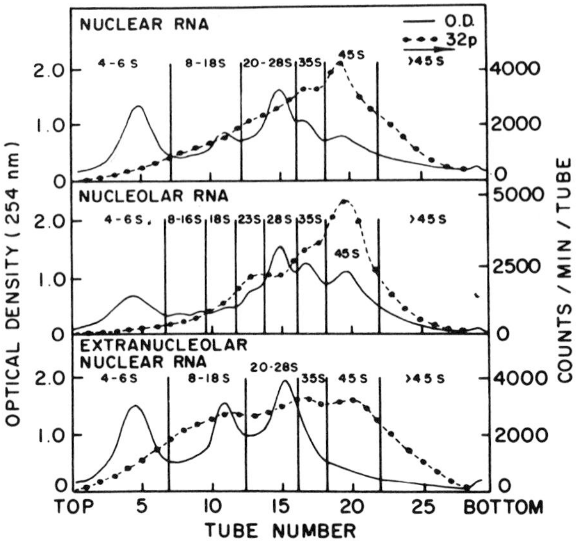

Fig. 1 Sedimentation profiles of nuclear, nucleolar, and extranucleolar nuclear RNA of rat liver. [^{32}P]orthophosphate, 2 mCi per rat, was injected intravenously 30 min prior to killing the animals. The solid line is optical density and the dotted line is radioactivity. (From Muramatsu *et al.*, 1966.)

its metabolism has become clarified, the designation HnRNA was introduced to indicate these RNA's are precursors of cytoplasmic messenger RNAs (mRNA). Cytoplasmic mRNA's are derived from HnRNA, but is not certain whether all the HnRNA species are informational or are precursors of cytoplasmic mRNA's.

The second group is the nucleolar RNA or pre-ribosomal RNA that is associated with nucleolar substructures and is characterized by a high G + C nucleotide composition in mammalian cells. Since the nucleolus is the synthetic site for ribosomal precursor RNA, the terms pre-rRNA, rapidly sedimenting, or HMW nucleolar RNA are interchangingly used to designate the precursors of the large ribosomal RNA's (two species, one each for the small and large ribosomal subunits).

There are marked structural and functional differences of eukaryotic organisms from prokaryotic organisms in cell division, growth, differentiation, specialization, and other biological phenomena including the actions of hormones and drugs. Intensive studies have been made on control mechanisms of gene action in eukaryotes. McCarthy and Hoyer (1964) and others reported tissue specific RNA's in studies on DNA–RNA hybridization; they suggested that selective gene transcription is responsible for specialization of cell functions (McCarthy and Church, 1970). Denis

(1966) and others demonstrated different RNA in embryonic development and differentiation. Britten and Kohne (1968) using DNA–DNA hybridization, demonstrated the existence of partial redundancies in many eukaryotic genes (DNA) and suggested that DNA is composed of three major classes: highly repetitious, moderately repetitious, and nonrepetitious.

Table I shows some differences of HMW RNA metabolism between prokaryotes and eukaryotes. One of the notable differences is the physical separation by the nuclear envelope which effectively separates the site of transcription (nucleus) from the site of translation (cytoplasm). The physical compartmentalization of information flow is paralleled with an evolution of complex ultrastructures of the nucleus and cytoplasm. Functionally the relation between the nucleus and cytoplasm (nucleocytoplasmic relationship) may play a role in control of gene action.

TABLE I
Differences of HMW RNA

Characteristic	Prokaryote	Eukaryote
Compartmentalization of template (chromosome) for RNA synthesis	One chromosome, not confined by membrane	Many chromosomes, confined by a nuclear envelope
Mode of information flow	Coupled transcription with translation	Uncoupled transcription (in nucleus) with translation (in cytoplasm)
Form of informational RNA	Polycistronic, synthesized as functionally active mRNA	Monocistronic and polycistronic(?), synthesized as HnRNA (pre-mRNA); processing before transport into cytoplasm
Form of noninformational RNA (RNA for ribosomal small and large subunits)	Synthesized as separate precursors	Synthesized as a giant ribosomal precursor; contains large spacer sequences

Both HMW informational (messenger) and noninformational (ribosomal) RNA species are produced as giant nuclear precursors which are processed to much smaller molecules. For noninformational RNA there is an initial precursor and intermediate precursors, most of which are separated. However, the numbers of intermediate species of informational RNA are not known.

II. Fractionation of Nuclear Substructures and Their Associated High-Molecular-Weight RNA

Few RNA species (perhaps tRNA) in the cell nucleus exist as free RNA molecules. Most RNA's are associated with proteins as ribonucleoproteins or ribonucleoprotein particles (RNP) associated with nuclear or nucleolar substructures. Examination of nuclear ultrastructure in interphase indicates a localization of RNA as RNP in perichromatin fibrils, perichromatin granules, interchromatin granules, granular bodies and coiled bodies in the extranucleolar portion of the nucleus (see Boutielle *et al.*, Volume I, Chapter 1) and fibrillar components and granular components in the nucleolar portion of the nucleus (Busch and Smetana, 1970). The current fractionation methods are not completely compatible with the morphological structures and further advancements to separate out these nuclear elements are required.

A. *Salt Fractionation*

The underlying principle of salt fractionation is to exploit the solubility characteristics of nuclear substructures. The most widely used methods employ sequential extractions of isolated nuclei with 0.15 *M* NaCl and subsequently with 2.0 *M* NaCl (Georgiev, 1967; Steele and Busch, 1966a, b; Busch *et al.*, 1968). The initial sequential extractions were primarily concerned with fractionation of various complex nuclear protein constituents. As the methodology of intact RNA extraction and of structural characterization was advanced, extensive studies have been made to identify the RNA associated with nuclear substructures (Fig. 2; Table II).

1. NUCLEAR SAP

The extraction with 0.15 *M* NaCl is designed to isolate the soluble phase through which there is exchange of metabolic substances and which operates throughout subnuclear compartments and through which influx and efflux of metabolites takes place between the nucleus and cytoplasm. However, other effects of the 0.15 *M* NaCl complicate the "nuclear sap" because the integrity of nuclear substructures is also subjected to alteration, and in turn structural alterations result in the release of loosely bound substances from the nuclear membrane, DNP fibrils, nucleolonemas, and other components. The nuclear sap fraction contains approximately 30–50% of total RNA of nuclei prepared by aqueous methods. The species of RNA extracted with 0.15 *M* NaCl include ribosomal RNA and

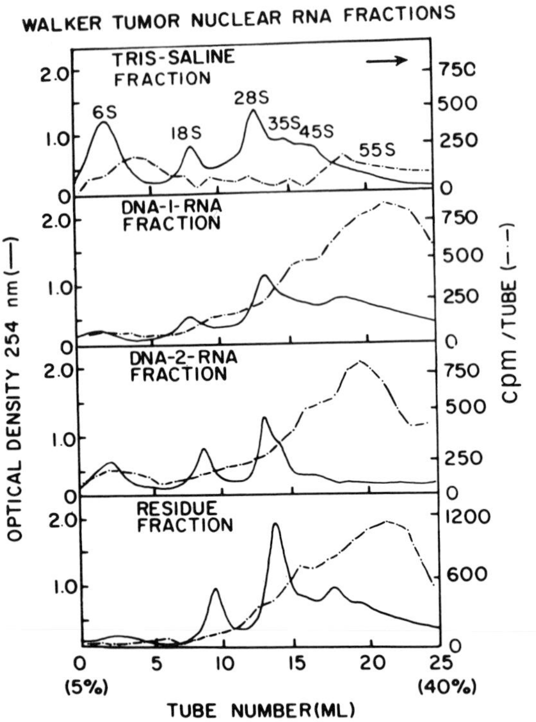

Fig. 2 Sedimentation profiles of RNA's from nuclear subfractions of Walker tumor cells; 1.5–2.0 mCi [^{32}P]orthophosphate was injected intravenously into each rat. After 20 min, the animals were sacrificed and nuclear preparations were made. The arrow shows the direction of sedimentation; the sucrose concentration ranged from 5 to 40% in the gradient. (From Busch *et al.*, 1968.)

TABLE II
^{32}P[a] Distribution of Nucleotides in Rapidly Sedimenting RNA

Fraction	A	U	G	C	$\frac{A+U}{G+C}$
		Chromatin (DNA-1-RNA)			
45 S	22.8	30.8	22.8	23.6	1.15
60 S	23.8	30.0	23.2	23.0	1.16
85 S	24.2	30.9	20.6	24.3	1.22
		Residue			
45 S	16.2	20.5	36.0	27.3	0.58
85 S	17.1	20.3	34.6	28.9	0.60

[a] Given as a 20-min pulse.

4–18 S RNA; the low-molecular-weight nuclear RNA (LMWN RNA) in this fraction is described in Chapter 5 of this volume. The rate of labeling of RNA in this fraction is intermediate between the rapidly labeled nuclear RNA and the cytoplasmic RNA.

When the pH of buffer saline extraction media is increased from 7 to 8, the yield of AU-rich RNA is considerably increased. Samarina *et al.* (1968) used pH 8 to isolate informosomes containing mRNA. The first step at pH 7.0 removes the nuclear sap and the second and third steps at pH 8 yield particles containing newly formed AU-rich RNA or DRNA (McParland *et al.*, 1972). Ultracentrifugation analysis showed that the 30 S RNP particles are composed of 12–18 S RNA and proteins (see Chapter 3 by Georgiev).

2. CHROMATIN FRACTION

The 2 *M* NaCl extract or the "chromatin fraction" (Busch *et al.*, 1968) contains almost all the HnRNA species. The chromatin fraction contains approximately 30% of the total nuclear RNA. Labeling with isotope precursors results in a rapid labeling of HMW nRNA, especially RNA with a sedimentation rate greater than 45 S RNA.

Sharma *et al.* (1969) extensively studied the RNA species associated with the chromatin fraction. The labeling and ultracentrifugation analysis showed a characteristic rapid labeling in the region containing larger than 45 S RNA which contains giant RNA molecules. Morphological studies indicated that their length of 5–8 μm is compatible with sedimentation coefficients of 65–85 S.

3. RESIDUE FRACTION

The fraction remaining after sequential treatment with dilute and concentrated salt solutions is designated as the "residue fraction." The residue fraction is constituted of unextracted chromatin and nucleolar residues which are enriched with RNP granules and nuclear membrane. Smetana *et al.* (1963) designated the residue fraction as the nuclear ribonucleoprotein network (NRN) and nucleoli; subsequently, Narayan *et al.* (1967) observed that RNA species associated are in interchromatin and perichromatin granules as well as nucleolar RNP. Both HnRNA and nucleolar HMW RNA were found in the residue fraction (Steele and Busch, 1966a); that HnRNA is also associated with the nuclear ribonucleoprotein network was found by actinomycin D treatment. The network is probably involved in the processing of HnRNA and pre-rRNA and their transport to the cytoplasm (Narayan *et al.*, 1967).

B. Phenol Fractionation

The introduction of phenol for RNA extraction by Kirby (1956) provided a means for preparing structurally and functionally intact RNA. Cellular RNA partitioned in a mixture of aqueous solution and phenol separates into three phases: most RNA in the aqueous phase, some RNA in the interphase, and little RNA in the phenol phase. Sibatani *et al.* (1959, 1960) and Yamana and Sibatani (1960) observed that the extractability of RNA into aqueous phase depends on a number of conditions, including pH and composition of the aqueous medium, composition of the phenol medium, and extraction temperature. Of these parameters, temperature is the most important, as shown by Georgiev and Mantieva (1962), who established a fractionation method for cellular RNA based on the differential thermal effect. Sodium dodecyl sulfate (SDS) is an important reagent to dissociate RNA from RNP complexes. Sibatani *et al.* (1959) observed that SDS liberates RNA from cellular constituents and subsequently Scherrer and Darnell (1962) and Hiatt (1962) demonstrated the isolation of structurally intact HMW nRNA. With a combination of the observations made by Georgiev and Mantieva (1962) and Sibatani *et al.* (1959), nuclear RNA can be fractionated into preribosomal RNA and HnRNA. The scheme recommended by Georgiev (1971) is (1) to extract at 55° for most of preribosomal RNA and some HnRNA, (2) at 65° for a stable fraction of HnRNA (mainly 18 S RNA), and (3) at 85° for rapidly labeled HnRNA. All the LMWN RNA is extracted at 55°C (see Chapter 3).

C. Mechanical Disruption of Nuclei

The nucleolar substructures have considerable tensile strength. When isolated nuclei are subjected to shearing forces, the integrity of extranucleolar portions is readily disrupted but the nucleoli are resistant to the mechanical disintegration. Various methods have been developed to isolate nucleoli (Muramatsu *et al.*, 1963; Busch and Smetana, 1970). At present time, two such methods are widely used for mammalian cells: the sonic oscillation method and the French press method. In one standard condition, nuclei are isolated in concentrated sucrose (2–2.4 *M*) containing 3.3 m*M* Ca^{2+} and subsequently disrupted by sonication in 0.25–0.35 *M* sucrose (Fig. 1). This method provides highly purified nucleoli which are characterized by morphologically intact ultrastructures and which retain most of the nucleolar components including DNA, RNA, proteins, and enzymes (Busch *et al.*, 1967; Siebert *et al.*, 1966).

The sonication method can also be used with nuclei isolated by the citric acid method (Busch and Smetana, 1970; Ro-Choi *et al.*, 1973b). This method may improve isolation of nucleoli from some types of cells for which the isolation of nuclei by the sucrose method is difficult and subsequent sonication is not effective.

Desjardins *et al.* (1966) used alternate compression and decompression of nuclei (with a French press) for mass isolation of nucleoli. As with the sonication method, these isolated nucleoli also retained the structural components.

Few attempts have been made to fractionate the disrupted extranucleolar portion of nuclei (Muramatsu *et al.*, 1966). New methods are required to fractionate the extranucleolar RNP particles of nuclei.

D. Enzymatic Digestion of Nuclei

The structural compactness of nucleolar RNP particles is maintained even after DNase treatment. This property of nucleolar RNP particles has been used to dissociate nucleoli from the nucleoplasm. For isolation of nucleoli, Penman *et al.* (1966) isolated nuclei with a detergent method and DNase was used to disrupt the nuclear ribonucleoprotein network. Subsequently, Vesco and Penman (1967) improved their method for isolation of nucleolar HMW RNA. Although used for HeLa cells, this method has not been applied to other mammalian cells. Evaluation of synthetic enzyme activity has not been performed on nucleoli isolated by the DNase method (Siebert *et al.*, 1966).

III. Heterogeneous Nucleoplasmic High-Molecular-Weight RNA (HnRNA)

In contrast to the discrete molecular sizes of nucleolar HMW RNA's, the nucleoplasmic RNA's are polydisperse. Because of the inherent physical properties and numbers of different species of HnRNA, fractionation has been very difficult.

A. Chromosomal Organization of HnRNA Cistrons

An estimation of genomic complexity shows that the genome in mammalian cells may contain as many as 7×10^6 informational cistrons in a total of 3×10^9 base pairs. The genome of bacterial cells contains 4×10^3 informational cistrons in a total of 4×10^6 base pairs (genome size). These cistrons are well organized into a defined set of chromosomes or chromatins (Judd *et al.*, 1972) and gene action is selectively initiated on chromosomal templates. Studies on isolated chromosomes indicate that the informational cistrons are uniformly distributed among all the chromosomes (Huberman and Attardi, 1967). Thus far, about 200 cistrons have been identified by a number of methods, including cytochemical techniques such as micromanipulation and *in situ* DNA–RNA hybridization (French and Kitzmiller, 1967; Steffensen and Wimber, 1972) and cytogenetic techniques such as linkage analysis of cell hybridization (Ruddle, 1972). DNA–RNA hybridization with HnRNA shows that at most 10% of the genome is transcribed in mouse L cells (Shearer and McCarthy, 1967) and HeLa cells (Soiero, 1968). In whole mouse embryo cells, approximately 30% of the genome was reported to be transcribed (Gelderman *et al.*, 1968). The extent of transcription of genes may be specific for any given cell type.

B. Heterogeneous RNA Transcriptional System

The synthetic transcriptional system for HnRNA requires RNA polymerase II (Table III). The usual template for this enzyme is nucleoplasmic DNA. In some cases, exogenous genes transcribed include DNA integrated from oncogenic DNA viruses (Georgiev *et al.*, 1972; Lindberg and Darnell, 1970; Oda and Dulbecco, 1968; Tonegawa *et al.*, 1970; Wall and Darnell, 1971; Westphal and Dulbecco, 1968) or from DNA complementary to oncogenic RNA viruses (Green *et al.*, 1971; Gulati *et al.*, 1972), which are products of reverse transcriptases (Baltimore, 1970; Temin and Mizutani, 1970).

TABLE III
Distribution of Nuclear RNA Polymerases [a]

	Enzyme class		
	I AI, AII	II BI, BII	III AIII
Localization	Nucleolus	Nucleoplasm	Nucleoplasm
Inhibitors	Cycloheximide	α-Amanitin	Rifamycin
Mn^{2+} 1–2 m*M*	++	+++	++
Mg^{2+} 2–8 m*M*	++	+	+
Mn^{2+}/Mg^{2+}	1–2	5–10	2.5
Ionic strength	0.04 *M*	0.1 *M*	0.2 *M*

[a] These values were taken from data of Roeder and Rutter (1970), Horgen and Griffin (1971), and Chambon *et al.* (1972).

Labeling with isotope precursors occurs in a broad sedimentation class ranging from 20 to 100 S with a maximal labeling of 30–50 S RNA classes (Yoshikawa-Fukuda *et al.*, 1964; Samarina *et al.*, 1965; Scherrer and Marcaud, 1965; Attardi *et al.*, 1966; Steele and Busch, 1966b; Warner *et al.*, 1966). The relationships of sedimentation coefficient and molecular size were studied by treatment of the RNA with heat or dimethylsulfoxide which showed that the RNA is a single polynucleotide chain (Sharma *et al.*, 1969). Direct electron microscopic visualization of the RNA also showed it was a single polynucleotide chain (Granboulan and Scherrer, 1969; Sharma *et al.*, 1969). Although it is commonly accepted that the larger HnRNA species are composed of single RNA chains with chain lengths ranging up to 30,000 nucleotides, treatment with formaldehyde (Mayo and DeKloet, 1971) resulted in a striking decrease of their sedimentation coefficients.

C. *Structural and Metabolic Heterogeneity*

The polydisperse property of HnRNA has complicated its structural analyses. Since the nucleotide composition is generally AU-rich regardless of molecular size, it is fortunate that advantage can be taken of the poly (A) sequences on the 3′-termini of these molecules.

1. POLY(A) SEQUENCES

The existence of 3′-terminal poly(A) sequences in most HnRNA (Canellakis *et al.*, 1970; Edmonds and Abrams, 1960; Edmonds and Caramela, 1969; Hadjvassiliou and Brawerman, 1966), was demonstrated by identification of two independent synthetic events in the synthesis of HnRNA (Jelinek *et al.*, 1973). The first event is sensitive to actinomycin D which blocks the transcriptional processes; the second is sensitive to cordycepin which inhibits the poly(A) synthesis (Penman *et al.*, 1970; Darnell *et al.*, 1971; Adesnik *et al.*, 1972; Mendecki *et al.*, 1972), catalyzed by the poly(A)-synthesizing enzyme. The poly(A) sequence has been isolated by treatment with pancreatic and T_1 RNases (Kates, 1970; Kates and Beeson, 1970; Lim and Canellakis, 1970; Darnell *et al.*, 1971; Edmonds *et al.*, 1971; Lee *et al.*, 1971), followed by analyses with a number of techniques. The chain length of poly(A) was determined to be 50–200 by electrophoretic mobility and other analyses. The 3′-end of the RNA molecules contains the poly(A) sequences (Molloy and Darnell, 1973).

It has been reported that the poly(A) sequence is the recognition site for a special protein which may be regulatory for protein synthesis (Kwan and Brawerman, 1972; Blobel, 1973). However, there is one exception where the poly(A) sequence is not an integral part of mRNA, i.e., mRNA for histones (Adesnik and Darnell, 1972; Greenberg and Perry, 1972). The chain lengths in poly(A) in mRNA have been reported to be shorter (50–170) than that reported for HnRNA (Mendecki *et al.*, 1972; Molloy and Darnell, 1973; Sheines and Darnell, 1973). It has been postulated that the poly(A) of HnRNA becomes shorter after its entry into the cytoplasm and association with ribosomes for protein synthesis.

2. SOME SPECIFIC STRUCTURAL CHARACTERISTICS

The 5′-end of HnRNA contains a triphosphoryl purine nucleotide (Chapter 3). In addition, there is one U-rich polyprimidine sequence which is not found in pre-rRNA's or rRNA's (Molloy *et al.*, 1972). Partial characterization of limited digestion products indicated the presence of loop regions (Jelinek and Darnell, 1972) in HnRNA and some nonconserved regions of HnRNA (Jelinek *et al.*, 1973). So far, no modified nucleotides have been found and therefore it is likely that HnRNA is composed only of the four major nucleotides.

D. The Concept of Cot and Analyses of Genomic Content of HnRNA

The complex population of HnRNA species is not easily resolved and the precise number of transcribed species in HnRNA is unknown. One of the analytical methods is based on hybridization. Britten and Kohne (1968) analyzed the genome by DNA–DNA hybridization techniques. Kinetic analyses showed that DNA renaturation is a complex second-order reaction. They introduced one important approach for definition of genomic organization by Cot values. These are defined as Co × t; Co is the initial DNA concentration in moles per liter (in terms of mononucleotides) and t is time of renaturation under standard conditions in seconds. Furthermore, $Cot_{1/2}$ was defined as $Co \times t_{1/2}$ where $t_{1/2}$ is expressed as time for 50% renaturation. The units of the Cot value are moles sec^{-1} $liter^{-1}$ or alternately as absorbance units (generally read at 260 nm) $hours^{-1}$ × 0.5. Figure 3 shows representative renaturation curves where $Cot_{1/2}$ corresponds to genomic complexity expressed as nucleotide pairs. The number of cistrons can be estimated when the size of genome is known. Figure 4 provides indications of genomic complexity which suggest the existence of three major characteristic sequences referred to as "highly repetitious," "intermediate," and "unique" sequences. DNA–RNA hybridization confirmed the existence of these different sequences (Ananieva *et al.*, 1968; Church and McCarthy, 1968; Melli and Bishop, 1969).

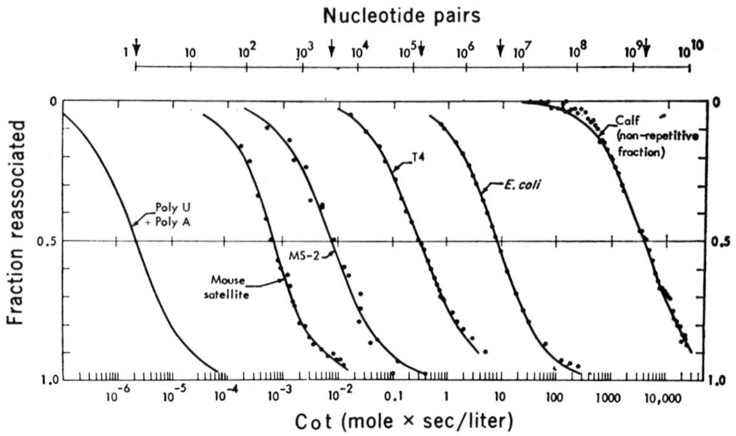

Fig. 3 Reassociation of double-stranded nucleic acids from various sources. The genome size is indicated by the arrows near the upper nomographic scale. Over a factor of 10^9, this value is proportional to the Cot required for half-reaction. The DNA was sheared and the other nucleic acids are reported to have approximately the same fragment size (about 400 nucleotides, single-stranded). (From Britten and Kohne, 1968.)

The time order of hybridization was found to be reiterated sequence (fastest) > intermediate sequence (variable) > unique sequences (slowest). Subsequently, the conditions for analyses of nonreiterated sequences were studied (McCarthy and Church, 1970). The kinetics of DNA–RNA hybridization show that HnRNA species are constituted of both reiterated sequences and unique sequences.

E. *Relationship of HnRNA to Cytoplasmic mRNA*

A number of comparative studies of HnRNA with cytoplasmic mRNA have provided evidence to define the precursor-product relationship between two RNA families. The selective labeling of HnRNA after treatment of cells with low doses of actinomycin D to inhibit nucleolar HMW RNA synthesis showed a flow of nuclear AU-rich RNA into the cytoplasm (see Chapter 3). Furthermore, sedimentation analyses of nonribosomal RNA associated with polysomes showed it had a more rapid labeling and a smaller size (6–25 S RNA) than HnRNA (Penman *et al.*, 1963). Nonribosomal RNA has template activity for protein synthesis (Brawerman *et al.*, 1963; DiGirolamo *et al.*, 1964) and HnRNA has similar activity (Cartouzou *et al.*, 1965; Jacob and Busch, 1967). However, mRNA is not always AU rich, as illustrated by the mRNA for silk fibroin, which contains 60% G + C (Suzuki and Brown, 1972).

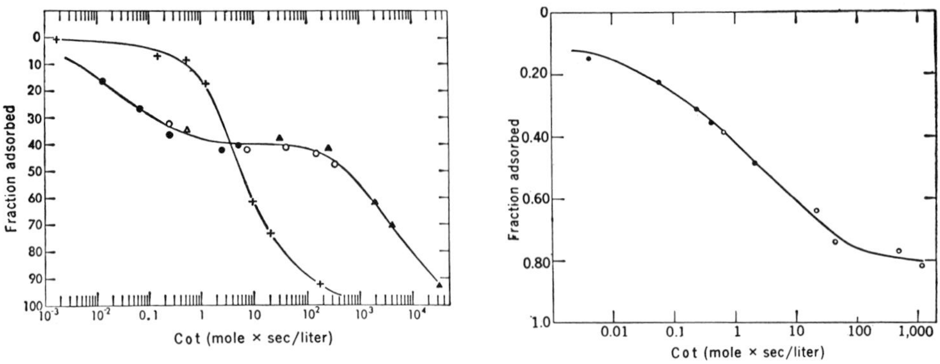

Fig. 4 (Left) The kinetics of reassociation of calf thymus DNA measured with hydroxyapatite. The DNA was sheared at 3.4 kb and incubated at 60°C in 0.12 *M* phosphate buffer. At various times, samples were diluted, if necessary (in 0.12 *M* phosphate buffer at 60°C), and pasesd over a hydroxyapatite column at 60°C. The DNA concentrations during the reaction were (μg/ml): open triangles, 2; closed circles, 10; open circles, 600; closed triangles, 8600. Crosses are radioactively labeled *Escherichia coli* DNA at 43 μg/ml present in the reaction containing calf thymus DNA at 8600 μg/ml. (Right) The kinetics of reassociation of salmon sperm DNA measured with hydroxyapatite. The DNA was sheared at 3.4 kb and incubated at 50°C in 0.14 *M* phosphate buffer. The samples were diluted into 0.14 *M* phosphate buffer at 50°C and passed over hydroxyapatite at 50°C. The DNA concentrations during the incubation were (μg/ml): closed circles, 8; open circles, 1600.

Hybridization competition studies indicate a structural homology of HnRNA and cytoplasmic mRNA (Scherrer, 1971; Scherrer and Marcaud, 1968; Scherrer *et al.*, 1970; Shearer and McCarthy, 1967; Soeiro and Darnell, 1970). The competition of HnRNA with cytoplasmic mRNA was stoichiometric but cytoplasmic RNA did not completely compete out HnRNA (20–40%). Poly(A) was found in both HnRNA and mRNA (Darnell *et al.*, 1971; Edmonds *et al.*, 1971; Kates, 1970; Lee *et al.*, 1971; Lim and Canellakis, 1970). Moreover, virus-specific RNA was detected both in HnRNA and mRNA in malignant transformed cells (Georgiev *et al.*, 1972; Green *et al.*, 1971; Gulati *et al.*, 1972; Lindberg and Darnell, 1970; Oda and Dulbecco, 1968; Tonegawa *et al.*, 1970; Wall and Darnell, 1971).

The assay of specific HnRNA using amphibian oocytes after injection provided evidence to support the precursor relationship of HnRNA to mRNA (Williamson *et al.*, 1973). However, not all the HnRNA contains poly(A) sequences and also not all the cytoplasmic mRNA's have poly(A) sequences (Adesnik and Darnell, 1972; Schochectman and Perry, 1972). It is also not clear how much HnRNA is precursor to cytoplasmic RNA and it remains to be determined whether HnRNA has other functions.

In eukaryotes, polycistronic mRNA has only recently been demonstrated (Daneholt, 1972). Moreover viral RNA associated with polysomes is 70 S RNA (Gulati *et al.*, 1972), suggesting that it is polycistronic in the cytoplasm. It is not clear that the giant HnRNA is a precursor of polycistronic mRNA of the cytoplasm. Thus far, only monocistronic mRNA have been purified from eukaryotic cells (Kuff and Roberts, 1967).

F. *Messenger RNA*

The term messenger RNA (mRNA) was used originally by Jacob and Monod (1961) and this term has been universally adopted to define mRNA as a polynucleotide that determines the sequence of amino acids in a polypeptide chain (Singer and Leder, 1966). Recently, a number of pure mRNA species, especially those derived from bacteria and virus, have been isolated and the primary structures of important parts of mRNA have been characterized in relation to the coding mechanism for protein synthesis (Sanger, 1971; Fiers *et al.*, 1971; Jukes and Gatlin, 1971). The structural studies show that (1) the translation of mRNA starts at the first initiation site which is some distance from the 5′-end, approximately 100 nucleotides apart; the initiation site is specified by an A—U—G codon; (2) the translation continues according to the assignment of genetic codons and even in the presence of the secondary structures due to hydrogen bonding, the internal A—U—G codons are not read as the initiator codon; (3) the translation ends at one of the termination codons, U—A—A, U—A—G, or U—G—A; The termination signal is repeated; and (4) subsequent to the termination, there is a region which is not translated. In the case of polycistronic mRNA, there are intercistronic sequences which are presumably functional for the proper recognition of regulatory proteins. However, only a few mRNA species have been isolated from eukaryotic cells and consequently the structural information is limited (Busch, 1974). There are technical difficulties in isolating pure mRNA species in satisfactory amounts. The difficulty is the complexity of isolation and purification procedures which involve mRNA associated with ribosomes (Perry and Kelley, 1968; Henshaw, 1968). The relative abundance of a particular mRNA is a limiting factor for its isolation. So far, no standard fractionation has been established and the present technology only allows isolation of special mRNA species from cells with specialized functions, usually with the aid of appropriate antibodies.

Assays of template activity for protein synthesis include the cell-free, *in vitro* assay and "injection" into oocytes. The former method is the conventional one and employs eukaryotic or *E. coli* ribosomes and protein initiation factors. This method has been widely used to characterize the synthetic products, although net protein synthesis has not been demonstrated. The injection method employs intact oocytes into which mRNA is injected and the product synthesized is assayed (Gurdon *et al.*, 1971). The advantage of this method is the small sample (pg) required. However, it has been found that the assay systems must be improved by the use of specific initiation factors which differ in various cell types (Wigle and Smith, 1972).

G. *Nuclear 8–18 S RNA*

Nuclear 8–18 S RNA is an intermediate-size class and composes 5–10% of total nuclear RNA (Figs. 1 and 2). The rate of labeling is less than that of HMW RNA and its overall nucleotide composition is GC rich (Muramatsu *et al.*, 1966; Ro-Choi *et al.*, 1973a). Although the RNA species have a relatively discrete sedimentation pattern, the analyses based on polyacrylamide gel electrophoresis show that 8–18 S RNA is a mixture of discrete and heterogeneous RNA species (Fig. 5). The fractionation pattern shows there are at least 40 different bands and unknown numbers of heterogeneous RNA species. All the discrete species have a high G + C content suggesting they are either precursors or degradation products of rRNA. However, further detailed fractionation based on Millipore techniques (Lee *et al.*, 1971) indicates that AU-rich RNA composes approximately 3% of 8–18 S RNA species.

The significance of the AU-rich RNA has been extensively studied by the phenol fractionation method (Georgiev, 1971), but the functions of the GC-rich RNA is not well defined. Since nuclear 8–18 S RNA contains nucleolar 8–18 S RNA (Muramatsu *et al.*, 1966), it is not surprising that 45 S nRNA and its products exhibit hybridization competition with nuclear 8–18 S RNA species. However, some 8–18 S RNA does not compete with nucleolar 45 S RNA; this RNA probably accounts for the template activity of 8–18 S RNA (Jacob and Busch, 1967).

H. *Posttranscriptional Modification of Transcriptional Unit of HnRNA*

To support the concept that HnRNA is the precursor of mRNA, the chemical intermediates must be defined. The addition of poly(A) to HnRNA and the cleavage reactions are posttranscriptional modifications, but the number of steps are not chemically defined. Since histone mRNA contains no poly(A) sequences, it is also uncertain whether alternate pathways exist for the formation of mRNA from HnRNA (Darnell *et al.*, 1973; Georgiev, 1971).

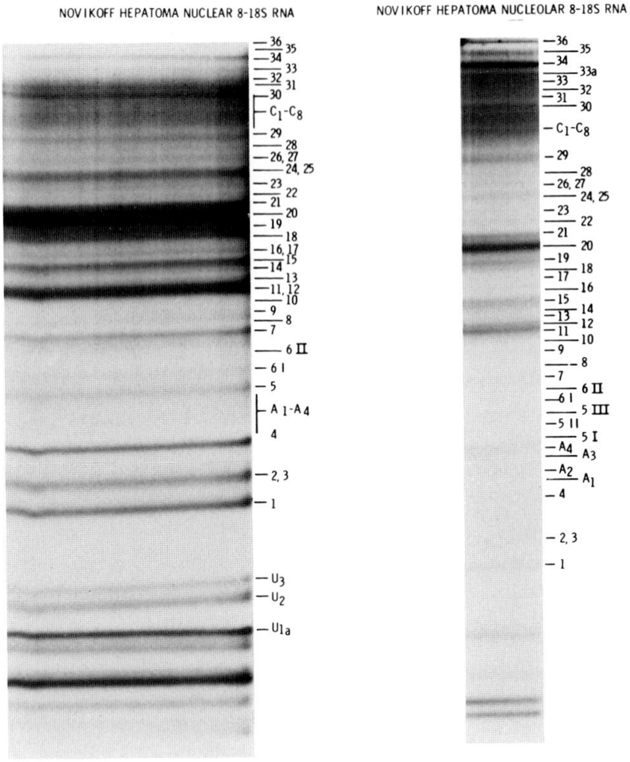

Fig. 5 Electrophoretic separation of nuclear and nucleolar 8–18 S RNA's of Novikoff hepatoma ascites cells. A 4.6% polyacrylamide slab gel was used at neutral pH in the presence of 6 *M* urea. (From Savage and Busch, 1973.)

I. *Regulatory Mechanisms and Models of the Heterogeneous Transcriptional Unit*

The mechanisms of information flow from DNA $\rightleftharpoons$ RNA $\longrightarrow$ protein include reverse transcription shown by cell transformation in which the viral RNA genome is incorporated into cell DNA as the "reversely" transcribed product.* Ultimate phenotypic expressions of eukaryotic operons are influenced by hormonal and other factors in cyto-

* Another example is suggested in gene amplification (Brown and Tochini-Valentini, 1972; Mahdavi and Crippa, 1972), which is implicated by the role of RNA-directed DNA polymerase (reverse transcriptase).

differentiation. The following provides examples of various steps that may be subject to control.

1. PRETRANSCRIPTIONAL CONTROLS

The number and kinds of transcribable genes are limited for any cell type presumably by nuclear proteins and possibly other factors (see Chapter 6 of this volume; Hearst and Botchan, 1970; Flamm, 1972; Rees and Jones, 1972). The unstable properties such as the number of genes for some transcribable genes are also controlled at the pretranscriptional levels. Redundant cistrons or families of redundant cistrons can be amplified in some cells by the activities of "master genes" (Callan, 1967).

2. TRANSCRIPTIONAL CONTROLS

The controls most studied are involved in the induction and repression of enzymes. Enzyme induction in mammalian cells produces five- to twentyfold changes in activity of specific enzymes in response to dietary alterations, hormone actions, or drug administration. There is extensive experimental evidence that a net increase of specific mRNA is associated with an increased rate of enzyme synthesis (Schimke and Doyle, 1970; Tomkins and Gelehrter, 1972).

3. POSTTRANSCRIPTIONAL CONTROLS

Posttranscriptional controls (Darnell *et al.*, 1973) may be regulatory, but so far no meaningful controls have been defined for poly(A) addition, selective cleavage, degradation, and maturation of precursor RNA into mRNA.

4. PRETRANSLATIONAL CONTROLS*

It is presumed that the pretranslational events include transport of the mRNA–protein complex from nucleoplasm to the cytoplasm through nuclear pores, and subequent assembly with ribosomes to form polysomes or an alternate pathway through cytoplasmic store in the form of informosomes (Spirin, 1969). Controls for mRNA–protein transport may exist in the nuclear RNP networks (Narayan *et al.*, 1967) or the nucleoli (Deak *et al.*, 1972). The stability of mRNA may vary, i.e., the half-life of mRNA in mammalian cells is reported to be 10–15 hr in liver (Revel and Hiatt, 1964) and brain (Appel, 1967) and 72 hr in reticulocytes (Marks *et al.*, 1962).

* See Chapter 3 of this volume.

a. Models of Gene Control. Inherent difficulties in experimental approaches to the eukaryotic genome result from the fact that it is 1000-fold more complex than the prokaryotic genome and the feasibility of genetic manipulation is limited. Some models proposed include the following.

1. Cascade regulation model: In this model (Scherrer and Marcaud, 1968), the heterogeneous transcriptional unit is constituted of polycistronic mRNA sequences and nonfunctional sequences. Its main feature is posttranscriptional modification in the processing mechanism and selective retention and rejection of particular segments of the transcriptional units and the pretranslational controls (Fig. 6). The model was supported by Church and McCarthy (1967) and Church *et al.* (1969).

2. Model of transcriptional and posttranscriptional controls: The proposed model suggests the importance of controls of transcription and post-transcriptional events (Darnell *et al.*, 1973). This model includes detailed mechanisms of HnRNA metabolism in relation to mRNA synthesis (Fig. 7).

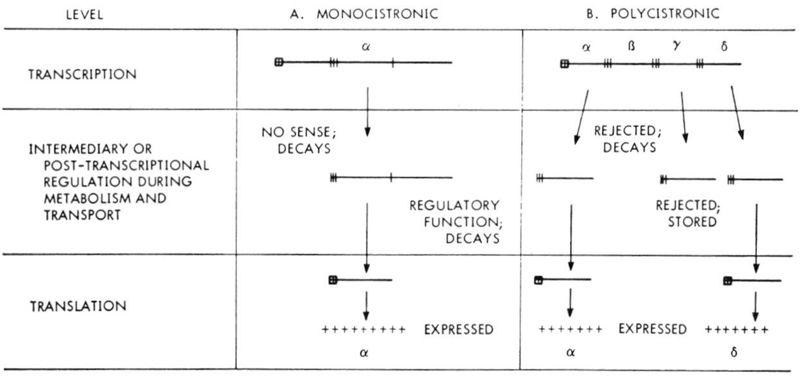

Fig. 6 Cascade model. If messenger-like RNA (MW 1–10 $\times$ 10^6 daltons) contains information of structural genes that correspond to polypeptides (α, β, γ, δ), two basically different arrangements can be considered: (A) Monocistronic transcriptional unit. The molecule contains only one cistron (α) which will be translated. The excess RNA contains no information but may have a structural or regulatory function prior to decay. (B) Polycistronic transcriptional unit. Several cistrons (α, β, γ, δ), linked together in the genome, are transcribed into a single molecule. It may correspond to an operon or to several independent cistrons which become separated during metabolism. Some of these individual cistrons reach the polysomes and are translated independently; some are immediately rejected and destroyed; others are rejected and stored at the intermediary level. The decay of the immediately rejected cistrons accounts for the observed nuclear turnover of mRNA. The polycistronic molecule map contain, in addition, sequences without structural information, such as those proposed for the monocistronic unit. (From Scherrer and Marcaud, 1968.)

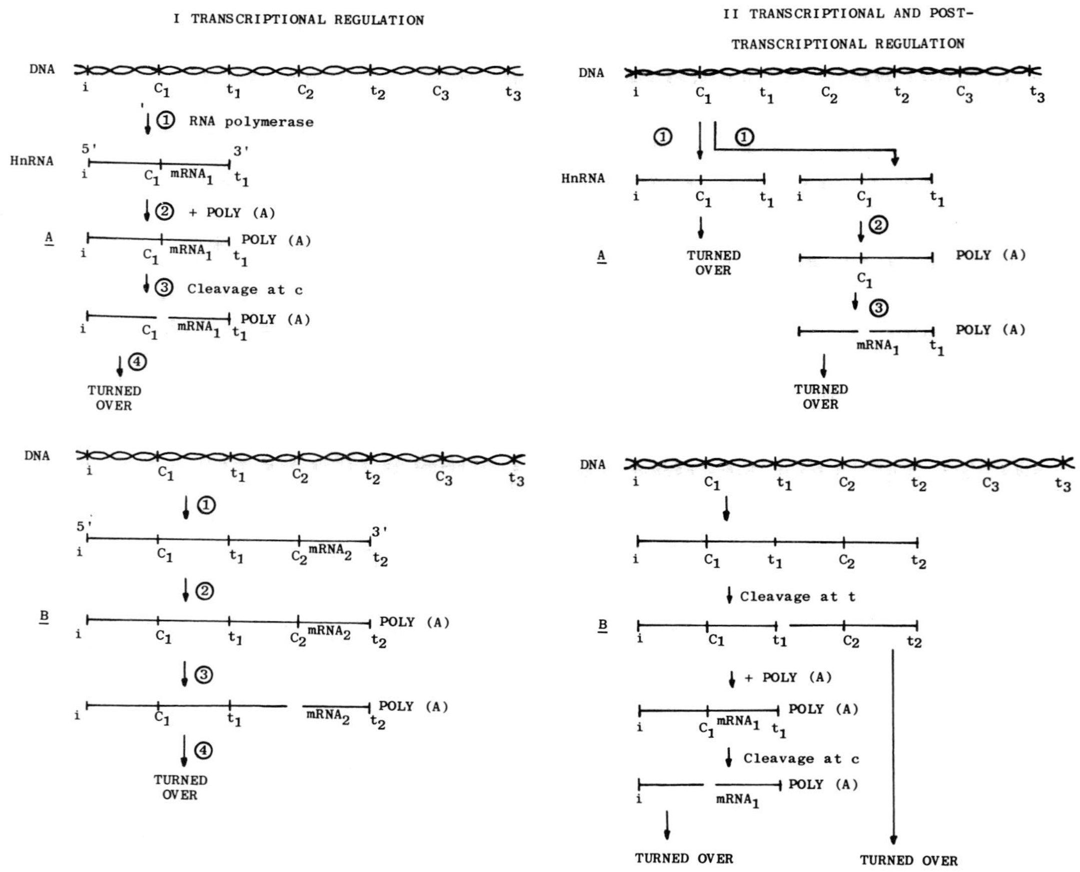
I TRANSCRIPTIONAL REGULATION
DNA
i
C1
t1
C2
t2
C3
t3
① RNA polymerase
HnRNA
5'
3'
mRNA1
② + POLY (A)
A
POLY (A)
③ Cleavage at c
④
TURNED OVER
B
mRNA2
II TRANSCRIPTIONAL AND POST-TRANSCRIPTIONAL REGULATION
Cleavage at t
+ POLY (A)
Cleavage at c
TURNED OVER

Fig. 7

3. The "gene battery" model: Based on the concepts of cytodifferentiation, a model was proposed (Britten and Davidson, 1969) to illustrate the mechanism of transcriptional controls in the "gene battery" framework. A "gene battery" is defined as a set of producer genes which is activated when a particular sensor gene activates its set of integrator genes (Fig. 8). The transcription of structural genes (producer gene) is regulated by a number of regulator genes (sensor gene, integrator gene, and receptor gene). The interrelationship among many genes is described in the form of gene circuitry which explains the control mechanism existing in the transcription processes (for example, hormone effect). One of the attractive features is an introduction of a functional role for redundant DNA sequences. A unique feature is the involvement of RNA for activation of receptor genes.

4. Eukaryotic Operon Model: This model has been presented in the preceding chapter (see Chapter 3 of this volume).

IV. Nucleolar High-Molecular-Weight RNA

For eukaryotes, HMW RNA's of both large and small ribosomal subunits are derived from a single giant transcriptional unit as the "primary ribosomal precursor RNA"; its genetic make-up is highly specialized (Birnstiel *et al.*, 1971; Busch and Smetana, 1970). The cistrons for the primary ribosomal precursor RNA (rDNA or ribosomal cistrons) are not randomly distributed in nuclear chromosomes but are clustered in nucleolus-organizing loci (NOR's). There are between 100 and 1000 rDNA cistrons. The sizes and number of the nucleoli are not constant but vary with the metabolic states of cells. For the synthesis of primary precursor RNA, the ribosomal cistrons are transcribed under common genetic control mechanisms and the synthetic reactions are catalyzed by a nucleolus specific RNA polymerase, nuclear RNA polymerase I (Table III).

Fig. 7 Four models of regulation of mRNA formation in eukaryotic cells. The symbols are: i for initiation of transcription, C for cleavage points at 5′-end of mRNA, t for termination at 3′-end of DNA-encoded region of mRNA (C_1, t_1, C_2, t_2, etc., indicate multiple sites on same molecule). The four steps in mRNA biosynthesis from HnRNA are (1) DNA-dependent transcription by RNA polymerase, (2) posttranscriptional addition of poly(A), (3) enzymatic cleavage at 5′-end of mRNA, (4) turnover of unused region(s) of HnRNA. Model IA shows transcriptional regulation at initiation site only. Model IB shows transcriptional regulation at initiation site only and termination site (t_1 is passed by in favor of t_2). Model IIA shows posttranscriptional regulation where one half HnRNA molecules yield an mRNA and one half are destroyed. Model IIB shows posttranscriptional regulation where specific cleavage at t_1 reveals the 3′-terminus of mRNA, for processing and $mRNA_2$ region is discarded. (From Darnell *et al.*, 1973.)

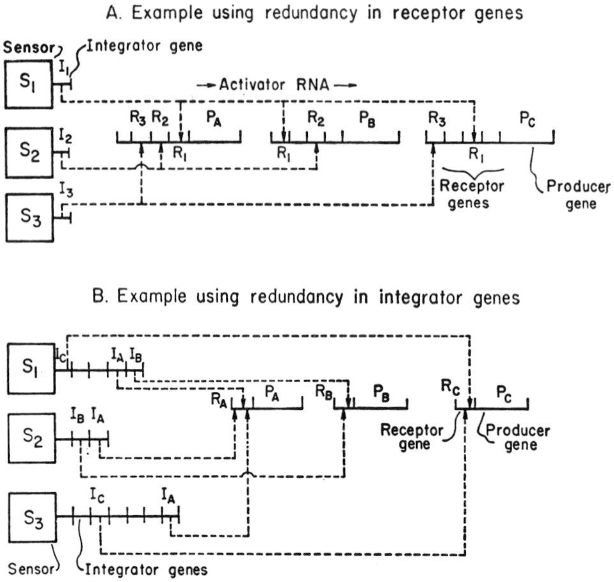

Fig. 8 Types of integrative system within the model. (A) Integrative system depending on redundancy among the regulator genes. (B) Integrative system depending on redundancy among the integrator genes. These diagrams schematize the events that occur after the three sensor genes have initiated transcription of their integrator genes. Activator RNA's diffuse (symbolized by dotted line) from their sites of synthesis—the integrator genes—to receptor genes. The formation of a complex between them leads to active transcription of the producer genes P_A, P_B, and P_C. (From Britten and Davidson, 1969.)

A. Nucleolar Organizer and Analyses of rDNA

The cluster of ribosomal RNA cistrons (DNA) is recognized in many chromosomes as a secondary constriction. The rDNA has a number of genetic and biochemical characteristics. Both *Xenopus laevis* and *Drosophila melanogaster* are excellent systems for characterization of rDNA (Birnstiel *et al.*, 1971; Brown and Gurdon, 1964; Ritossa and Spiegelman, 1965).

1. In general, the number of ribosomal cistrons is proportional to the number of nucleolar organizers; the number of nucleoli vary depending on metabolic states of the cell.

2. Various mutants can be produced by genetic manipulation which generates the complete or partial deletion of nucleolar organizer.

3. The genetic behavior of the nucleolar organizer is characterized by classic genetics as a single Mendelian factor.

4. The rDNA of some nucleolus organizers can be separated as a satellite DNA and analyzed quantitatively by DNA–RNA hybridization.

Mapping of rDNA shows various sequences; 18 S and 28 S rRNA cistrons and transcribable and nontranscribable spacers (Birnstiel *et al.*, 1971; Dawid *et al.*, 1970). The tandem arrangement and gene linkage between 18 S rRNA and 28 S rRNA cistrons were demonstrated by the hybridization properties of rDNA with two rRNA's and by the isolation of primary pre-rRNA. The spacers corresponding to the nonhybridizable sequences occupy more than 50% of rDNA. The transcribable spacers were identified by the characterization of the primary pre-rRNA, from which the nonribosomal sequences are lost during formation of 18 S and 28 S rRNA. The spacers were demonstrated by the periodity of regions with and without growing RNA chains transcribed on rDNA (Miller and Hamkalo, 1972; Scheer *et al.*, 1974) and also by the periodity of regions containing thermally stable and unstable rDNA sequences (Brown *et al.*, 1972).

A distinct feature of the control mechanisms for rRNA synthesis is genetic amplification by which enormous amounts of ribosomes can be formed in a particular cell (see Volume I, Chapter 2). The redundancy of ribosomal cistrons indicates a special genetic mechanism operates for rDNA production in oocytes.

B. *Ribosomal Transcriptional Unit and Its Derived RNA Species*

In contrast to cells of amphibia and insects, which are most suitable for study of rDNA, mammalian cells provide a number of advantages for study of nucleolar RNA because three experimental problems have been resolved: (1) mass isolation of morphologically intact nucleoli, (2) preparation of structurally intact HMW RNA, and (3) fractionation of HMW RNA. The first successful demonstration of nucleolar HMW RNA was achieved in rat liver cells (Muramatsu *et al.*, 1966) and HeLa cells (Penman *et al.*, 1966).

1. METABOLIC IDENTIFICATION OF THE PRIMARY RIBOSOMAL TRANSCRIPTIONAL UNIT AND ITS DERIVED SPECIES

Although cytochemical evidence that the nucleolus is the site for ribosomal RNA synthesis was obtained initially, direct evidence for nucleolar function was provided by metabolic studies on isolated nucleoli (Busch and Smetana, 1970; Darnell, 1968; Perry, 1967). The first direct evidence, based on studies of interrelationships among nucleolar HMW RNA's and cytoplasmic rRNA's, was the kinetic profile of isotope flux from nucleoli into the cytoplasm. The sequence of labeling with various isotopic precursors ([^{32}P]orthophosphate, [^{14}C]- or [^{3}H]nucleoside, [^{14}C]- or [^{3}H]-methionine, and others) showed that of the nucleolar HMW RNA's and cytoplasmic rRNA's, the first labeled RNA species (the primary ribosomal transcriptional unit) was 45 S nRNA or its larger oligomers (Busch and Smetana, 1970).

The combined use of isotopic precursors and selective inhibitors of nucleolar RNA synthesis (low doses of actinomycin D and cycloheximide) provided a more detailed analysis of rates of isotope flow among the RNA molecules. From these kinetic studies, the metabolic results are (Muramatsu *et al.*, 1966; Penman *et al.*, 1970; Liau and Perry, 1969):

a. The primary transcriptional unit of nucleolar HMW RNA is 45 S RNA or its oligomers.

b. Other species of nucleolar HMW RNA's are derived from 45 S RNA as a result of posttranscriptional modification or processing that is catalyzed by endonuclease(s) and possibly exonuclease(s).

c. Cytoplasmic rRNA's are products of nonconservative processing of 45 S RNA in which approximately 50% of the molecule is lost for the formation of rRNA.

d. The kinetic pathways from 45 S RNA to rRNA are two irreversible, branched, elementary steps for which first-order kinetics were applied to determine the rates (Choi *et al.*, 1971).

$$\text{45 S nRNA} \rightarrow \text{41 S nRNA(?)} \begin{cases} \nearrow \text{35 S nRNA} \rightarrow \text{28 S nRNA} \rightarrow \text{28 S rRNA} \\ \searrow \text{23 S nRNA} \rightarrow \text{18 S rRNA} \end{cases}$$

e. All of the RNA species exist as RNP.

Of the other nRNA larger than 45 S nRNA (Hidvegi *et al.*, 1971; Tiollais *et al.*, 1971), the largest species identified was 85 S RNA. The labeling pattern suggested there are larger precursors than 45 S RNA. Furthermore, chase studies with actinomycin D indicated that 85 S RNA is a precursor of 45 S RNA. The 85 S RNA may be an oligomer of 45 S RNA that is metabolically very labile (Quagliarotti *et al.*, 1970).

2. INTERRELATIONSHIP OF STRUCTURAL HOMOLOGY

The first structural evidence for the homology among nucleolar HMW RNA's and rRNA's was based on the analyses of nucleotide composition and modified nucleotides (Choi and Busch, 1970; Egawa *et al.*, 1971; Muramatsu *et al.*, 1966; Smith *et al.*, 1967; Darnell, 1968; Wagner *et al.*, 1967; Jeanteur *et al.*, 1968). Recently, more refined methods were applied to compare the structural homology among pre-rRNA's and rRNA. DNA–RNA hybridization and hybridization competition provided evidence (Jeanteur and Attardi, 1969; Quagliarotti *et al.*, 1970) for structural interrelationships and precursor–product relationships (Fig. 9).

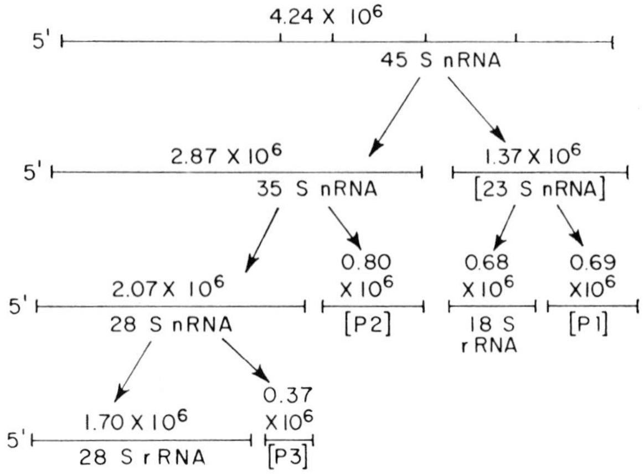

Fig. 9 Proposed cleavage of 45 S nucleolar RNA (nRNA) and its products to form the 18 S and 28 S rRNA and other polynucleotide fragments. (From Quagliarotti *et al.*, 1970.)

DNA–RNA hybridization has been used to compare pre-rRNA's and rRNA's (Fig. 10) of various species (Brown *et al.*, 1972). Based on DNA–RNA hybridization competition the rRNA's of *Xenopus laevis* and *Xenopus mulleri* are the same; the pre-RNA's are different. These findings were interpreted as showing the sequences for rRNA's are genetically stable and the spacer sequences are genetically unstable.

C. Importance of Ribosomal RNA's

Bacterial rRNA has been used as a model system since protein synthesis on ribosomes was initially proposed (Watson, 1964). Recent advances in the study of ribosome structure and function have further defined the mechanism of protein synthesis, as well as a series of ribosome states and cycles (Lucas-Lenard and Lipman, 1971; Pestka, 1971; Kurland, 1972). The ribosome contains structural components that are interdependent (pleiotropic) for its functional integrity (Kurland, 1972). To understand the importance of such structural components, two complementary approaches have been made: one analyzing ribosomal proteins (Nomura, 1970; Kurland, 1972) and the other, rRNA's (Fellner, 1969).

1. Direct RNA–Protein Interactions

The reconstitution of ribosomal subunits demonstrates an assembly pathway with specific RNA–protein interactions (Nomura, 1970). Analyses of 21 proteins of the 30 S subunit indicated that 6 proteins (Schaup

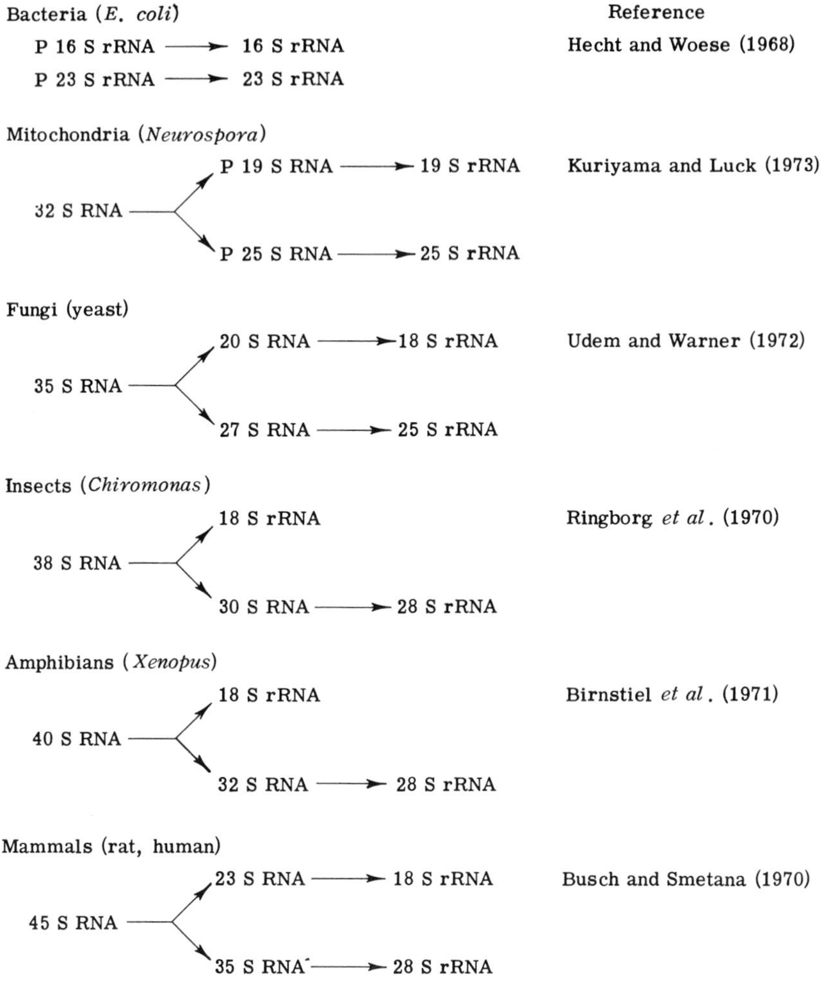

Fig. 10 Schematic representation of rRNA synthesis in various systems. The symbol P refers to precursor, which signifies the molecules slightly larger than the mature rRNA.

et al., 1970, 1971; Zimmerman *et al.*, 1972) form a stable complex with 16 S rRNA. Detailed studies based on the determination of RNA sequences showed that 900 nucleotides at the 5′-end combine with 5 proteins and 500–600 nucleotides at the 3′-end combine with one protein (Zimmerman *et al.*, 1972). It was also found that the protein-binding sites are in the highly helical regions of the RNA. Studies with 38 proteins of the 50 S subunit demonstrated that 8 proteins form a complex with 23 S rRNA (Stöffler *et al.*, 1971).

2. REQUIREMENT FOR STRUCTURAL INTEGRITY OF rRNA'S FOR RIBOSOME FUNCTION

Disruption of covalent linkages of RNA results in an impairment of ribosome function. Treatment of ribosomes or subunits with endonucleases produced a decreased binding of tRNA and mRNA (Lee and Quintanilla, 1972). The size of fragments produced by T_1 RNase was much larger than with pancreatic RNase, suggesting there are unshielded portions of rRNA's (Santer and Szekely, 1971). The importance of the 3′-end was demonstrated by treatment of ribosome with colicin E_3 which cleaves approximately 50 nucleotides (E_3 fragment) from the 3′-end (Bowman *et al.*, 1971; Senior and Holland, 1971). Without the E_3 fragment, the ribosome did not function well.

3. ALTERATIONS OF RIBOSOME FUNCTION BY NUCLEOTIDE MODIFICATION

Treatment of the 30 S subunit with kethoxal, which specifically binds to guanine, showed that modification of 10 guanine residues inhibited ribosome function; 6–7 of these were involved in binding to tRNA and mRNA (Noller and Chaires, 1972). The resistance to antibiotics such as Kasugamycin was reported to result from alterations in the normal patterns of methylation in 16 S rRNA or 23 S rRNA (Helser *et al.*, 1970, 1971).

D. *Primary Structure of High-Molecular-Weight rRNA's and Pre-rRNA's*

1. EVOLUTIONARY COMPLEXITY OF rRNA AND PRE-rRNA

There are a number of evolutionary differences in the molecular organization of ribosomes between prokaryotes and eukaryotes that are associated with the structures of rRNA's and pre-RNA's. One is the sedimentation coefficients of the ribosomal subunits which are 30 S and 50 S for the small and large subunits of prokaryotes and 40 S and 60 S for eukaryotes. For the mitoribosomes, the values are correspondingly lower.

a. Molecular Weight and Composition. Physical studies of rRNA provide evidence of a nonrandom increase of their molecular weights during evolution (Loening *et al.*, 1969; Attardi and Amaldi, 1970; Perry *et al.*, 1970). The rRNA molecular sizes were more varied for the large subunit (23 S–28 S, 10^6–1.9×10^6 daltons) than for the small subunit (16 S–18 S, 0.6×10^6–0.7×10^6 daltons). The nonrandom increase was paralleled with an increasing G + C content, up to 70% for HeLa cells, which is more prominent in 28 S than 18 S rRNA (Attardi and Amaldi, 1970). These parallel trends were related to increased stability of the secondary structures in rRNA's. They may also be correlated with the numbers and types of ribosomal proteins. The increasing size of pre-rRNA's may also reflect a nonrandom increase in size of the nonconserved portions of pre-rRNA's or spacers (Loening *et al.*, 1969; Perry *et al.*, 1970) (Table IV).

b. Modified Nucleotides. Some modified nucleotides are found in rRNA's (Table V). Both bacterial and eukaryotic rRNA contain ribose and base methylated nucleotides.

2. STRUCTURAL DEFINITION OF PRECURSOR–PRODUCTS RELATIONSHIP

The feasibility of the nucleotide sequencing of HMW RNA in general has been enhanced by new techniques and the determination of total sequence has begun for bacterial rRNA (Fellner, 1969) and mammalian rRNA's and pre-rRNA (Busch *et al.*, 1972).

a. Products of Complete Alkaline Hydrolysis. The distribution analyses of alkali-stable oligonucleotides showed there are structural differences of rRNA's (Tamaoki and Lane, 1968) and pre-rRNA's (Wagner *et al.*, 1967; Choi and Busch, 1970; Egawa *et al.*, 1971; Maden *et al.*, 1972). There are 16 kinds of alkali-resistant dinucleotides and one trinucleotide (Um—Gm—Up) in 28 S rRNA and pre-rRNA's. Analyses of 2′-*0*-methylation patterns showed that purine nucleotides are predominantly methylated. The nonconserved portions of 45 S rRNA are apparently not methylated. The 5′-terminal oligonucleotide is Am—Gm—Cm—Ap.

b. Products of Complete Endonuclease Digestions. The comparative studies of digestion products with pancreatic RNase (Jeanteur *et al.*, 1968; Seeber and Busch, 1971), T_1 RNase (Salim *et al.*, 1970; Inagaki and Busch, 1972a), combinations of T_1 RNase and U_2 RNase (Nazar and Busch, 1973), as well as pancreatic RNase and T_1 RNase (Birnboim and Coakley, 1971), provided evidence for precursor–product relationship of HMW RNA and rRNA. Table VI shows the longer oligonucleotides which are found in the rRNA's and pre-rRNA's. One of the most important findings was the identification of specific fragments (P and Q) in the spacer portions within 45 S RNA (Inagaki and Busch, 1972b).

c. The Secondary Structures. Since the dynamics of ribosomes are involved in the scheme of ribosome states and cycles, there is no doubt that it is important to begin studies on the conformational changes of rRNA's during the ribosome function. The extensive physical studies, including X-ray diffraction, indicate that the secondary and tertiary structures of rRNA's are very similar in ribosomes and as free molecules in solution (Attardi and Amaldi, 1970).

Recently, enzymatic approaches were taken to probe the secondary

TABLE IV
Evolution of the Ribosomal Transcriptional Unit

Organism	Ribosomal transcriptional unit (daltons $\times 10^{-6}$)	32 S intermediate (daltons $\times 10^{-6}$)	Large rRNA (daltons $\times 10^{-6}$)	Small rRNA (daltons $\times 10^{-6}$)	% of precursor conserved	References
Mammal (HeLa)	4.4	2.2	1.75	0.7	54	Loening *et al.* (1969)
Amphibian (*Xenopus*)	2.5	1.6	1.5	0.7	88	Loening *et al.* (1969)
Plant (*Phaseolus*)	2.3	1.4	1.3	0.7	88	Loening *et al.* (1969)
Rodent (mouse)	4.19	2.16	1.70	0.65	56	Perry *et al.* (1970)
Marsupial (potoroo or rat kangaroo)	4.19	2.16	1.70	0.65	56	Perry *et al.* (1970)
Bird (fowl)	3.92	1.98	1.61	0.63	57	Perry *et al.* (1970)
Reptile (iguana)	2.74	1.58	1.51	0.62	78	Perry *et al.* (1970)
Amphibian (frog)	2.76	1.65	1.58	0.61	79	Perry *et al.* (1970)
Fish (trout)	2.70	1.60	1.55	0.65	81	Perry *et al.* (1970)
Insect (*Drosophila*)	2.85	1.60	1.40	0.65	72	Perry *et al.* (1970)
Plant (tobacco)	2.76	1.50	1.29	0.66	71	Perry *et al.* (1970)

TABLE V
Modified Nucleosides in rRNA

	A	U	G	C	Reference
Escherichia coli	m^6A	ψ	m^1G	m^4C	Nichols and Lane (1966);
	m^6_2A		m^2G	m^5C	
	m^2A		m^7G		
	Am	Um	Gm	Cm,m^4Cm	Fellner (1969)
Eukaryotes (HeLa cell)	m^1A	ψ	m^1G	m^3C	Amaldi and Attardi (1968); Iwanami and Brown (1968)
	m^6A	m^3U	m^2G		
	m^6_2A		m^2_2G		
			m^7G		
	Am	Um	Gm	Cm	

structures of rRNA's and were very useful in isolating the helical regions. The use of limited digestion with T_1 RNase show the presence of stable secondary structures in 28 S rRNA (Gould, 1966; Wikman *et al.*, 1969). Further studies showed that pre-rRNA's also contain such structures and suggested that secondary structures of 45 S RNA contain discretely independent regions for 28 S rRNA and other portions (Wikman *et al.*, 1969; Kanamaru *et al.*, 1972). These stable regions were isolated and the largest stable fragment isolated was named the "B3" fragment. The structures of such regions are undefined but by analogy with the studies on bacterial ribosomes and their RNA sequences (Zimmerman *et al.*, 1972; Kurland, 1972) the B3 fragment may have multiple loops that serve as binding sites for ribosomal proteins. One portion is defined (Fig. 11).

A-C-C-C-C-C-U-C-U-C-C-U-U-U-C-C-G-C-C-C-G-
G-G-C-C-C-G-C-C-C-C-U-C-C-U-C-U-C-C-C-G-C-
G-G-G-G-C-C-C-C-G-C-C-G-U-C-C-C-C-G-C-G-U-
C-G-U-C-G-C-C-G-U-G-G-U-C-C-C-C-C-C-U-C-U-
C-C-U-C-U-U-C-C-C-G-U-C-C-Gp

Fig. 11 Nucleotide sequence of B_3-9 subcomponent isolated from 28 S rRNA of Novikoff hepatoma cells. (From Kanamaru *et al.*, in press).

E. Topography of Primary Transcriptional Unit

When the total nucleotide sequence of 45 S RNA is known, uncertainties existing about the mechanism of posttranscriptional modification (nonconservative processing) will be clarified. Moreover, the detailed structure of genes (rDNA) complementary to 45 S RNA will be defined. Nevertheless, several suggestions have been made about the location of

TABLE VI

Comparisons of Sequences of rRNA's and Pre-rRNA's of Novikoff Hepatoma Ascites Cells

Sequence of complete digestion products	Ribosomal RNA's		Nucleolar RNA's			Reference
	18 S rRNA	28 S rRNA	28 S rRNA	35 S rRNA	45 S rRNA	
Pancreatic RNase products						
(A—G)(G)(G)(G)A—A—G—G—Up	−	+	+	+	+	Seeber and Busch (1971)
(A—A—A—G)(A—G)(G)(G)(G)Up	−	+	+	+	+	
(A—A—G)(A—G)(Cm—G)(G)(G)(G)Up	−	+	+	+	+	
(A—A—A—A—G)(A—G)(G)(G)(G)—A—A—A—Cp	−	+	+	+	+	
T_1 RNase products						
C—U—C—C—Gm—U—A—U—U—C—A—A—U—U—A—Gp	−	+	+	+	+	Choi and Busch (1974)
Am—Gm—Cm—A—A—A—U—U—C—A—U—A—U—U—C—A—A—A—C—Gp	−	+	+	+	+	
Um—Gm—U—U—U—C—A—C—C—C—A—U—A—U—C—A—A—U—A—A—C—Gp—C—A—Gp	−	+	+	+	+	
A—A—A—U—A—C—C—A—Cm—U—A—C—U—U—C—C—A—U—C—Gp	−	+	+	+	+	Inagaki and Busch (1972a)
A—A—C—C—U—A—U—C—U—U—C—A—U—C—U—C—A—A—A—C—U—U—U—A—A—A—U—Gp	−	+	+	+	+	
Spacer sequences						
C—C—C—A—C(C_9U_8)—G (P)	−	−	−	+	+	Inagaki and Busch (1972b)
$A_2C_{11}U_5G$ (Q)	−	−	−	−	+	
U_2 RNase and T_1 RNase products						
C—C—C—C—C—C—U—C—C—C—Ap	−	+	+	+	+	Nazar and Busch (1974)
C—U—C—U—C—C—C—C—C—C—Gp	−	+	+	+	+	
C—C—U—C—U—C—C—C—U—C—C—Gp	−	+	+	+	+	
C—C—C—C—C—C—C—U—U—U—C—C—Gp	−	+	+	+	+	
C—C—C—U—C—C—C—U—U—U—C—C—Gp	−	+	+	+	+	
C—C—C—C—U—C—C—U—C—U—C—C—C—Gp	−	+	+	+	+	
C—C—C—C—C—U—C—U—C—C—U—U—C—C—Gp	−	+	+	+	+	
C—C—C—C—U—C—U—C—U—U—C—C—C—C—Gp	−	+	+	+	+	
C—C—C—C—C—U—C—U—C—C—U—U—U—C—C—Gp	−	+	+	+	+	
U—U—C—C—C—C—C—C—U—C—U—C—C—U—C—U—U—C—C—C—Gp	−	+	+	+	+	
C—C—C—U—U—C—U—U—C—C—Ap	+	−	−	−	+	
C—C—C—C—C—C—U—U—C—C—C—Gp	+	−	−	−	+	

18 S rRNA, 28 S rRNA, and nonribosomal segments (the transcribed spacers) within the transcriptional unit. These are based mainly on biochemical techniques (Fig. 12).

Apart from the molecular sizes which are known to vary according to cell type, the main difference is the different polarity of 18 S rRNA and 28 S rRNA within the transcriptional unit. In the *Xenopus* model, the topography was constructed by comparison of complementary RNA of rDNA with 18 S rRNA and 28 S rRNA (Reeder and Brown, 1970). The different size classes of complementary RNA were produced progressively on rDNA templates by *E. coli* RNA polymerase. DNA–RNA hybridization competition indicated a high efficacy of competition of the initially formed products with 18 S rRNA and an increasing competition efficacy of subsequent products with 28 S rRNA.

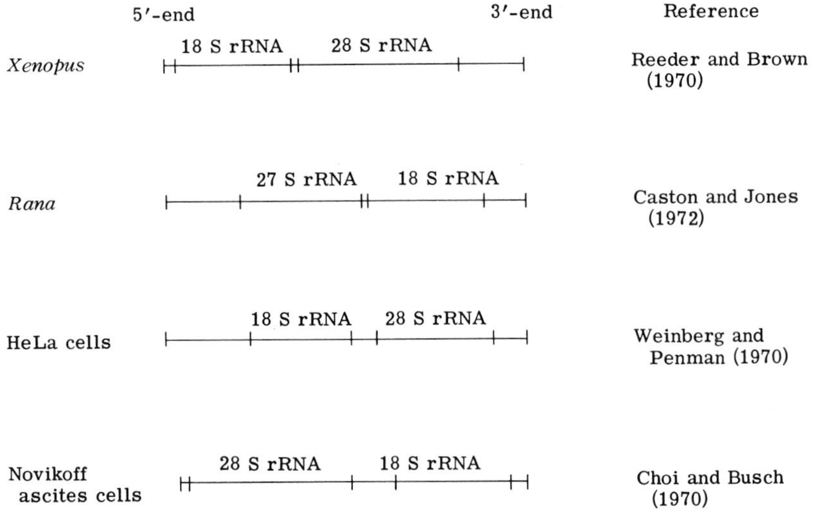

Fig. 12 Polarity of rRNA's within ribosomal transcriptional units from various systems.

In the model for *Rana* (frog), the time sequence of pre-rRNA synthesis (Caston and Jones, 1972) shows that 28 S rRNA is more proximal to the 5′-end than 18 S rRNA. Cordycepin (3′-deoxyadenosine) was used to study the transcriptional products and the appearance of cytoplasmic rRNA's were compared in HeLa cells (Siev *et al.*, 1969). This drug produces a premature termination of transcriptional processes and the appearance of cytoplasmic 18 S rRNA was not inhibited.

In the model of Novikoff hepatoma ascites cells, the structural analyses of the 5′-end were made among all the nucleolar HMW RNA's and cytoplasmic rRNA's (Choi and Busch, 1970; Egawa *et al.*, 1971). No triphosphoryl moiety was found among HMW RNA's. The 5′-ends of 18 S rRNA and 28 S rRNA were different. Among the nucleolar 45 S RNA, 32 S RNA, 28 S RNA, and cytoplasmic 28 S rRNA, the 5′-ends are identical (Choi and Busch, 1970), i.e., (Am—Gm—Cm—Ap) (Nazar and Busch, 1974).

F. *Regulatory Mechanism of Pre-rRNA Metabolism and Model of Ribosome Metabolism*

The overall gene expression of ribosomal cistrons includes a series of metabolic events which occur in the nucleolus, nucleoplasm, and cytoplasm where the final gene products are ribosomes. The sequence of events is closely linked and factors governing any step are likely to influence the dynamics of gene expression.

1. PRETRANSCRIPTIONAL CONTROLS

The gene amplification in oocytes of amphibians and insects is a good example of positive control. The mechanism of amplification is by gene duplication, either by DNA-dependent or RNA-dependent DNA polymerase (Brown and Tochini-Valentini, 1972; Mahdavi and Crippa, 1972). Another positive control mechanism is gene "magnification" observed in "bobbed" mutants of *Drosophila* (Ritossa, 1968), which occurs in flies over a few generations.

2. TRANSCRIPTIONAL CONTROLS

In the course of synthesis of primary transcriptional unit, three coordinated reactions are involved: (1) transcription specified by RNA polymerase I; this process is very sensitive to low doses of actinomycin D; (2) transcriptional modification operated by the enzymes for base methylation, ribose methylation, base rearrangement for ψ-formation, and cleavage or trimming of the 5′-end portion of the polynucleotide being transcribed; and (3) assembly of ribosomal proteins with transcribed polynucleotides. Since the nucleolar pool of ribosomal proteins is small (Pederson and Kumar, 1971), an inhibition of ribosomal protein synthesis in the cytoplasm induces a reduced rate of ribosome formation (Maden, 1971). Furthermore, the transcription and subsequent processing are affected. These findings indicate that there is a closely linked cytonucleolar regulation involved in the feedback controls between the cytoplasmic protein synthesis and intranucleolar ribosome assembly.

Many hormones, including estrogens, androgens, growth hormones, thyroid hormones, and hydrocortisone, enhance the synthesis of nucleolar HMW RNA (Busch and Smetana, 1970).

3. POSTTRANSCRIPTIONAL CONTROLS

One of the posttranscriptional controls may include posttranscriptional modification catalyzed by endonuclease(s) (Prestayko *et al.*, 1973) and possibly exonuclease(s) (Perry and Kelley, 1972). The requirement of structural integrity of 45 S RNA was demonstrated by the use of base analogs such as 8-azaguanine (Perry, 1964), toyocamycin (Tavitian *et al.*, 1968), and tubercidin (Tavitian *et al.*, 1968). The incorporation of such analogs into 45 S RNA was reported to interrupt the processing at various steps. Adenosine analog such as cordycepin causes a premature termination of 45 S RNA and interferes with the formation of 32 S RNA (Siev *et al.*, 1969).

Puromycin causes a selective degradation of 32 S RNA and possibly 23 S RNA or 18 S RNA (Soeiro *et al.*, 1968; Warner *et al.*, 1966). The overall effect is little or no appearance of 18 S rRNA and 28 S rRNA in the cytoplasm. Cycloheximide causes a rapid degradation of 28 S RNA or 18 S RNA (Willems *et al.*, 1969).

Viral infection (poliovirus-infected HeLa cells) produces an accumulation of 41 S RNA and 36 S RNA in nucleoli (Weinberg and Penman, 1970). In early mitosis, the processing of 45 S RNA ceases (Fan and Penman, 1971).

Although the 23 S precursor was found (Egawa *et al.*, 1971), 18 S rRNA has not been found in isolated nucleoli. There may be an additional processing in 23 S RNA of the nuclear ribonucleoprotein network. Other controls may involve maturation of RNP in which additional RNA (5.5 S RNA) is added to RNP containing 28 S RNA and tightly hydrogen bonded (Pene *et al.*, 1968; Prestayko *et al.*, 1970).

4. INTERCOMPARTMENTAL TRANSPORT

There are two processes of intercompartmental transports: nucleolar–nucleoplasmic transport and nucleoplasmic–cytoplasmic transport. Actinomycin D (Girard *et al.*, 1964) and camptothecin (Wu *et al.*, 1972) are known to inhibit selectively the transport of 28 S RNA into the cytoplasm.

G. *An Extended Model of Nucleolar RNA Metabolism*

Based on the previous discussion, the original model of molecular RNA metabolism proposed by Perry (1967) is extended (Fig. 13). As progress is being made actively in a number of laboratories, this extended model serves as a working scheme for the mechanism of ribosomal RNA synthesis.

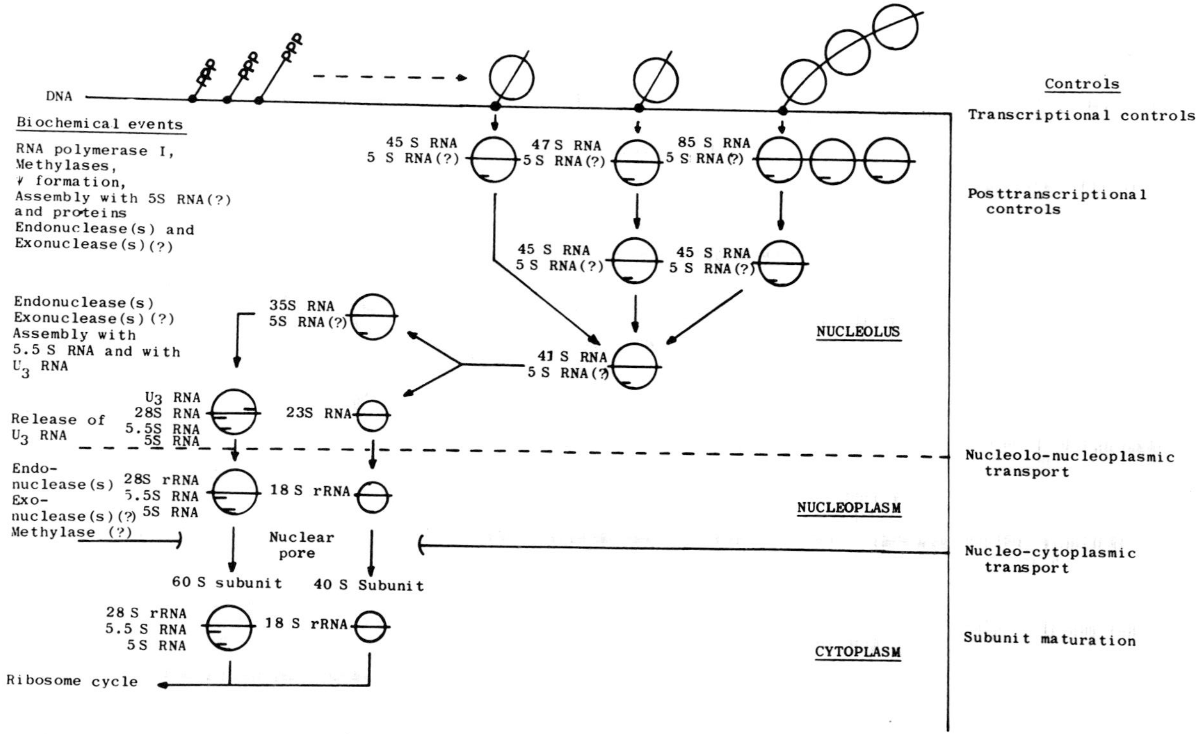

Fig. 13 An extended model of ribosome genesis. (From Perry, 1967; Darnell, 1968; Busch and Smetana, 1970; Prestayko *et al.*, 1970; Weinberg and Penman, 1970; Egawa *et al.*, 1971; Hidvegi *et al.*, 1971; Tiollais *et al.*, 1971.)

ACKNOWLEDGMENT

The authors wish to express their appreciation to *Science* for permission to reproduce Figures 3, 4 and 7, copyright 1973 by the American Association for the Advancement of Science.

REFERENCES

Adesnik, M., and Darnell, J. E. (1972). *J. Mol. Biol.* **67**, 397.

Adesnik, M., Salditt, W., Thomas W., and Darnell, J. E. (1972). *J. Mol. Biol.* **71**, 21.

Amaldi, F., and Attardi, G. (1968). *J. Mol. Biol.* **33**, 737.

Ananieva, L. N., Kozlov, Y. V., Ryskov, A. P., and Georgiev, G. P. (1968). *Mol. Biol.* **2**, 588.

Appel, S. H. (1967). *Nature (London)* **213**, 1253.

Attardi, G., and Amaldi, F. (1970). *Ann. Rev. Biochem.* **39**, 183.

Attardi, G., Parnas, H., Huang, M. I. H., and Attardi, B. (1966). *J. Mol. Biol.* **20**, 145.

Baltimore, D. (1970). *Nature (London)* **226**, 1209.

Birnboim, H. C., and Coakley, B. V. (1971). *Biochem. Biophys. Res. Commun.* **42**, 1169.

Birnstiel, M. (1967). *Ann. Rev. Plant Physiol.* **18**, 25.

Birnstiel, M. L., Chipcase, M., and Speirs, J. (1971). *Progr. Nucl. Acid Res. Mol. Biol.* **11**, 351.

Blobel, G. (1973). *Proc. Nat. Acad. Sci. U.S.* **70**, 924.

Bowman, C. M., Dahlberg, J. E., Ikemura, T., Konisky, J., and Nomura, M. (1971). *Proc. Nat. Acad. Sci. U.S.* **68**, 964.

Brawerman, H., Gold, L., and Eisenstadt, J. (1963). *Proc. Nat. Acad. Sci. U.S.* **50**, 630.

Britten, R. J., and Davidson, E. H. (1969). *Science* **165**, 349.

Britten, R. J., and Kohne, D. E. (1968). *Science* **161**, 529.

Brown, D. D., and Gurdon, J. B. (1964). *Proc. Nat. Acad. Sci. U.S.* **51**, 139.

Brown, D. D., Wensink, P. C., and Jordan, E. (1972). *J. Mol. Biol.* **63**, 57.

Brown, R. D., and Tocchini-Valentini, G. P. (1972). *Proc. Nat. Acad. Sci. U.S.* **69**, 1746.

Burdon, R. H. (1971). *Progr. Nucl. Acid Res. Mol. Biol.* **11**, 33.

Busch, H. (1974). *In* "The Molecular Biology of Cancer" (H. Busch, ed.), p. 187. Academic Press, New York.

Busch, H., and Smetana, K. (1970). "The Nucleolus." Academic Press, New York.

Busch, H., Desjardins, R., Grogan, D., Higashi, K., Jacob, S. T., Muramatsu, M., Ro, T. S., and Steele, W. J. (1967). *Nat. Cancer Inst. Monogr.* **23**, 193.

Busch, H., Arendell, J. P., Morris, H. P., Neogy, R. K., and Schwartz, S. M. (1968). *Cancer Res.* **28**, 280.

Busch, H., Choi, Y. C., Daskal, I., Inagaki, A., Olson, M. O. J., Reddy, R., Ro-Choi, T. S., Shibata, H., and Yeoman, L. C. (1972). *In* "Gene Transcription in Reproductive Tissue" (E. Diczfalusy, ed.), pp. 33–63. Karolinska Inst., Stockholm.

Callan, H. G. (1967). *J. Cell Sci.* **2**, 1.

Canellakis, N., Canellakis, E. S., and Lim, L. (1970). *Biochim. Biophys. Acta* **209**, 128.

Cartouzou, G., Mante, S., and Lissitzky, S. (1965). *Biochem. Biophys. Res. Commun.* **20**, 212.

Caston, J. D., and Jones, P. H. (1972). *J. Mol. Biol.* **69,** 19.
Chambon, P., Gissinger, F., Kedinger, C., Mandel, J. L., Meilhac, M., and Muret, P. (1972). *In* "Gene Transcription in Reproductive Tissue" (E. Diczfalusy, ed.), pp. 222–246. Karolinska Inst., Stockholm.
Choi, Y. C., and Busch, H. (1970). *J. Biol. Chem.* **245,** 1954.
Choi, Y. C., and Busch, H. (1974). In preparation.
Choi, Y. C., Mauritzen, C. M., Taylor, C. W., and Busch, H. (1971). *Physiol. Chem. Phys.* **3,** 116.
Church, R. B., and McCarthy, B. J. (1967). *Proc. Nat. Acad. Sci. U.S.* **58,** 1548.
Church, R. B., and McCarthy, B. J. (1968). *Biochem. Genet.* **2,** 55.
Church, R. B., Luther, S. W., and McCarthy, B. J. (1969). *Biochim. Biophys. Acta* **190,** 30.
Daneholt, B. (1972). *Nature (London) New Biol.* **240,** 229.
Darnell, J. E. (1968). *Bacteriol. Rev.* **32,** 262.
Darnell, J. E., Jelinek, W. R., and Molloy, G. R. (1973). *Science* **181,** 1215.
Darnell, J. E., Wall, R., and Tushinski, R. J. (1971). *Proc. Nat. Acad. Sci. U.S.* **68,** 1321.
Dawid, I. B., Brown, D. D., and Reeder, R. H. (1970). *J. Mol. Biol.* **51,** 341.
Deak, I., Sidebottom, E., and Harris, H. (1972). *J. Cell Sci.* **11,** 379.
Denis, H. (1966). *J. Mol. Biol.* **22,** 285.
Desjardins, R., Smetana, K., Grogan, D., Higashi, K., and Busch, H. (1966). *Cancer Res.* **26,** 97.
DiGirolamo, A., Henshaw, E. C., and Hiatt, H. H. (1964). *J. Mol. Biol.* **8,** 479.
Edmonds, M., and Abrams, R. (1960). *J. Biol. Chem.* **235,** 1142.
Edmonds, M., and Caramela, M. G. (1969). *J. Biol. Chem.* **244,** 1314.
Edmonds, M., Vaughan, M. H., and Nakazato, H. (1971). *Proc. Nat. Acad. Sci. U.S.* **68,** 1336.
Egawa, K., Choi, Y. C., and Busch, H. (1971). *J. Mol. Biol.* **56,** 565.
Fan, H., and Penman, S. (1971). *In* "Drugs and Cell Regulation" (E. Mihich, ed.), pp. 79–98, Academic Press, New York.
Fellner, P. (1969). *Eur. J. Biochem.* **11,** 12.
Fiers, W., Contreras, R., DeWachter, R., Haegeman, G., Merregaert, J., Min Jou, W., and Vandenberghe, A. (1971). *Biochimie* **53,** 495.
Flamm, W. G. (1972). *Int. Rev. Cytol.* **32,** 2.
French, W. L., and Kitzmiller, J. B. (1967). *Amer. Zool.* **7,** 782.
Gelderman, A. H., Rahe, A. V., and Britten, R. J. (1968). *Carnegie Inst. Yearbook* **67,** 320.
Georgiev, G. P. (1967). *Progr. Nucl. Acid Res. Mol. Biol.* **6,** 259.
Georgiev, G. P. (1971). *Current Top. Develop. Biol.* **7,** 1.
Georgiev, G. P., and Mantieva, V. L. (1962). *Biochim. Biophys. Acta* **61,** 153.
Georgiev, G. P., Ryskov, A. P., Coutelle, C., Mantieva, V. L., and Avakyan, E. R. (1972). *Biochim. Biophys. Acta* **259,** 259.
Girard, M., Penman, S., and Darnell, J. E. (1964). *Proc. Nat. Acad. Sci. U.S.* **51,** 205.
Gould, H. (1966). *Biochemistry* **5,** 1103.
Granboulan, N., and Scherrer, K. (1969). *Eur. J. Biochem.* **9,** 1.
Green, M., Rokutanda, H., and Rokutanda, M. (1971). *Nature (London)* **230,** 229.
Greenberg, J. R., and Perry, R. P. (1972). *J. Mol. Biol.* **72,** 91.
Gulati, S. C., Axel, R. and Spiegelman, S. (1972). *Proc. Nat. Acad. Sci. U.S.* **69,** 2020.

Gurdon, J. B., Lane, C. D., Woodland, H. R., and Marbaix, G. (1971). *Nature (London)* **233,** 177.
Hadjivassiliou, A., and Brawerman, G. (1966). *J. Mol. Biol.* **20,** 1.
Hearst, J. E., and Botchan, M. (1970). *Ann. Rev. Biochem.* **39,** 151.
Helser, T. L., Davies, J. E., and Dahlberg, J. E. (1970). *Nature (London) New Biol.* **235,** 6.
Helser, T. L., Davies, J. E., and Dahlberg, J. E. (1971). *Nature (London) New Biol.* **233,** 12.
Henshaw, E. C. (1968). *J. Mol. Biol.* **36,** 401.
Hiatt, H. (1962). *J. Mol. Biol.* **5,** 217.
Hidvegi, E. J., Prestayko, A. W., and Busch, H. (1971). *Physiol. Chem. Phys.* **3,** 17.
Horgen, P. A., and Griffin, D. H. (1971). *Proc. Nat. Acad. Sci. U.S.* **68,** 338.
Huberman, J. A., and Attardi, G. (1967). *J. Mol. Biol.* **29,** 487.
Inagaki, A., and Busch, H. (1972a). *J. Biol. Chem.* **247,** 3327.
Inagaki, A., and Busch, H. (1972b). *Biochem. Biophys. Res. Commun.* **49,** 1398.
Iwanami, Y., and Brown, G. M. (1968). *Arch. Biochem. Biophys.* **126,** 8.
Jacob, F., and Monod, J. (1961). *J. Mol. Biol.* **3,** 318.
Jacob, S. T., and Busch, H. (1967). *Biochim. Biophys. Acta* **138,** 249.
Jeanteur, P., Amaldi, F., and Attardi, G. (1968). *J. Mol. Biol.* **33,** 757.
Jeanteur, P., and Attardi, G. (1969). *J. Mol. Biol.* **45,** 305.
Jelinek, W., and Darnell, J. E. (1972). *Proc. Nat. Acad. Sci. U.S.* **69,** 2537.
Jelinek, W., Adesnik, M., Salditt, M., Sheiness, D., Wall, R., Molloy, G., Philipson, L., and Darnell, J. E. (1973). *J. Mol. Biol.* **75,** 515.
Judd, B. H., Shen, M. W., and Kaufman, T. C. (1972). *Genetics* **71,** 139.
Jukes, T. H., and Gatlin, L. (1971). *Progr. Nucl. Acid Res. Mol. Biol.* **11,** 303.
Kanamaru, R., Choi, Y. C., and Busch, H. (1972). *Physiol. Chem. Phys.* **4,** 103.
Kates, J. (1970). *Cold Spring Harbor Symp. Quant. Biol.* **35,** 743.
Kates, J., and Beeson, J. (1970). *J. Mol. Biol.* **50,** 19.
Kirby, K. S. (1956). *Biochem. J.* **64,** 405.
Kuff, E. L., and Roberts, N. E. (1967). *J. Mol. Biol.* **26,** 211.
Kurland, C. G. (1972). *Ann. Rev. Biochem.* **41,** 377.
Kwan, S., and Brawerman, G. (1972). *Proc. Nat. Acad. Sci. U.S.* **69,** 3247.
Lee, J. C., and Quintanilla, I. V. (1972). *Biochemistry* **11,** 1357.
Lee, R., Mendecki, J., and Brawerman, G. (1971). *Proc. Nat. Acad. Sci. U.S.* **68,** 1331.
Liau, M. C., and Perry, R. P. (1969). *J. Cell Biol.* **42,** 277.
Lim, L., and Canellakis, E. S. (1970). *Nature (London)* **227,** 710.
Lindberg, V., and Darnell, J. E. (1970). *Proc. Nat. Acad. Sci. U.S.* **65,** 1089.
Loening, V. E. (1968). *Ann. Rev. Plant Physiol.* **19,** 37.
Loening, V. E., Jones, K. W., and Birnstiel, M. L. (1969). *J. Mol. Biol.* **45,** 353.
Lucas-Lenard, J., and Lipman, F. (1971). *Ann. Rev. Biochem.* **40,** 409.
Maden, B. E. H. (1971). *Progr. Biophys. Mol. Biol.* **22,** 127.
Maden, B. E. H., Lees, C. D., and Salim, M. (1972). *FEBS Lett.* **28,** 293.
Mahdavi, V., and Crippa, M. (1972). *Proc. Nat. Acad. Sci. U.S.* **69,** 1749.
Marks, P. A., Burka, E. R., and Schlessinger, D. (1962). *Proc. Nat. Acad. Sci. U.S.* **48,** 2163.
Mayo, V. S., and DeKloet, S. R. (1971). *Biochim. Biophys. Acta* **247,** 74.
McCarthy, B. J., and Church, R. B. (1970). *Ann. Rev. Biochem.* **39,** 131.
McCarthy, B. J., and Hoyer, B. H. (1964). *Proc. Nat. Acad. Sci. U.S.* **52,** 915.
McParland, R., Crooke, S. T., and Busch, H. (1972). *Biochim. Biophys. Acta* **269,** 78.
Melli, M., and Bishop, J. D. (1969). *J. Mol. Biol.* **40,** 117.

Mendecki, J., Lee, S. Y., and Brawerman, G. (1972). *Biochemistry* **11,** 792.
Miller, O. L., Jr., and Hamkalo, B. (1972). *Int. Rev. Cytol.* **33,** 1.
Molloy, G. R., and Darnell, J. E. (1973). *Biochemistry* **12,** 2324.
Molloy, G. R., Thomas, W. L., and Darnell, J. E. (1972). *Proc. Nat. Acad. Sci. U.S.* **69,** 3684.
Muramatsu, M., Smetana, K., and Busch, H. (1963). *Cancer Res.* **23,** 510.
Muramatsu, M., Hodnett, J. L., and Busch, H. (1966). *J. Biol. Chem.* **241,** 1544.
Narayan, K. S., Steele, W. J., Smetana, K., and Busch, H. (1967). *Exp. Cell Res.* **46,** 65.
Nazar, R. N., and Busch, H. (1973). *Biochim. Biophys. Acta* **299,** 428.
Nazar, R. N., and Busch, H. (1974). *J. Biol. Chem.* **249,** 919.
Nichols, J. L., and Lane, B. G. (1966). *Can. J. Biochem.* **44,** 1633.
Noller, H. F., and Chaires, J. B. (1972). *Proc. Nat. Acad. Sci. U.S.* **69,** 3115.
Nomura, M. (1970). *Bacteriol. Rev.* **34,** 228.
Oda, K., and Dulbecco, R. (1968). *Proc. Nat. Acad. Sci. U.S.* **60,** 525.
Pederson, T., and Kumar, A. (1971). *J. Mol. Biol.* **61,** 655.
Pene, J. J., Knight, E., and Darnell, J. E. (1968). *J. Mol. Biol.* **28,** 491.
Penman, S., Scherrer, K., Becker, I., and Darnell, J. E. (1963). *Proc. Nat. Acad. Sci. U.S.* **49,** 654.
Penman, S., Smith, I., and Holtzman, E. (1966). *Science* **154,** 786.
Penman, S., Rosbash, M., and Penman, M. (1970). *Proc. Nat. Acad. Sci. U.S.* **67,** 1878.
Perry, R. P. (1964). *Nat. Cancer Inst. Monogr.* **14,** 73.
Perry, R. P. (1967). *Progr. Nucl. Acid Res. Mol. Biol.* **6,** 219.
Perry, R. P., and Kelley, D. E. (1968). *J. Mol. Biol.* **35,** 37.
Perry, R. P., and Kelley, D. E. (1972). *J. Mol. Biol.* **70,** 265.
Perry, R. P., Cheng, T.-Y., Freed, J. J., Greenberg, J. R., Kelley, D. E., and Tartoff, K. D. (1970). *Proc. Nat. Acad. Sci. U.S.* **65,** 609.
Pestka, S. (1971). *Ann. Rev. Microbiol.* **25,** 487.
Prestayko, A. W., Tonato, M., and Busch, H. (1970). *J. Mol. Biol.* **47,** 505.
Prestayko, A. W., Lewis, B. C., and Busch, H. (1973). *Biochim. Biophys. Acta* **319,** 323.
Quagliarotti, G., Hidvegi, E., Wikman, J., and Busch, H. (1970). *J. Biol. Chem.* **245,** 1962.
Reeder, R. H., and Brown, D. D. (1970). *J. Mol. Biol.* **51,** 361.
Rees, H., and Jones, R. N. (1972). *Int. Rev. Cytol.* **32,** 53.
Revel, M., and Hiatt, H. H. (1964). *Proc. Nat. Acad. Sci. U.S.* **51,** 810.
Ritossa, F. M. (1968). *Proc. Nat. Acad. Sci. U.S.* **60,** 509.
Ritossa, F. M., and Spiegelman, S. (1965). *Proc. Nat. Acad. Sci. U.S.* **53,** 737.
Ro-Choi, T. S., Choi, Y. C., Savage, H. E., and Busch, H. (1973a). *In* "Methods in Cancer Research" Vol. IX (H. Busch, ed.), p. 72. Academic Press, New York.
Ro-Choi, T. S., Smetana, K., and Busch, H. (1973b). *Exp. Cell Res.* **79,** 43.
Roeder, R. G., and Rutter, W. J. (1970). *Proc. Nat. Acad. Sci. U.S.* **65,** 675.
Ruddle, F. H. (1972). *Advan. Hum. Genet.* **3,** 173.
Salim, M., Williamson, R., and Maden, B. E. (1970). *FEBS Lett.* **12,** 109.
Samarina, O. P., Asriyan, I. S., and Georgiev, G. P. (1965). *Proc. Acad. Sci. U.S.S.R.* **163,** 1510.
Samarina, O. P., Lukanidin, E. M., Molnar, J., and Georgiev, G. P. (1968). *J. Mol. Biol.* **33,** 251.
Sanger, F. (1971). *Biochem. J.* **124,** 833.
Santer, M., and Szekely, M. (1971). *Biochemistry* **10,** 1841.

Schaup, H. W., Green, M., and Kurland, C. G. (1970). *Mol. Gen. Genet.* **109,** 193.
Schaup, H. W., Green, M., and Kurland, C. G. (1971). *Mol. Gen. Genet.* **112,** 1.
Scheer, U., Trendelenburg, M. F., and Franke, W. W. (1974). *Exp. Cell Res.* (in press).
Scherrer, K., and Darnell, J. E. (1962). *Biochem. Biophys. Res. Commun.* **7,** 486.
Scherrer, K., and Marcaud, L. (1965). *Bull. Soc. Chim. Biol.* **47,** 1697.
Scherrer, K., and Marcaud, L. (1968). *J. Cell Physiol. Suppl.* 1 **72,** 181.
Scherrer, K., Spohr, G., Granboulan, N., Morel, C., Grosclaude, J., and Chezzi, C. (1970). *Cold Spring Harbor Symp. Quant. Biol.* **35,** 539.
Schimke, R. T., and Doyle, D. (1970). *Ann. Rev. Biochem.* **39,** 929.
Schochetman, G., and Perry, R. P. (1972). *J. Mol. Biol.* **63,** 591.
Seeber, S., and Busch, H. (1971). *J. Biol. Chem.* **246,** 7151.
Senior, B. W., and Holland, I. B. (1971). *Proc. Nat. Acad. Sci. U.S.* **68,** 958.
Sharma, O. K., Arendell, J. P., Hidvegi, E. J., Marks, F., Prestayko, A., Smetana, K., and Busch, H. (1969). *Physiol. Chem. Phys.* **1,** 185.
Shearer, R. W., and McCarthy, B. J. (1967). *Biochemistry* **6,** 283.
Sheiness, D., and Darnell, J. E. (1973). *Nature (London) New Biol.* **241,** 265.
Sibatani, A., Yamana, K., Kimura, K., and Okagaki, T. (1959). *Biochim. Biophys. Acta* **33,** 590.
Sibatani, A., Yamana, K., Kimura, K., and Takahagi, T. (1960). *Nature (London)* **186,** 215.
Siebert, G., Villalobos, J., Jr., Ro, T. S., Steele, W. J., Lindenmayer, G., Adams, H., and Busch, H. (1966). *J. Biol. Chem.* **241,** 71.
Siev, M., Weinberg, R., and Penman, S. (1969). *J. Cell Biol.* **41,** 510.
Singer, M., and Leder, P. (1966). *Ann. Rev. Biochem.* **35,** 195.
Smetana, K., Steele, W. J., and Busch, H. (1963). *Exp. Cell Res.* **31,** 198.
Smith, S. J., Higashi, K., and Busch, H. (1967). *Cancer Res.* **27,** 849.
Soeiro, R., Vaughan, M., and Darnell, J. E. (1968). *J. Cell Biol.* **36,** 91.
Soeiro, R. (1968). *In* "Regulatory Mechanisms for Protein Synthesis in Mammalian Cells." Academic Press, New York.
Soeiro, R., and Darnell, J. E. (1970). *J. Cell Biol.* **44,** 467.
Spirin, A. S. (1969). *Eur. J. Biochem.* **10,** 20.
Steele, W. J., and Busch, H. (1966a). *Biochim. Biophys. Acta* **129,** 54.
Steele, W. J., and Busch, H. (1966b). *Biochim. Biophys. Acta* **119,** 501.
Steffensen, D. M., and Wimber, D. E. (1972). *In* "Nucleic Acid Hybridization in the Study of Cell Differentiation" (H. Ursprung, ed.), pp. 47–63. Springer-Verlag, New York and Berlin.
Stoffler, G., Daya, L., Rak, K. H., and Garret, R. A. (1971). *J. Mol. Biol.* **62,** 411.
Suzuki, Y., and Brown, D. D. (1972). *J. Mol. Biol.* **63,** 409.
Tavitian, A., Uretsky, S. C., and Acs, S. (1968). *Biochim. Biophys. Acta* **157,** 33.
Temin, H. M., and Mizutani, S. (1970). *Nature (London)* **226,** 1211.
Tiollais, P., Galibert, F., and Bairon, M. (1971). *Proc. Nat. Acad. Sci. U.S.* **6,** 1117.
Tomkins, G. M., and Gelehrter, T. D. (1972). "Biochemistry Actions of Hormones" (G. Litwack, ed.), Vol. II, pp. 1–20. Academic Press, New York.
Tonegawa, S., Walter, G., Bernardini, A., and Dulbecco, R. (1970). *Cold Spring Harbor Symp. Quant. Biol.* **35,** 823.
Udem, S. A., and Warner, J. R. (1972). *J. Mol. Biol.* **65,** 213.
Vesco, C., and Penman, S. (1967). *Biochim. Biophys. Acta* **169,** 188.
Wagner, E. K., Penman, S., and Ingram, V. M. (1967). *J. Mol. Biol.* **29,** 371.
Wall, R., and Darnell, J. E. (1971). *Nature (London) New Biol.* **232,** 73.

Warner, J. R., Soeiro, R., Birnboim, H. C., Girard, M., and Darnell, J. E. (1966). *J. Mol. Biol.* **19,** 349.
Warner, J. R., Girard, M., Latham, H., and Darnell, J. E. (1966). *J. Mol. Biol.* **19,** 373.
Watson, J. D. (1964). *Bull Soc. Chem. Biol.* **46,** 1399.
Weinberg, R. A., and Penman, S. (1970). *J. Mol. Biol.* **47,** 169.
Westphal, H., and Dulbecco, R. (1968). *Proc. Nat. Acad. Sci. U.S.* **59,** 1158.
Wigle, D. T., and Smith, A. E. (1972). *Nature (London) New Biol.* **242,** 136.
Wikman, J., Howard, E., and Busch, H. (1969). *J. Biol. Chem.* **244,** 5471.
Willems, M., Penman, M., and Penman, S. (1969). *J. Cell Biol.* **41,** 177.
Williamson, R., Drewienkiewicz, C. E., and Paul, J. (1973). *Nature (London) New Biol.* **241,** 66.
Yamana, K., and Sibatani, A. (1960). *Biochim. Biophys. Acta* **41,** 295.
Yoshikawa-Fukada, M., Fukada, T., and Kawada, Y. (1964). *Biochim. Biophys. Acta* **103,** 383.
Zimmerman, R. A., Muto, A., Fellner, P., Ehresmann, C., and Branlant, C. (1972). *Proc. Nat. Acad. Sci. U.S.* **69,** 1282.

5

Low-Molecular-Weight Nuclear RNA's

Tae Suk Ro-Choi and Harris Busch

I. Introduction*

All forms of cellular life contain many types of transfer RNA, messenger RNA, and two types of ribosomal RNA (18 S RNA and 28 S RNA). The main biochemical function of the various forms of RNA's is to provide a link between genetic information contained in DNA sequences and the primary structure of proteins. These RNA's are synthesized in the nucleus on DNA templates by DNA-dependent RNA polymerases of various forms (Chambon *et al.*, 1970; Ro and Busch, 1964; Roeder and Rutter, 1969) and are transferred to the cytoplasm where they serve specific function. Low-molecular-weight RNA with sedimentation coefficients from 4 S to 8 S was long considered to be only transfer RNA.

With the discovery of a new type of low-molecular-weight RNA, namely, the 5 S RNA of the ribosomes, by Rosset and Monier (1963), the search for other types of low-molecular-weight RNA was stimulated. With the successful application of polyacrylamide gel electrophoresis to RNA biochemistry, many more new species of low-molecular-weight RNA have been discovered. These include a class of low-molecular-weight nuclear RNA and virus-associated low-molecular-weight RNA's. These low-molecular-weight nuclear RNA's (LMWN RNA) either are specifically localized to the nucleolus or to the extranucleolar portion of the nucleus of mammalian and other cells (Egyhazi *et al.*, 1969; Enger and Walters, 1970; Hellung-Larsen *et al.*, 1971; Hellung-Larsen and Fredriksen, 1972; Hodnett and Busch, 1968; Larsen *et al.*, 1967, 1969; Loening, 1967; Moriyama *et al.*, 1969; Nakamura *et al.*, 1968; Ro-Choi *et al.*, 1970, 1971, 1972; Zapisek *et al.*, 1969). Some RNA's smaller in size than these which contain dihydropyrimidines have also been reported to be present in the nucleus, specifically in association with the chromatin fraction (Bonner and Widholm, 1967; Huang and Bonner, 1965; Shih and Bonner, 1969). Low-molecular-weight RNA's associated with cytoplasmic membranes have also been reported. Although the functions of most of the LMWN RNA's are not defined, they are a unique class of RNA species with specific characteristics (Table I).

* Abbreviations used: LMWN RNA, low-molecular-weight nuclear ribonucleic acid; LMW RNA, low-molecular-weight ribonucleic acid; Nu, nuclei; No, nucleoli; Rib, ribosome; A, adenosine; U, uridine; G, guanosine; C, cytidine; ψ, pseudouridine; T, ribothymidine; hu, dihydrouridine; A′, ^{3}H derivative of adenosine (nucleoside was oxidized with sodium periodate and reduced with [^{3}H]KBH_4 on its ribose moiety); U′, ^{3}H derivative of uridine; G′, ^{3}H derivative of guanosine; C′, ^{3}H derivative of cytidine; ψ', ^{3}H derivative of pseudouridine; m^{6}A′, ^{3}H derivative of N^6-methyladenosine; m^{2}G′, ^{3}H derivative of N^2-methylguanosine, $m_3^{2,2,7}$G′, ^{3}H derivative of N^2,N^2-dimethyl-7-methylguanosine; Gly, glycerol; B, background; AMP, adenylic acid; UMP, uridylic acid; CMP, cytidylic acid; GMP, guanylic acid; CEV, cucumber exocortis disease virus; PSTV, potato spindle tuber viroid.

TABLE I
General Characteristics of LMWN RNA

1. Size ranges: 100–300 nucleotides; total number per cell, 1–2 × 10^6 molecules (Weinberg and Penman, 1968)
2. Stable half-lives of up to one cell cycle (Weinberg and Penman, 1969)
3. Specific localization: (*a*) nucleolus-associated RNA, U-3 RNA; (*b*) nucleoplasmic RNA: 4.5 S $RNA_{I,II,\text{ and }III}$; 5 S RNA_{III}, U-1 RNA, and U-2 RNA (Busch *et al.*, 1971); (*c*) cytoplasm: absent
4. Cell distribution: present in all eukaryotic cells studied, including human cells (HeLa cells, KB cells, lymphocytes, fibroblast), rat cells (liver cells, Novikoff hepatoma cells), hamster cells (Chinese hamster ovary cells), mouse cells (3T3 cells, Yoshida ascites cells, Ehrlich ascites cells, L cells), *Xenopus laevis* cells, sea urchin eggs, and *Tetrahymena pyriformis*
5. Some exist in ribonucleoprotein complexes (Enger and Walters, 1970; Rein, 1971)
6. Exhibit specificity of hybridization with nuclear and nucleolar DNA (Sitz and Busch, 1973)
7. Specific sequences; some have an unusual distribution and content of modified nucleosides (Reddy *et al.*, 1972; Ro-Choi *et al.*, 1971, 1972)

The progress in structural analysis of various RNA's from viral, bacterial, and mammalian cells contributes to the understanding of important processes regulating transcription and translation. Although at the present time the tertiary structure of only one RNA ($tRNA^{ala}$) is known, the sequences of many tRNA molecules, isolated mostly from microorganisms, has been established. Such studies have already uncovered important structure–function relationships, for example, with regard to the location of the anticodon triplet of tRNA, and details of its interaction with messenger RNA code letters. Some progress has been made in aminoacyl synthetase specificity and structural differences in natural and mutant tRNA.

The sequence analysis of bacteriophage MS2 messenger RNA and its translation into protein sequences is another remarkable contribution. More recently, ribosomal RNA and its precursors have been subjects for the analysis of their primary sequences with particular respect to the relationship of their structures to the orientation of their accompanying proteins which provide the functional mosaics for protein synthesis (Fuller and Hodgson, 1967; Gould, 1967; Sanger, 1971). However, sequences of only four RNA's from mammalian cells have been determined: 5 S RNA from KB cells (Forget and Weissman, 1969); nuclear 4.5 S RNA_I from Novikoff hepatoma cells (Ro-Choi *et al.*, 1972); $tRNA^{ser}$ from rat liver (Ginsberg *et al.*, 1971); nuclear U-2 RNA (Shibata *et al.*, 1973a, b); and one viral RNA of mammalian cell origin, VA-RNA (Ohe and Weissman, 1971). Among these RNA's, 4.5 S RNA_I and U-2 RNA are localized in the cell nucleus. Moreover, these studies have set an important pattern in the use of physical and biochemical methods for analysis of the structure of various RNA species.

II. Classification and Number of Low-Molecular-Weight Nuclear RNA's

One of the criteria of classification is the size of these RNA molecules. Figure 1 shows representative gel patterns of low-molecular-weight nuclear and nucleolar RNA. There are six major groups of LMWN RNA's: 4 S RNA (tRNA), 4.5 S RNA, 5 S RNA, U-1 RNA, U-2 RNA, and U-3 RNA. The nomenclature from several laboratories for these RNA's is shown in Table II.

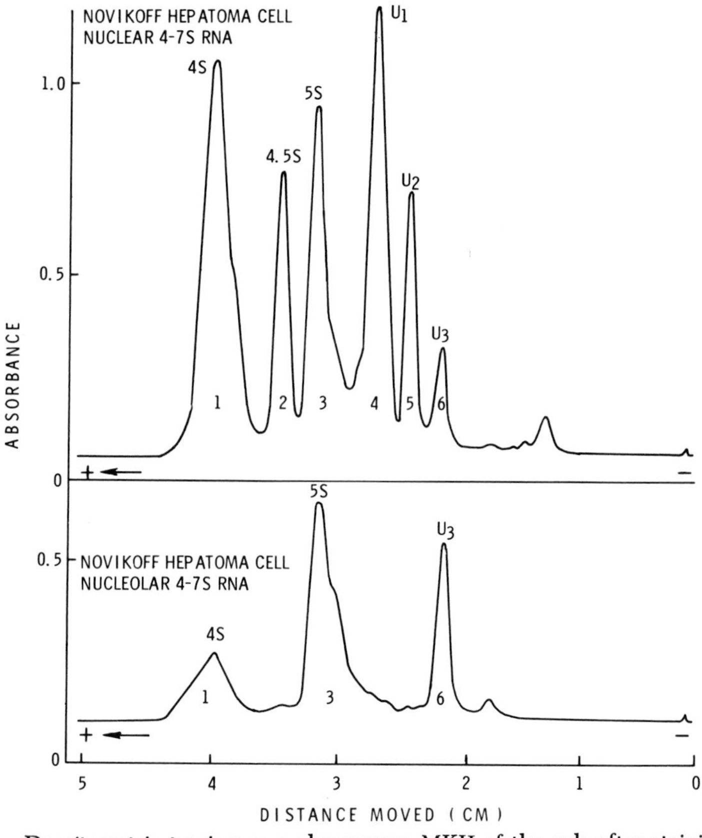

Fig. 1 Densitometric tracing on a chromoscan MKII of the gels after staining with methylene blue. Electrophoresis of the 4–8 S RNA from nuclei and nucleoli of Novikoff hepatoma cells was carried out on 8% polyacrylamide gels. The buffer system used is 0.04 *M* Tris acetate, 0.02 *M* sodium acetate, 1 m*M* EDTA at pH 7.2. The arabic numbers are arranged from the fastest moving band (1) to the slowest moving band (6).

TABLE II
Species of LMWN RNA of Mammalian Cells

Band no.	1	2	3	4	5	6	Reference
Species	4 S RNA	4.5 S RNA I, II, and III	5 S RNA I, II, and III	U-1 RNA 5.5 S (28 S) U-1a and U-1b	U-2 RNA	U-3 RNA	Busch *et al.* (1971) Ro-Choi *et al.* (1971)
	I	H	G and G′	F and D	C	A	Weinberg and Penman (1968)
	II	III	IV	VI	VII		Zapisek *et al.* (1969)
Approximate chain length	78	97	121	138–150	180	185	Larsen *et al.* (1969)
	80	100	121	125–150	165	180	Weinberg and Penman (1968)
	80(a)	96(b)	120(c)	170(e)	196(d)		(a) Zachau (1967) (b) Ro-Choi *et al.* (1972) (c) Forget and Weissman (1969) (d) Shibata *et al.* (1973a,b) (e) Reddy *et al.* (1974)
Estimated no. of molecules/nucleus	5×10^5	3×10^5	6×10^5	1×10^6	5×10^5	2×10^5	Weinberg and Penman (1968)

The complexity of LMWN RNA is also evident from the 3′-terminal analysis of these RNA's. Table III shows the ^3H distribution in the major 3′-nucleosides. The 4 S (tRNA) and 5 S RNA (mainly 5 S rRNA) have A and U as their 3′-terminus. The predominant terminals of 4.5 S $RNA_{I\ and\ II}$ and U-1 RNA are U, U, and G, respectively. Both U-2 and U-3 had more complex terminals (Table III.) As will be discussed in Section V, these six groups of RNA's can be subfractionated into several subspecies using DEAE–Sephadex column chromatography. Although it is possible that these molecules exhibit internal microheterogeneity, as in the case of tRNA, evidence that this is the case has not yet been obtained. The possibility that there might be an artifactual breakdown product of high-molecular-weight RNA has been ruled out, as summarized in Table IV.

III. Localization of Low-Molecular-Weight RNA in the Cell

Analysis of LMW RNA from fractionated subcellular organelles shows specific localization of these RNA's. Figure 2A shows the sucrose density gradient of RNA extracted from nucleoli, nuclei, and ribosomes. The

TABLE III
3′-Terminal Nucleoside Analysis of LMWN RNA of Novikoff Hepatoma Cell Nuclei

	Radioactivity (%) in nucleoside derivatives			
RNA species	A′	U′	G′	C′
4 S	*89*	—	—	—
4.5 S (total)	11	80	—	—
4.5 S RNA_I	—	*87*	—	—
4.5 S RNA_{II}	13	*80*	—	—
5 S	11	*75*	—	—
U–1	—	13	*78*	—
U–2	*61*	—	—	28
U–3	*54*	23	10	14

The mixture of these LMWN RNA's were labeled with $[^3H]KBH_4$ after oxidation with $NaIO_4$ at the 3′-terminal end. The low-molecular-weight nuclear RNA was fractionated by preparative gel electrophoresis (7 or 8% gel) into six major components (4 S, 4.5 S, 5 S, U-1, U-2, and U-3). Each fraction was hydrolyzed with 0.3 *N* KOH and analyzed for 3′-nucleosides.

TABLE IV
Evidence That LMWN RNA's Are Native Molecular Species[a]

1. The polyacrylamide gel patterns of LMWN RNA's are reproducible for individual cell types and are similar from cell type to cell type including liver, spleen, fibroblasts, ovary, kidney, and tumors (Hellung-Larsen *et al.*, 1971; Moriyama *et al.*, 1969; Rein and Penman, 1969)
2. The nucleotide compositions of specific RNA fractions are constant and differ from those of HMW RNA
3. The RNase content of Novikoff hepatoma nuclear preparations is low and the residual RNase activity is inhibited at the low pH of citric acid used for these nuclear isolations
4. When labeled 18 S RNA and 28 S rRNA were added to the preparations, no degradation was observed during isolation of nuclear RNA either with hot or cold phenol extractions
5. The nuclear RNA bands are found in preparations of RNA obtained from whole cells so they cannot be artifacts of nuclear isolation techniques
6. The special nuclear RNA bands are equally well extracted with either hot or cold phenol
7. The labeling patterns of the nuclear RNA with either [^{14}C]sodium formate or [^{14}C]methylmethionine differ from those of tRNA, 5 S RNA, or HMW RNA.
8. Unique and specific linear sequences of nucleotides have been demonstrated for 4.5 S RNA_I, U-1, and U-2 RNA (Ro-Choi *et al.*, 1972; Shibata *et al.*, 1973a,b)

[a] From Busch *et al.* (1971); Moriyama *et al.* (1969).

RNA in the shaded areas (4–8 S RNA) was reprecipitated and analyzed by 8% analytical polyacrylamide gel electrophoresis (Fig. 2B). Whole nuclear RNA contains the greatest variety of these RNA's and the cytoplasmic sap contains the smallest number of bands. The fast-moving RNA in the cytoplasmic sap is largely composed of transfer or amino acid acceptor RNA which has the highest mobility of any of the types of LMW RNA in these groups. The LMW RNA of the ribosomal fraction contains at least two bands in the 4 S region: a dark band of 5 S RNA that has a slower migration rate and another dark band that follows the 5 S RNA (Fig. 2B). Probably there are only three main types of low-molecular-weight RNA in the ribosomes; i.e., 4 S transfer RNA (with a trailing band), 5 S RNA (with a trailing band), and a third component which will be referred to as 5.5 S (28 S) RNA (Pene *et al.*, 1968; Prestayko *et al.*, 1971).

In the whole nuclear RNA as obtained either from Novikoff hepatoma, normal liver, or other tissues, there is a dense band of 4.5 S RNA between the 4 S and 5 S bands, and a distinct region containing two dense bands referred to as U-1 RNA. These closely approximated two bands are followed by a single dense band which is U-2 RNA. The smaller U-3 band which follows the U-2 band is concentrated in the nucleolus (Figs. 1 and 2B).

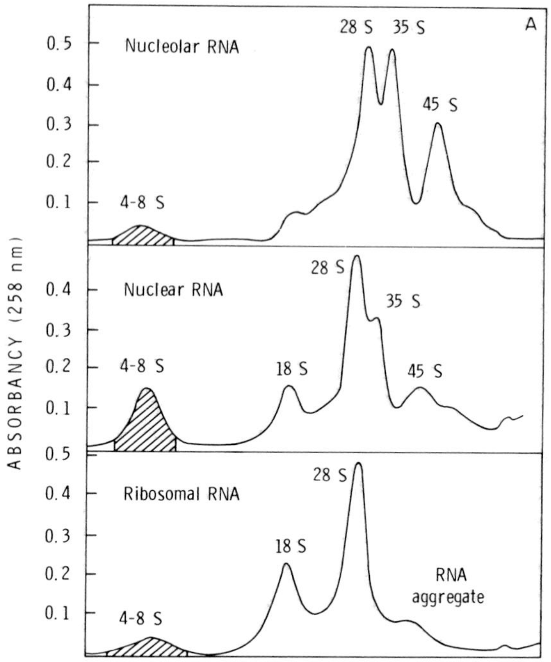

Fig. 2 (A) Sucrose density gradient centrifugation profiles of Novikoff hepatoma nucleolar, nuclear, and ribosomal RNA. RNA was extracted with SDS–phenol at 65°C from each organelle. The 4–8 S RNA (shaded) was used for further studies. (B) Polyacrylamide gel (8%) electrophoretic patterns of 4–8 S RNA from different cell fractions. Slot 1: nuclear 4–8 S RNA; slot 2: ribosomal 4–8 S RNA; slot 3: crude mitochondrial 4–8 S RNA; and slot 4: soluble cytoplasmic sap 4–8 S RNA.

Nucleolar low-molecular-weight RNA has a much simpler pattern on gel electrophoresis than whole nuclear RNA. Like the ribosomal RNA, the nucleolar RNA has only three major electrophoretic components; however, there are differences in these components, i.e., the nucleolus has 4 S RNA, 5 S RNA, and a slower moving U-3 RNA.

Of the six classes of LMW RNA (4 S, 4.5 S, 5 S, U-1, U-2, and U-3) 4 S RNA is present in all of the subcellular fractions. Although more than 99% is in the cytoplasm, it is present in the nucleoplasm and the nucleolus (Table V). In HeLa cells, 4.5 S RNA was found both in the nucleoplasm and nucleolus in the ratio of 2:1 (Table V); in the Novikoff hepatoma cells, 4.5 S RNA was found exclusively in the nucleoplasm. Whether this difference is a real difference in cell types or due to different isolation procedures (Busch, 1967; Penman *et al.*, 1966) is not clear at the present time. The 5 S RNA was distributed throughout the cell organelles, although the greatest concentration was found in the ribosomes and nu-

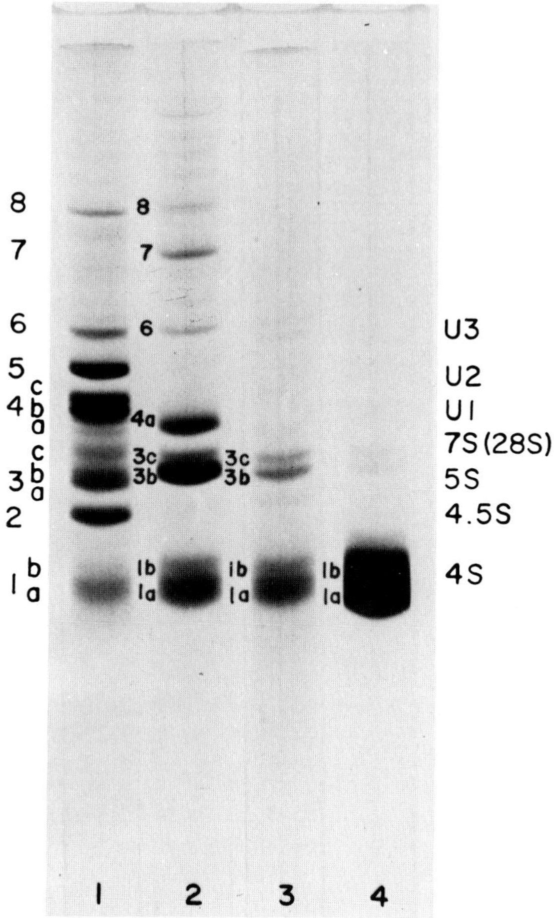

Fig. 2B

cleoli. Nuclear 5 S RNA contains another nuclear-specific 5 S RNA designated as 5 S RNA_{III} (see Section V) which is different from ribosomal 5 S RNA.

The U-1 and U-2 RNA's are not found either in the nucleolus or cytoplasm. These RNA's are exclusively localized in the nucleoplasm. The salt fractionation of the liver nuclei indicates that 65–75% of the total LMWN RNA is associated with the chromatin fraction (Table VI). The U-2 RNA (7 S U-rich RNA) was exclusively in the chromatin fraction in the nucleus. Whether 4.5 S RNA is in the nuclear sap or structural elements of the chromatin or in the nuclear ribonucleoprotein network (Narayan *et al.*, 1966) has not been established. The possibility that 4.5 S RNA is associated with the nuclear ribonucleoprotein network or the inner layer of

TABLE V
Estimated Numbers of Molecules of Various Types of RNA in Nuclei, Nucleoplasm, and Cytoplasm[a]

Species of RNA	Estimated Number of molecules/cell		
	Cytoplasm	Nucleoplasm	Nucleolus
U-3			2×10^5
U-2		4×10^5	1×10^5
U-1		1×10^6	7×10^4
5.5 S (28 S)	5×10^6	5×10^4	1×10^4
5 S	5×10^6	3×10^5	3×10^5
4.5 S		2×10^5	1×10^5
4 S tRNA	1×10^8	2×10^5	
18 S	5×10^6	6×10^4	
28 S	5×10^6	7×10^4	
35 S			4×10^4
45 S			1×10^4

[a] Modified from Weinberg and Penman (1968).

TABLE VI
Distribution of 7 S U-Rich RNA in Rat Liver Nucleus[a]

Fraction	% 4–8 S of total nuclear 4–8 S	% U-rich RNA	% U-rich RNA of total nuclear U-rich RNA
Tris–saline	15–20	10–15	5
Chromatin	65–75	60–65	90
Residue	10–15	—	—
Whole nuclear 4–7 S	100	48	100

[a] From Prestayko and Busch (1968).

the nuclear membrane is suggested by the finding that the chromatin fraction obtained by 2 *M* NaCl extraction after 0.14 *M* NaCl extraction contains only a small amount of 4.5 S RNA. The U-3 and 8 S nucleolar RNA was found to be associated with preribosomal RNP particles isolated from the nucleolus (Busch *et al.*, 1971). Moreover, it was found that part of the U-3 RNA is hydrogen bonded to 28 S nucleolar RNA. The U-3 nucleolar RNA is transitionally associated with the 28 S RNA molecule in the nucleolus since no U-3 RNA is associated with 28 S RNA of the ribosomal subunits or the whole ribosome.

A. *Nucleolar Low-Molecular-Weight RNA*

Sucrose gradient centrifugation techniques for the separation of RNA extracted from isolated nucleoli separated a class of RNA designated as the 4–8 S RNA. This LMW RNA is of interest for two reasons: (1) Its rate of labeling is slow by comparison with the more rapidly sedimenting nucleolar RNA, and (2) its nucleotide composition differs from that of the more rapidly sedimenting RNA. In addition, the nucleolar low-molecular-weight RNA has a different overall nucleotide composition in the fractions obtained from the tumor and the liver nucleoli (Table VII). Accordingly, it was of interest to isolate and fractionate the 4–8 S nucleolar RNA (Nakamura *et al.*, 1968). An initial fractionation was carried out by Sephadex G-100 gel filtration. Later, using polyacrylamide gel electrophoresis, the same result was obtained. This RNA was also separated into three major fractions (4 S, 5 S, and U-3) and two minor fractions [5.5 S (28 S) and 8 S RNA] (Fig. 3).

TABLE VII
Nucleotide Compositions of 4–8 S RNA

Type	AMP	UMP	GMP	CMP	(AMP + UMP)/(GMP + CMP)
Cytoplasmic RNA's					
4 S RNA[a]	18	21	32	27	0.66
5 S RNA[b]	18	23	32	27	0.71
Novikoff nucleolar					
4–7 S RNA[c]	19	26	32	22	0.85
Walker nucleolar					
4–7 S RNA[c]	19	22	34	25	0.70
Liver nucleolar					
4–7 S RNA[d]	20	29	30	21	0.96

[a] Hodnett and Busch (1968); Steele and Busch (1967).
[b] Galibert *et al.* (1966a,b, 1967).
[c] Nakamura *et al.* (1968).
[d] Muramatsu *et al.* (1966).

1. NUCLEOLAR 4 S RNA

The amino acid acceptor activity of nucleolar 4 S RNA was only 60% that of the nuclear 4 S RNA (Table VIII). The major and minor (base-modified) nucleoside content of 4 S RNA's (Table IX) determined by the method of Randerath and Randerath (1971) showed that the compositions

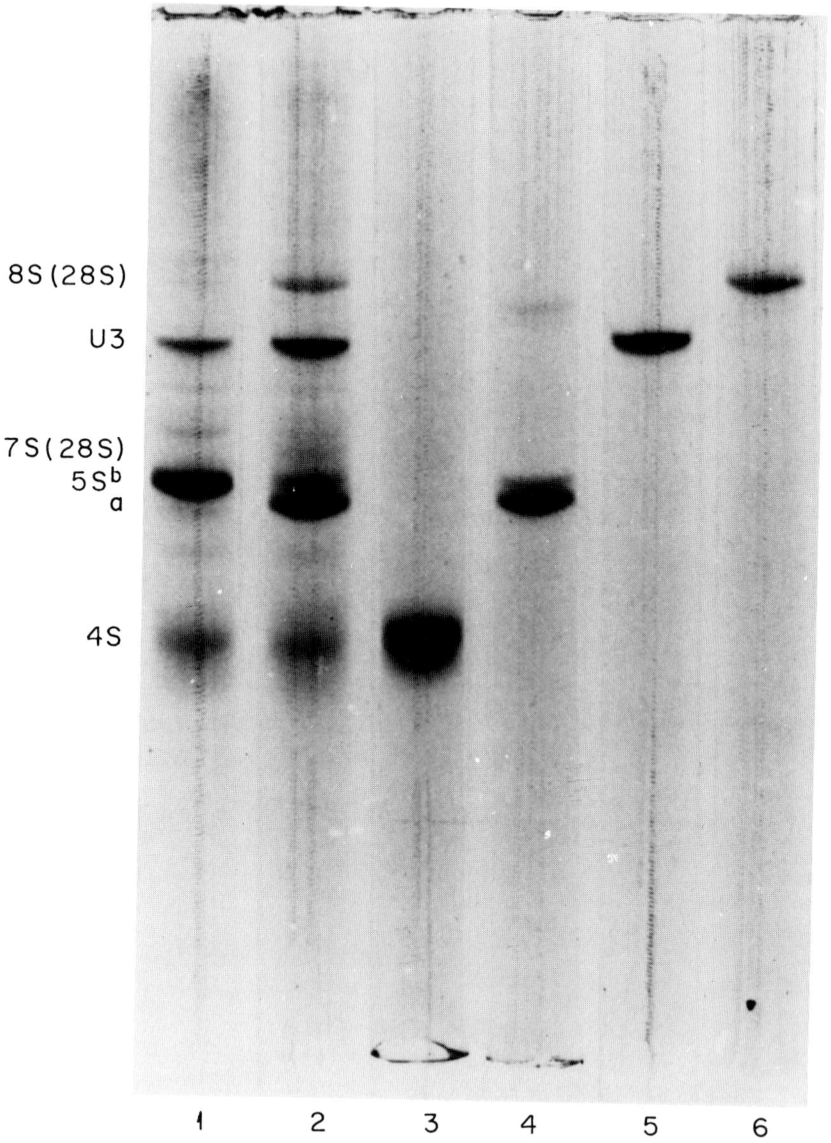

Fig. 3 Polyacrylamide gel (8%) electrophoresis patterns of 4–8 S RNA extracted from Novikoff hepatoma cell nucleoli at 25°C (slot 1) and 65°C (slot 2). Slots 3–6 are reruns of the RNA extracted from the gel bands corresponding to 4 S, 5 S, U-3, and 8 S (28 S), respectively.

TABLE VIII
Amino Acid-Acceptor Capacities of 4 S RNA's[a]

	Acceptance ($\mu\mu$moles/A_{260})	
tRNA	Valine	Leucine
Novikoff cytoplasmic	100	85
Novikoff nuclear	71	42
Novikoff nucleolar	37	28
Rat liver cytoplasmic	63	31

[a] From Ritter and Busch (1971).

TABLE IX
Nucleoside Composition of 4 S RNA's from Nucleoli, Nuclei, Ribosomes, and Cell Sap[a]

	% Total nucleosides			
Nucleoside	Nucleoli	Nuclei	Ribosomes	Cell sap
Major				
U	19[b]	17	17	17
A	20[b]	19	19	19
C	24	26	26	26
G	27	27	27	27
	91	89	89	89
Modified				
m^1A	0.7	0.9	0.9	0.9
m^6A	0.2	0.2	0.2	0.2
m^3C	0.2	0.2	0.2	0.2
m^5C	1.2	1.2	1.3	1.3
m^1G	0.6	0.7	0.6	0.7
m^2G	0.6[b]	1.2	1.1	1.2
m_2^2G	0.4	0.5	0.5	0.5
m^7G	0.6	0.6	0.5	0.6
T	0.4	0.5	0.4	0.5
ψ	2.8	2.9	2.9	3.0
hU	1.6[b]	2.3	2.5	2.5
	9.5	11.1	11.2	11.6

[a] From Reddy *et al.* (1972).
[b] These values are significantly different for the nucleolar 4 S RNA.

of 4 S RNA from nuclei, ribosomes, and the cell sap are essentially the same. Of the total nucleosides of the nuclear, cytoplasmic, and ribosomal 4 S RNA's, 11.3% were modified. The nucleolar 4 S RNA differed from these RNA's in its higher content of U and A. In the nucleolar 4 S RNA, the modified nucleosides only accounted for 9.5% of the total and there was significantly less dihydro-U and N^2-methyl-G. The reason for the differences between the nucleolar 4 S RNA and 4 S RNA's from ribosomes, cell sap, or nuclei is not clear. It is possible that the 4 S RNA from nucleoli contains RNA other than tRNA since nucleolar 4 S RNA has only 60% of the amino acid acceptor activity of nuclear or cytoplasmic 4 S RNA (Table VIII). It is also possible that the modification of precursor nucleosides to dihydro-U and N^2-methyl-G in nucleoli is a slower process compared with other modifications so that nucleolar 4 S RNA contains a greater proportion of immature tRNA. Whether the low content of dihydro-U and N^2-methylguanosine in nucleolar 4 S RNA is related to its low amino acid acceptor activity remains to be answered. There is evidence that in some tRNA's, the recognition site of aminoacyl-tRNA synthetase may reside at the dihydro-U loop and its stem (Dudock *et al.*, 1969, 1970, 1971). Interestingly, N^2-methyl-G appears to be specifically localized to the dihydro-U stem in several tRNA's which were sequenced (Zachau, 1967). However, its low rate of synthesis, as determined by kinetics of labeling with [^{32}P]-labeled inorganic phosphate (^{32}Pi) and its different composition rule out a role for this RNA as a precursor of nuclear and cytoplasmic tRNA. In any event, nucleolar 4 S RNA is different from the 4 S RNA of other parts of the cell with respect to amino acid acceptor activity and content of modified nucleosides.

2. U-3 RNA

Although relatively little is known or understood about the structure of U-3 RNA by comparison with the other LMW RNA's, these molecules, like those in the 8 S (28 S) RNA group, are of special interest because of their specific localization to the nucleolus. In addition, these U-3 RNA's and 8 S RNA's were found to be associated with nucleolar 28 S RNA along with 5.5 S (28 S) RNA. Several molecular species appear to be present in the U-3 and 8 S (28 S) RNA's since the major 3′-terminal nucleosides of U-3 RNA are adenosine and uridine and the major 3′-terminal nucleosides of 8 S (28 S) RNA are cytidine and uridine. However, from the preliminary sequence analysis of U-3 RNA, it appears to have one main chain of U-3 RNA with microheterogeneity of the 3′-end of the molecule.

Interestingly, neither nucleolar 45 S RNA nor 35 S RNA contains the

U-3 or 8 S RNA components that are present in the nucleolar 28 S RNA. However, these RNA species (U-3 and 8 S RNA) do not appear to be precursors of cytoplasmic RNA's; therefore their function must be restricted to the nucleus and, more specifically, to the nucleolus. The finding that 5.5 S (28 S) RNA is present in the molar ratio of approximately 1:1 to nucleolar 28 S RNA suggests that it binds to nucleolar 28 S RNA in a relatively specific site. On the other hand, the molar ratio of 8 S (28 S) and U-3 (28 S) to nucleolar 28 S RNA is only approximately 1:2, suggesting that only some of the nucleolar 28 S RNA's are bound to these molecules. These results suggest that these small molecules may serve a role in processing of nucleolar 28 S RNA and hence are associated only for a limited time. Since the nucleolar 35 S and 45 S RNA's are not bound to any of these molecular species, it seems that, in the course of processing of nucleolar 35 S RNA to nucleolar 28 S RNA, the 5.5 S (28 S), U-3 (28 S), and 8 S (28 S) RNA are added from preexisting pools.

It is also interesting to note that species differences in LMWN RNA mobilities on gel electrophoresis were most significant in the region of U-3 RNA. Rein and Penman (1969) reported that U-3 RNA from 3T3 cells migrated faster than that of HeLa cells. Similar differences were found for U-3 RNA of Ehrlich ascites cells and HeLa cells by Hellung-Larsen *et al.* (1971).

B. Extranucleolar Nuclear Low-Molecular-Weight RNA

1. 4.5 S RNA

Almost all studies on LMWN RNA species (Table II) have shown the presence of a nucleus specific RNA that has an intermediate electrophoretic mobility between 4 S and 5 S RNA. Although the function of this RNA is not known, its localization differs from that of some other nuclear RNA's inasmuch as it is not found in the nucleolus and hence is limited in location to the chromatin or nucleoplasm (Fig. 1). As shown in Fig. 2A, this RNA was not found in any of the cytoplasmic fractions. The purified 4.5 S RNA from Novikoff hepatoma cell nuclei has been separated into three distinct molecular species using DEAE–Sephadex column chromatography (Fig. 4). The same results were obtained by Hellung-Larsen and Frederiksen (1972) with Ehrlich ascites tumor cells. The 4.5 S RNA region was separated into three peaks on 13% gels. However, in KB cells (Larsen *et al.*, 1970, 1972), HeLa cells (Weinberg and Penman, 1968, 1969), and CHO cells this region was a single monodisperse peak even on 15% gel. The LMW RNA from 3T3 cells (Rein and Penman, 1969) showed heterogeneity in this region. The significance of this difference is not

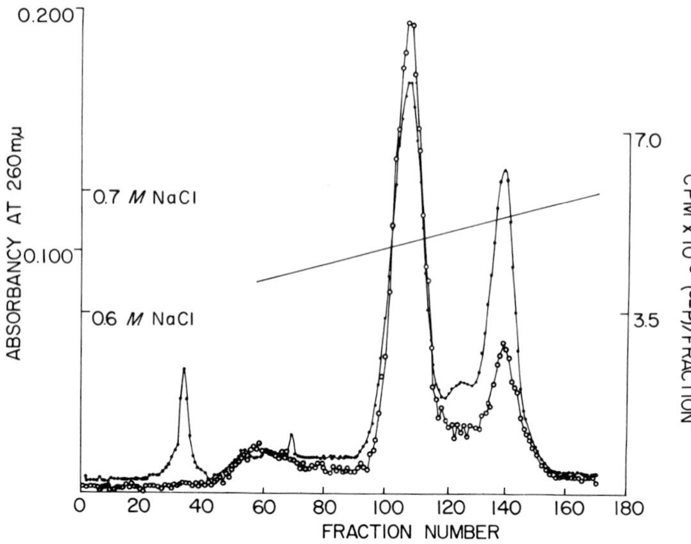

Fig. 4 Chromatography of 4.5 S RNA on DEAE Sephadex A-50. Unlabeled 4.5 S RNA was mixed with [^{32}P]labeled 4.5 S RNA and a linear gradient of 0.6–0.8 *M* NaCl in 0.02 *M* sodium acetate, pH 5.1, was used for the fractionation. Absorbancy, ●——●; radioactivity, ○——○.

known at the present time. The three fractions obtained from DEAE–Sephadex column chromatography migrated as a single band with the same mobilities as the original 4.5 S RNA on 8% analytical gel (Fig. 5). Analysis of nucleotide compositions of 4.5 S RNA (Table X) showed that they contained approximately equal quantities of each of the four nucleotides except for 4.5 S RNA_{III}.

The RNA in DEAE–Sephadex fraction III, or 4.5 S RNA_{III}, has a significantly lower content of CMP than 4.5 S RNA_{I}. In 4.5 S RNA_{I}, the nucleotide ratio AMP + UMP/GMP + CMP was approximately 1.0 and the purine to pyrimidine ratio was also approximately 1.0. In 4.5 S RNA_{III}, the purine to pyrimidine ratio was 1.2–1.3. The 4.5 S RNA_{I} does not have any modified nucleotides and its chain length is 96 (see Section VI). An analogous RNA with a chain length of approximately 100 and no modified nucleotides is present in *Escherichia coli* (Griffin, 1971). This RNA is not a tRNA precursor as shown by the structural differences from any of known tRNA's. However, rapidly labeled RNA migrating similarly to this 4.5 S RNA was reported to be present in the cytoplasm and to be a precursor of tRNA (Bernhardt and Darnell, 1969). The 4.5 S RNA_{II} has not been characterized in detail due to the amount of this RNA present in the cell (Fig. 4). However, by the method of microanalysis of modified nucleoside developed by Randerath and Randerath (1971), this RNA contains no modified nucleosides (Reddy *et al.*, 1972).

In synchronized HeLa cell system in tissue culture, the synthesis of these LMWN RNA's was the same throughout the cell cycle (Weinberg

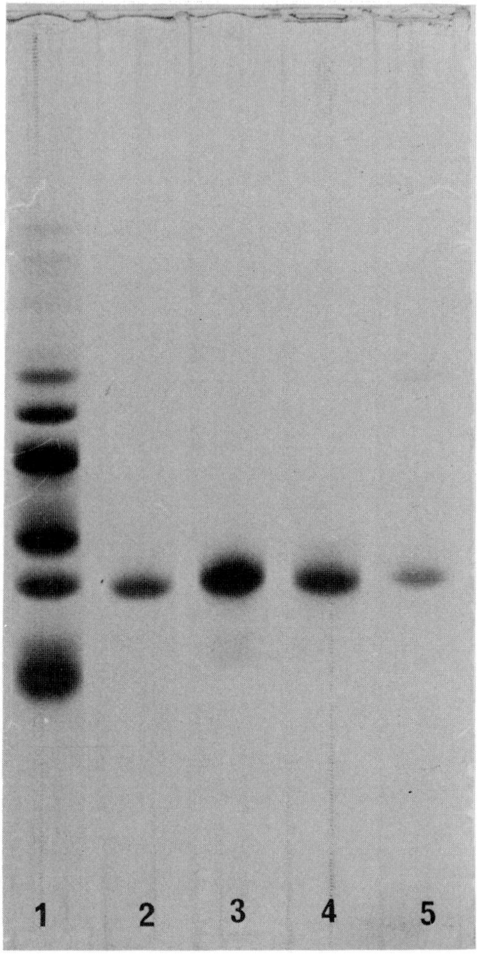

Fig. 5 Polyacrylamide gel (8%, pH 7.2, and 0.5 × 7 cm) electrophoretic pattern of 4.5 S RNA fractions obtained from DEAE–Sephadex A-50 column chromatography. Slot 1: total 4–8 S nuclear RNA; slot 2: purified 4.5 S RNA applied on DEAE–Sephadex; slot 3: 4.5 S RNA_{I} (peak 1, Fig. 4); slot 4: 4.5 S RNA_{II} (peak II, Fig. 4); and slot 5: 4.5 S RNA_{III} (peak III, Fig. 4). The direction of migration is from top to bottom.

TABLE X
Nucleoside Composition of LMW RNA's from Nuclei, Nucleoli, and Ribosomes

Type	U	A	C	G	ψ	m^6A	m^2G	$m^{2,2,7}G$	Alkali-stable oligonucleotides
Nuclei									
4.5 S_I	24	25	25	26	0.1				
4.5 S_{II}	26	25	25	24					
4.5 S_{III}	23	25	21	26	2.9	0.8	0.8		AmpAp GmpAp GmpGp
5 S_I	24	19	25	31	0.4				
5 S_{II}	24	20	26	30					
5 S_{III}	31	24	20	23	1.7				UmpUp,GmpCp
U-1A	25	19	27	30					
U-2	26	22	24	23	5.5			0.5	CmpUp,Cmpψp XmpGp,GmpAp UmpAp,GmpAp GmpGmpCp $m_3^{2,2,7}G$,AmpUmpCp
U-3	25	20	25	29	0.9			0.4	
Nucleoli									
5 S	24	20	27	30					
5.5 S	23	21	27	30	0.4				
U-3	24	20	25	29	1.1			0.4	
Ribosomes									
5 S	24	20	27	30					
5.5 S	23	20	27	30	0.4				

and Penman, 1969). Most of this RNA has a half-life of 24 hr or longer. This indicates that this RNA might be synthesized only once in the cell cycle. Similarities in the 5′- and 3′-ends of this molecule to those of λ and ϕ80 LMW RNA which is required for the DNA synthesis suggest that 4.5 S RNA_I may have a similar function in Novikoff hepatoma cells (see p. 202).

2. 4.5 S RNA_{III}

The 4.5 S RNA_{III} is another extranucleolar nuclear species not found in the cytoplasm. The 4.5 S RNA_{III} is a unique RNA species with adenosine-3′,5′-diphosphate (pAp) as the 5′-terminal and 2′-*O*-methyluridine as the 3′-terminal nucleoside. It has a very low T_m of 37.5° compared to 65° for the 4.5 S RNA_I. As shown in Table X, this RNA has a higher A and a lower C content than 4.5 S RNA_I. Analysis of the alkali-resistant dinucleotides showed that 4.5 S RNA_{III} contains four alkali-resistant dinucleotides, i.e., Am-Ap, Gm-Ap, and 2 Gm-Gp. Analysis of base-modified nucleosides of 4.5 S RNA_{III} (Fig. 6; Table X) indicates that this RNA contains ψ, N^2-methyl-G, and N^6-methyl-A, which account for 2.9, 0.8, and 0.8% of the total nucleosides, respectively. Since the approximate chain length of this RNA is 93 (El-Khatib *et al.*, 1970), the number of residues

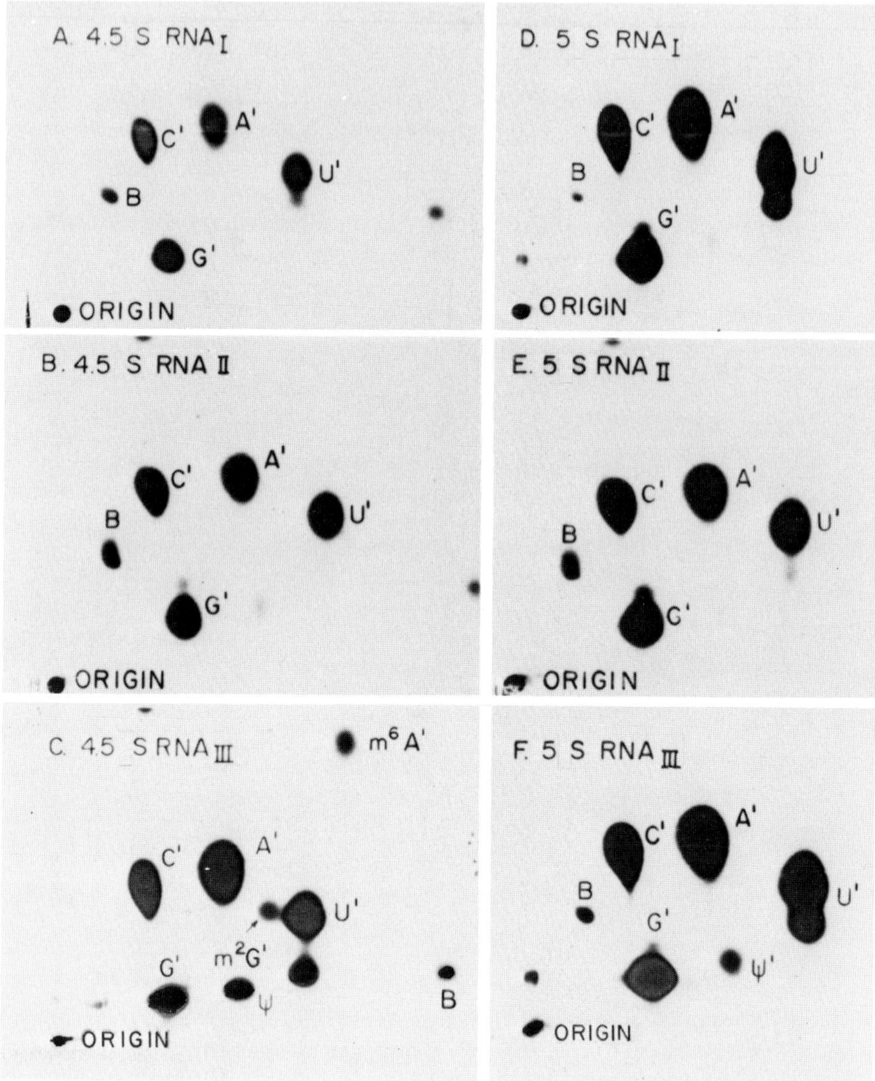

Fig. 6 Fluorographs of two-dimensional thin layer chromatography. Separation of ^3H nucleoside derivatives from LMWN RNA's. A, 4.5 S RNA_I; B, 4.5 S RNA_{II}; C, 4.5 S RNA_{III}; D, 5 S RNA_I; E, 5 S RNA_{II}; and F, 5 S RNA_{III} (see footnote, p. 152, for definitions of symbols appearing on figure).

of ψ, N^2-methylguanosine, and N^6-methyladenosine is 3, 1, and 1, respectively. The presence of N^6-methyladenosine is unique to this RNA since no other RNA of mammalian origin, including tRNA (Randerath, 1971; Reddy *et al.*, 1972) contains this N^6-methyladenosine. As indicated earlier, this is the only RNA which has a 2′-*O*-methylated 3′-end. As in the case of 4.5 S RNA_I, its function is still unknown.

3. 5 S RNA

Three types of nuclear 5 S RNA have been found in Novikoff hepatoma cells (Ro-Choi *et al.*, 1971) and HeLa cells (Weinberg and Penman, 1969). These include the two ribosomal isomers that are rapidly synthesized and a nucleus-specific species of 5 S RNA, which is best isolated by chromatography of whole nuclear 5 S RNA on DEAE-Sephadex columns (Fig. 7).

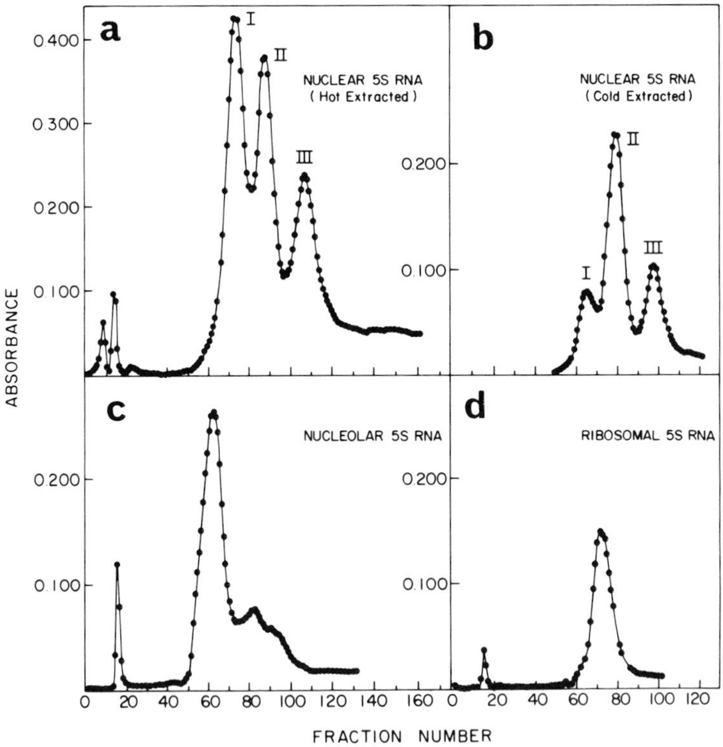

Fig. 7 Chromatographic pattern of 5 S RNA's from various fractions on DEAE–Sephadex A-50. A linear gradient of 0.6–0.8 *M* NaCl in 0.02 *M* sodium acetate, pH 5.1, was used for the fractionation. (a) Nuclear 5 S RNA extracted with SDS phenol at 65°C; (b) nuclear 5 S RNA extracted with SDS phenol at 23°C; (c) nucleolar 5 S RNA extracted with SDS phenol at 65°C; (d) ribosomal 5 S RNA extracted with SDS phenol at 65°C.

In comparison of the three types of 5 S RNA from Novikoff hepatoma cell nuclei with nucleolar 5 S RNA and 5 S rRNA, nuclear 5 S RNA_{I} and 5 S RNA_{II} were found to be conformational isomers of 5 S rRNA. This was shown by nucleotide composition (Table X) and "fingerprinting" RNase digestion products (see Section VI). When the RNA was extracted by the hot SDS–phenol procedure, 38, 34, and 28% of the nuclear 5 S RNA were found in the 5 S RNA_{I}, 5 S RNA_{II}, and 5 S RNA_{III}, respectively (Fig. 7). The proportion of 5 S RNA_{I} was markedly decreased when the RNA was extracted at 23°C (Fig. 7). When nucleolar or ribosomal 5 S RNA was chromatographed in the same way, only one main peak was found. The nucleotide compositions of nuclear 5 S RNA_{I} and 5 S RNA_{II}, as well as 5 S rRNA and nucleolar 5 S RNA, are virtually identical (Table X). The composition of nuclear 5 S RNA_{III} differs in its higher AMP and UMP content and lower CMP content. The AMP + UMP/GMP + CMP ratios for the ribosomal-type 5 S RNA's are 0.74–0.80 but that of 5 S RNA_{III} is 1.24.

Recently, a precursor for bacterial 5 S RNA was found by Jordan *et al.* (1970) and was shown to be different at the 5′-end by 2–3 nucleotides. However, its characteristics differ significantly from those of the 5 S RNA_{III} of mammalian cells. The 5 S RNA_{I} and 5 S RNA_{II} do not contain any modified nucleotides which is in agreement with those reported for KB cell ribosomal 5 S RNA. However, 5 S RNA_{III} contains two alkali-stable dinucleotides (Um—Up and Gm—Cp) and two ψ residues. As shown by the absence of this 5 S RNA_{III} in the nucleolus and cytoplasm, this is also a unique nucleoplasmic RNA. The function of this RNA is also unknown. The reason why nucleolar and ribosomal 5 S RNA have only one peak (Fig. 7) compared to ribosomal-type 5 S RNA (5 S $RNA_{I,II}$) peaks in nuclear 5 S RNA is uncertain. Although 5 S RNA has been reported to exist in two conformational states that are interconvertible by heat (Aubert *et al.*, 1968; Jordan *et al.*, 1970), the finding of only one 5 S peak in the nucleolar and ribosomal RNA fractions suggests that there may be other restrictions on conformation in these molecules in these sites. Whether this is related to the maturation process of ribosomes is unclear at the present time.

4. U-1 NUCLEAR RNA

Among the earliest findings in studies on the LMWN RNA's was the presence of a dense, double-banded zone, referred to as U-1 RNA, that has an electrophoretic mobility lower than that of 5 S RNA (Fig. 2). Although the two bands in the U-1 RNA were not adequately separable by gel electrophoresis or chromatography on DEAE–Sephadex, recently it became possible to isolate one of the two bands referred to as U-1a by chromatography on benzoylated DEAE (BD)–cellulose. This band has a 3′-terminal guanosine (>90%) which is unique to the U-1a RNA. Another interesting feature of this molecule is the apparent lack of a 5′-terminal phosphate since no pNp 5′-terminal nucleoside diphosphate has been found on alkaline hydrolysis. The other 3′-terminal nucleosides, including adenosine and uridine which are present in the U-1 RNA, are not found in the purified U-1a band. Although it seems probable that there is a minimum number of four bands in the U-1 region, two are minor bands of which one is the 5.5 S (28 S) RNA separated by heating from the 28 S RNA; the other minor band has not been characterized. These U-1a and U-1b bands contain isomers of one nucleoplasmic RNA; it is not found in the nucleolus or cytoplasm. The tentative structure of this RNA (Ro-Choi *et al.*, 1974) is

($m_3^{2,2,7}$G,Am,Um)—A—C—U—U—A—C—C—U—G—G—C—A—G—G—G—
A—G—A—U—A—C—C—A—U—G—A—U—C—A—C—G—A—A—G—G—
U—G—U—U—U—U—C—U—C—U—C—C—A—G—G—G—C—G—U—G—
G—U—C—U—A—U—C—C—A—U—U—G—A—G—G—C—G—C—Am—
C—U—C—C—G—U—G—G—A—U—G—C—U—G—A—C—C—C—C—U—
G—C—G—A—U—U—U—C—C—U—C—C—A—A—A—U—G—C—G—G—
G—A—A—A—C—U—C—G—A—C—U—G—C—A—U—A—A—U—U—U—
G—U—G—G—U—A—G—U—G—C—G—G—G—G—A—C—U—G—U—U—
C—G—C—G—C—U—C—C—U—C—U—C—G_{OH}.

5. U-2 NUCLEAR RNA

The U-2 RNA fraction was the first LMWN RNA that was obtained in a high state of purity by preparative gel electrophoresis (Moriyama *et al.*, 1969; Hodnett and Busch, 1968). Like U-1 RNA, U-2 RNA is localized to the extranucleolar portion of the nucleus, and its function is also uncertain. According to the data of Clason and Burdon (1969), it is probably only synthesized once during the cell cycle and hence may serve as a structural or initiating element in gene readouts. The nucleotide compositions of these RNA's in the liver and tumor were virtually identical (Busch and Smetana, 1970). Although both U-1 and U-2 RNA have high uridylic acid contents, their nucleotide compositions differ, as shown in Table X.

An unusual feature of U-2 RNA is that it contains 10 2′-O-methylated nucleotides; 6 are in the alkalai-stable dinucleotides Cm—Up, Cm—ψp, m^6Am—Gp, Gm—Gp, Um—Ap, and Gm-Ap. Two are in the alkali-stable trinucleotide Gm—Gm—Cp and the others are in the 5′-terminal tetranucleotide ($m_3^{2,2,7}$G, Am—Um—Cp). U-2 RNA contains 10 ψ-residues and one residue of $m_3^{2,2,7}$G (trimethyl-G) which was not found in bacterial or mammalian tRNA. Fluorographs of the separation of ^{3}H derivatives of nucleosides obtained from nuclear U-2 and nucleolar U-3 RNA are shown in Fig. 8. An unusual spot which was not present in 4 S RNA but was found in both U-2 and U-3 RNA was identified as a trialcohol derivative of N^2,N^2-dimethyl-7-methylguanosine. Trimethylguanosine, first found in LMW RNA from Chinese hamster ovary cells by Zapisek *et al.* (1969), was synthesized according to the method described by Saponara and Enger (1969), who defined the structure of this nucleoside; it was labeled according to the method of Randerath and Randerath (1971). Figure 8F shows the fluorograph of the position of this labeled trimethyl-G derivative, as well as its purity.

Analysis of the 3′-terminal nucleosides after alkaline hydrolysis revealed that this RNA has 62% A and 28% C (Table II). After T_1 RNase digestion of [^{3}H]labeled U-2 RNA, two oligonucleotides were released, one with a cytidine and the other with an adenosine terminal. These two terminal oligonucleotides differ in length from one another by only one nucleotide. The sequence analysis of U-2 RNA (see Section VI) confirmed that this RNA has one common 5′-end and two 3′-ends which differ by one nucleotide.

C. *Low-Molecular-Weight RNA Found in All Cell Fractions*

1. 4 S RNA

The 4 S RNA which is generally referred to as transfer RNA is present in all types of cells and even in viruses. It is the most numerous and widely distributed low-molecular-weight RNA species. Although more than 99% is in the soluble sap of the cell, a considerable amount of tRNA was found in the nucleolus, nucleus, and ribosomes. Much of the current research on nucleic acid structure has centered on tRNA. This is understandable since the tRNA molecule is relatively small and individual molecular species can be readily isolated.

a. Nuclear 4 S RNA. The 4 S RNA of the nucleus is primarily amino acid acceptor or tRNA. Although there has been some question as to whether this RNA might differ significantly from other tRNA of the cell, studies by Ritter and Busch (1971) have not revealed any notable difference. Cochromatography on hydroxylapatite, BD–cellulose, and DEAE–Sephadex was employed for comparison of nuclear and cytoplasmic Val- and Leu-tRNA's (Fig. 9). Apparently, there is no nucleus-specific Val- or Leu-tRNA in the Novikoff hepatoma cells.

Whether nuclear 4 S RNA participates in protein synthesis in the nucleus by the same mechanism as cytoplasmic tRNA in the ribosomes is not clear at the present time. Since no authentic protein-synthesizing ma-

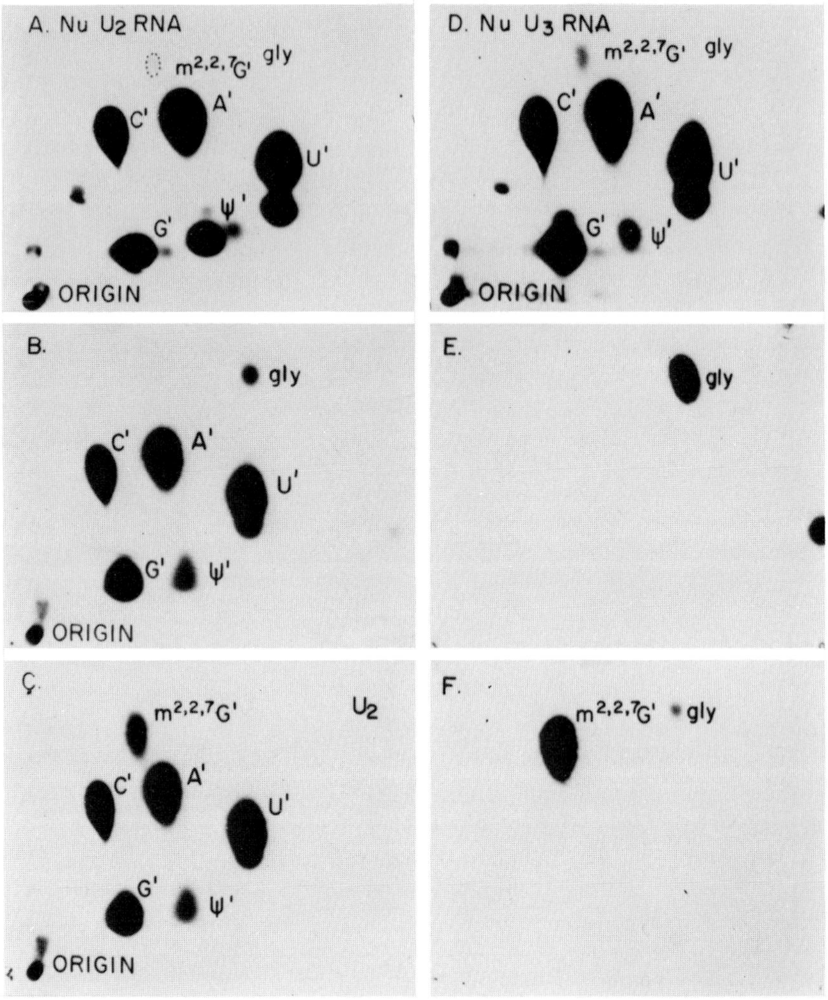

Fig. 8 Fluorographs of two-dimensional, thin-layer chromatography. Separation of ^{3}H nucleoside derivatives from nuclear U-2 RNA and nucleolar U-3 RNA. A, nuclear U-2 RNA; B, sample A heated at 95° for 2 hr in 10% piperidine; C, sample A cochromatographed with [^{3}H]trialcohol derivative of N^2,N^2-dimethyl-7-methyl guanosine; D, nucleolar U-3 RNA; E, ^{3}H derivative of N^2,N^2-dimethyl-7-methyl guanosine heated at 95° for hr in 10% piperidine; and F, ^{3}H derivative of N^2,N^2-dimethyl-7-methyl guanosine.

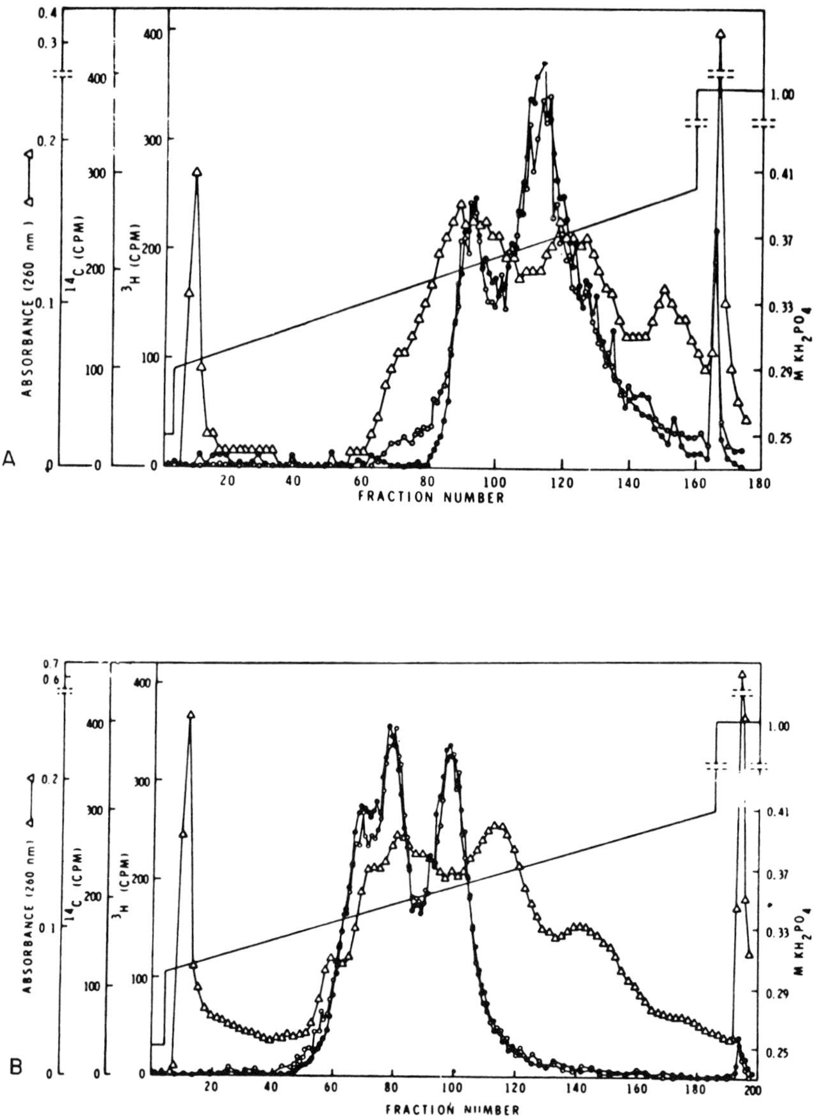

Fig. 9 (A) Cochromatography of Novikoff hepatoma cytoplasmic [^{14}C]Val-tRNA (●——●) and Novikoff hepatoma nuclear [^{3}H]Val-tRNA (○——○) on a hydroxylapatite column. (B) Cochromatography of Novikoff hepatoma cytoplasmic [^{14}C]Leu-tRNA (●——●) and Novikoff hepatoma nuclear [^{3}H]Leu-tRNA (○——○) on a hydroxylapatite column. (From Ritter and Busch, 1971.)

chinery is found in the nucleus, it is doubtful that nuclear 4 S RNA acts as aminoacyl transfer RNA. The nuclear function of these tRNA's remain to be answered. Pre-tRNA was reported to exist in the cytoplasm by Bernhardt and Darnell (1969), but no such precursors have been found in the nucleus. Since the amino acid-accepting activity of nuclear 4 S RNA was essentially the same as that of the cytoplasmic tRNA, it appears that the newly synthesized tRNA is converted to an active form as soon as it is formed or in close proximity to the site of its formation. The proximity of methylation and modification reactions to biosynthetic sites has also been found for high-molecular-weight nucleolar precursors of ribosomal RNA (Muramatsu and Fujisawa, 1968; Choi and Busch, 1970).

b. Cytoplasmic 4 S RNA. Cytoplasmic 4 S RNA's are essentially all transfer RNA's. The amino acid acceptor activity and content of modified nucleosides of Novikoff hepatoma were essentially the same as reported values for tRNA from human tissue and tRNA from rat liver and brain (Randerath, 1971; Randerath and Randerath, 1971). However, some differences have been found for the major and minor nucleoside compositions of 4 S RNA from mitochondria (Randerath *et al.*, 1973). Although its amino acid acceptor activity is not known, this RNA has a high A + U/G + C ratio with an unusually high A content and fewer modified nucleosides (hypomethylation).* No differences were found between ribosome-bound 4 S RNA and soluble sap 4 S (Table IX). The function of cytoplasmic tRNA is its amino acid transfer activity for the protein synthesis on ribosomes. Whether tRNA is involved in the regulatory processes in the cytoplasm is not clear at the present time. Although the concept of transcriptional control of Jacob and Monod (1961) for genetic regulation is well accepted, the alternative model, suggested by several workers and summarized by Strehler *et al.* (1967), postulated that selective translation of messages is a guiding force in differential protein synthesis. Gonano (1967) reported that the tRNA molecule responsible for inserting the serine residue into position 3 of the α-chain of hemoglobin was not the same molecule that inserted serine into either position 49 or 52 of the chain. Similarly, Weisblum *et al.* (1967) found that $tRNA_I^{arg}$ and $tRNA_{II}^{arg}$, purified from yeast tRNA, inserted arginine residues into different positions of the α-chain of the hemoglobin. Weisblum *et al.* (1965) demonstrated that two species of *E. coli* $tRNA^{leu}$ distributed leucine differently into various peptides of the α-chain of rabbit hemoglobin. These results suggest that a relatively rare molecular species of tRNA responsible for inserting amino acid into one location of a given protein molecule could become the rate-limiting factor in the synthesis of this molecule.

* In addition, two unknown nucleosides were found in mitochondrial 4 S RNA which was not present in cytoplasmic 4 S RNA (Table XI).

TABLE XI
Comparison of Base Composition (Mole%) of Mitochondrial and Cytoplasmic 4 S RNA from Rat Liver[a]

Nucleoside	Mitochondrial 4 S RNA	Cytoplasmic 4 S RNA
U	20.89	13.80
A	26.58	17.15
C	22.39	25.93
G	20.80	27.74
m^1A	1.01	1.38
m^5U	0.21	0.60
hU	1.21	2.82
m_2^2G	0.37	0.61
m^2G	0.97	1.37
m^1G	0.60	0.79
m^3C	0.16	0.34
m^5C	1.05	2.13
ψ	2.82	3.83
I	0.17	0.30
m^7G	0.37	0.79
m^6A	0.18	0.45
X_1[b]	0.11	Absent
X_2[b]	0.13	Absent

[a] Unpublished data of K. Randerath, E. Randerath, and L. S. Y. Chia.

[b] One of two unidentified modified nucleosides specific for mitochondrial tRNA. These compounds are absent from cytoplasmic and nuclear tRNA.

This form of control would occur even though an excess was available of other tRNA molecules capable of transferring amino acid (Hampel *et al.*, 1972).

2. CYTOPLASMIC 5 S RNA

The 5 S ribosomal RNA has been shown to be associated with large ribosomal subunits from all bacterial (Comb *et al.*, 1965; Elson, 1961; Galibert *et al.*, 1966a, b; Rosset and Monier, 1963; Schleich and Goldstein, 1966), animal (Bachvaroff and Tongur, 1966; Brown and Littna, 1966; Comb and Katz, 1964; Comb *et al.*, 1965; Galibert *et al.*, 1965, 1966a, b; Knight and Darnell, 1967; Watson and Ralph, 1967; Ro-Choi *et al.*, 1970), plant (Chakravorty, 1969; Li and Fox, 1969), and lower eukaryotic cells (Comb *et al.*, 1965; Marcot-Queiroz *et al.*, 1965) examined so far. The 5S rRNA from three cell types, *E. coli* (Brownlee *et al.*, 1968), *Pseudomonas fluorescens* (DuBuy and Weissman, 1971) and KB cells (Forget and Weissman, 1968, 1969) have been completely sequenced (see Section VI). It is present in 1:1 molar ratio to large ribosomal sub-unit RNA (28 S or 23 S RNA).

Although it is clear that 5 S RNA is a component of ribosomal RNA and its synthesis is coordinated with 18 S and 28 S rRNA, it has been shown that 5 S RNA cistrons are located at different sites from 18 S and 28 S RNA cistrons. In anucleolate *Xenopus* embryos which do have 5 S RNA genes (Brown and Weber, 1968), there is no 5 S RNA synthesis. There is coordinated accelerated synthesis of all three RNA species in amphibian oocytes, where no amplification of the 5 S genes is detectable (Brown and Dawid, 1968). In addition, RNA–DNA hybridization experiments along with labeling kinetics have shown that there are different numbers of cistrons for 5 S RNA compared to those of high-molecular-weight rRNA (Table XII). These results oppose the existence of a common precursor of 5 S RNA and high-molecular-weight RNA. The existence of a pool of 5 S RNA in *E. coli* (Morell and Marmur, 1968) as well as in HeLa cells (shown by kinetics of labeling) substantiates the absence of a common precursor for the 5 S RNA and high-molecular-weight rRNA. Recently, Brown *et al.* (1971) purified the 5 S DNA from *Xenopus laevis* and have shown that this DNA has a very low GC (33–35%) content; each repeat contains about 500,000 daltons. If one gene for 5 S RNA were present in each repeat, it should comprise about 17% of 5 S DNA; 83% of each repeating unit is considered to be "spacer" DNA.

TABLE XII
Genome Redundancy for Ribosomal RNA Components and 4 S RNA[a]

Organism	18 S + 28 S RNA	5 S rRNA	tRNA
HeLa cells	280	2000	1260
Rabbit	250	—	—
Chick	100	—	—
Rat liver	300	—	—
Novikoff hepatoma cell	300	—	—
Xenopus laevis	450–610	9000–24,000	6500–1150
Drosophila melanogaster	130	—	860
Tobacco	3450	—	—
Pea	4500	—	—
Neurospora crassa	125	—	—
Bacillus subtilis	8–9	4–5	42
E. coli	5–6	11	50–60

[a] The values are the number of cistrons per genome (from Attardi and Amaldi, 1970).

Although its specific function remains unknown, its removal results in loss of ribosomal activity (Aubert *et al.*, 1967). Three possible functions have been postulated: (1) Formation of peptidyl 5 S RNA in the peptide elongation reaction in protein synthesis; this was postulated on the basis of space-filling model building studies without experimental evidence. (2) Function as a tRNA binding site on ribosomes; this was based on the presence of sequences complementary to the "universal" tRNA G—T—ψ—C—G sequence which has been considered as a possible ribosome-binding site. This possibility was suggested also by Jordan (1971) from studies on 5 S RNA conformation by partial ribonuclease hydrolysis. It was found that the sequence of the most accessible region by RNase is complementary to the universal transfer RNA G—T—ψ—C—G sequence. (3) An essential structural component which provides crucial protein-binding sites. The modification of the 3′-terminal of 5 S RNA by sodium periodate oxidation and reduction by potassium borohydride had no effect on its function. In addition, reconstituted 50 S subunits in the absence of 5 S RNA greatly reduced activity and lacked 4 out of 39 possible 50 S ribosomal proteins that are present in particles containing 5 S RNA. Recently, Blobel (1971) was able to isolate a single 5 S RNA-binding protein from rat liver and rabbit reticulocyte ribosomes. This protein has a molecular weight of about 35,000.

3. 5.5 S RNA

In animal cells a 28 S rRNA-associated, low-molecular-weight RNA is found. This component has also been designated as 7 S RNA (Knight and Darnell, 1967) or 5.5 S RNA (Weinberg and Penman, 1968). Such an RNA component has not been found in *E. coli.* However, the postribosomal supernatant of *E. coli* contains a 6 S RNA which recently has been sequenced (Brownlee, 1971). No function has been suggested so far for the 5.5 S RNA and the 6 S RNA. The 5.5 S RNA (7 S RNA) in mammalian cells was found to be associated with 28 S ribosomal RNA by hydrogen bonding. This RNA was not released from 28 S rRNA when RNA was extracted at 23°C or lower. However, heating released this RNA from 28 S rRNA. RNA with the same mobility was also found to be present in nuclear and nucleolar low-molecular-weight RNA from Novikoff hepatoma cells (Ro-Choi *et al.*, 1970) and HeLa cells (Weinberg and Penman, 1968). Moreover, this RNA was found to be associated with nucleolar 28 S RNA when RNA was extracted at room temperature (Prestayko *et al.*, 1970). The 35 S and 45 S nucleolar precursors of ribosomal RNA were not found to be associated with 5.5 S RNA. Although mechanisms of association of this RNA with 28 S RNA has been proposed, the precise role of this RNA in relation to maturation of ribosomes is uncertain at the present time. Interestingly, Hellung-Larsen and Frederiksen (1972) have reported the presence of an LMW RNA associated with 25 S rRNA in *Tetrahymena pyriformis*. A component designated as "T_0" was found to be associated with ribosomal RNA and liberated by phenol extraction at 55°. The RNA T_0 is associated with 25 S rRNA and constitutes about 1.5% of 25 S rRNA. The release of T_0 from 25 S RNA by heat also caused a conversion of 25 S to 17 S RNA and suggested that the 25 S RNA is composed *in vivo* of two 17 S RNA molecules. It is interesting that the 25 S-associated T_0 component of *Tetrahymena* migrated slower (about 6 S) on gels than does 28 S RNA-associated 5.5 S RNA of Ehrlich ascites cells; therefore it probably has a larger molecular weight whereas the ribosomal RNA components of 25 S RNA, 17 S RNA, and "4.9 S" RNA migrate faster than the corresponding components from Ehrlich ascites cells and probably have lower molecular weights.

IV. Low-Molecular-Weight RNA in Various Tissues

Up to 10 different LMWN RNA components in addition to tRNA and 5 S RNA have been found in Novikoff hepatoma cells of rats, some predominantly in the nucleolus and others in the nucleoplasmic fraction (Ro-Choi *et al.*, 1970). This specific class of LMWN RNA has been demonstrated in various tissues including human cells (HeLa cells, KB cells lymphocytes), Chinese hamster cells, 3T3 cells, rat liver, Yoshida and Ehrlich ascites cells, L cells, sea urchin eggs, *Tetrahymena pyriformis*, and *Xenopus laevis* (Enger and Walters, 1970; Hellung-Larsen *et al.*, 1971; Hellung-Larsen and Frederiksen, 1972; Rein and Penman, 1969; Weinberg and Penman, 1968).

The gel electrophoretic patterns of LMWN RNA components of mammalian cell origin (HeLa cells, Novikoff hepatoma cells, KB cells, Chinese hamster cells, Ehrlich and Yoshida ascites cells, and human lymphocytes) appear to be nearly identical, although minor differences between some species have been reported. Using a method of double-labeling of LMWN RNA, it was found that the precise mobilities of three of the species—U-1(D), U-2(C), and U-3(A)—are different in mouse cells (3T3 and L cells) and human cells (HeLa cells and WI_{38} cells) (Rein and Penman, 1969). In comparison of LMWN RNA from Ehrlich, Yoshida, L5178Y, and HeLa cells, it was found that the gel electrophoretic pattern of Ehrlich, Yoshida, and L5178Y ascites cells were identical whereas there were differences in the components of U-3(A) and 4.5 S(H) RNA between HeLa cells and Ehrlich ascites cells. Three components in 4.5 S RNA were demonstrated in the Novikoff hepatoma cells and Ehrlich ascites cells whereas in HeLa cells there was only one component even in 15% gels (Hellung-Larsen *et al.* 1971; Rein and Penman, 1969; Ro-Choi *et al.* 1970).

These LMW RNA's have been found in cells other than mammalian cells (Table XIII). Low-molecular-weight RNA's associated with various viruses were also reported (Kaper and West, 1972; Larsen *et al.*, 1972; Lebowitz *et al.*, 1971; Marcaud *et al.*, 1971; Semancik and Weathers, 1972). One associated with adenovirus 2-infected KB cells (VA RNA), two λ phage LMW RNA's and one ϕ80-associated, LMW RNA have been sequenced (see Section VI). Although the functions of these viral low-molecular-weight RNA's are not completely understood, some plant low-molecular-weight RNA (Semancik and Weathers, 1972) has been shown to be infectious.

In the three "self-terminating" RNA species in phage (λ phage or ϕ80)-

TABLE XIII
Existence of LMW RNA Other Than tRNA

Species	Nuclei	Ribosomes	Soluble sap	Reference
E. coli	—	5 S RNA	4.5 S RNA 6 S RNA	Hindley (1967) Brownlee (1971)
Tetrahymena pyriformis	5 S RNA T_1, T_2, T_3, and T_4 RNA	4.9 S RNA 6 S RNA(To)	5 S RNA	Hellung-Larsen and Frederiksen (1972)
Xenopus laevis	H, 5 S RNA (G + G′) D,C,B, and A[a]	5 S RNA 5.5 S RNA	5 S RNA	Rein and Penman (1969)
Mammalian cells[a]				
Viruses				
Adenovirus 2 (KB cell); VA-RNA (6 S RNA)				Ohe and Weissman (1970, 1971)
Avian leukosis virus (9 S RNA)				Obara *et al.* (1971)
Rous sarcoma virus (7 S RNA)				Bishop *et al.* (1970)
Visna virus; 5–7 S RNA				Harter *et al.* (1971)
CEV (exocortis disease virus); 7–10 S RNA				Semancik and Weathers (1972)
PSTV (potato spindle tuber viroid); 4–6 S RNA				Diener and Smith (1971)
Qβ RNA; 6 S RNA				Kacian *et al.* (1971)
λ phage (*E. coli*); 6 S RNA				Lebowitz *et al.* (1971)
λ phage (*E. coli*); 4 S RNA				Dahlberg and Blattner (1973)
ϕ80 (*E. coli*) smaller than 4 S RNA				Pieczenik *et al.* (1972)

[a] For nomenclature, see Table II.

infected *E. coli* that have been sequenced, all have common 5′- and 3′-ends which are pppGp and —U—U—U—U—(U—U—A—A), respectively (Lebowitz *et al.*, 1971; Pieczenik *et al.*, 1972; Dahlberg and Blattner, 1973). The RNA species sequenced by Dahlberg and Blattner (1973) has been suggested to function as a primer for DNA synthesis (Hayes and Szybalski, 1973).

V. General Procedure for the Purification of Low-Molecular-Weight RNA

A. *Preparation of RNA*

Tissue organelles are prepared according to the method described previously. The nucleoli can be isolated by three different procedures.

1. Sucrose-Ca^{++}–sonication method (Busch, 1967). This method is the most commonly used method and was developed in this laboratory. The method involves isolation of nuclei by modified Chauveau method which employs 2.0–2.4 *M* sucrose containing 3.3 m*M* Ca^{++}. The isolated nuclei are suspended in 0.34 *M* sucrose and sonicated for 45–60 sec. The sonicate was then overlayered on 0.88 *M* sucrose and centrifuged to obtain clean nucleoli. Sonication can be replaced by compression and decompression method.

2. Citric acid–sonication method (Ro-Choi *et al.*, 1973b). Recently developed method which could be used for the chemical analysis of nucleoli. The methods involve isolation of nuclei by citric acid method and the nuclei were suspended in 0.02 M sodium acetate at pH 5.1. The suspension was then sonicated and nucleoli were obtained by differential centrifugation. The nucleoli isolated by this method do not have RNase activity, thus permitting a better chemical analysis of nucleoli. The nuclei and other tissue organelles (ribosomes, mitochondria and cell sap) are isolated by the method reported ("Methods in Enzymology," Vol. 12, 1967).

3. Salt fractionation (Penman *et al.*, 1966). This method consists of high salt extraction of nuclei with concomitant treatment with DNase. The nuclei isolated by detergent method was suspended in 0.5 *M* NaCl and treated with DNase. This was then centrifuged to obtain "nucleolar" fraction and nucleoplasm.

The extraction of RNA is best done by the SDS–phenol method described previously. The 0.3% SDS was used either at 65° or 23°C. The RNA extracted by this method includes over 95% of RNA present in the tissue organelles; they are mixtures of high- and low-molecular-weight RNA (Steele and Busch, 1967).

B. Fractionation of RNA by Sucrose Gradient Centrifugation

This is one of the simplest and most widely used techniques for the fractionation of RNA. A sucrose gradient is essentially a supporting and stabilizing medium through which macromolecules are moved by centrifugal force. The separation is based on sedimentation velocity (molecular weight). The RNA as extracted from cell organelles (Section V, A) is a mixture of high- and low-molecular-weight RNA's. The range of gradient and composition of buffer containing sucrose varies for specific purposes. The standard buffer system used for sucrose gradient centrifugation is 0.02 *M* sodium acetate, pH 5.1 containing 0.1 *M* NaCl and 1 m*M* EDTA. The gradient is usually from 5 to 45%. Although resolution in the high-molecular-weight region where the molecular weight differences are 10^6 is reasonably acceptable, the resolution at the low-molecular-weight region where the molecular weight ranges from 24,000 to 100,000 is very poor. Accordingly, the LMW RNA isolated (Fig. 2A) is a mixture of tRNA and other LMW RNA species.

C. Polyacrylamide Gel Electrophoresis

This is one of the most successful methods for the fractionation of low-molecular-weight RNA. Details of the procedure were described previously (Ro-Choi *et al.*, 1973a). Three main procedures have been used for the fractionation of RNA: (*a*) analytical disc gel electrophoresis, (*b*) preparative gel electrophoresis (continuous electrophoresis and elution), and (*c*) electrophoresis on slab gels.

Analytical disc gel electrophoresis has been most widely used. The RNA labeled with ^{32}P and ^{14}C or ^{14}C and ^{3}H may be coelectrophoresed and the radioactivity distribution studied by counting the gel slices. For staining the RNA, 50–100 μg of RNA is required for a single disc gel of 0.5–0.7 cm diameter. The gels can be stained with methylene blue and scanned by chromoscan for band density. The gels used contained from 6 to 15% polyacrylamide (Figs. 1 and 2B; Adesnik and Levinthal, 1970).

Preparative gel electrophoresis is needed for further chemical analysis of RNA fractions. One of the first RNA's that was purified by this method is U-2 RNA (Hodnett and Busch, 1968). Subsequently the 4.5 S RNA and 5 S RNA were purified and analyzed (Ro-Choi *et al.*, 1970, 1971, 1972). The same method was used by Saponara and Enger (1969) for the separation of LMW RNA from Chinese hamster cells in tissue culture. With slab gel electrophoresis (8–15% slab gels), [^{32}P]labeled nuclear 4–8 S RNA with high specific activity (total of 10^{10-11} cpm) was separated; an autoradiogram is shown in Fig. 10. The slab gel is the most useful method

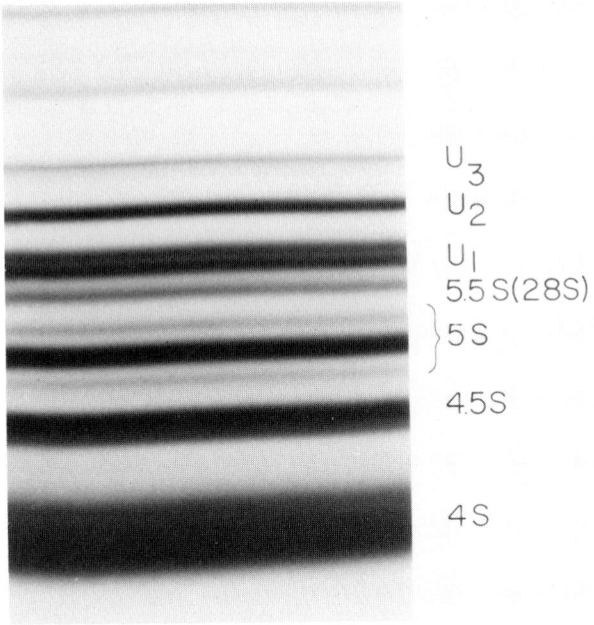

Fig. 10 Autoradiograph of [^{32}P]labeled 4–8 S RNA separated on 10% polyacrylamide slab gel.

for isolation of [^{32}P]labeled RNA; this gel can be easily subjected to autoradiography. For the sequence analysis of U-2 RNA, this method was used routinely; the gel band was cut out and RNA was extracted. This method is particularly important for the initial separation of partial hydrolysis products of RNA (Ro-Choi *et al.*, 1972; Fig. 11). A two-dimensional gel electrophoretic technique which was developed by DeWachter and Fiers (1972) is now proving to be of considerable value.

D. Column Chromatography

Although the RNA purified by gel electrophoresis at neutral pH is a single band on analytical gel electrophoresis, subsequent chromatography on DEAE–Sephadex is sometimes useful for the further purification of LMWN RNA. The 4.5 S RNA and 5 S RNA were separated into three distinct fractions of RNA (Figs. 4 and 7). MAK column chromatography has been used to separate some of the LMW RNA. Frederiksen *et al.* (1971) isolated two components designated as X_1 and X_2 from Ehrlich ascites tumor cells, Yoshida ascites tumor cells, and L5178 Y lymphoblast ascites cells. Three species of low-molecular-weight RNA were isolated

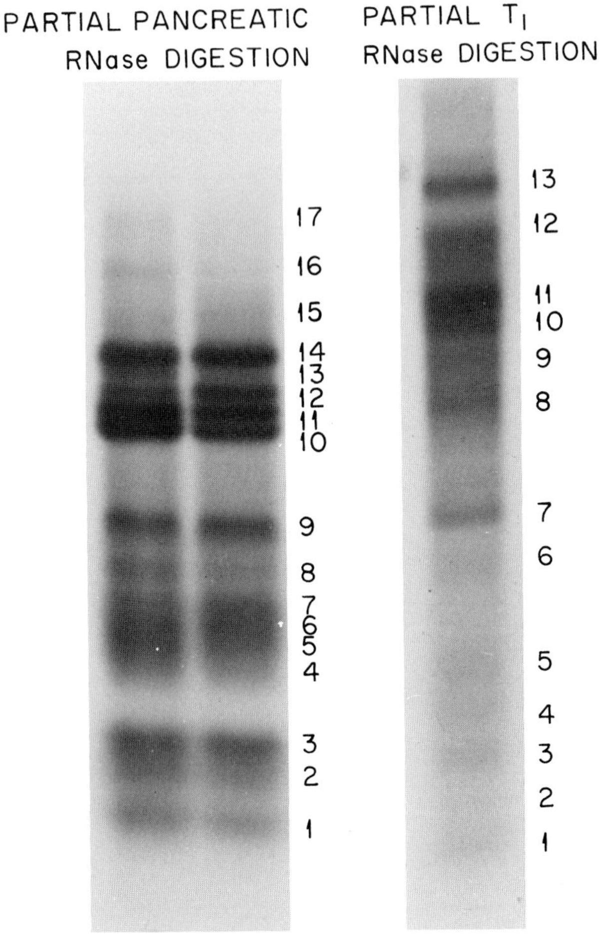

Fig. 11 Autoradiograph of electrophoresis of limited digestion products of 4.5 S RNA_I on a 16% polyacrylamide slab gel. Digestion was carried out at enzyme to substrate ratios of 1:10,000 at 0° or 4°C for 10–15 min for the pancreatic RNase and T_1 RNase digestion, respectively. The hydrolyzates were applied onto the gel directly and electrophoresis was performed at constant voltage of 250 V for 36 hr at 4°.

from *Euglena gracilis* by Spiess and Richter (1970), using MAK column chromatography. Initial separation of low-molecular-weight nuclear and nucleolar RNA was carried out by gel filtration on Sephadex G-100 (Hodnett and Busch, 1968; Moriyama *et al.*, 1969; Nakamura *et al.*, 1968). Although the separation of LMWN RNA was not satisfactory due to the complexity of the mixture, the separation of nucleolar LMW RNA and ribosomal LMW RNA was quite satisfactory in view of the fact that they contain only three components.

For the separation of tRNA many different methods have been reported (series of detailed methods are described in "Methods in Enzymology," Vol. XII). For better separation of tRNA from other RNA, precipitation with cadmium and zinc salts (Raj *et al.*, 1968; Raj and Narasinga Rao, 1969; Premsagar and Narasinga Rao, 1969), lithium extraction, and ammonium sulfate precipitation (Avital and Elson, 1969) have been applied. To prevent degradation of RNA by RNase, precipitation with isopropanol (Sein *et al.*, 1969), dimethylsulfate (Gutcho, 1968), and cetavlon (Legocki *et al.*, 1967) were introduced. In addition, bentonite, sodium dodecyl sulfate, or diethyl pyrocarbonate were also used (Rogg *et al.*, 1969; Abadom and Elson, 1970; Robins and Raacke, 1968). Although countercurrent distribution method was used by Holley *et al.* (1965) to purify $tRNA^{phe}$ for their sequence work, a breakthrough in column chromatography of tRNA was achieved by benzoylated DEAE–cellulose (Gillam *et al.*, 1967, 1968) and by the reversed-phase column chromatographic method (Weiss and Kelmers, 1967; Griffin and Black, 1971). MAK column chromatography (Rushizky, 1969; Goldin and Kaiser, 1969), polyacrylamide gel column chromatography, and electrophoresis on lipophilic membranes were also used to purify tRNA (Winsten, 1969).

VI. Sequence Analysis of Low-Molecular-Weight RNA

The present techniques for the sequence analysis of RNA can be divided into two groups: (1) the method used by Holley *et al.* (1965) in which the analysis is based on the optical properties of its constituents, and (2) the method developed by Sanger and Brownlee (1967); their analysis is based on radioactivity of labeled RNA. This latter procedure is extremely useful, especially in the case of RNA derived from mammalian sources. The extremely small amounts of pure RNA available from mammalian cell constituents prevent the use of spectrophotometric methods for the sequence analysis.

The development of high-level labeling technique of tumor cells with [^{32}P]orthophosphate (Mauritzen *et al.*, 1970) enabled us to sequence some of the rare species of RNA molecules, namely, 4.5 S RNA_I (Ro-Choi *et al.*, 1972) and U-2 RNA (Shibata *et al.*, 1973a, b) of Novikoff hepatoma cell nuclei. Only five RNA's have been sequenced of mammalian cell origin; they are $tRNA^{ser}$ from rat liver by Ginsberg *et al.* (1971); 5 S rRNA from KB cells tissue culture system by Forget and Weissman (1969); VA RNA from KB cells infected with adenovirus 2 (Ohe and Weissman, 1971), in addition to two nuclear RNA's mentioned above. Except for the sequence of tRNA^{ser}, all four of the sequences were determined by the method of Brownlee *et al.* (1968).

A. *4 S RNA*

Transfer RNA's are the smallest of the biologically active nucleic acids, and it has been possible to characterize these macromolecules in terms of nucleotide sequences as well as other properties. Since the sequence analysis of Holley *et al.* (1965), 45 tRNA's or more have been sequenced. However, only one mammalian cell tRNA was sequenced, the tRNA^{ser} from rat liver (Fig. 12). Although the primary structure established by base sequence analysis makes it possible to arrange the molecule in the "cloverleaf" model, there were some differences in comparison to serine tRNA from yeast. All the double helical regions are composed of the classic A·U and G·C base pairs (Fig. 12). In fact, most of the differences between yeast and liver serine tRNA, both of which are good substrates for liver serine tRNA synthetase, are found within these double helical regions.

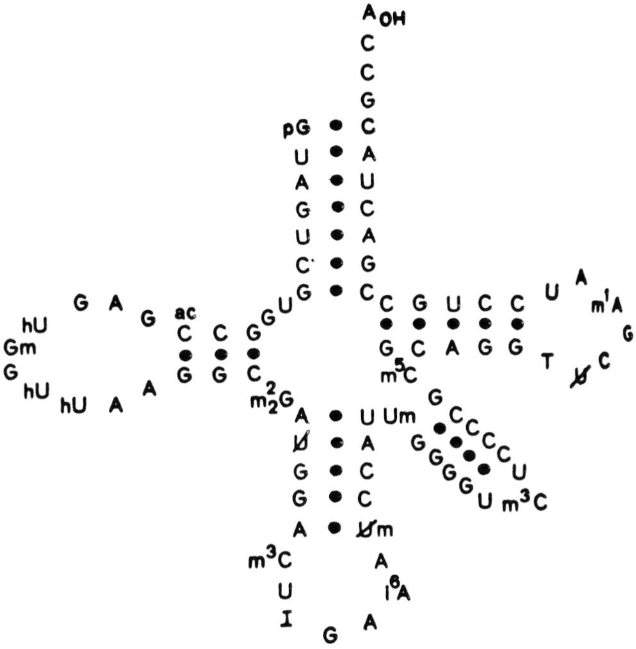

Fig. 12 Cloverleaf model of rat liver serine tRNA$_{I}$. (From Ginsberg *et al.*, 1971.)

B. 4.5 S RNA

The 4.5 S RNA was initially obtained from nuclear 4–8 S RNA of Novikoff hepatoma cells by electrophoresis on preparative polyacrylamide gels. Chromatography on DEAE–Sephadex resulted in the separation of three 4.5 S RNA components. Of these, the 4.5 S RNA_I had a purity of approximately 90% as shown by terminal analysis, i.e., the 3′-terminal nucleoside of 4.5 S RNA_I is uridine (87%) and its 5′-terminal nucleoside 3′, 5′-diphosphate is pGp (90%). Highly labeled 4.5 S RNA_I was obtained from cells incubated with [^{32}P]orthophosphate *in vitro* (Mauritzen *et al.*, 1970). Twenty milliliters (packed volume) of cells were incubated with 500 mCi of [^{32}P]orthophosphate in 1 liter of medium for 9 hr. The purified 4.5 S RNA_I from these cells had a specific activity of 100 μCi/mg RNA and a total of 10–30 $\times 10^6$ cpm in each experiment.

The RNA was then digested with T_1 RNase or pancreatic RNase and fractionated by two-dimensional electrophoresis (Fig. 13). Each spot was cut out and eluted with triethylammonium bicarbonate and analyzed further for their individual sequence. Table XIV shows the catalogue of complete pancreatic and T_1 RNase digestion products. Partial digestion of 4.5 S RNA_I was carried out at the enzyme substrate ratio of 1:10,000 (w/w) and products were separated either on 16% polyacrylamide slab gel (Fig. 11) or by chromatography on DEAE–Sephadex A-25 at pH 7.4. The gel bands or fractions from neutral column were repurified by chromatography on DEAE–Sephadex A-25 at pH 3.3. Partial T_1 RNase digestion produced two large fragments designated at PT_1—2-B and PT_1—2-C that are 3′- and 5′-fragments. The complete sequence of 4.5 S RNA_I is shown in Fig. 14. The secondary structure was constructed with maximal base pairing with high stability numbers (Tinoco *et al.*, 1971) and to fit the partial hydrolytic products of this RNA molecule with limited enzyme digestions with pancreatic RNase or T_1 RNase (Fig. 15).

This RNA has 96 (or 97) nucleotides and has no modified nucleotides. One of the striking features of this molecule is the regional difference in the content of purines and pyrimidines. The 5′-end is rich in purines and the 3′-end is rich in pyrimidines. The center of the molecule has a high content of AMP residues since half of the total adenylic acid residues are clustered in the center between residue 37 and 55. The 5′-terminus was found to be pGp when the RNA was labeled *in vivo*, whereas a mixture of pGp, ppGp, and pppGp was found when the RNA was labeled *in vitro*. Like a number of other RNA species, there was some heterogeneity of the 3′-terminal of 4.5 S RNA_I. Two types of 3′-terminal fragments were found which differed in their content of UMP, namely, —C—A—C—C—U—U—U—U_{OH} and —C—A—C—C—U—U—U—U—

TABLE XIV

Molar Ratios of Oligonucleotides Produced by Complete Digestion of 4.5 S RNA_I with Pancreatic RNase and T_1 RNase

Oligo-nucleo-tide number[a]	Sequence	Experi-mental molar ratio	Theo-retical molar ratio	Oligo-nucleo-tide number[a]	Sequence	Experi-mental molar ratio	Theo-retical molar ratio
1	Cp	15.3	15	1	Gp	7.8	7
2	Up	12.2	12	2	A—Gp	3.2	3
3	A—Cp	4.1	4	3	C—C—Gp	1.1	1
4	A—Up	1.5	1	4	C—U—Gp,U—C—Gp	2.2	1, 1
5	G—Up	2.1	2	5	A—U—Gp	1.1	1
6	A—A—Cp	1.0	1	6	C—U—A—Gp	1.2	1
7	A—G—Cp	2.6	3	7	U—U—C—Gp	1.0	1
8	G—G—Cp	2.1	2	8	C—U—C—A—Gp	1.1	1
9	G—G—Up	1.2	1	9	U—U—A—A—A—Gp	1.1	1
10	A—G—G—Cp	1.0	1	10	U—U—C—C—C—A—Gp	1.1	1
11	A—A—A—A—A—Up	0.7	1	11	C—A—C—C—A—C—Gp	0.7	1
12	A—A—A—G—G—Cp	0.8	1	12	C—A—C—C—U—U—U—U_{OH}	0.8	1
13	A—A—G—A—G—Up	1.0	1	13A	pppGp	0.4	1
14	G—A—G—A—G—G—A—Up	1.0	1	13B	C—A—C—C—U—U—U—U—U_{OH}	0.4	
14A	pppG—G—Up	0.6	1	13C	U—C—C—U—U—C—C—A—Gp	0.9	1
				14	C—C—A—A—A—A—A—U—A—A—C—A—C—C—U—A—U—A—A—Gp	0.7	1

[a] The numbers refer to the oligonucleotide shown in Fig. 13. The experimental molar ratios were calculated from the percentage of the total activity in each spot, and the number of nucleotides in the oligonucleotide. The radioactivity was determined by the Cerenkov method (Clausen, 1968) after the spots were cut out according to their autoradiographic pattern. The values are averages of seven experiments. Theoretical molar ratios refer to the moles of each oligonucleotide in the final structure.

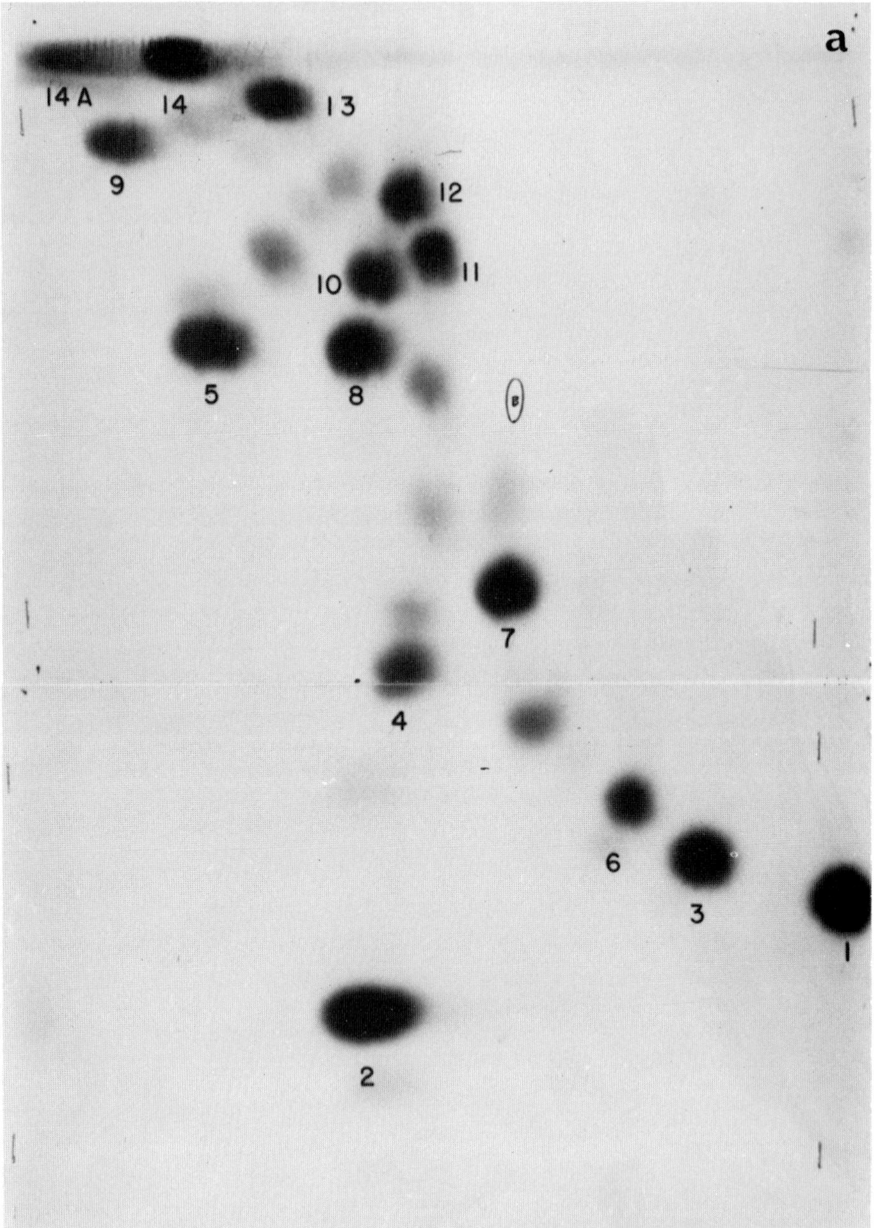

Fig. 13a

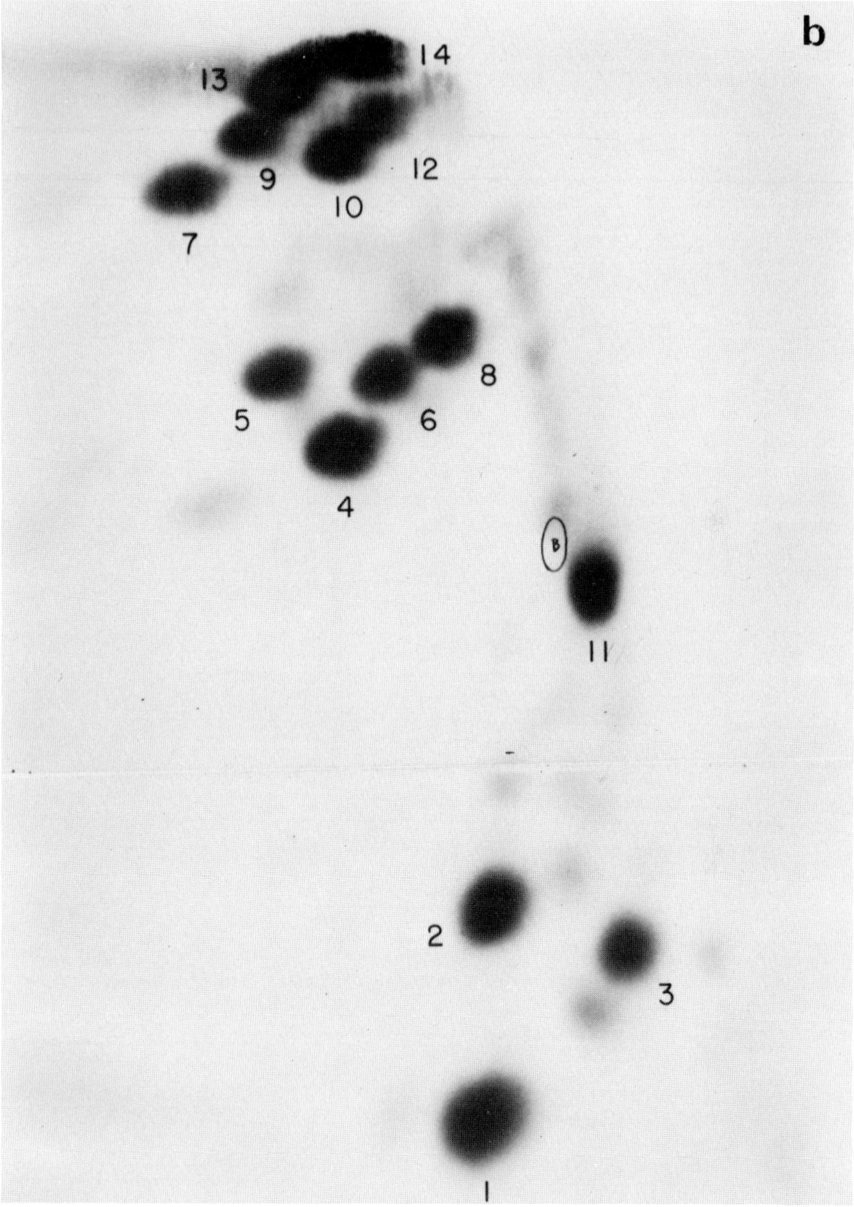

Fig. 13 Autoradiographs of two-dimensional electrophoretic patterns of (a) pancreatic RNase digest and (b) T_1 RNase digest of [^{32}P]labeled 4.5 S RNA_I of Novikoff hepatoma cell nuclei. Electrophoresis was from right to left on a cellulose acetate strip at pH 3.5 (5% acetic acid, 0.01% pyridine) in 7 *M* urea for the first dimension and top to bottom on DEAE paper at pH 1.7 (7% formic acid) for the second dimension.

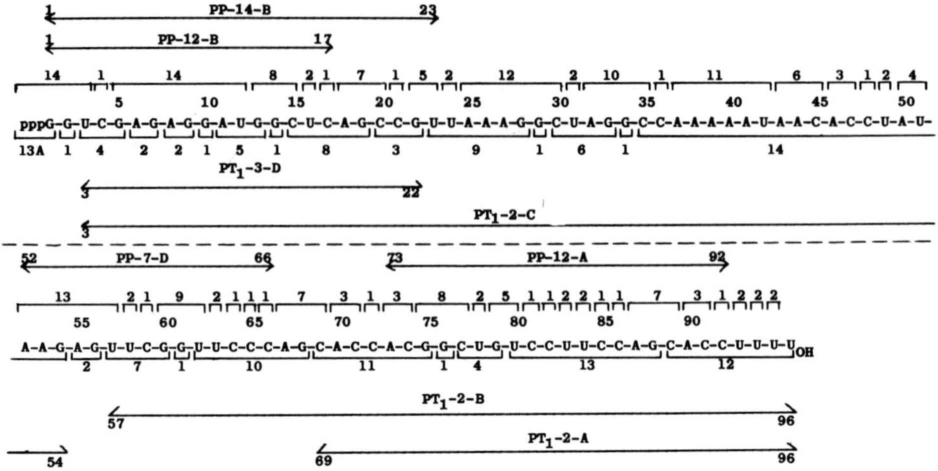

Fig. 14 Total primary sequence of 4.5 S RNA$_I$ of Novikoff hepatoma cell nuclei. The bars above the sequence represent fragments obtained by partial pancreatic RNase digestion, and the bars below the sequence represent fragments obtained by partial T_1 RNase digestion.

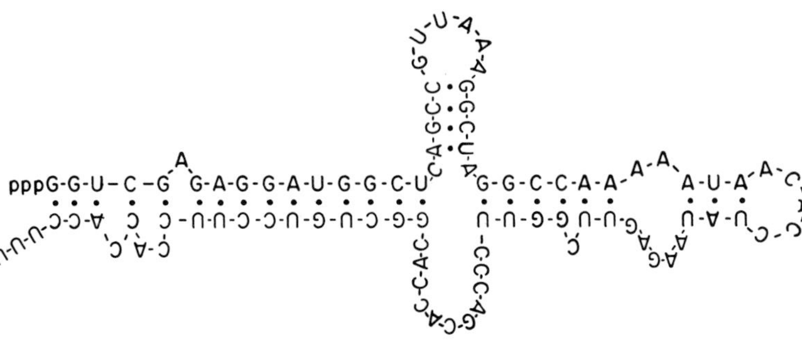

Fig. 15 One of the possible secondary structures of 4.5 S RNA$_I$ of Novikoff hepatoma cell nuclei. The secondary structure was constructed to have maximal base pairing with high stability numbers and to fit the partial hydrolytic products of this RNA molecule with limited enzyme digestions with pancreatic RNase or T_1 RNase.

U_{OH}. The former accounted for two thirds of the terminals and the latter accounted for one-third. This type of heterogeneity has been previously found in KB cell and mouse liver 5 S rRNA (Forget and Weissman, 1969; Williamson and Brownlee, 1969) and adenovirus VA RNA (Ohe and Weissman, 1970, 1971).

It is of interest that some "self-terminating" RNA's, e.g., VA RNA (Lebowitz *et al.*, 1971), ϕ80 RNA (Pieczenik *et al.*, 1972), oop RNA from λ phage-infected *E. coli* (Dahlberg and Blattner, 1973) have similar 3′-

ends with respect to —U—U—U—U-tracts. The oop RNA has been suggested to be a primer RNA for λ phage DNA synthesis (Figs. 14 and 16).

C. 5 S RNA

The sequences of 5 S rRNA have been determined from three cell types, e.g., KB cells (Forget and Weissman, 1969), *E. coli* (Brownlee *et al.*, 1968), and *P. fluorescens* (DuBuy and Weissman, 1971) (Fig. 17) and partial sequences were reported in Novikoff hepatoma cells (Ro-Choi *et al.*, 1971) and mouse liver cells (Williamson and Brownlee, 1969). *Escherichia coli* 5 S RNA was found to be 120 nucleotides long, while KB cell 5 S RNA existed in two forms, 120 and 121 nucleotides long, respectively. In *E. coli* 5 S RNA (Fig. 17), various sequences are repeated twice in the molecule, and the whole molecule can be divided into two halves which display a considerable similarity. On the other hand, KB cell 5 S RNA, which has a base composition very similar to that of *E. coli* 5 S RNA, has a different nucleotide sequence with only a limited amount of homology to its *E. coli* counterpart.

Both nucleotide sequences permitted extensive base pairing between the 5′- and 3′-ends of the molecule and there is more than one possible arrangement of base-paired regions among the internal nucleotide sequences. The 5′-terminal nucleotide is uridylic acid and the 3′-terminal nucleoside is uridine in both *E. coli* and *P. fluorescens* 5 S RNA. Although these two sequences have marked similarities, they do not fit exactly to

pppG-U-U-G-A-U-A-G-A-U-C-C-A-G-U-A-A-U-G-A-C-C-U-C-A-G-A-A-

C-U-C-A-U-C-C-U-G-G-A-U-U-U-G-U-U-G-C-C-G-C-C-U-U-A-G-A-A-

C-G-C-U-C-G-G-U-U-G-C-C-G-G-G-C-G-U-U-U-U-U-U-A-OH

pppG-U-U-A-C-U-U-G-C-C-G-U-A-A-C-A-C-A-G-U-G-U-

G-A-U-A-A-G-G-G-C-C-U-U-U-G-A-A-G-U-U-A-C-C-G-C-

U-U-C-A-A-G-G-G-C-U-U-U-U-U_{OH}

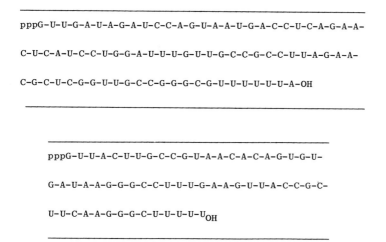

Fig. 16 Primary sequences of λ phage oop RNA (top) and φ80 (bottom) LMW RNA (*E. coli* systems). (From Dahlberg and Blattner, 1973; Pieczenik *et al.*, 1972.)

A

10 20 30

pU-G-C-C-U-G-G-C-G-G-C-C-G-U-A-G-C-G-C-G-G-U-G-G-U-C-C-C-A-C-C-U-G-A-C-C-C-C-

40 50 60 70

A-U-G-C-C-G-A-A-C-U-C-A-G-A-A-G-U-G-A-A-A-C-G-C-C-G-U-A-G-C-G-C-C-G-A-U-G-G-U-

80 90 100 110 120

A-G-U-G-U-G-G-G-G-U-C-U-C-C-C-C-A-U-G-C-G-A-G-A-G-U-A-G-G-G-A-A-C-U-G-C-C-A-G-G-C-A-U_{OH}

B

10 20 30

pU-G-U-U-C-U-G-U-G-A-C-G-A-G-U-A-G-U-G-G-C-A-U-U-G-G-A-A-C-A-C-C-U-G-A-U-C-C-C-

40 50 60 70

A-U-C-C-C-G-A-A-C-U-C-A-G-A-G-G-U-G-A-A-A-C-G-A-U-G-C-A-U-C-G-C-C-G-A-U-G-G-U-

80 90 100 110 120

A-G-U-G-U-G-G-G-G-U-U-U-C-C-C-C-A-U-G-U-C-A-A-G-A-U-C-U-C-G-A-C-C-A-U-A-G-A-G-C-A-U_{OH}

C

10 20 30

pG-U-C-U-A-C-G-G-C-C-A-U-A-C-C-A-C-C-C-U-G-A-A-C-G-C-G-C-C-C-G-A-U-C-U-C-G-U-

40 50 60 70

C-U-G-A-U-C-U-C-G-G-A-A-G-C-U-A-A-G-C-A-G-G-G-U-C-G-G-G-C-C-U-G-G-U-U-A-G-U-A-C-U-

80 90 100 110 120

U-G-G-A-U-G-G-G-A-G-A-C-C-G-C-C-U-G-G-G-A-A-U-A-C-C-G-G-G-U-G-C-U-G-U-A-G-G-C-U-U_{OH}

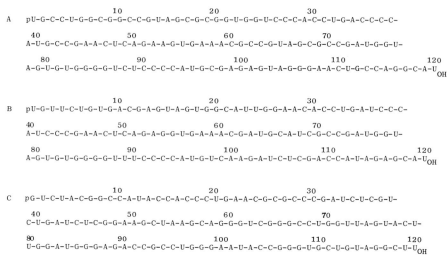

Fig. 17 Comparison of primary sequences of 5 S RNA from (A) *E. coli*, (B) *P. fluoroscense*, and (C) KB cells.

any of the proposed models for the secondary structure of other 5 S RNA's. It is interesting that two forms of 5 S RNA, differing by one nucleotide in only one position, have been found in an *E. coli* strain in about equal amounts (Brownlee *et al.*, 1968). The sequence shown in Fig. 17 is common to strains MRE 600 and CA265. However, strain MRE 600 contains a second form in which the G in position 13 is changed to a U and in strain CA265 the C in position 12 is changed to an A. A similar finding was observed in the sequence of 5 S rRNA from *P. fluorescens*. In *P. fluorescens*, alternative sequences were reported at position 7 and 19. In addition, the 5 S RNA precursor in exponentially growing *E. coli* cells contains the 5′-terminal sequences of pAUUUG(I), pUUUG(II), and pUUG(III), in addition to the mature form of pUG which is found in mature 5 S RNA (Monier *et al.*, 1969; Galibert *et al.*, 1970a,b). These precursor molecules were also found when maturation was arrested by drug treatment or amino acid starvation (Galibert *et al.*, 1967).

Although the fact that precursor 5 S RNA_I (p 5 S RNA_I) has a 5′-terminal adenylic acid residue which appears to satisfy the requirements of *E. coli* RNA polymerase for initiation (Maitra *et al.*, 1967), no di- nor triphosphate group has been found at its 5′-end (Feunteun *et al.*, 1972). This observation is in agreement with the finding that the 5 S RNA cistrons are located at the 3′-end of a long transcriptional unit (Pace *et al.*, 1970). Whether there are similar precursors in *P. fluorescens* is not known at present. From careful analysis of maturation of 5 S RNA precursors in

E. coli, the following pathway has been proposed by Feunteun *et al.* (1972):

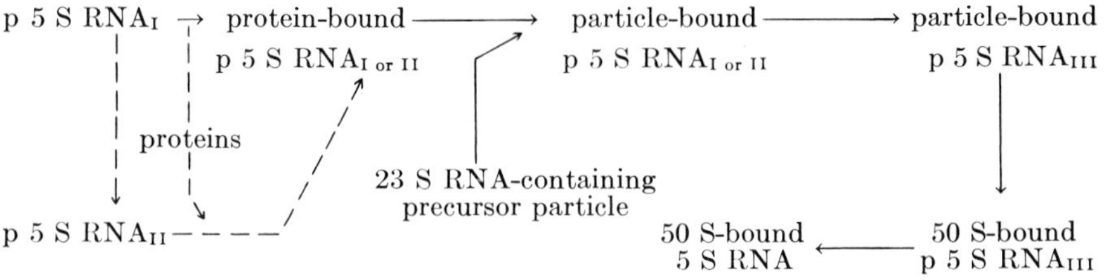

The formation of p 5 S RNA_{III} appears to be a result of endonucleolytic cleavage from p 5 S RNA_I and p 5 S RNA_{II} is a by-product under conditions of inhibition of protein synthesis.

It is interesting that in mammalian cell ribosomal 5 S RNA (KB cell, mouse liver cell, and Novikoff hepatoma cell), minor microheterogeneity was found at the 3′-end where only one nucleotide is different. All of these 5 S RNA's from mammalian cell appear to have identical sequences, namely, pGp as the 5′-terminal and uridine as the 3′-terminal. Mixtures of pppGp, ppGp, and pGp as 5′-terminals were found in HeLa cells as well as Novikoff hepatoma cells (Hatlen *et al.*, 1969; Ro-Choi et al., 1971). The existence of pppGp and ppGp as 5′-ends in 5 S rRNA indicates that this RNA is a primary product of the transcription. Whether the microheterogeneity at the 3′-end indicates processing of this RNA from the 3′-end (Perry and Kelley, 1972) is not yet clear.

The nucleus-specific 5 S RNA_{III} observed in Novikoff hepatoma cells (Ro-Choi *et al.*, 1971), as well as in HeLa cells, (Weinberg and Peuman, 1968), differs from other 5 S RNA in its higher content (31%) of uridylic acid and modified nucleotides (Table X). The 5 S RNA_{III} contains two alkali-stable dinucleotides, UmU and GmC, and 2 ψ-residues. Its 3′-terminals are cytidine and uridine, as compared with uridine alone for ribosomal 5 S RNA. One interesting structural feature of this RNA is that the "fingerprints" obtained after RNase digestions are unique. Pancreatic RNase digestion produced AAC, AAU, AAAU, (AG)C, (AAG)U, and (AAAG)U from 5 S RNA_{III} but not from 5 S RNA_I or 5 S RNA_{II}. Moreover, T_1 RNase digestion produced $(U_2C_2)G$, (CUA)G, $(C_4A_3U_8)G$, [(UmU) $(C_2A_3U_5)$-(AGmC)] $(CAU_2)G$, $(C_2AU_2)G$, and $(C_3A_3U_2)G$ from 5 S RNA_{III} but not from 5 S RNA_I and 5 S RNA_{II}. Of these, the most interesting is $(C_4A_3U_8)G$, which contains eight uridylic acid residues and in itself accounts for the much higher U content of this RNA as compared with 5 S RNA_I or 5 S RNA_{II}. Since 5 S RNA_{III} is found only in the whole nuclear RNA and not in nucleolar or ribosomal RNA, it too is presumably associated with the chromatin fraction. Recently Kanehisa *et al.* (1972) have found a 5 S RNA that is associated with chick liver chromatin; this RNA contains 28% uridylic acid. It was also shown that this RNA is stimulatory for RNA synthesis when chick liver chromatin was used as a template for *E. coli* RNA polymerase.

D. U-2 RNA

From preliminary experiments on 3′- and 5′-terminal analysis of the RNA, it was thought this RNA might contain more than one species of RNA. There are two 3′-termini, adenosine and cytidine, in a ratio of 2:1 (Table III). However, this RNA was not subfractionated by DEAE–Sephadex column chromatography and electrophoresis on polyacrylamide gels at pH 3.3. The primary sequence of U-2 RNA is shown in Fig. 18; the interesting feature of this RNA is its abundance of modified nucleotides. There are ten 2′-*0*-methylated nucleotides. This RNA contains one residue of N^2, N^2, 7-trimethyl guanylic acid which was found first in LMW RNA from Chinese hamster cells (Saponara and Enger, 1969). This unusual nucleotide is localized at the 5′-end fragment of U-2 RNA. Another interesting feature of this RNA is that it contains 12 residues of ψ, one of which is adjacent to 2′-*0*-methylated cytosine, and is therefore re-

```
Nm-Am-Um-m^{2,2,7}_3 G-C-G-C-Gm-Gm-C-U-C-C-ψ-ψ-C-ψ-U-U-U-U-Gm-G-C-U-A-
                                   10                   20

A-m'Gm-A-U-C-A-m^6 Am-G-U-G-ψ-A-G-ψ-A-ψ-Cm-ψ-G-ψ-ψ-U-C-Um-A-U-C-A-
        30                        40                        50

G-U-ψ-U-U-A-A-ψ-A-U-Cm-U-U-C-G-A-U-A-C-G-U-C-C-U-C-U-A-U-C-C-G-
        60                  70                   80

A-G-G-A-C-A-A-U-A-ψ-U-A-ψ-U-A-A-A-U-G-G-A-U-U-U-U-U-G-G-A-A-C-U-
        90              100                  110

A-G-G-A-G-U-U-G-G-A-A-U-A-G-G-A-G-C-U-U-G-C-U-C-C-G-U-C-C-A-C-C-
  120                130                 140

U-C-A-C-G-C-A-U-C-G-A-C-C-U-G-G-U-A-U-U-G-C-G-C-A-G-U-A-C-C-C-U-
150                160                 170                  180

C-A-G-G-A-A-C-G-G-U-G-C-A-C-C-A_OH
              190
```

Fig. 18 Primary nucleotide sequence of U-2 RNA of Novikoff hepatoma cell nuclei.

leased as Cmpψp by alkaline hydrolysis. This RNA has the highest ψ content of any RNA reported so far.

All of the modified nucleotides are clustered at the 5′-one-third of the molecule except for two ψ-residues which are located within two-thirds of the molecule from the 3′-end. The 3′-half of the molecule does not have any modified nucleotides and is rich in pyrimidines. Some micro-heterogeneity was found at the positions with ψ-residues, i.e., 20–30% unmodified U was found at positions 42 and 62. The fact that all modified nucleotides are clustered at the 5′ end of the molecule suggests that this RNA might be in specific complexes with other macromolecules or specific conformations for the specific modifying enzymes. One of the possible secondary structures of 3′-two-thirds of U-2 nuclear RNA is shown in Fig. 19. No attempt has been made to construct a secondary structure of one-third at the 5′-end due to the presence of unusual bases. Preliminary results (Hermolin, J., personal communication) suggest that this RNA is associated with proteins as an RNP complex. The 3′-end of U-2 RNA may be protected by its secondary structure and attached proteins. The 5′-end might be more self-protective by virtue of the unusual bases. On the other hand, the highly methylated 5′-end has a hydrophobicity which may have a functional role. According to the data of Clason and Burdon (1969), it is synthesized just prior to S phase. However, Weinberg and Penman (1969) found that this RNA is synthesized throughout the cell cycle and its half-life is approximately 25 hr.

Although tRNA has the largest variety of modified nucleotides and the highest percent of total modified nucleotides, their localization is in the looped regions of the molecule. They have a universal structural homology (cloverleaf model) regardless of their origins. In the series of LMWN RNA, each species has its own unique sequence and there are no sequence homologies at levels greater than hexanucleotides.

VII. Structure and Function of Low-Molecular-Weight RNA

Since most of the structural and functional relationships of RNA have been studied using tRNA, this section will focus on tRNA structure and function relationship.

A. *Universal Characteristics of tRNA Structure*

The cloverleaf arrangement of tRNA proposed by Holley *et al.* (1965) appears to fit all the tRNA sequenced from bacterial mammalian cell origin. Some of the characteristics of tRNA are listed on Table XV.

N_m-A_m-U_m-$m_3^{2,2,7}$G-C-G-C-G_m-G_m-C-U-C-C-ψ-ψ-C-ψ-

U-U-U-U-G_m-G-C-U-A-A-G'_m-A-U-C-A-A'_m-G-U-G-

ψ-A-G-ψ-A-ψ-C_m-ψ-G-ψ-ψ-U-C-U_m-A-U-C-A-G-

U-ψ-U-U-A-A-ψ-A-U-C_m-U-U-C-G-

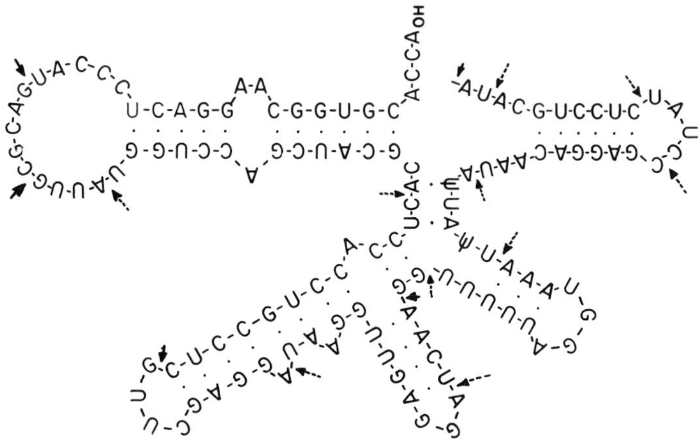

Fig. 19 One of the possible secondary structures of two-thirds of the 3′-end of U-2 RNA of Novikoff hepatoma cell nuclei. The secondary structure was constructed to have maximal base pairing with high stability numbers and to fit the partial hydrolytic products of this RNA molecule with limited enzyme digestions with pancreatic RNase or T_1 RNase. Solid arrows (———→) indicate cleavage sites by partial T_1 RNase digestion; (— — →) indicate cleavage sites by partial pancreatic RNase digestion.

B. *Partial Cleavage and Function of tRNA*

tRNA has been dissected by specific cleavage reactions; the isolated large fragments were reconstituted to regain a totally or partially biologically active molecule (Imura *et al.*, 1969b; Mirzabekov *et al.*, 1969d, e; Oda *et al.*, 1969; Seno *et al.*, 1969a, b).* Such studies have been used to

* Cleavage into two halves and reconstitution was also done for the ribosomal 5 S RNA from *E. coli* (Jordan and Monier, 1971). The reconstituted 5 S RNA from two fragments, I (nucleotide Nos. 42–120) and II (nucleotide Nos. 1–41), has been shown to have identical fingerprints and conformation to that of native 5 S RNA. Moreover, this reconstituted 5 S RNA bound to reconstituted 50 S ribosomal subunits with essentially the same efficiency as native 5 S RNA.

TABLE XV
Characteristics of tRNA

1. Chain length of the molecule: 80 ± 5 nucleotides
2.

	Loops		
	Dihydrouridine loop	Anticodon loop	ψ loop
Size	Variable	17 nucleotides	17 nucleotides
Stem	3–4 base pairs	5 base pairs A · ψ pair	5 base pairs Rich in G:C pair
Loop	8–12 nucleotides	7 nucleotides	7 nucleotides
Modified nucleotides	DihU	Hypermodified nucleotides	"Universal" penta-nucleotides G—T—ψ—C—G
Function	Aminoacyl tRNA synthetase recognition site ?	Binding site in the presence of triplet codons	Ribosome binding site ? 5 S RNA binding site ?

3. Amino acid acceptor stem and extra arm
   Amino acid acceptor stem: (a) 7 base pairs at —C—C—A stem
   (b) —C—C—A is added posttranscriptially
   (c) Similar in various tRNA
   Extra arm: Variable in base compositions as well as in length
4. The position of first anticodon nucleotide is at 35 ± 1 from the 5′-end of the molecule
5. Thymidine of the "universal" pentanucleotide is at 23 ± 1 from the 3′-end of the molecule
6. The length from the 5′-end to anticodon is approximately 80 Å
7. The numbers of nucleotides between m^5C to T, m^2G to 5′-end, and m^2G to the first anticodon are 5–6, 9–10, and 9, respectively

define the specifying parts of the molecule (Clark *et al.*, 1968; Mirzabekov *et al.*, 1968, 1970; Seno *et al.*, 1969a,b). Heterologous fragments of tRNA's from different species can be hybridized (Mirzabekov *et al.*, 1969c, e; Thiebe and Zachau, 1969) and the hybrids retain part of the biological function. Amino acid acceptor activity was studied for fragments or various combinations of partial molecules in an attempt to find recognition sites for aminoacyl tRNA synthetases. In yeast $tRNA^{ala}$, an isolated amino acid-accepting stem fragment retains some specificity for the amino acid-accepting activity (Imura *et al.*, 1969a, b; Mirzabekov *et al.*, 1969a, b). Although certain parts of the dihydrouridine loop and arm can be removed without loss of amino acid acceptor activity in *E. coli* $tRNA^{met}$, the removal of eight nucleotides in this region renders the molecule inactive (Seno *et al.*, 1971).

C. *Specificity of Aminoacyl tRNA Synthetases*

Since each tRNA has a specific anticodon for a specific amino acid, it seemed possible that the anticodon might specify which amino acid should be accepted. However, single breaks at the anticodon produced no change in amino acid acceptor activity. The $tRNA_1^{tyr}$ yeast and $tRNA_1^{tyr}$ and $tRNA_2^{tyr}$ from *E. coli* have the same anticodon (GUA) but they can be aminoacylated only by the homologous aminoacyl tRNA synthetase. Also, $tRNA^{tyr}$ from *E. coli* Su_{III}^+ has anticodon CUA but this tRNA can be aminoacylated by the same tyrosyl tRNA synthetase as the wild-type $tRNA^{tyr}$, which has GUA as anticodon. From these studies, the anticodon has been eliminated as a recognition site. The fact that the sequences of $tRNA^{phe}$ from wheat, and $tRNA^{val}$ from *E. coli,* all of which are phenylalanylated by yeast phenyalanyl tRNA synthetase have the highest sequence homology in the dihydro-U loop and its stem (Dudock *et al.*, 1970) suggested that this area contained recognition sites. However, at present several lines of evidence suggest that this may not be true for all of the tRNA's studied.

The amino acid acceptor region is another site that has been proposed to contain tRNA synthetase recognition sites. In yeast $tRNA^{ala}$, after UV-irradiation, the acceptor activity was decreased markedly; the UV targets were the fifth, sixth, and seventh nucleotide from the 3′-end of the molecule. In addition, fragments including the acceptor stem can distinguish the appropriate synthetase and amino acid (Chambers, 1969; Schulman and Chambers, 1968) and thus may be considered the recognition site. At present it seems likely that the aminoacyl tRNA synthetase recognizes the appropriate tRNA molecule by its exterior shape.

D. *Mutant tRNA*

The relation between nucleotide sequence, structure, and function of tRNA has been studied for a number of mutants. These include the mutants that read the UAG terminator codon as an amino acid; the amber (UAG) suppressor of Su_I^+ serine tRNA, Su_{II}^+ glutamine tRNA, and Su_{III}^+ tyrosine tRNA, and the suppressor of the UGA nonsense codon which is a tryptophan tRNA (Hirsh, 1971; Hirsh and Gold, 1971). The wild type of Su_{III} gene (Su_o^-) is the structural gene for one of the tyrosine tRNA's of *E. coli* which recognizes the codon UAU and UAC. The Su_{III}^+ tRNA recognizes only the amber codon UAG. The mutation to Su^- can occur not only by reversion to Su_o^- but also by changes in the tRNA sequence which produces a defective tRNA. Some mutants of the Su_{III} tyrosine suppressor transfer RNA (Abelson *et al.*, 1970; Smith *et al.*,

1970a,b) differ from the wild-type Su_{III}^+ tRNA molecule by a single base change. Three mutant $tRNA^{tyr}$'s have G→A substitutions at positions 15, 17, and 31, respectively. A mutant tRNA containing an A residue in place of G in the "dihydrouracil loop" appears to be defective in a step after the acylated tRNA is bound to the ribosome. There was no change in kinetics of aminoacylation or the ribosome binding of the tRNA. Another mutant tRNA having an A residue in place of a G in the "anticodon" stem has been shown to be defective in its apparent affinity for the tyrosyl tRNA synthetase. In two temperature-sensitive mutants, single base substitutions at the amino acid acceptor stem and dihydrouracil arm were observed which produced mispaired bases from a G·C to an A·C pair. These do not have suppressor activity *in vivo.* However, second-site revertants of these mutant to wild-type suppressors produce a tRNA with an A·U pair at this position. These changes from G·C to A·U pairs do not alter the K_m of aminoacylation of tRNA with tyrosine tRNA synthetase, which suggests that these regions are not involved in recognition by the synthetase.

Suppressor $tRNA^{trp}$ for the nonsense codon UGA was unchanged in the anticodon CCA which is complementary to the tryptophan codon, UGG. This RNA had changes in the dihydrouracil arm where A was substituted for G to make an A·U pair. Tryptophan tRNA from several related Su^- strains denatured on isolation but that from the suppressor strain is stable.

These studies show that even a single change in nucleotide sequence can produce a marked functional difference. Although the functions of LMWN RNA's are not known at the present time, it will be of great interest to study the effect of mutant changes in nucleotide sequences on cellular function.

VIII. Conclusions

A number of species of LMW RNA are present in many types of cells. The only known function of the low-molecular-weight RNA is the amino acid acceptor and transfer activity of tRNA. Possible involvements of LMW RNA in gene regulatory functions are: (1) gene activation, (2) derivation from intercistronic or untranslated portion of structural gene products, and (3) structural component of chromatin.

1. Gene activator RNA: Some models for gene regulation in higher organisms include RNA (Britten and Davidson, 1969; Georgiev, 1972). Recently, a stimulatory activity was found for certain species of low-molecular-weight chromatin RNA in *in vitro* RNA-synthesizing system (Kanehisa *et al.,* 1972; Tanaka and Kanehisa, 1972). These results suggest

that these RNA might function as specific or nonspecific "derepressor RNA".

2. Primer for nucleic acid synthesis: From structural analysis of 4.5 S RNA_I and U-2 and preliminary results on U-1 and U-3 RNA's, there is no sequence homology in these molecules at the levels greater than hexa-nucleotides. This indicates that these species of low molecular weight RNAs are not nonspecific degradation products but are unique individual molecules. Thus, it is reasonable to speculate that each individual molecule might have its own unique function. 4.5 S RNA_I has pppGp as a 5′-end and U—U—U—U_{OH} at the 3′-end. The existence of pppGp indicates that this RNA is a primary transcriptional product. RNA's with 5′- and 3′-ends similar to 4.5 S RNA_I have been sequenced in bacterial systems. These include 4 S RNA (81 nucleotides long: oop RNA) from λ phage-infected *E. coli* (Dahlberg and Blattner, 1973), 6 S RNA synthesized *in vitro* using λ DNA and *E. coli* RNA polymerase (Lebowitz *et al.*, 1971) and smaller than 4 S RNA (62 nucleotides long) from ϕ80 infected *E. coli* system (Pieczenik *et al.*, 1972).

The oop RNA in λ phage-infected *E. coli* has been demonstrated to act as a primer for the DNA synthesis (Hayes and Szybalski, 1973). It is possible that 4.5 S RNA_I might have similar functions in Novikoff hepatoma cell nuclei.

The lack of a 5′-phosphate in U-1 and U-2 RNA suggests these are not direct transcriptional products but are processed during or after transcription. The processing might be produced by simple phosphatase activity or endo- or exonucleolytic activity. Other interesting features of these molecules now known are unusual modifications of the 5′-end in U1, U2 and U3 RNA which contain $N^{2,2,}$-7-trimethyl-G. If modification occurred before the processing, these hypermodified 5′-ends might serve as recognition sites for specific processing enzyme(s).

In addition, the number of these low-molecular-weight nuclear RNA's roughly corresponds to that of functional genes in mammalian cells (3×10^6). This finding suggests these RNA's in complexes with specific proteins may localize at specific loci such as the initiation or termination site(s) of gene transcriptional unit(s) and direct the RNA polymerase activity.

3. Structural elements: The 5 S rRNA has been sequenced for three different species. Despite all the information available at the present time, no definite function(s) is assigned to this RNA. One clear fact is that this RNA is essential for the functional 50 S subunit reconstitution. This RNA might serve as a critical steric locus for important protein-binding sites. A similar function was proposed for the nucleolar U-3 RNA for the correct and proper binding sites for the structural or processing protein(s)

for the immature ribosomal particles (Prestayko *et al.*, 1970). An analogous situation might exist in the genetic transcriptional unit where these LMWN RNA's may provide critical binding sites for different histones, nonhistone proteins, or even for RNA polymerase or DNA polymerase. Another interesting aspect of low-molecular-weight RNA is the existence of such RNA in certain viruses. Adenovirus 2 produced 6 S RNA and CEV and PSTV are infectious low-molecular-weight RNA's in plants. It will be of interest to know what kind of interrelationship exists between low-molecular-weight RNA of oncogenic viruses and low-molecular-weight RNA of host cells. If LMWN RNA is important in regulatory functions of gene activity, replacing one of these RNA's with viral low-molecular-weight RNA might lead to disturbances in gene regulation and give rise to important reactions in oncogenesis or gene integration.

REFERENCES

Abadom, P. N., and Elson, D. (1970). *Biochim. Biophys. Acta* **199**, 528.

Abelson, J. N., Gefter, M. L., Barnett, L., Landy, A., Russell, R. L., and Smith, J. D. (1970). *J. Mol. Biol.* **47**, 15.

Adesnik, M., and Levinthal, C. (1970). *J. Mol. Biol.* **48**, 187.

Attardi, G., and Amaldi, F. (1970). *Ann. Rev. Biochem.* **39**, 183.

Aubert, M., Monier, R., Reynier, M., and Scott, J. F. (1967). *Proc. FEBS Meeting, 4th* **3**, 151.

Aubert, M., Scott, J. F., Reynier, M., and Monier, R. (1968). *Proc. Nat. Acad. Sci. U.S.* **61**, 292.

Avital, S., and Elson, D. (1969). *Biochim. Biophys. Acta* **179**, 297.

Bachvaroff, R. J., and Tongur, V. (1966). *Nature (London)* **211**, 248.

Bernhardt, D., and Darnell, J., Jr. (1969). *J. Mol. Biol.* **42**, 43.

Bishop, J. M., Levinson, W. E., Sullivan, D., Fanshier, L., Quintrell, N., and Jackson, J. (1970). *Virology* **42**, 927.

Blobel, G. (1971). *Proc. Nat. Acad. Sci. U.S.* **68**, 1881.

Bonner, J., and Widholm, J. (1967). *Proc. Nat. Acad. Sci. U.S.* **57**, 1379.

Britten, R. J., and Davidson, E. H. (1969). *Science* **165**, 349.

Brown, D. D., and Dawid, I. B. (1968). *Science* **160**, 272.

Brown, D. D., and Littna, E. (1966). *J. Mol. Biol.* **20**, 95.

Brown, D. D., and Weber, C. S. (1968). *J. Mol. Biol.* **34**, 661.

Brown, D. D., Wensink, P. C., and Jordan, E. (1971). *Proc. Nat. Acad. Sci. U.S.* **68**, 3175.

Brownlee, G. G. (1971). *Nature (London) New Biol.* **229**, 147.

Brownlee, G. G., Sanger, F., and Barrell, B. G. (1968). *J. Mol. Biol.* **34**, 379.

Busch, H. (1967). *Methods Enzymol.* (L. Grossman and K. Moldave, eds.), Vol. **12**, 448–464.

Busch, H., and Smetana, K. (1970). "The Nucleolus." Academic Press, New York.

Busch, H., Ro-Choi, T. S., Prestayko, A. W., Shibata, H., Crooke, S. T., El-Khatib, S. M., Choi, Y. C., and Mauritzen, C. M. (1971). *Persp. Biol. Med.* **15**, 117.

Chakravorty, A. K. (1969). *Biochim. Biophys. Acta* **179**, 67.

Chambers, R. W. (1969). *J. Cell Physiol.*, Suppl. 1, **74**, 179.

Chambon, P., Gissinger, F., Mandel, J. L., Jr., Kedinger, C., Guiazdowski, M., and Meihlac, M. (1970). *Cold Spring Harbor Symp. Quant. Biol.* **35**, 693.
Choi, Y. C., and Busch, H. (1970). *J. Biol. Chem.* **245**, 1954.
Clark, B. F. C., Dube, S. K., and Marcker, K. A. (1968). *Nature (London)* **219**, 484.
Clason, A. E., and Burdon, R. H. (1969). *Nature (London)* **223**, 1063.
Clausen, T. (1968). *Anal. Biochem.* **22**, 70.
Comb, D. G., and Katz, S. (1964). *J. Mol. Biol.* **8**, 790.
Comb, D. G., Sarkar, N., DeVallet, J., and Pinzino, C. J. (1965). *J. Mol. Biol.* **12**, 509.
Dahlberg, J. E., and Blattner, F. R. (1973). *Fed. Proc.* 664. Abs. (2539).
DeWachter, R., and Fiers, W. (1972). *Anal. Biochem.* **49**, 184.
Diener, T. O., and Smith, D. R. (1971). *Virology* **46**, 498.
DuBuy, B., and Weissman, S. M. (1971). *J. Biol. Chem.* **246**, 747.
Dudock, B. S., Katz, G., Taylor, E., and Holley, R. W. (1969). *Proc. Nat. Acad. Sci. U.S.* **62**, 941.
Dudock, B. S., Diperi, C., and Michael, M. S. (1970). *J. Biol. Chem.* **245**, 2465.
Dudock, B. S., Diperi, C., Scileppi, K., and Rezselbach, R. (1971). *Proc. Nat. Acad. Sci. U.S.* **68**, 681.
Egyhazi, E., Daneholt, B., Edstrom, J. E., Lambert, B., and Ringborg, U. (1969). *J. Mol. Biol.* **44**, 517.
El-Khatib, S. M., Ro-Choi, T. S., Choi, Y. C., and Busch, H. (1970). *J. Biol. Chem.* **245**, 3416.
Elson, D. (1961). *Biochim. Biophys. Acta* **53**, 232.
Enger, M. D., and Walters, R. A. (1970). *Biochemistry* **9**, 3551.
Feunteun, J., Jordan, B. R., and Monier, R. (1972). *J. Mol. Biol.* **70**, 465.
Forget, B. G., and Weissman, S. M. (1968). *J. Biol. Chem.* **243**, 5709.
Forget, B. G., and Weissman, S. M. (1969). *J. Biol. Chem.* **244**, 3148.
Frederiksen, S., Tonnesen, T., and Hellung-Larsen, P. (1971). *Arch. Biochim. Biophys.* **142**, 238.
Fuller, W., and Hodgson, A. (1967). *Nature (London)* **215**, 817.
Galibert, F., Larsen, C. J., Lelong, J. C., and Boiron, M. (1965). *Nature (London)* **207**, 1039.
Galibert, F., Larsen, C. J., Lelong, J. C., and Boiron, M. (1966a). *Bull. Soc. Chim. Biol.* **48**, 21.
Galibert, F., Lelong, J. C., Larsen, C. J., and Boiron, M. (1966b). *J. Mol. Biol.* **21**, 385.
Galibert, F., Lelong, J. C., and Larsen, C. J. (1967). *C. R. Acad. Sci. Paris* **265**, 279.
Galibert, F., Eladari, M. E., Hampe, A., and Boiron, M. (1970a). *Eur. J. Biochem.* **13**, 281.
Galibert, F., Eladari, M. E., Larsen, C. J., and Boiron, M. (1970b). *Eur. J. Biochem.* **13**, 273.
Georgiev, G. P. (1972). *Current Top. Develop. Biol.* **7**, 1.
Gillam, I., Millward, S., Blew, D., von Tigerstrom, M., Wimmer, E., and Tener, G. M. (1967). *Biochemistry* **6**, 3043.
Gillam, I., Blew, D., Warrington, R. C., von Tigerstrom, M., and Tener, G. M. (1968). *Biochemistry* **7**, 3459.
Ginsberg, T., Rogg, H., and Staehelin, M. (1971). *Eur. J. Biochem.* **21**, 249.
Goldin, M., and Kaiser, I. I. (1969). *Biochem. Biophys. Res. Commun.* **36**, 1013.
Gonano, F. (1967). *Biochemistry* **6**, 977.
Gould, H. J. (1967). *J. Mol. Biol.* **29**, 307.
Griffin, A. C., and Black, D. D. (1971). *Methods Cancer Res.* **6**, 189.
Griffin, B. E. (1971). *FEBS Lett.* **15**, 165.

Gutcho, S. (1968). *Biochim. Biophys. Acta* **157,** 76.
Hampel, A. E., Saponara, A. G., Walters, R. A., and Enger, M. D. (1972). *Biochim. Biophys. Acta* **269,** 428.
Harter, D. H., Schlom, J., and Spiegelman, S. (1971). *Biochim. Biophys. Acta* **240,** 435.
Hatlen, L. E., Amaldi, F., and Attardi, G. (1969). *Biochemistry* **8,** 4989.
Hayes, S., and Szybalski, W. (1973). *Fed. Proc.* 529, Abs. (1746).
Hellung-Larsen, P., and Fredriksen, S. (1972). *Biochim. Biopyhs. Acta* **262,** 290.
Hellung-Larsen, P., Fredriksen, S., and Plesner, P. (1971). *Biochim. Biophys. Acta* **254,** 78.
Hindley, J. (1967). *J. Mol. Biol.* **30,** 125.
Hirsh, D. (1971). *J. Mol. Biol.* **58,** 439.
Hirsh, D., and Gold, L. (1971). *J. Mol. Biol.* **58,** 459.
Hodnett, J. L., and Busch, H. (1968). *J. Biol. Chem.* **243,** 6336.
Holley, R. W., Apgar, J., Everett, G. A., Marquisee, M., Merrill, S. H., Penswick, J. R., and Zamir, A. (1965). *Science* **147,** 1462.
Huang, R. C., and Bonner, J. (1965). *Proc. Nat. Acad. Sci. U.S.* **54,** 960.
Imura, N., Schwam, H., and Chambers, R. W. (1969a). *Proc. Nat. Acad. Sci. U.S.* **62,** 1203.
Imura, N., Weiss, G. B., and Chambers, R. W. (1969b). *Nature (London)* **222,** 1147.
Jacob, F., and Monod, J. (1961). *J. Mol. Biol.* **3,** 318.
Jordan, B. R. (1971). *J. Mol. Biol.* **55,** 423.
Jordan, B. R., Feunteun, J., and Monier, R. (1970). *J. Mol. Biol.* **50,** 605.
Jordan, B. R., and Monier, R. (1971). *J. Mol. Biol.* **55,** 423.
Kacian, D. L., Mills, D. R., and Spiegelman, S. (1971). *Biochim. Biophys. Acta* **238,** 212.
Kanehisa, T., Tanaka, T., and Kano, Y. (1972). *Biochim. Biophys. Acta* **277,** 584.
Kaper, J. M., and West, C. K. (1972). *Prep. Biochem.* **2,** 251.
Knight, E., and Darnell, J. E. (1967). *J. Mol. Biol.* **28,** 491.
Larsen, C. J., Galibert, F., Lelong, J. C., and Boiron, M. (1967). *C. R. Acad. Sci. Paris* **D264,** 1523.
Larsen, C. J., Galibert, F., Hampe, A., and Boiron, M. (1969). *Bull. Soc. Chim. Biol.* **51,** 649.
Larsen, C. J., Lebowitz, P., Weissman, S. M., and DuBuy, B. (1970). *Cold Spring Harbor Symp. Quant. Biol.* **35,** 35.
Larsen, C. J., Ravicovitch, R. E., Bazilier, M., Mauchauffe, M., Robin, J., and Boiron, M. (1972). *C. R. Acad. Sci. Paris* **274,** 1396.
Lebowitz, P., Weissman, S. M., and Radding, C. M. (1971). *J. Biol. Chem.* **246,** 5120.
Legocki, A. B., Szymkowiak, A., Pech, K., and Pawelkiewicz, J. (1967). *Acta Biochem. Pal.* **14,** 323.
Li, P. H., and Fox, R. H. (1969). *Biochim. Biophys. Acta* **182,** 255.
Loening, U. E. (1967). *Biochem. J.* **102,** 251.
Maitra, U., Nakata, Y., and Hurwitz, J. (1967). *J. Biol. Chem.* **242,** 4908.
Marcaud, L., Portier, M. M., Kourilsky, P., Barrell, B. G., and Gros, F. (1971). *J. Mol. Biol.* **57,** 247.
Marcot-Queiroz, J., Julien, J., Rosset, R., and Monier, R. (1965). *Bull. Soc. Chim. Biol.* **47,** 183.
Mauritzen, C. M., Choi, Y. C., and Busch, H. (1970). *Methods Cancer Res.* **6,** 253.
Mirzabekov, A. D., Grünberger, D., and Bayev, A. A. (1968). *Biochim. Biophys. Acta* **166,** 68.

Mirzabekov, A. D., Kazarinova, L. Ya, Lastity, D., and Bayev, A. A. (1969a). *FEBS Lett.* **3**, 268.
Mirzabekov, A. D., Kazarinova, L. Ya., and Bayev, A. A. (1969b). *J. Mol. Biol.* **3**, 879.
Mirzabekov, A. D., Kazarinova, L. Ya., Lastity, D., and Bayev, A. A. (1969c). *J. Mol. Biol.* **3**, 909.
Mirzabekov, A. D., Lastity, D., and Bayev, A. A. (1969d). *FEBS Lett.* **4**, 281.
Mirzabekov, A. D., Levina, E. S., and Bayev, A. A. (1969e). *FEBS Lett.* **5**, 218.
Mirzabekov, A. D., Lastity, D., Levina, E. S., and Bayev, A. A. (1970). *FEBS Lett.* **7**, 95.
Monier, R., Feunteun, J., Forget, B., Jordan, B. R., Reynier, M., and Varrichio, F. (1969). *Cold Spring Harbor Symp. Quant. Biol.* **34**, 139.
Morell, P., and Marmur, J. (1968). *Biochemistry* **7**, 1141.
Moriyama, Y., Hodnett, J. L., Prestayko, A. W., and Busch, H. (1969). *J. Mol. Biol.* **39**, 335.
Muramatsu, M., Hodnett, J. L., and Busch, H. (1966). *J. Biol. Chem.* **241**, 1544.
Muramatsu, M., and Fujisawa, T. (1968). *Biochim. Biophys. Acta* **157**, 476.
Nakamura, T., Prestayko, A. W., and Busch, H. (1968). *J. Biol. Chem.* **243**, 1368.
Narayan, K. S., Smetana, K., Steele, W. J., and Busch, H. (1966). *Proc. Amer. Ass. Cancer Res.* **7**, 52.
Obara, T., Bolognesi, D. P., and Bauer, H. (1971). *Int. J. Cancer* **7**, 535.
Oda, K., Kimura, F., Harada, F., and Nishimura, S. (1969). *Biochim. Biophys. Acta* **179**, 97.
Ohe, K., and Weissman, S. M. (1970). *Science* **167**, 879.
Ohe, K., and Weissman, S. M. (1971). *J. Biol. Chem.* **246**, 6991.
Pace, B., Peterson, R. L., and Pace, N. R. (1970). *Proc. Nat. Acad. Sci. U.S.* **65**, 1097.
Pene, J. J., Knight, E., and Darnell, J. E. (1968). *J. Mol. Biol.* **33**, 609.
Penman, S., Smith, I., and Holtzman, E. (1966). *Science* **154**, 786.
Perry, R. P., and Kelley, D. E. (1972). *J. Mol. Biol.* **70**, 265.
Pieczenik, G., Barrell, B. G., and Gefter, M. L. (1972). *Arch. Biochem. Biophys.* **152**, 152.
Premsagar, K. D. A., and Narasinga Rao, M. S. (1969). *Biochim. Biophys. Acta* **182**, 394.
Prestayko, A. W., and Busch, H. (1968). *Biochim. Biophys. Acta* **169**, 327.
Prestayko, A. W., Tonato, M., and Busch, H. (1970). *J. Mol. Biol.* **47**, 505.
Prestayko, A. W., Tonato, M., Lewis, B. C., and Busch, H. (1971). *J. Biol. Chem.* **246**, 182.
Raj, B. K., and Narasinga Rao, M. S. (1969). *Biochemistry* **8**, 1277.
Raj, B. K., Premsagar, K. D. A., and Narasinga Rao, M. S. (1968). *Biochem. Biophys. Res. Commun.* **31**, 723.
Randerath, K. (1971). *Cancer Res.* **31**, 658.
Randerath, K., and Randerath, E. (1971). *In* "Procedures in Nucleic Acid Research" (G. L. Cantoni and D. R. Davies, eds.), Vol. 2, p. 796. Harper, New York.
Randerath, K., Randerath, E., and Chia, L. S. Y. (1973). Unpublished data.
Reddy, R., Ro-Choi, T. S., Henning, D., Shibata, H., Choi, Y. C., and Busch, H. (1972). *J. Biol. Chem.* **247**, 7245.
Reddy, R., Henning, D., Ro-Choi, T. S., and Busch, H. (1974). In preparation.
Rein, A. (1971). *Biochim. Biophys. Acta* **232**, 306.
Rein, A., and Penman, S. (1969). *Biochim. Biophys. Acta* **190**, 1.
Ritter, P. O., and Busch, H. (1971). *Physiol. Chem. Phys.* **3**, 411.
Ro, T. S., and Busch, H. (1964). *Cancer Res.* **24**, 1630.

Robins, H. I., and Raacke, I. D. (1968). *Biochem. Biophys. Res. Commun.* **33,** 240.
Ro-Choi, T. S., Moriyama, Y., Choi, Y. C., and Busch, H. (1970). *J. Biol. Chem.* **245,** 1970.
Ro-Choi, T. S., Reddy, R., Henning, D., and Busch, H. (1971). *Biochem. Biophys. Res. Commun.* **44,** 963.
Ro-Choi, T. S., Reddy, R., Henning, D., Takano, T., Taylor, C. W., and Busch, H. (1972). *J. Biol. Chem.* **247,** 3205.
Ro-Choi, T. S., Choi, Y. C., Savage, H., and Busch, H. (1973a). *Methods Cancer Res.* **9,** 71.
Ro-Choi, T. S., Smetana, K., and Busch, H. (1973b). *Exp. Cell Res.* **79,** 43.
Ro-Choi, T. S., Reddy, R., Raj, N. B., Henning, D., and Busch, H. (1974). *Am. Assoc. Cancer Res.* (Abstract).
Roeder, R. G., and Rutter, W. J. (1969). *Nature (London)* **224,** 234.
Rogg, H., Wehrli, W., and Staehelin, M. (1969). *Biochim. Biophys. Acta* **195,** 13.
Rosset, R., and Monier, R. (1963). *Biochim. Biophys. Acta* **68,** 653.
Rushizky, G. W. (1969). *Anal. Biochem.* **29,** 459.
Sanger, F. (1971). *Biochem. J.* **124,** 833.
Sanger, F., and Brownlee, G. G. (1967). *Methods Enzymol.* **12,** 361.
Saponara, A. G., and Enger, M. D. (1969). *Nature (London)* **223,** 1365.
Schleich, T., and Goldstein, J. (1966). *J. Mol. Biol.* **15,** 136.
Schulman, L. H., and Chambers, R. W. (1968). *Proc. Nat. Acad. Sci. U.S.* **61,** 308.
Sein, K. T., Becarevic, A., and Kanazir, D. (1969). *Anal. Biochem.* **28,** 65.
Semancik, J. S., and Weathers, L. G. (1972). *Nature (London) New Biol.* **237,** 242.
Seno, T., Kobayashi, M., and Nishimura, S. (1969a). *Biochim. Biophys. Acta* **182,** 280.
Seno, T., Kobayashi, M., and Nishimura, S. (1969b). *Biochim. Biophys. Acta* **190,** 285.
Seno, T., Kobayashi, I., Fukuhara, M., and Nishimura, S. (1971). *Fed. Eur. Biochem. Soc. Lett.* **7,** 343.
Shibata, H., Reddy, R., Henning, D., Ro-Choi, T. S., and Busch, H. (1973a). *Mol. Cell Biochem.* (in press).
Shibata, H., Reddy, R., Ro-Choi, T. S., Henning, D., and Busch, H. (1973b). In preparation.
Shih, T. Y., and Bonner, J. (1969). *Biochim. Biophys. Acta* **182,** 30.
Sitz, T., and Busch, H. (1973). In preparation.
Smith, J. D., Anderson, K., Cashmore, A., Hooper, M. L., and Russell, R. C. (1970a). *Cold Spring Harbor Symp. Quant. Biol.* **35,** 21.
Smith, J. D., Barnett, L., Brenner, S., and Russell, R. L. (1970b). *J. Mol. Biol.* **54,** 1.
Spiess, E., and Richter, A. (1970). *Arch. Mikrobiol.* **75,** 37.
Steele, W. J., and Busch, H. (1967). *Methods Cancer Res.* **3,** 61.
Strehler, B. L., Hendley, D. D., and Hirsch, G. P. (1967). *Proc. Nat. Acad. Sci. U.S.* **57,** 1751.
Tanaka, T., and Kanehisa, T. (1972). *J. Biochem.* **72,** 1273.
Thiebe, R., and Zachau, H. G. (1969). *Biochem. Biophys. Res. Commun.* **36,** 1024.
Tinoco, I., Jr., Uhlenbeck, O. C., and Levine, M. D. (1971). *Nature (London)* **230,** 362.
Watson, J. D., and Ralph, R. K. (1967). *J. Mol. Biol.* **26,** 541.
Weinberg, R. A., and Penman, S. (1968). *J. Mol. Biol.* **38,** 289.
Weinberg, R. A., and Penman, S. (1969). *Biochim. Biophys. Acta* **190,** 10.
Weisblum, B., Gonano, F., Von Ehrenstein, G., and Benzer, S. (1965). *Proc. Nat. Acad. Sci. U.S.* **53,** 328.
Weisblum, B., Cherayil, J. D., Bock, R. M., and Soll, D. (1967). *J. Mol. Biol.* **28,** 275.

Weiss, J. F., and Kelmers, A. D. (1967). *Biochemistry* **6**, 2507.
Williamson, R., and Brownlee, G. G. (1969). *FEBS Lett.* **3**, 306.
Winsten, W. A. (1969). *Biochim. Biophys. Acta* **182**, 402.
Zachau, H. G. (1967). *FEBS Symp. 4th* p. 169.
Zapisek, W. F., Saponara, A. G., and Enger, M. D. (1969). *Biochemistry* **8**, 1170.

PART II

Nuclear Proteins

6

Nuclear Proteins

Mark O. J. Olson and Harris Busch

I. Introduction

Studies on nuclear proteins of a variety of cell types have been predicated on the concept that these proteins are somehow responsible for the overall phenotypic definition of the cell, for the initiation, continuation, and cessation of growth and cell division, and for the moment-to-moment regulation of specific genes. Although it has long been postulated that specific proteins are responsible for "gene control" in mammalian systems (Bonner and Ts'o, 1964; Busch, 1965; Hnilica, 1972; Stedman and Stedman, 1950), only recently has evidence been presented to support this idea. In part, the experimental difficulties in testing this concept were related to the inadequacies of the methods available for definition of the number, types, and functions of nuclear proteins. Fortunately, these problems are now being rapidly overcome.

The nuclear proteins are divisible into two broad classes: (*a*) histones, and (*b*) nonhistone proteins (NHP)* which include the nuclear enzymes. The histones are basic and tightly bound to DNA predominantly by multiple ionic bonds. Because of their small number, common features of structure, and relatively limited modifications, they are generally considered to be primarily structural proteins although special regulatory functions cannot be ruled out. Generally, it has been considered that if histones have a gene regulatory role, it involves the fixation of the specific phenotype of cells rather than their responses to various stimuli or environmental changes. On the other hand, cogent evidence has been provided that the nonhistone proteins are involved in gene derepression in eukaryotic cells.

Many of the nonhistone nuclear proteins are bound to RNA either in particles (Volume I, Chapter 2) or lengthy strands such as those of the granular and fibrillar elements of the nucleolus (Volume I, Chapter 2). In addition, many of the RNP particles of the nucleus are either completed ribosomes on the outer layer of the nuclear envelope or preribosomal particles which contain many of the ribosomal proteins. Thus, in the definition of the NHP involved in gene control, it is essential to differentiate those that have structural functions as part of the nucleolar and other particulate elements from those that may be associated specifically with DNA. The nuclear enzymes involved in synthesis of DNA and RNA contain subunits that must also be distinguished from NHP that may have gene control functions.

In recent years, extensive progress has been made in nuclear protein chemistry, particularly of the histones, and this has served as a base for

* Abbreviations used: NHP, nonhistone protein(s); RNP, ribonucleoprotein; DNP, deoxyribonucleoprotein; BME, β-mercaptoethanol; SDS, sodium dodecyl sulfate.

further consideration of their interrelationships with DNA. Moreover, vast improvements in methods for isolation, separation, and analysis of the nonhistone proteins (see p. 244) have given rise to the hope that further comprehension of their functions may be forthcoming. In view of this progress, it is the plan of this chapter to deal with the histones first and then with the NHP, about which less is understood and yet much is being learned at present. At least, at this point, with the aid of two-dimensional electrophoretic separation methods, it is possible to estimate the numbers and types of nuclear proteins (Fig. 1) which may soon be defined functionally. Gene control proteins may soon be identified and hopefully those involved in neoplasia may be differentiated from those that are not.

II. General Characteristics of Histones

Histones were first observed by Miescher (Miescher, 1897), who found them associated with the DNA ("soluble nuclein") of unripe salmon testes. Kossel, whose primary interest was protamines, named them histones (Kossel, 1884). Since the late 1940's when it was first proposed that his-

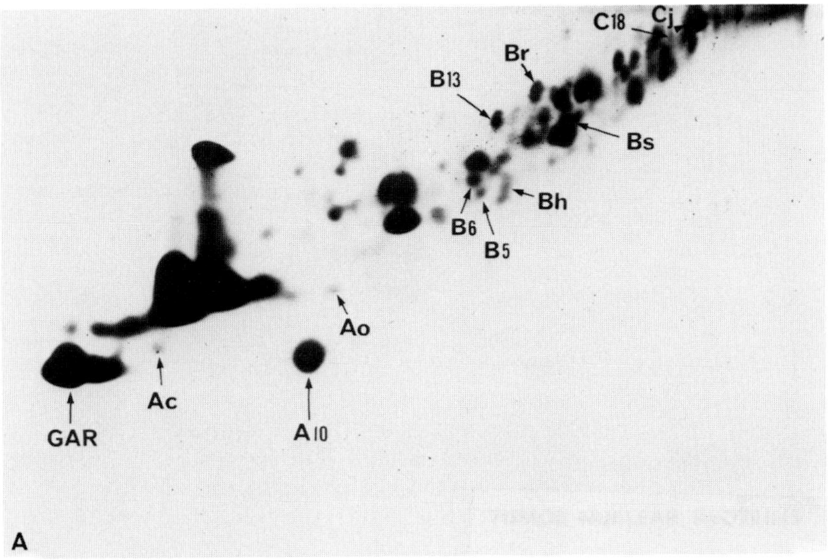

Fig. 1 Two-dimensional electrophoresis of 0.4 *N* H_2SO_4-extracted nuclear proteins of normal liver and Novikoff hepatoma. (A) Novikoff hepatoma nuclear proteins. (B) Diagram of A. (C) Normal liver nuclear proteins. (D) Diagram of C. (From Yeoman *et al.*, 1973.)

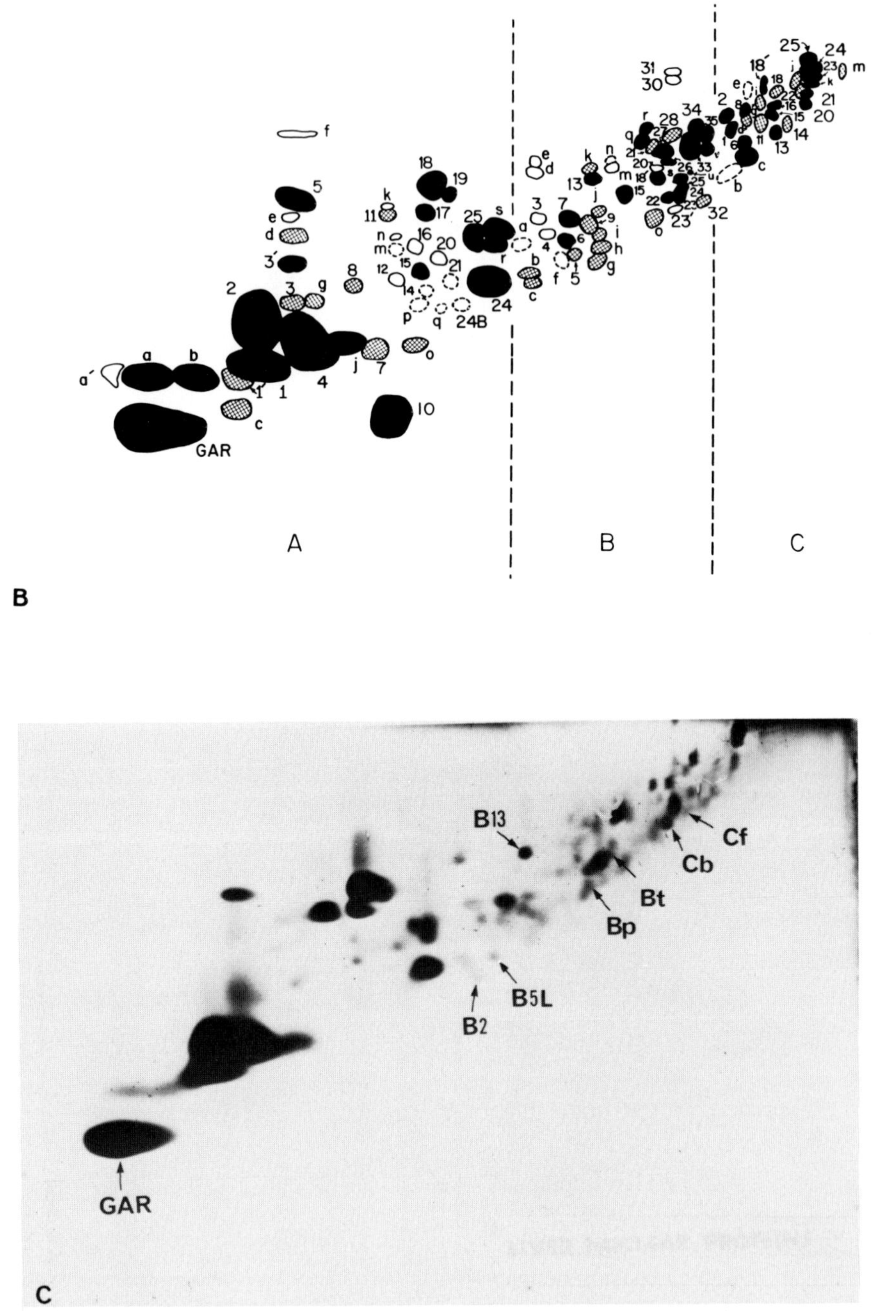

Figs. 1B and C

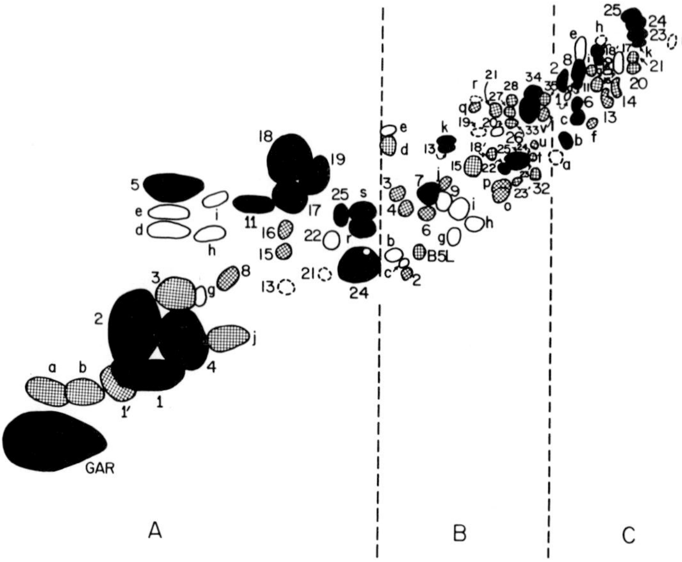

D

Fig. 1D

tones might have a role in gene regulation (Stedman and Stedman, 1950) the histones of a number of species have been purified and chemically defined.

A. *Definition of Histones*

The histones are the basic proteins associated with DNA in the somatic cells of all eukaryotes but not in prokaryotes (Bloch, 1969; Busch, 1965; Johns, 1971).* Histones have a number of common characteristics, including a content of more than 20% basic amino acids, no tryptophan, and small amounts of cysteine or cystine (the arginine-rich histone F_3 is the only one that contains cysteine).

B. *Number and Types of Histones*

Although the actual number of unique species of histones was not clear until recently (Bonner and Ts'o, 1964), the development of high resolution chromatographic and electrophoretic systems showed that in most tissues there are five classes of histones. By polyacrylamide gel electrophoresis (Panyim and Chalkley, 1969a), which is widely used as

* In some recent studies, small amounts of VLR (F_1, I) histones have been found in cytoplasmic ribosomes (Gurley *et al.*, 1973).

a tool for the analysis of histones (Fig. 2), the fastest moving band is the GAR (F2a1, IV) histone (Table I) followed by the triplet of bands of the SLR (F2b, IIb2), AL (F2a2, IIb1), and AR (F3, III) histones. The slowest moving band is the VLR (F1, I) histone. Some properties of calf thymus histones, which are the most characterized, are shown in Table I.

Although there are five major classes of histones in most tissues, there is some heterogeneity within these groups that results from partial chemical modification of specific amino-acid side chains in some histones. For example, the GAR (F2a1, IV) histone of calf thymus separates into two subfractions in certain electrophoretic systems; one band is the parent histone molecule and the other contains one acetyl group on a lysine residue (Panyim et al., 1971). This type of chemical modification—acetylation, methylation, and phosphorylation—greatly increases the number of variants of histones (see Section on histone modification).

In addition to heterogeneity from modification of amino acids, some histone classes contain subfractions which differ from one another in amino acid sequence. Four VLR histone subfractions of calf thymus have been separated by chromatography on amberlite IRC-50 (Kinkade and Cole, 1966a). Three of these are distinguishable by sequence differences of a few amino-acid residues (Kinkade and Cole, 1966b). The number of VLR subfractions varies from tissue to tissue (Bustin and Cole, 1968). Rabbit thymus contains five subfractions while rabbit liver has three and chicken liver contains only two.

Some histones are unique to certain tissues and are found in addition to the five major classes. These include histone V or F2c in avian erythrocytes (Hnilica, 1964; Neelin *et al.*, 1964) and histone T, which is found in the cells of rainbow trout testes (Wigle and Dixon, 1971).

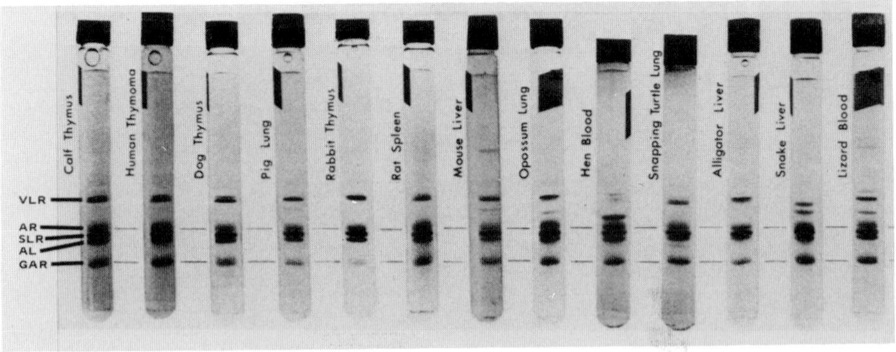

Fig. 2 Polyacrylamide gel electrophoresis patterns of vertebrate histones. (From Panyim and Chalkley, 1969b.)

TABLE I
Major Histone Fractions from Calf Thymus

Class	Fraction	No. of sub-fractions	Lys (%)	Arg (%)	Total residues	Molecular weight	NH_2 terminal	COOH terminal
Lys-rich	Very lysine-rich, VLR (F1, I)	3–4	26.8	1.8	212	19,500–21,000	Ac-Ser	Lys
Intermediate	Slightly lysine-rich, SLR (F2b, IIb2)	1	16.0	6.4	125	13,800	Pro	Lys
	Arginine- and lysine-rich, AL (F2a2, IIb1)	1	10.8	9.3	129	14,000	Ac-Ser	Lys
Arg-rich	Arginine-rich, AR (F3, III)	3	9.6	13.3	135	15,300	Ala	Ala
	Glycine- and arginine-rich, GAR (F2a1, IV)	2	9.8	13.7	102	1,300	Ac-Ser	Gly

A special class of basic proteins, the protamines, is found in the spermatocytes of higher animals. Most information regarding these proteins has come from studies on fish spermatozoa; here they are called protamines (Kossel, 1928). In rainbow trout, for example, there is complete replacement of histone by protamine during spermatogenesis (Marushige and Dixon, 1969). The protamines are characterized by molecular weight of 4000–4500, an arginine content of approximately 70%, and a lack of acidic amino acids (Felix, 1960; Bloch, 1969; Phillips, 1971). The structures of several protamines have been defined by Ando and his associates (Ando *et al.* 1962, 1966, 1967, 1969) as follows:

Clupeine Y-I

10
ALA-ARG-ARG-ARG-ARG-SER-SER-SER-ARG-PRO-ILE-ARG-ARG-ARG-
20
ARG-PRO-ARG-ARG-ARG-THR-THR-ARG-ARG-ARG-ARG-ALA-GLY-
30
ARG-ARG-ARG-ARG

Clupeine Y-II

10
PRO-ARG-ARG-ARG-THR-ARG-ARG-ALA-SER-ARG-PRO-VAL-ARG-
20
ARG-ARG-ARG-PRO-ARG-ARG-VAL-SER-ARG-ARG-ARG-ARG-ALA-
30
ARG-ARG-ARG-ARG

Clupeine Z

10
ALA-ARG-ARG-ARG-ARG-SER-ARG-ARG-ALA-SER-ARG-PRO-VAL-ARG-
20
ARG-ARG-ARG-PRO-ARG-ARG-VAL-SER-ARG-ARG-ARG-ARG-ALA-
30
ARG-ARG-ARG-ARG

Salmine A-I

10
PRO-ARG-ARG-ARG-ARG-SER-SER-SER-ARG-PRO-VAL-ARG-ARG-
20
ARG-ARG-ARG-PRO-ARG-VAL-SER-ARG-ARG-ARG-ARG-ARG-ARG-
30
GLY-GLY-ARG-ARG-ARG-ARG

Iridine Ia

10
PRO-ARG-ARG-ARG-ARG-SER-SER-SER-ARG-PRO-VAL-ARG-ARG-
20
ARG-ARG-ARG-PRO-ARG-ARG-VAL-SER-ARG-ARG-ARG-ARG-ARG-
30
ARG-GLY-GLY-ARG-ARG-ARG-ARG

Iridine Ib

10
PRO-ARG-ARG-ARG-ARG-ARG-ARG-SER-SER-SER-ARG-PRO-ILE-ARG-
20
ARG-ARG-ARG-PRO-ARG-ARG-VAL-SER-ARG-ARG-ARG-ARG-ARG-
30
GLY-GLY-ARG-ARG-ARG-ARG

Iridine II

10
PRO-ARG-ARG-ARG-ARG-SER-SER-SER-ARG-PRO-VAL-ARG-ARG-
20
ARG-ARG-ALA-ARG-ARG-VAL-SER-ARG-ARG-ARG-ARG-ARG-ARG-
30
GLY-GLY-ARG-ARG-ARG-ARG

Much less is known about the nature of basic proteins of mammalian spermatocytes. A protein rich in arginine from bovine spermatozoa was recently isolated and sequenced (Coelingh *et al.*, 1972). In addition a lysine- and arginine-rich protein has been isolated from mouse testis (Lam and Bruce, 1971). A similar protein is found in rat testis (Kistler *et al.*, 1972).

C. Occurrence of Histones

The presence of histones has been confirmed in all multicellular organisms examined (Johns, 1971). In unicellular organisms, histones appear to be present in those species that have a well-defined nucleus such as *Physarum polycephalum* (Mohberg and Rusch, 1969) but have been reported to be absent in *Neurospora crassa** (Dwivedi *et al.*, 1969) and certain other fungi (Leighton *et al*,. 1971). It is not certain what level of organization in the evolutionary scale brought about the appearance of histones.

A recent intriguing finding is that histones are present in some viruses. They have been found in polyoma virus (Frearson and Crawford, 1972) of the rat and in the SV40 virus of the monkey (Estes *et al.*, 1972). Although the function of the histones is not known, it is of interest that both of these are DNA-containing viruses.

III. The Structures of Histones

The primary amino-acid sequences of most of the histones have been defined (Figs. 3–7). The glycine- and arginine-rich (GAR) or histone IV

* Recently, Hsiang and Cole (1973) isolated two slightly lysine-rich histones from *Neurospora crassa* chromatin during exponential growth.

```
                                           10
AcetylSer-Gly-Arg-Gly-Lys-Gly-Gly-Lys-Gly-Leu-

                                          20
Gly-Lys-Gly-Gly-Ala-Lys(Ac)-Arg-His-Arg-Lys(Me)-

                                       30
Val-Leu-Arg-Asp-Asn-Ile-Gln-Gly-Ile-Thr-Lys-Pro-

                          40
Ala-Ile-Arg-Arg-Leu-Ala-Arg-Arg-Gly-Gly-Val-Lys-

                    50
Arg-Ile-Ser-Gly-Leu-Ile-Tyr-Glu-Glu-Thr-Arg-Gly-

            60
Val-Leu-Lys-Val-Phe-Leu-Glu-Asn-Val-Ile-Arg-Asp-

    70                                     80
Ala-Val-Thr-Tyr-Thr-Glu-His-Ala-Lys-Arg-Lys-Thr-

                                       90
Val-Thr-Ala-Met-Asp-Val-Val-Tyr-Ala-Leu-Lys-Arg-

                          100
Gln Gly Arg Thr Leu Tyr Gly Phe Gly Gly COOH.
```

Fig. 3 Sequence of the GAR (F2a1, IV) histone from calf thymus. Basic clusters in this figure and succeeding ones are in italics. (From Ogawa *et al.*, 1969.)

(Fig. 3) was the first to be sequenced (DeLange *et al.*, 1969a; Ogawa *et al.*, 1969). When the sequence of the GAR histone from pea seedling was compared with that of calf thymus (DeLange *et al.*, 1969b), only two conservative amino-acid substitutions were found, an isoleucine substituted for a valine at position 60 and an arginine for a lysine at residue 77. This finding shows that in the course of evolution most of the histone sequence was conserved and most mutations were probably lethal.

The SLR (F2b, IIb2) histone (Fig. 4) has little sequence homology to the GAR histone (Iwai *et al.*, 1970). However, in both sequences, there is clustering of basic residues (underlined) and the center of the molecule is largely hydrophobic.

In the complete sequence of amino acids (DeLange *et al.*, 1972, 1973) of the AR histone (F3, III), the distribution of the amino acids (Fig. 6) is similar to that of the GAR histone in that the center of the molecule is largely hydrophobic. The total sequence of the AR (F3, III) histone from chicken erythrocytes is highly homologous to the calf thymus AR histone with one less cysteine replaced by a serine residue and only one or two other substitutions (Brandt and Von Holt, 1972). Similarly, in carp testes AR (F3, III) histone, one cysteine residue is replaced by a serine (Hooper

5 10
H-Pro-Gln-Pro-Ala-Lys-Ser-Ala-Pro-Ala-Pro-Lys-Lys-

15 20 25
Gly-Ser-Lys-Ala-Val-Thr-Lys-Lys-Ala-Gln-Lys-Lys-Asp-

30 35
Gly-Lys-Lys-Arg-Lys-Arg-Ser-Arg-Lys-Glu-Ser-Tyr-Ser-

40 45 50
Val-Tyr-Val-Tyr-Lys-Val-Leu-Lys-Gln-Val-His-Pro-Asp-

55 60
Thr-Gly-Ile-Ser-Ser-Lys-Ala-Met-Gly-Ile-Met-Asn-Ser-

65 70 75
Phe-Val-Asn-Asp-Ile-Phe-Glu-Arg-Ile-Ala-Gly-Glu-Ala-

80 85 90
Ser-Arg-Leu-Ala-His-Tyr-Asn-Lys-Arg-Ser-Thr-Ile-Thr-

95 100
Ser-Arg-Glu-Ile-Gln-Thr-Ala-Val-Arg-Leu-Leu-Leu-Pro-

105 110 115
Gly-Glu-Leu-Ala-Lys-His-Ala-Val-Ser-Glu-Gly-Thr-Lys-

120 125
Ala Val Thr Lys Tyr Thr Ser Ser Lys OH.

Fig. 4 Sequence of the SLR (F2b, IIb2) from calf thymus. (From Iwai *et al.*, 1970.)

10
Ac-Ser-Glu-Ala-Pro-Ala-Glu-Thr-Ala-Ala-Pro-Ala-Pro-Ala-

20
Glu-Lys-Ser-Pro-Ala-Lys-Lys-Lys-Lys-Ala-Ala-Lys-Lys-Pro-

30 40
Gly-Ala-Gly-Ala-Ala-Lys-Arg-Lys-Ala-Ala-Gly-Pro-Pro-Val-

50
Ser-Glu-Leu-Ile-Thr-Lys-Ala-Val-Ala-Ala-Ser-Lys-Glu-Arg-

60
Asn-Gly-Leu-Ser-Leu-Ala-Ala-Leu-Lys-Lys-Ala-Leu-Ala-Ala-

70
Gly-Gly-Tyr-

Fig. 5 The amino terminal amino-acid sequence of the VLR (F1, I) histone of rabbit thymus, fraction 3. (From Rall and Cole, 1971.)

9 10
H_2N-Ala-Arg-Thr-Lys-Gln-Thr-Ala-Arg-Lys$(CH_3)_{0-2}$-Ser-

14 20
Thr-Gly-Gly-Lys(Ac)-Ala-Pro-Arg-Lys-Gln-Leu-Ala-Thr-

23 27 30
Lys(Ac)-Ala-Ala-Arg-Lys$(CH_3)_{0-2}$-Ser-Ala-Pro-Ala-Thr-

40
Gly-Gly-Val-Lys-Lys-Pro-His-Arg-Thr-Arg-Pro-Gly-Thr-

50
Val-Ala-Leu-Arg-Glu-Ile-Arg-Arg-Tyr-Gln-Lys-Ser-Thr-

60 70
Glu-Leu-Leu-Ile-Arg-Lys-Leu-Pro-Phe-Gln-Arg-Leu-Val-

80
Arg-Glu-Ile-Ala-Gln-Asp-Phe-Lys-Thr-Asp-Leu-Arg-Phe-

90
Gln-Ser-Ser-Ala-Val-Met-Ala-Leu-Gln-Glu-Ala-Cys-Glu-

100 110
Ala-Tyr-Leu-Val-Gly-Leu-Phe-Glu-Asp-Thr-Asn-Leu-Cys-

120
Ala-Ile-His-Ala-Lys-Arg-Val-Thr-Ile-Met-Pro-Lys-Asp-

130 135
Ile-Gln-Leu-Ala-Arg-Arg-Ile-Arg-Gly-Glu-Arg-Ala-COOH.

Fig. 6 The amino-acid sequence of the AR (F3, III) histone from calf thymus. (From DeLange *et al.*, 1972.)

et al., 1973). The sequence is identical to calf thymus AR (F3, III) histone at all other positions.

The amino terminal sequences of several fractions of the VLR (FI, I) histone from rabbit thymus were also determined (Rall and Cole, 1971). One of these (fraction 3) is shown in Fig. 5. It is of interest that unlike the other histones the amino terminal half of this molecule has a composition not unlike common enzymes such as ribonuclease or lysozyme (Bustin and Cole, 1970). However, the carboxyl terminal half of this histone is very rich in lysine and proline.

The most recently completed histone sequence (Fig. 7) is that of the AL (F2a1, IIb1) histone (Olson *et al.*, 1972; Sugano *et al.*, 1972; Yeoman *et al.*, 1972). Strikingly, the first nine residues of the AL and GAR histones are identical except for the glutamine at position 6:

```
           1                   5                  10
AL:   AcetylSer-Gly-Arg-Gly-Lys-Gln-Gly-Gly-Lys-Ala-Arg-Ala-Lys — —
GAR: AcetylSer-Gly-Arg-Gly-Lys — Gly-Gly-Lys-Gly-Leu-Gly-Lys-Gly-Gly
          15                  20                  25
AL:   -Ala-Lys-Thr-Arg-Ser-Ser-Arg-Ala-Gly-Leu-Gln-Phe-Pro-Val-Gly-Arg-
GAR: -Ala-Lys — Arg-His-Arg-Lys-Val-Leu-Arg-Asp-Asn-Ile-Gln-Gly-Ile-
```

This finding suggests these regions of the AL and GAR histones have similar functions (Olson *et al.*, 1972). It is of interest that three histones have the same amino terminal, *N*-acetylserine, and three have a carboxyl terminal lysine.

IV. Histones and Chromatin Structure

Chromatin is the functional form of the eukaryotic chromosome in which DNA is found in close association with various other macromolecules. The structure of chromatin is dependent on the cell type, its cycle stage, its age, and the method of extraction. Metaphase chromatin contains 13–17% DNA, 8–15% RNA, and 68–79% protein (Comings, 1972). Most studies were performed on interphase chromatin which represents chromatin fibers largely in an extended form. Interphase chromatin contains about 25% DNA, less RNA (3–4%), and about 70% protein. The histone to NHP ratio ranges from 0.25:1 to approximately 1:1 and varies from cell type to cell type (Busch, 1965; Busch and Steele, 1964; Comings, 1972; DuPraw, 1968; Georgiev, 1969; Hnilica, 1972; Klyszejko-Stefanowicz and Polanowska, 1971; MacGillivray *et al.*, 1972b; Schjeide, 1970).

Chromatin apparently has ordered geometrical features in the nucleus (Bartley and Chalkley, 1968; Panyim *et al.*, 1971; Wagner and Spelsberg, 1971). It has been suggested that some increases in RNA synthesis are associated with structural modifications in chromatin that make new sites of DNA available for transcription (Kleiman and Huang, 1971; Von Hippel and McGhee, 1972; Wong *et al.*, 1972; also, see Chapter 4).

Although the structure of chromatin is not well defined, a number of recent studies have provided some insight into how histones fit into the chromatin matrix (Fig. 8) and how they interact with DNA (see Chapter 4). Whether histones bind to the large or small groove of the DNA double helix has been an open question. The elucidation of the amino-acid sequence of the GAR histone (F2a1, IV) allowed construction of space-filling models in which the amino terminal region of the GAR histone (Shih and Bonner, 1970a; Sung and Dixon, 1970) was fitted as an α-helix in the large groove of the DNA double helix (Fig. 8). The α-amino groups of lysine residues in this model aligned to form salt bridges with the phos-

```
        1                                             10
AcetylSer-Gly-Arg-Gly-Lys-Gln-Gly-Gly-Lys-Ala-Arg-Ala-Lys-

    15                  20                  25
Ala-Lys-Thr-Arg-Ser-Ser-Arg-Ala-Gly-Leu-Gln-Phe-Pro-Val-

        30                  35                  40
Gly-Arg-Val-His-Arg-Leu-Leu-Arg-Lys-Gly-Asn-Tyr-Ala-Glu-

            45                  50                  55
Arg-Val-Gly-Ala-Gly-Ala-Pro-Val-Tyr-Leu-Ala-Ala-Val-Leu-

                60                  65
Glu-Tyr-Leu-Thr-Ala-Glu-Ile-Leu-Glu-Leu-Ala-Gly-Asn-Ala-

 70                  75                  80
Ala-Arg-Asp-Asn-Lys-Lys-Thr-Arg-Ile-Ile-Pro-Arg-His-Leu-

    85                  90                  95
Gln-Leu-Ala-Ile-Arg-Asn-Asp-Glu-Glu-Leu-Asn-Lys-Leu-Leu-

       100                 105                 110
Gly-Lys-Val-Thr-Ile-Ala-Gln-Gly-Gly-Val-Leu-Pro-Asn-Ile-

           115                 120                 125
Gln-Ala-Val-Leu-Leu-Pro-Lys-Lys-Thr-Glu-Ser-His-His-Lys-

           129
Ala-Lys-Gly-Lys-COOH.
```

A

***Fig.* 7** (A) Sequence of the AL (F2a2, IIb1) histone from calf thymus. (From Yeoman *et al.*, 1972.) (B) Molecular model of the AL histone.

phate residues in DNA. A similar arrangement is possible with the AL (F2a2, IIb1) histone (Olson *et al.*, 1972). Since histone I does not inhibit the binding of actinomycin D, it may also bind to the major groove of DNA (Olins, 1969; Pietsch, 1969).

Numerous physical studies including circular dichroism studies showed that the GAR histone assumes a more α-helical structure when bound to DNA than when free in solution (Shih and Fasman, 1971). Conversely, when the lysine-rich histone is bound to DNA, the physical behavior is that of a protein in the extended form (Shih and Fasman, 1972). Proton magnetic resonance studies on the SLR (F2b, IIb2) histone support the idea that basic regions are the primary sites of interaction (Bradbury *et al.*, 1971). Nonbasic regions seem to have secondary structures that may be sites for histone–histone interactions. Thermal denaturation profiles of various DNA histone complexes indicate that the effects of each histone on DNA structure are quite specific (Ansevin and Brown, 1971). Similar studies on the interaction of half-molecules of the SLR (F2b, IIb2) histone

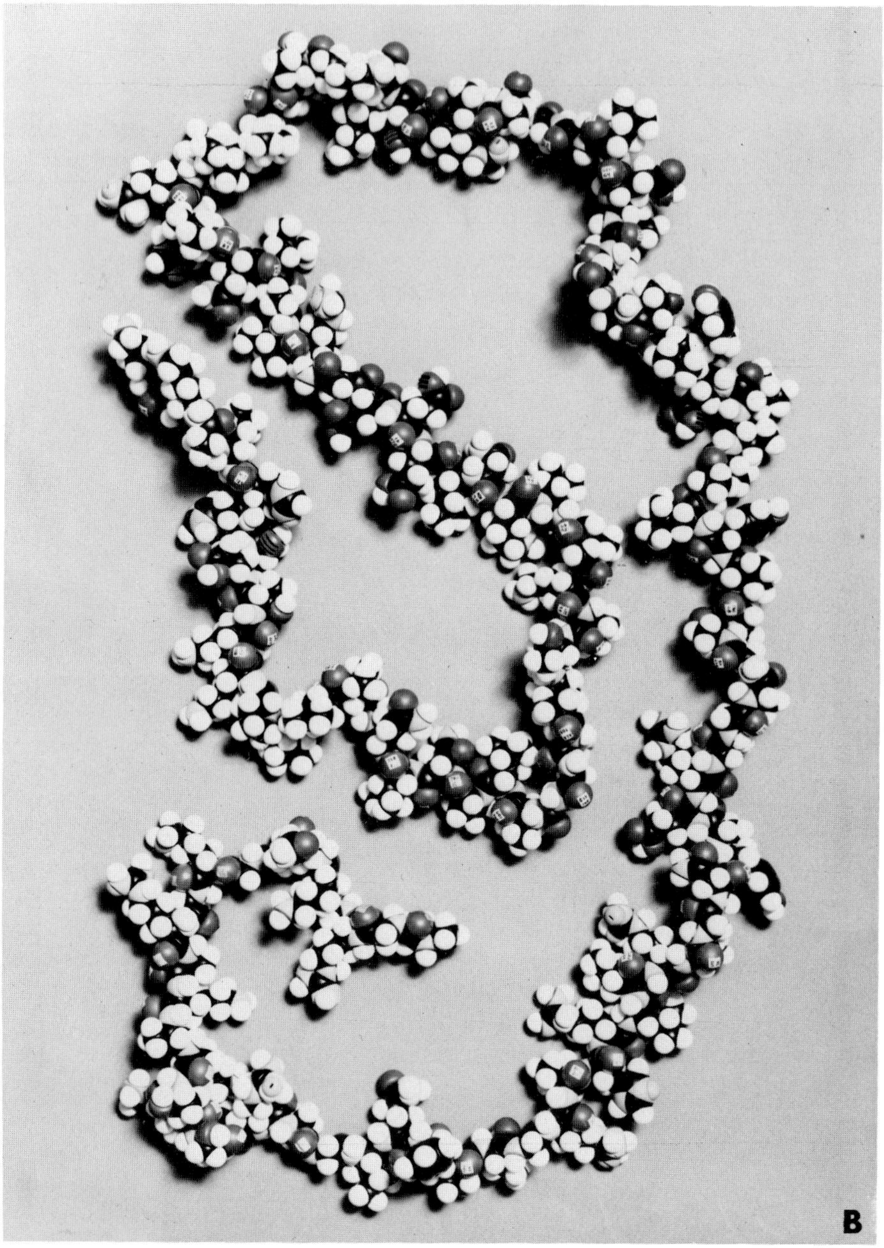
B

Fig. 8 DNA–histone molecular model showing dense packing of atoms in the complex. Left: GAR (F2a1, IV) histone of the large groove of DNA. Right: "naked" DNA double helix.

with DNA show that the amino terminal half of this molecule stabilizes the DNA against thermal denaturation better than the carboxyl terminal portion (Li and Bonner, 1971).

Various histones differ in their apparent manner of attachment to the chromatin matrix. Histones appear to interact with DNA both by electrostatic and hydrophobic types of bonds (Bartley and Chalkley, 1972). The lysine-rich groups of histones (F1, F2b, and F2a2) are more easily extracted with NaC1 than are the arginine-rich (F2a1 and F3) histones (Bonner *et al.*, 1968b). If deoxycholate, which apparently specifically binds to hydrophobic regions, is used as a dissociating agent, the arginine-rich and slightly lysine-rich histones are selectively removed while the very lysine-rich histones are last to be extracted (Hadler *et al.*, 1971).

Recently Clark and Felsenfeld (1972) were able to remove selectively from chromatin all histones except the GAR (F2a1, IV) and the AR (F3, III) histones using urea–calcium chloride–phosphate buffers. After digestion of this complex with staphylococcal nuclease, it was found that these two arginine-rich histones remained firmly bound to intact segments of DNA which were rich in guanine and cytosine. Thus, there may be some general specificity of interaction of types of histones for compositions of DNA.

Although the number of basic amino acids nearly equals the number of negatively charged nucleotides in chromatin (Vendrely *et al.*, 1960), there is evidence to suggest that the proteins are unevenly distributed on the DNA. Titration of chromatin with polylysine, nuclease digestion, and exchange of protein between radioactively labeled and unlabeled DNA indicate that one-third to one-half of the DNA is "open" and not covered by protein (Comings, 1972; Itzhaki, 1970). Other studies have shown that the accessibility of DNA in chromatin was such that up to 71.5% of the DNA was digested (Mirsky and Silverman, 1972).

The proteins associated with chromatin have basic regions, presumably contributed mostly by histones, that are exposed and are not tightly bound to DNA. Modification of rabbit liver chromatin with acetic anhydride reached a limiting value of acetylation of 54 amino-acid residues/100 base pairs (Simpson, 1972). A subsequent study employing limited digestion of chromatin with trypsin showed that 55 peptide bonds per 100 DNA base pairs were susceptible to cleavage (Simpson, 1972). Tryptic digestion did not alter thermal denaturation profiles appreciably, indicating that sites of DNA–histone interaction were not changed. However, other physical characteristics such as flow dichroism, specific viscosity, and circular dichroism of the modified chromatin appeared to be identical to those of free DNA. This supports the idea that chromatin has a supercoil conformation and that the histones act to promote supercoiling. The center regions of histones may interact with one another to produce this effect while the ends of the proteins bind to DNA. X-ray diffraction studies also support a supercoil model for nucleohistone (Pardon and Wilkins, 1972).

The DNA double helix in an uncovered form would tend to form a rigid, rodlike structure which would not fit well inside the cell nucleus. It is, therefore, logical that molecules in large quantities such as histones would be necessary to promote packing of the DNA in a suitable structure such as a supercoil.

V. Modified Amino Acids in Histones

Figures 3–7 show that a number of amino acids in histones are modified by phosphorylation, acetylation, and methylation. The presence of these modifying groups introduces microheterogeneity into some of the histone fractions. The structures of the modified amino acids are shown in Table II.

TABLE II
Chemical Modifications of Histones

Modifying group	Structure of modified amino acid	Histones modified	References
Phosphoryl	Phosphoserine[a] 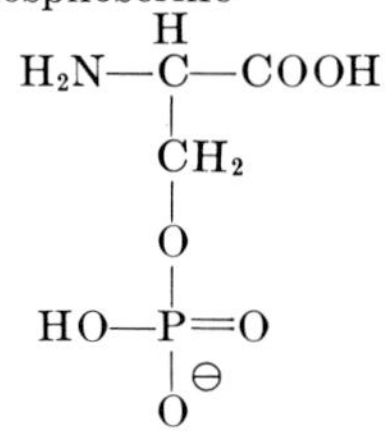	GAR (IV)	Marushige *et al.* (1969)
		AL (IIb$_1$)	Marushige *et al.* (1969) Sung *et al.* (1971)
		AR (III)	Hayashi and Iwai (1970) Marushige *et al.* (1969)
		VLR (I)	Balhorn *et al.* (1971) Hayashi and Iwai (1970) Kleinsmith *et al.* (1966) Langan (1969) Marushige *et al.* (1969) Sherod *et al.* (1970)
		SLR (IIb$_2$)	Hayashi and Iwai (1970) Marushige *et al.* (1969)
		V	Adams *et al.* (1970) Murray and Milstein (1967)
Acetyl[b]	α-*N*-acetylserine 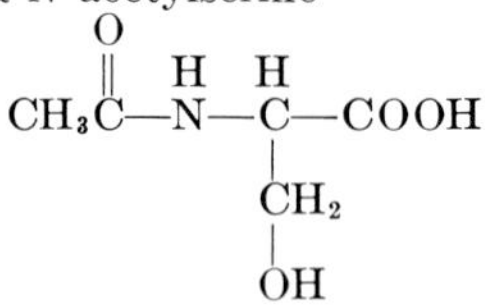	AL, GAR	Phillips (1963) Phillips (1968)
		VLR	Rall and Cole (1971)
	ε-*N*-acetyllysine 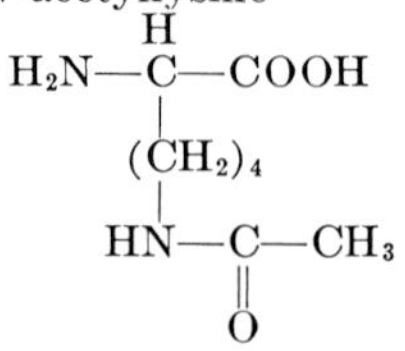	GAR and AR (best acceptors)	Panyim and Chalkley (1969) Pogo (1966, 1968) Pogo *et al.* (1969) Candido and Dixon (1971) Candido and Dixon (1972) DeLange *et al.* (1969) DeLange *et al.* (1972) Vidali *et al.* (1968)
		Other histones AL, SLR, VLR	Candido and Dixon (1971) Candido and Dixon (1972) Vidali *et al.* (1968)

TABLE II (*Continued*)

Modifying group	Structure of modified amino acid	Histones modified	References
Methyl	ε-*N*-mono- and dimethyllysine[c] 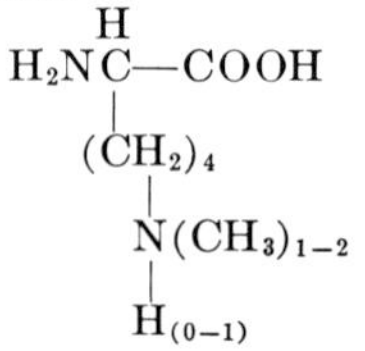	GAR, AR (major)	DeLange *et al.* (1969) DeLange *et al.* (1972) Ogawa *et al.* (1969) Sekeris *et al.* (1967)
	3-Methylhistidine 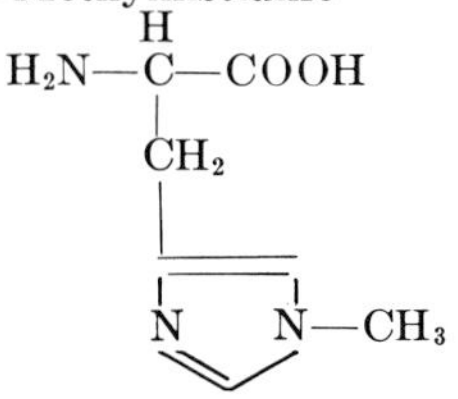	I and V	Gershey *et al.* (1969)
	ω-*N*-methylarginine[d] 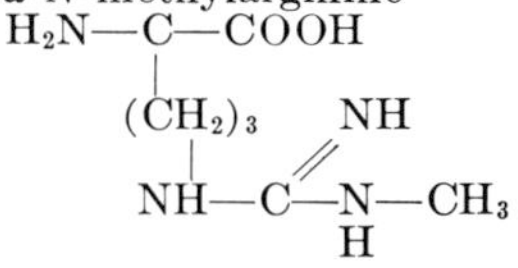	Crude histone	Paik and Kim (1970)

[a] Phosphothreonine has been reported to be present but was not observed in sequence studies (Kleinsmith *et al.*, 1966; Ord and Stocken, 1966).
[b] The presence of O-acetyl groups has been noted in some instances (Gallwitz and Sekeris, 1969; Nohara *et al.*, 1966, 1968; Pogo, 1968; Pogo *et al.*, 1969).
[c] Trimethyllysine has been reported but not confirmed in peptides derived from histones (Hempel *et al.*, 1968 a, b; Paik and Kim, 1970).
[d] *In vitro* studies have shown that histones can be enzymatically modified to contain 2-methylguanidinomethylarginine (Kaye and Sheratzky, 1969).

A. Phosphorylation of Histones

Phosphate is esterfied to the hydroxyl of serine and possibly to a lesser extent to threonine; under various conditions, phosphate is found in all of the major histones. The placement of a negatively charged phosphate moiety on a positively charged protein might be expected to change the conformation of that protein or to lessen its binding affinity for DNA. Therefore, phosphorylation could act to (*a*) "uncover" specific genes which are repressed by histones, (*b*) modify the structure of chromatin in preparation for DNA synthesis and cell division, or (*c*) aid in the removal of the histones from DNA to facilitate replacement by another type of protein. Table III shows a number of correlations of histone phosphorylation with biological events.

TABLE III
Phosphorylation of Histones

Possible function of phosphorylation	Biological event or system	Site of phosphorylation	References
Gene activation	Glucagon stimulation	VLR (F1, I) histones	Langan (1968) Langan (1969) Langan (1969)
Cell division	Regenerating liver	VLR (F1, I) histones	Balhorn *et al.* (1971); Buckingham and Stocken (1970); Ord and Stocken (1968); Stevely and Stocken (1968)
		AL (F2a2, IIb1) histone	Sung *et al.* (1971)
		All histone fractions	Gutierrez-Cernosek and Hnilica (1971)
	Morris hepatomas	VLR (F1, I) histones	Balhorn *et al.* (1972)
	Cultured Ehrlich ascites and Morris hepatoma cells	VLR (F1, I) histones	Sherod *et al.* (1970) Balhorn *et al.* (1972)
	Cultured Chinese hamster ovary cells	VLR (F1, I) and SLR (F2b, IIb2)	Shelton and Allfrey (1970); Shepherd *et al.* (1971)
Replacement of histones	Spermiogenesis in trout testes	All major fractions	Marushige *et al.* (1969)

A specific kinase of liver catalyzes the *in vitro* incorporation of phosphate into histones and this activity is stimulated by adenosine 3′5′-cyclic phosphate (cyclic AMP) (Langan, 1968). *In vivo* experiments demonstrated that a specific serine residue in a lysine-rich histone is phosphorylated in liver cells treated with dibutyryl cyclic AMP (Langan, 1969a). Glucagon and, to a lesser extent, insulin stimulated this cyclic AMP-mediated phosphorylation of a lysine-rich histone in liver (Langan, 1969b).

The level of glucagon-stimulated phosphorylation is on the order of one percent of the total lysine-rich histones (Langan, 1969a). This is consistent with the activation of a small percentage of the genome of nondividing cells. However, this is the only well-defined instance which correlates histone phosphorylation with gene activation; the exact molecular events are still obscure.

A second type of phosphorylation has been observed which seems to be associated with cell division in regenerating liver and cell cultures (Table

II). With high-resolution electrophoretic techniques for the study of lysine-rich histone phosphorylation (Balhorn *et al.*, 1971; Sherod *et al.*, 1970), one subfraction (the major one) of VLR (f1, I) histones reaches its maximum phosphorylation level 29 hours after partial hepatectomy (Balhorn *et al.*, 1971). The rapidly dividing Ehrlich ascites tumor cells (Sherod *et al.*, 1970) or Morris hepatoma cells in culture (Balhorn *et al.*, 1972b) have a much higher level of phosphorylation in the lysine-rich histones than normal tissues (Balhorn *et al.*, 1971). When these studies were extended to a series of Morris hepatomas with varying growth rates (Balhorn *et al.*, 1972a), a positive correlation was found between growth rate and the degree of phosphorylation of one of the VLR (f1, I) histone subfractions (Fig. 9). Phosphorylation of the VLR (f1, I) histone may be an obligatory event in DNA synthesis or cell replication since the F1 histone is phosphorylated only during DNA syntheses (Balhorn *et al.*, 1972c). Since this type of phosphorylation is largely present in rapidly replicating

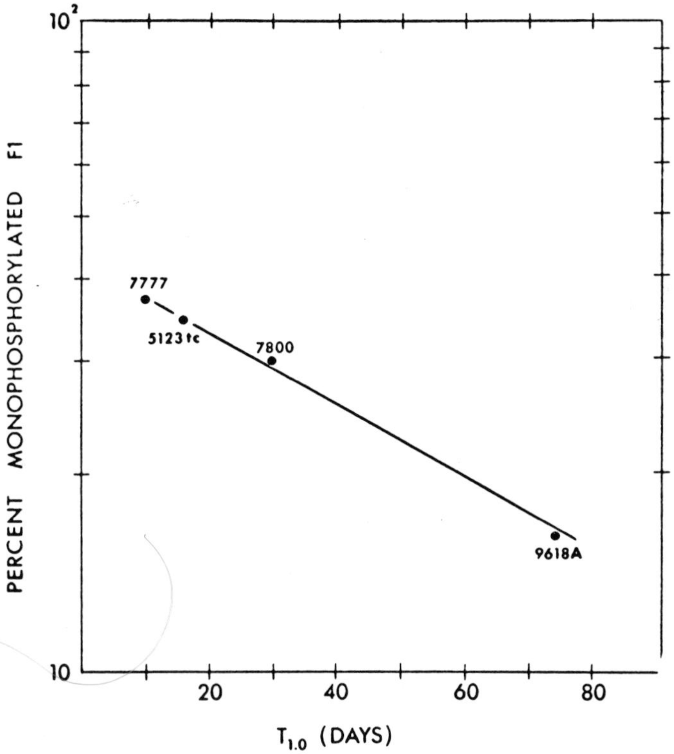

Fig. 9 Phosphorylation of VLR (F1, I) histone correlated with growth rate. T_{10} is the time required for a transplanted tumor to grow to 1 gm. The numbers on the data points indicate the type of Morris hepatoma studied. (From Balhorn *et al.*, 1972.)

tissues and essentially absent in stationary tissues, this modification is probably related more to cell division than gene expression.

A third type of histone phosphorylation occurs in the transformation of somatic tissues into sperm cells. This process is accompanied, first, by an increase in histone content and a decrease in nonhistone proteins of trout testis and, second, by a complete replacement of the histones by the highly basic protamines. A marked decrease in template activity of DNA results from this process. In this replacement process, preformed histones are phosphorylated (Marushige and Dixon, 1969). This phosphorylation may be related to the removal of histones from DNA in preparation for replacement by protamine or may serve as a signal for specific histone proteases (Marushige *et al.*, 1969). However, it is equally plausible that phosphorylation and subsequent dephosphorylation are important in the "annealing" of highly basic portions of histones onto the DNA double helix (Candido and Dixon, 1972; Louie and Dixon, 1972).

B. *Acetylation of Histones*

Two distinct types of acetylation are found in the histones: (*a*) acetylation of amino terminal moieties and (*b*) "internal" acetylation of the ϵ-amino groups of lysine. The amino terminal serine residues of the GAR (F2a1, IV), AL (F2a2, IIb1) and lysine-rich (F1, I) histones are N-acetylated (Phillips, 1963, 1968a; Rall and Cole, 1971). During the lifetime of the histone, the α-NH_2-acetyl groups do not appear to turn over appreciably (Gershey *et al.*, 1968; Wilhelm and McCarty, 1970). In the F2a (GAR and AL) histones, *N*-acetylseryl–tRNA has been reported to serve as the initiator amino acid for synthesis of these proteins (Liew *et al.*, 1970). Chain initiation in these histones seems to be a unique case since *N*-formylmethionine or methionine has been found to be the initiator in other systems.

The acetyl groups of the ϵ-amines of lysine are metabolically active, turn over at varying rates in different tissues, and are found in fractional molar ratios at specific sites in histone molecules. These acetyl groups have been found in all histones except the VLR (F1, I) histones.

C. *Physiological Significance of Histone Acetylation*

The physiological significance of histone acetylation is uncertain. Acetylation of the free ϵ-amino groups of lysine eliminates the positive charge of this group. As with phosphorylation, this modification should be capable of changing the charge characteristics, the conformation or the mode of binding of the histone to DNA.

Some examples of correlation of acetylation with events in cell cycle or gene expression have been reported (Allfrey, 1969). In various calf organs (Panyim and Chalkley, 1969b), the rapidly dividing tissues contain roughly equal amounts of the acetylated form of the GAR histone (F2a1, IV) and the nonacetylated parent histone. Slowly dividing tissues contain considerably more of the unmodified GAR histone. In regenerating liver, an increase in acetylation of arginine-rich histones was reported to precede the increase in RNA synthesis (Pogo, 1968; Pogo *et al.*, 1969). Similar changes occur in human lymphocytes stimulated with phytohemagglutinin (Pogo, 1966). However, turnover rates of acetyl groups in arginine-rich histones do not vary appreciably with growth rate in a series of Morris hepatomas and liver (Byvoet and Morris, 1971). Moreover, in Novikoff hepatoma the acetyl groups of histones turn over much more slowly than in normal liver. Although this effect could be explained by the low level of histone deacetylase found in Novikoff hepatoma (Libby, 1970), it does not support the concept of an important control function for acetylation.

Although there are temporal relationships between histone acetylation and various events in the cell cycle or gene activation, no direct relationship has been proved. Acetylation may be one of several processes that must occur before a specific gene segment is selected for activation (Allfrey, 1971). Alternatively, partial acetylation of specific residues of histones may serve to neutralize the repulsive effects of closely spaced amino groups during histone binding to DNA (Louie and Dixon, 1972). During the slow annealing of histones onto the DNA double helix, acetylases and deacetylases may serve to properly orient the amino groups of the histones.

D. Methylation of Histones

ϵ-Methylation of lysine (Murray, 1964) in histones is highly specific. The GAR (F2a1, IV) histone of calf thymus is methylated at lysine 20 (DeLange *et al.*, 1969a); 75% of the molecules contain a dimethyl- and 25% a monomethyllysine (Ogawa *et al.*, 1969). Pea seedlings do not contain methylated residues in the GAR histone (DeLange *et al.*, 1969b).

In addition to methylated lysine, methylated histidine and arginine have been found (Gershey *et al.*, 1969). 3-Methylhistidine is present in duck erythrocyte histone fractions I and V. Calf thymus histones contain small amounts of methylarginine while rat liver histones contain significant amounts of this derivative (Paik and Kim, 1970c).

In contrast to phosphorylation and acetylation reactions which occur in S phase or early in the cell cycle (Shepherd *et al.*, 1971a, b), methylation of histones seem to occur late in the cell cycle (Sedwick *et al.*, 1972;

Tidwell *et al.*, 1968). In addition, the turnover rate of methyl groups seems to be similar to that of the parent histones, indicating that methylation is essentially an irreversible process (Byvoet *et al.*, 1972).

E. *Function of Methylation of Histones*

Because of the small size of the methyl group and the relatively small change in the p*K* of the ε-amino group by introduction of the methyl moiety, any change in the general characteristics of the protein must be subtle. Changes in conformation or in susceptibility to specific proteolysis may result. Methylation may play a role in the condensation of chromatin late in the cell cycle prior to mitosis (Gershey *et al.*, 1969; Tidwell *et al.*, 1968).

VI. Histone Synthesis and Metabolism

A. *Site of Histone Synthesis*

Although it seemed possible that histone synthesis occurred in the nucleus (Allfrey, 1954; Allfrey *et al.*, 1964; Flamm and Birnstiel, 1964; Reid and Cole, 1964; Reid *et al.*, 1968), evidence for cytoplasmic synthesis of histones first emerged from cytochemical studies on developing grasshopper spermatids (Bloch and Brack, 1964). Later, HeLa cells synchronized by selective detachment of cells in mitosis were found to synthesize histones on isolated small polysomes during S phase. A rapidly labeled 7–9 S mRNA fraction (see Chapter 5) was associated with these polysomes (Borun *et al.*, 1967).

Cytoplasmic histone synthesis was confirmed by studies on HeLa cell microsome preparations (Gallwitz and Mueller, 1969a, b, 1970) and on sea urchin embryo polyribosomes (Kostraba and Wang, 1971; Nemer and Lindsay, 1969). Messenger RNA coding for histones has been isolated from HeLa cells and translated into protein by cell free extracts of mouse ascites tumor (see Chapter 5).

Although there is now good evidence that histones are synthesized in the cytoplasm in several cell types, it is not yet clear whether all synthesis occurs in the cytoplasm. Since some cytoplasmic contamination of nuclear preparations may occur and vice versa, more definitive studies are needed to finally establish whether or not histone synthesis occurs in the nucleus.

B. *Time of Synthesis*

The synthesis of DNA and histones generally appear to be closely coupled in the S phase. Early histochemical and autoradiographic studies indicated that histones and DNA were synthesized concurrently in several tissues including liver and mouse fibroblasts (Bloch and Godman, 1955), in *Euplotes eurytomus* (Prescott, 1966), and in HeLa cells and Chinese hamster cells in tissue culture (McClure and Hnilica, 1970; Robbins and Borun, 1967). Histone synthesis stopped when DNA synthesis was inhibited by hydroxyurea in HeLa cell microsomal preparations (Gallwitz and Mueller, 1970). In some studies, only part of the histone synthesis in HeLa cells is parallel to DNA synthesis (Sadgopal and Bonner, 1969), and different histone fractions have differences in their degrees of dependence on the synthesis of DNA (Chalkley and Maurer, 1965). The degree of concurrence of synthesis of DNA and histones may vary from cell to cell depending on the physiological or developmental state of each cell. Rapidly proliferating and relatively undifferentiated cell types seem to have less correlation between histone and DNA synthesis. Significant amounts of histone synthesis, however, have been reported during G_2, M, and G_1 phases in addition to the major amount of synthesis during S phase (Gurley and Hardin, 1968).

Thus, there is general agreement that the bulk of histone synthesis occurs concurrently with DNA synthesis; some histone synthesis and turnover may occur during other times of the cell cycle independent of DNA synthesis. Operationally, a small degree of asynchrony in cell populations could account for apparent independence from DNA synthesis.

C. Histone Turnover

The DNA–histone complex is metabolically stable as shown by the lack of significant differences in turnover rates of DNA and histones (Byvoet, 1966). In Chinese hamster cells in tissue culture all histones except the lysine-rich group were metabolically stable. However, during inhibition of DNA synthesis, all the histones did turn over slowly and the lysine-rich histones turned over faster than all other types of histones. In HeLa cells and mouse mastocytes there was essentially no turnover of histones unrelated to new DNA synthesis over eight generations (Hancock, 1969).

In general, histone turnover apears to be very slow in the systems studied to date. Any observed turnover of histones may be attributed to cell death, cell replacement, or possibly gene regulation.

D. Differential Rates of Histone Synthesis

There are conflicting reports of differential rates of synthesis of various histone subfractions. Differences in rates of biosynthesis of histone fractions were found in liver and hepatoma (Evans *et al.*, 1962; Holbrook *et al.*, 1960, 1962). In Walker tumor [^{14}C]lysine was incorporated faster in the arginine-rich histones than in the VLR histones. In other studies, [^{14}C]leucine incorporation into lysine-rich histones was greatest in tissues undergoing cell division while incorporation was more active in the arginine-rich histones in cells not synthesizing DNA at a significant rate (Chambon *et al.*, 1973).

During pregnancy and lactation, the rates of synthesis for lysine-rich and arginine-rich histones were greater than for slightly lysine-rich fractions (Stellwagen and Cole, 1969b). In mouse mammary gland explants, a mixture of cortisol, prolactin, and insulin first depressed the incorporation of lysine into one of the lysine-rich subfractions and later elevated incorporation into another. There is need for care in interpretation of these results because of possible pool effects and possible contamination with rapidly labeled NHP in preparations used for these metabolic studies (Hnilica, 1972).

VII. Functions of Histones

The histones are found in the chromatin of all eukaryotic organisms but not in the prokaryotes. Along with the nuclear membrane, the presence of histones is a feature that distinguishes the eukaryotes from prokaryotes. Although the proposal that histones repress genome segments was supported by *in vitro* studies on transcription, neither the precise roles they play in specific genetic regulation nor as structural elements of chromatin are well defined.

A. *Histones in Genetic Restriction*

Studies on pea seedlings first showed that histones added to DNA dramatically reduced its transcription by RNA polymerase (Huang and Bonner, 1962). However, the insolubility of the DNA–histone complex is such that this addition of histones simply resulted in unavailability of DNA as a template. Subsequent studies have shown that stepwise removal of histones from chromatin increases its template activity (Allfrey *et al.*, 1963; possibly also because of its greater solubility (Johns and Hoare, 1970; Sonnenberg and Zubay, 1965).

Since histone synthesis is closely coupled to DNA synthesis, it has been presumed that histones are either needed immediately to repress segments of newly synthesized DNA or that they are necessary for functional chromatin structure. Histones may also be required to suppress further synthesis of DNA (Hnilica, 1972; Wang, 1968) or to prevent its cleavage by DNase (Clark and Felsenfeld, 1971).

B. Histone Specificity

Even if such evidence indicated participation of histones in gene repression, the problem of the specificity of histones for such repression is a difficult one. There are simply not enough unique species of histones to serve the function of specifically repressing the multitude of genes that need to be repressed in each cell type. Further, there is no evidence that specific histones combine with specific genes (Shih and Bonner, 1970b).

This is not to mean that histones exhibit no tissue or species specificity. Although the electrophoretic patterns of histones are generally very similar from tissue to tissue, the relative amounts of each type of histone do vary within a given organism (Panyim *et al.*, 1971). Furthermore, some tissues contain unique species of histones. A notable case is found in avian erythrocytes where there is a unique histone (F2c, V) that is rich in lysine and serine (Hnilica, 1964; Neelin *et al.*, 1964). This histone appears in chromatin during maturation of erythrocytes in birds (Dick and Johns, 1969), and the nucleus becomes genetically inactive. The appearance of histone V is believed to be associated with the complete repression of the erythrocyte genome.

Tissue specificity of histones has been reported for lung which has little self-replication; an extra histone (1°) is present in lung that is absent in rapidly dividing cells (1°) such as ascites tumors (Panyim and Chalkley, 1969c). This histone is similar to the lysine-rich histones but has a greater electrophoretic mobility.

In mouse mammary gland explants, insulin, cortisol, and prolactin cause a reduced incorporation of lysine into one of the lysine-rich histone subfractions and an increased incorporation into another (Hohman and Cole, 1971). It has been suggested that this restructuring of newly synthesized chromatin could alter the state of differentiation of these cells.

Another instance of a relation between histones and gene regulation is the observation that one fraction of the VLR (F1, I) histones of rabbit thymus (fraction 3) is readily phosphorylated by a specific histone kinase whereas fraction 4 does not respond to the enzyme (Langan *et al.*, 1971). The serine residue which is phosphorylated in fraction 4 is replaced by an alanine residue in fraction 3. It is conceivable that the chromatin of a cell type programed to respond to phosphorylation could have a different distribution of the fractions 3 and 4 of the VLR histones in the chromatin matrix.

In support of earlier chromatographic evidence of VLR histone species and tissue specificity are recent immunochemical studies. Stollar and Ward (1970) were able to prepare antisera against the five histone classes of calf thymus by injecting the RNA–histone complexes into rabbits. Using this method on isolated calf thymus VLR histone subfractions, Bustin and Stollar (1972) showed considerable organ and species specificity. For instance, antisera to calf thymus VLR subfraction III reacted best with calf thymus VLR subfraction III, moderately well with whole calf thymus VLR, and poorly with calf liver whole VLR histones. It is of interest that the antigenic determinants reside in the highly basic carboxyl terminal portion of the VLR histone (Bustin and Stollar, 1973). An important question regarding these studies is whether the specificity observed is related to functional properties of the histone subfractions or if it is a reflection of evolutionary drift in the histone structure.

The above examples point to a special role that seems to be played by the VLR histones. Although the role played by most histones in gene expression appears to be somewhat limited, it is possible that they act in concert with the nonhistone proteins. Because of the large amounts of histones present in the chromatin, their function would appear to be largely structural although a regulatory role cannot yet be excluded.

VIII. Nonhistone Nuclear Proteins

In view of the uncertainties about the role of the histones as gene regulatory proteins, studies were initiated in the authors' and other laboratories on the possible gene regulatory role of the NHP more than a decade ago (Busch, 1965).* One proposal for possible gene regulation by the NHP was that these proteins might be capable of forming complexes with histones and displacing them from DNA (Fig. 10); another was that such proteins bind directly to DNA (Busch, 1965; Chaudhuri *et al.*, 1972; MacGillivray *et al.*, 1972b; Stellwagen and Cole, 1969a; Ursprung *et al.*, 1968). Although early studies were made on turnover, composition, and terminals of these proteins in mammalian tissues (Busch, 1965), the initial studies of their roles as gene regulators were made with prokaryotic cells. Recently, it became possible to identify and characterize the "*lac* repressor" of *Escherichia coli* (Gilbert and Müller-Hill, 1966), the λ phage repressor (Alberga *et al.*, 1971; Barret and Dingle, 1971; Riggs *et al.*, 1968), a repres-

* There are now many reviews on the subject of nonhistone nuclear proteins (Allfrey, 1969, 1970, 1971; Baserga and Stein, 1971; Bonner, 1965; Bonner *et al.*, 1968a,b, Busch, 1962, 1965, Busch *et al.*, 1964, 1971, 1972; Busch and Davis, 1958; Busch and Smetana, 1970; Busch and Steele, 1964; Comings, 1972; Diczfalusy, 1972; Dijkstra and Weide, 1972; Georgiev, 1969; Hearst and Botchan, 1970; Hnilica, 1967, 1972; Keller, 1972; Klyszejko-Stefanowicz and Polanowska, 1971; MacGillivray *et al.*, 1972b; Malamud, 1971; Mandal, 1968; Maskos, 1971; Spelsberg *et al.*, 1971b; Stein, 1972; Stein and Baserga, 1972; Stein *et al.*, 1974; Tobey *et al.*, 1971).

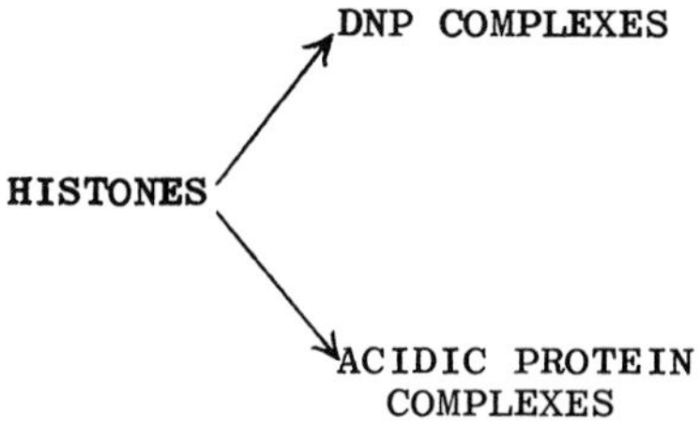

$$\text{DNP} \underset{k_2}{\overset{k_1}{\rightleftharpoons}} (\text{HIST})^{+} \quad + \quad (\text{DNA})^{-}$$

$$(\text{HIST-ACIDIC PROTEIN}) \underset{k_4}{\overset{k_3}{\rightleftharpoons}} (\text{HIST})^{+} \quad + \quad (\text{ACIDIC-PROTEIN})^{-}$$

Fig. 10 Possible schemes of interactions of DNA, histones, and acidic proteins.

sor for the tryptophan operon (Barret and Dingle, 1971; Zubay *et al.*, 1972), and factors involved in control of the bacterial RNA polymerases (Jones and Berg, 1966).

The studies on bacterial systems had the remarkable advantages of the availability of temperature sensitive mutants that permitted rapid advances in structural and functional analyses of the lac repressor. This protein has been partially sequenced (Adler *et al.*, 1972; Fig. 11) and there is some understanding of its mechanism of binding to DNA. One of the most interesting features of the *lac* repressor is its very tight binding to its operator DNA; its dissociation complex is 10^{-12} *M* (Jobe *et al.*, 1972). Although DNA-binding proteins have been found in a va-

H_2N-Met-Lys-Pro-Val-Thr-Leu-Tyr-Asp-Val-Ala-Glu-Tyr-Ala-

Gly-Val-Ser-Tyr-Gln-Thr-Val-Ser-Arg-Val-Val-Asp-Gln-Ala-

Ser-His-Val-Ser-Ala-Lys-Thr-Arg-Glu-Lys-Val-Glu-Ala-Ala-

Met-Ala-Glu-Leu-Asx-Tyr-Ile-Pro-Asx-

Fig. 11 The amino-acid sequence of the N-terminal region of the *lac* repressor. (From Adler *et al.*, 1972.)

riety of mammalian cells, and it is possible that they control growth and possibly other cellular activities such as rRNA synthesis (Crippa, 1970), clear correlation of their binding, mRNA synthesis, and specific cell functions have not yet been provided despite some evidence that such events occur in a number of endocrine glands (see Section H).

A. *NHP and Chromatin*

On the basis of evidence from bacterial systems, it is generally assumed that gene repressors have specific binding sites on DNA. However, unlike the bacterial systems where most of the DNA-binding proteins appear to have repressor function, in mammalian cells it appears that the active chromatin has a higher content of NHP than inactive chromatin and that NHP involved in gene control generally have derepressor rather than repressor functions. In current studies, the systems for binding analysis of the NHP are similar to those developed for the prokaryotes (Gross, 1968; Klein and Bonhoeffer, 1972; Losick, 1972; Von Hippel and McGhee, 1972) and include affinity chromatography which has been useful for the steroid-binding proteins and other binding proteins (Cuatrecasas, 1971, 1972; Wilchek, 1972).

Nuclei and chromatin preparations of active tissues contain larger amounts of NHP than corresponding preparations of inactive tissues (Stein and Baserga, 1972). However, it seems likely that much of this protein may be synthetic enzymes, proteins of nuclear ribonucleoprotein complexes engaged in transport of preribosomal RNP particles, or proteins of preribosomal RNP particles. There appear to be much larger numbers of NHP with faster turnover rates than histones (Steele and Busch, 1963, 1964; Stein and Baserga, 1970, 1972) in nuclei and nucleoli that would be appropriate for regulator molecules. It seems likely that gene control proteins of the chromatin are present in very small numbers. Despite remarkable improvements in methods for identification and analysis of the proteins of chromatin (Fig. 12), quantitatively improved methods will be necessary to identify them satisfactorily. In studies on gene regulatory proteins, it is also important to note that some may not be "nucleus specific" but may be "cytonucleoproteins" that travel to and fro from the nucleus to the cytoplasm and are only fleetingly chromatin bound (Goldstein, 1967; Goldstein and Prescott, 1967, 1968; Paine and Feldherr, 1972). Some of these proteins may be in the saline (0.15 *M* NaCl) extracts of cell nuclei (Fig. 13) (Bakey and Sorof, 1969; Bakey *et al.*, 1969; Busch, (1965) or in the extracts of nuclei performed with physiological concentrations of KCl, $CaCl_2$, and $MgCl_2$ (Kellermayer and Busch, 1973).

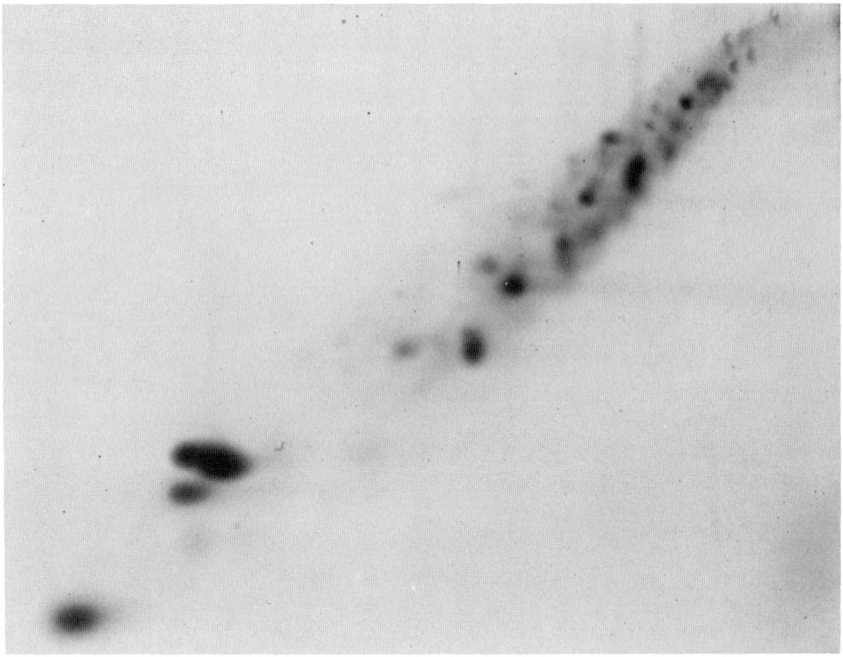

Fig. 12 Two-dimensional electrophoresis of chromatin associated NHP remaining after extraction with 0.4 *N* H_2SO_4. (Courtesy of Dr. L. C. Yeoman and Mr. Charles Taylor.)

B. Extrinsic Effects on NHP

Additional evidence for a role of the NHP in nuclear function is their change in various biological events. Recently, interactions have been reported for specific NHP and carcinogens, drugs, and plant and mammalian hormones (Alberga *et al.*, 1971; Albert, 1972; Barker, 1971; Elgin *et al.*, 1971; Epifanova, 1971; Hamilton and Teng, 1969; Hardin *et al.*, 1970; Jensen and DeSombre, 1972; Jungmann and Schweppe, 1972b; Kostraba and Wang, 1971; Lotlikar and Paik, 1971; O'Malley *et al.*, 1972b; Ruddon and Rainey, 1971; Stein, 1972; Sung *et al.*, 1971; Venis, 1972). Moreover, the NHP vary with different stages of the cell cycle (Baserga and Stein, 1971; Malamud, 1971; Stein and Baserga, 1972; Tobey *et al.*, 1971) and with age (Lukanidin *et al.*, 1972; Phytila and Sherman, 1968; Von Hahn, 1971; Zhelabovskaya and Berdyshev, 1972). On the basis of evidence from prokaryotes, induced changes in biologically active tissues, and their large number, the NHP apparently are important control molecules.

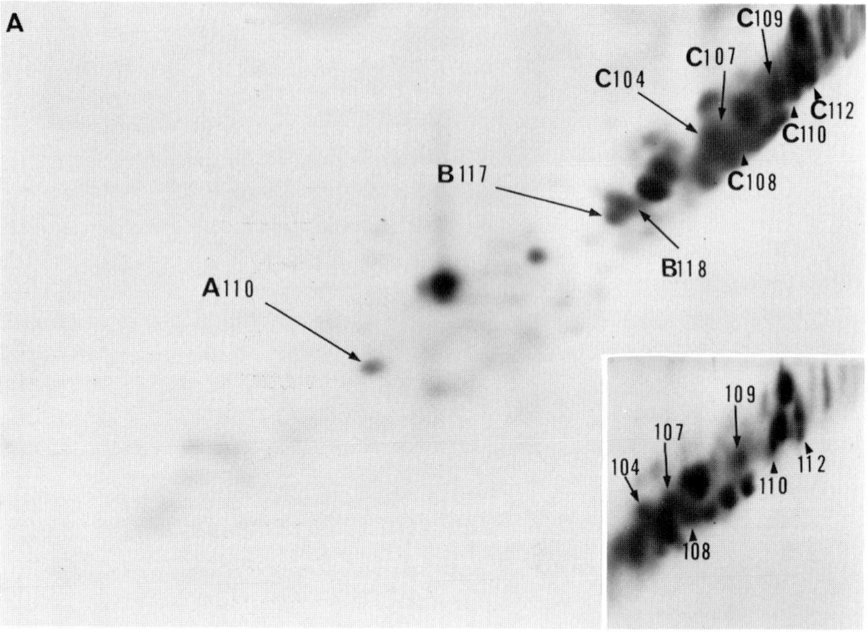

Fig. 13 Two-dimensional gel electrophoresis of soluble nuclear proteins. (A) Nuclear proteins extracted from rat liver nuclei with a buffer containing 150 m*M* KCl, 3.8 m*M* $CaCl_2$, 12 m*M* $MgCl_2$, 10 m*M* Tris–HCl, pH 7.2 (concentrations of K^+, Ca^{2+}, and Mg^{2+} similar to normal cytoplasmic concentrations). (Insert) Expanded resolution of C region of the gel. (B) Diagram of A. (Courtesy of Dr. Miklos Kellermayer.)

C. Sources of NHP

The nonhistone nuclear proteins include the nuclear enzymes and other proteins of the nucleus except for the histones. Therefore, the source material for studies on these proteins may be the whole nucleus or parts of it such as chromatin, the nucleoplasm, or nucleoli. Many of the studies have been performed on isolated chromatin, because its nonhistone proteins are presumably closely associated with DNA and evidence for their functional specificity has been provided.

The nucleoprotein complex of chromatin may be isolated as a residue from whole cells after extraction with dilute saline followed by differential centrifugation through sucrose density gradients (Bonner *et al.*, 1968a) to remove cytoplasmic and soluble nuclear components. Alternatively, chromatin may be prepared similarly from isolated cellular components such as nuclei (Palau *et al.*, 1967; Shaw and Huang, 1970) or nucleoli

B

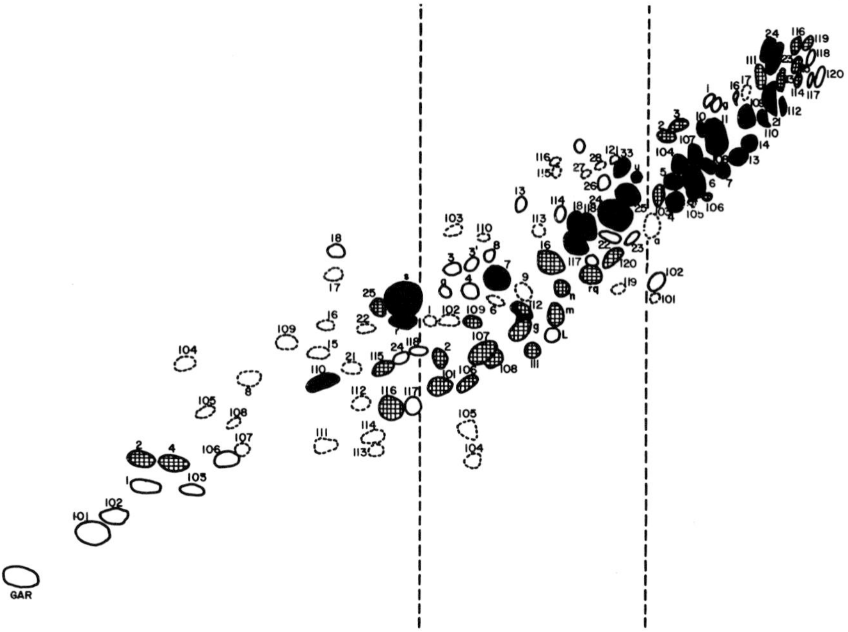

(Wilhelm *et al.*, 1972a), to avoid contamination by cytoplasmic proteins which tend to adhere to chromatin strands (Johns and Forrester, 1969).

D. Isolation of NHP

Because methods for isolation of NHP are relatively primitive (Table IV) by comparison to procedures for histones, many methods have been employed (Birnie, 1972; Busch, 1965; Busch and Steele, 1964; Comings, 1972; Dounce, 1971; Dounce *et al.*, 1972; Elgin *et al.*, 1971; Hnilica, 1972; Klyszejko-Stefanowicz and Polanowska, 1971; MacGillivray *et al.*, 1972a; Mandal, 1968; Schjeide, 1970; Sharma and Sharma, 1972; Spelsberg *et al.*, 1971b; Stellwagen and Cole, 1969a). In general, the nonhistone proteins may be dissociated from the nucleoprotein complex and solubilized by high concentrations of NaCl (2–3 *M*) and urea (5–7 *M*) (Arnold and Young, 1972; Bekhor *et al.*, 1969; Shaw and Huang, 1970; Zbarsky and Georgiev, 1959) or guanidine hydrochloride in concentrations up to 5 *M* (Levy *et al.*, 1972). Separation of histones from nonhistones may be achieved by ion exchange chromatography (Levy *et al.*, 1972; MacGillivray *et al.*, 1972), selective precipitation of the DNA–histone complex by 0.15 *M* NaCl (Patel and Wang, 1965) or preextraction of the histones with dilute mineral acid (Elgin and Bonner, 1970). Proteins that are difficult to solubilize

TABLE IV
Summary of Principal Methods for Obtaining NHP

Method[a,b]	Purpose of method	Reference
Bring to 1.5 *M* NaCl, ppt NHP with H_2SO_4	Solution and dissociation of DNP with high salt, ppt NHP from DNP	Zbarsky and Georgiev (1959)
Suspend in large volumes of water, shear with Waring blender	Use of very dilute solutions for solubilization of DNP; NHP will ppt as gel	Zubay *et al.* (1972)
Bring to 1 *M* NaCl and disperse with Dounce homogenizer; dialyze against water	Dissociation with 1 *M* NaCl, removal of the DNA–histone complex after dialysis	Wang (1966)
Dissociated DNP; treat with Bio-Rex 70; discard resin	Bio-Rex 70 resin binds and removes histones	Langan (1966)
Apply solubilized NHP to column of DNA cellulose	Affinity chromatography removes specific DNA binding proteins[c]	Alberts *et al.* (1968)
Wash nuclei with 0.35 *M* NaCl prior to extraction	Removes saline-soluble NHP and cytoplasmic contamination	Johns and Forrester (1969)
Extract directly with 8 *M* urea, 0.4 *M* guanidine HCl, 0.1% BME; remove nucleic acids by ultracentrifugation; apply to Bio-Rex 70 column	Extraction and dissociation with high urea, separation of histones and NHP on column	Levy *et al.* (1972)
Bring to 3 *M* NaCl with 7 *M* urea, DNA ppt	Avoidance of pH extremes by dissociation in high salt urea; a "mild" procedure	Shaw and Huang (1970)
Bring to 3 *M* NaCl separate DNA on Bio-gel A-50	Similar to above	Shaw and Huang (1970)
Bring to 2 *M* NaCl and 0.05 *M* $AlCl_3$	$AlCl_3$ will ppt DNA + NHP; leaves histones in solution; a "mild" procedure	Loeb (1968)
Extract nuclei directly with 8 *M* urea, 0.05 *M* phosphate, pH 8.0	One-step procedure to solubilize and extract NHP	Gronow and Griffiths (1971)
Bring to 2 *M* NaCl with 5 *M* urea and then 0.0135 $LaCl_3$	High salt solubilizes and dissociates DNP; $LaCl_3$ causes DNA to ppt leaving protein in solution	Yoshida and Shimura (1972)
Extract nuclei directly with 0.1 *M* Salyrgan; sediment through sucrose	Salyrgan solubilizes chromatin including all NHP; a "mild" procedure	Dijkstra and Weide (1972)
Extract nuclei or chromatin with 0.4 *N* H_2SO_4; extract buffer suspended residue with phenol	Phenol layer contains phosphoprotein fraction	Teng *et al.* (1971)

[a] Starting material may be whole cells, nuclei, or chromatin. The treatment may dissociate and solubilize any one or combination of NHP, histones, and DNA.

[b] The starting material is usually in dilute or isotonic salt solutions at this point.

[c] Affinity chromatography also can be used to isolate specific steroid-binding NHP proteins (see text). (From Olson *et al.*, 1974a.)

or to dissociate from DNA may be extracted with sodium dodecyl sulfate (SDS) from preextracted chromatin or nuclei (Levy *et al.*, 1972; Wang, 1966; Wilhelm *et al.*, 1972). Phenol extraction has been used to obtain a class of residual proteins which contain covalently bound phosphate (Frearson and Kirby, 1964; Teng *et al.*, 1971).

E. Fractionation and General Characteristics of NHP

Because of the limited resolution and sensitivity of present methodology, the precise number of NHP species is not known for any tissue. In rat liver, the number of chromosomal NHP as observed after one-dimensional SDS gel electrophoresis is from 14 to 27 bands, with 70% of the protein found in 10 to 15 bands (Elgin and Bonner, 1972). Similar or greater numbers of proteins were found in the rabbit, spleen, kidney, and liver chromatin NHP fraction by SDS gels (Levy *et al.*, 1972). In chromatin it appears that there is a limited number of the major NHP present. However, numerous minor protein bands not visible by staining techniques were detected by radioactive labeling (Borun and Stein, 1972; Levy *et al.*, 1973; Stein and Borun, 1972). Recent studies on 0.4 *N* sulfuric acid extracts of nucleoli of rat liver and Novikoff hepatoma have revealed the presence of 96 distinct protein spots on two-dimensional polyacrylamide gel electrophoresis, as shown in Fig. 14 (Orrick *et al.*, 1973; Yeoman *et al.*, 1973a). While 7–10 of these spots are histones, most of the proteins visualized are nonhistone proteins. This technique offers a vastly improved method for cataloguing and surveying the NHP of various tissues.

The NHP of rat chromatin have been reported to range in molecular weight from about 5,000 to 100,000 under dissociating conditions (Elgin and Bonner, 1970, 1972), although many are probably larger in their undenatured state. In HeLa cells chromosomal NHP had molecular weights from 15,000 to 180,000; 85% of the proteins had molecular weights greater than 40,000 (Bhorjee and Pederson, 1972).

Amino-acid analyses carried out on various preparations have demonstrated that these mixtures are generally rich in glutamic and aspartic acids or their amides. Since the NHP are generally retained by anion exchange resins at approximately neutral pH, most are probably acidic. By isoelectric focusing, the isoelectric point of the rat liver chromatin NHP was reported to be 6–10 (Arnold and Young, 1972; Gronow and Griffiths, 1971).

Very little structural information is available on purified NHP. Recently, two nonhistone proteins, θI and ϵ, were isolated from rat liver chromatin (Elgin and Bonner, 1972). Protein ϵ had a single N-terminal glycine. A major protein of Novikoff hepatoma nucleoli has been isolated in a state

A

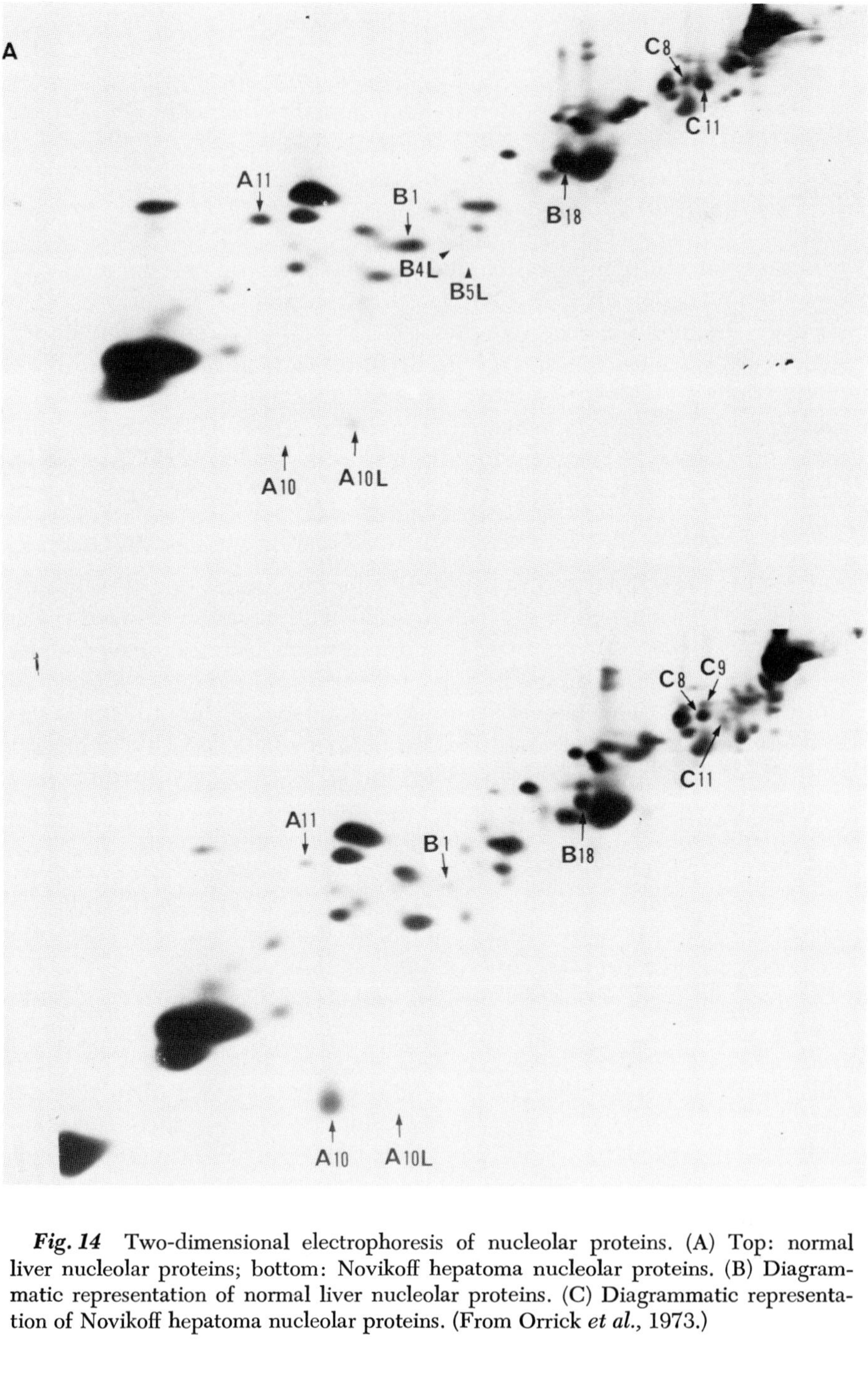

Fig. 14 Two-dimensional electrophoresis of nucleolar proteins. (A) Top: normal liver nucleolar proteins; bottom: Novikoff hepatoma nucleolar proteins. (B) Diagrammatic representation of normal liver nucleolar proteins. (C) Diagrammatic representation of Novikoff hepatoma nucleolar proteins. (From Orrick *et al.*, 1973.)

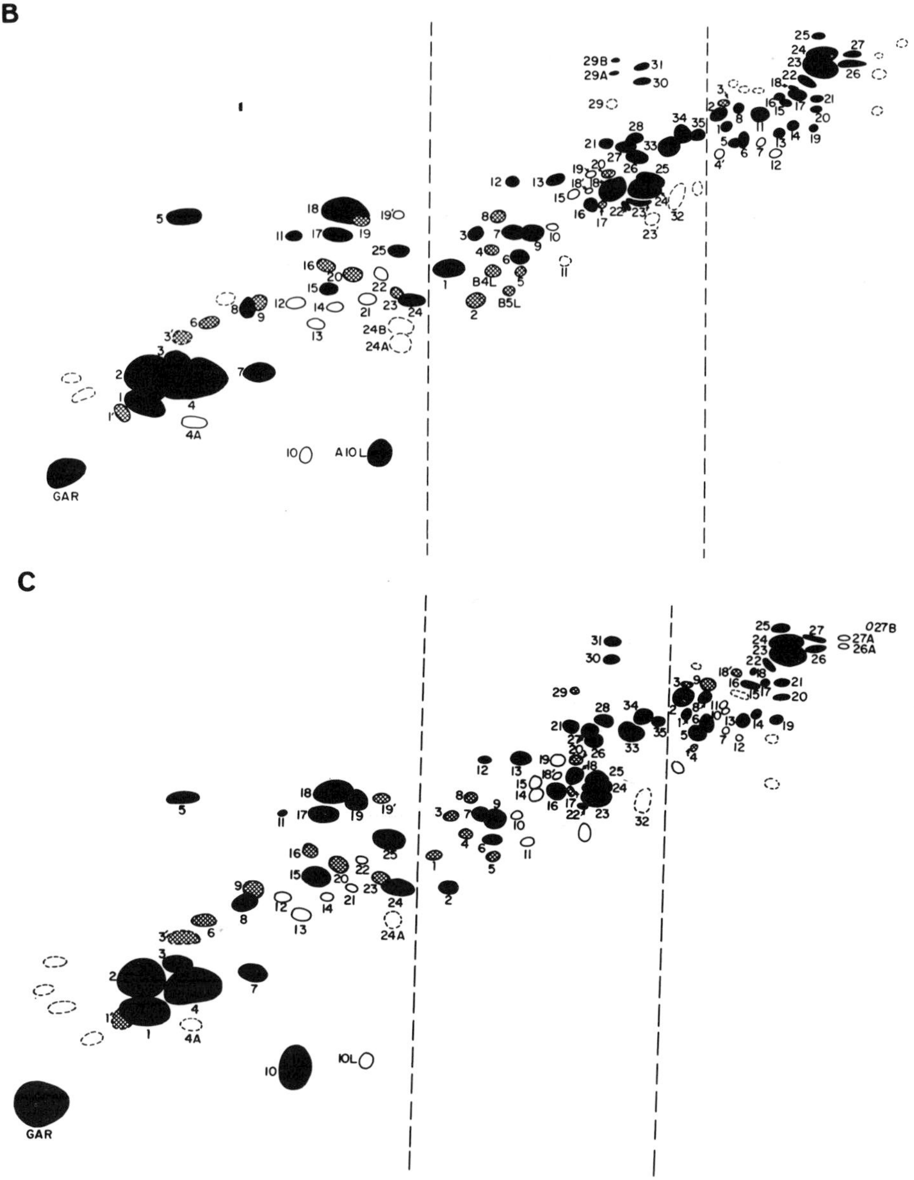

of high purity (Knecht and Busch, 1971); it has an amino terminal serine, a molecular weight of 60,000, and was found as a single band on one-dimensional gel electrophoresis. Comparison of the amino acid compositions of these proteins are shown in Table V. Improved isolation and characterization of all the NHP are required.

TABLE V
Amino-Acid Compositions of NHP

	Nucleolar extract[a]	Nucleolar band 15[a]	Chromosomal acidic proteins[b]	NHPθI[c]	NHP_{ϵ}[c]
Lysine	7.3	10.2	6.2	4.2	13.3
Histidine	2.4	1.0	2.2	1.3	1.3
Arginine	6.0	4.3	9.8	2.7	5.3
Aspartic acid	10.0	11.2	9.4	9.0	10.6
Threonine	5.1	5.3	4.9	5.5	3.7
Serine	7.6	6.2	5.2	13.3	9.1
Glutamic acid	13.8	15.9	13.6	12.8	17.2
Proline	5.3	4.7	4.4	6.1	4.8
Glycine	7.6	10.6	8.3	16.4	9.5
Alanine	7.0	9.8	8.4	9.0	8.2
Half cystine				0.0	0.0
Valine	6.0	5.6	6.4	4.2	2.5
Methionine	2.1	0.8	1.3	1.5	1.4
Isoleucine	4.1	3.3	4.9	2.7	2.8
Leucine	8.9	6.9	8.6	4.8	4.5
Tyrosine	2.4	1.2	1.8	2.6	2.8
Phenylalanine	3.6	3.1	3.2	3.9	2.9
Tryptophan	ND	ND	ND	ND	ND
Total A	23.8	27.1	23.0	21.8	27.8
Total B	15.7	15.5	18.2	8.2	19.9
A/B	1.5	1.75	1.26	2.7	1.4
Molecular weight		66,000		49,000	31,000

[a] From Knecht and Busch (1971).
[b] From Patel (1972).
[c] From Nohara *et al.* (1966)

F. Organ and Species Specificity

Unlike the histones which have similar electrophoretic profiles when obtained from various species and tissues, the nonhistone proteins have been found to vary from organ to organ and from species to species (Davis *et al.*, 1972; Elgin and Bonner, 1970, 1972; Gabel, 1972; Kleinsmith *et al.*, 1970; LeStourgeon and Rusch, 1971; Shelton and Neelin, 1971; Teng *et al.*, 1971; Wang, 1972; Wilhelm *et al.*, 1972a). However, surveys of various tissues in the rat, chicken, and pea indicate that many of the nonhistone proteins are common to all organs of these species (Elgin and Bonner, 1970). One-dimensional polyacrylamide gel electrophoresis of NHP of normal lymphocyte leukemic cells and Burkitt lymphoma cells showed that in the tumors there was an increase in the higher molecular weight NHP (Weinsenthal and Ruddon, 1972). However, many of the differences observed appear to be due to high protease activity in the normal cell nuclei (Weinsenthal and Ruddon, 1973). Although very few tissue or species differences were found in comparing the NHP of several mouse and bovine tissues (MacGillivray *et al.*, 1972), the corresponding proteins from mouse brain had markedly different electrophoretic characteristics than other mouse tissues. As noted earlier, discrepancies in these reports may result from the presence of ribosomal and preribosomal proteins in the NHP as well as proteins of the informosomes (see Chapter 3).

Recently the NHP of normal rat liver and Novikoff hepatoma ascites cells were compared by the two-dimensional polyacrylamide gel electrophoresis technique (Yeoman *et al.*, 1973b). The majority of the components observed were common to both tissues. However two proteins found in hepatoma, CG and CH, were not found in normal liver; conversely, two spots in normal liver, Bi and Bs, were absent in the hepatoma. In addition, quantitative differences in concentrations of a number of proteins were observed in comparing these two tissues.

In particular, the nuclear phosphoproteins have been reported to have species specificity, both in their patterns of phosphorylation and electrophoretic mobilities (Platz *et al.*, 1970; Teng *et al.*, 1971). Additional evidence of species specificity was obtained by an examination of their ability to form complexes with DNA of various species (Lukanidin *et al.*, 1971; Teng *et al.*, 1971). The greatest degree of binding of the nuclear phosphoproteins was found with the DNA from the same species. Rat phosphoproteins complexed best with rat DNA, to a lesser extent with mouse DNA, and not at all with calf, human, or dog DNA (Teng *et al.*, 1971). The electrophoretic profiles of DNA-binding phosphoproteins from rat liver and rat kidney are shown in Fig. 15.

Immunochemical studies have demonstrated specificity of the antigenic properties of the nonhistone protein. NHP–DNA complexes from chick oviduct produced high levels of complement fixation with rabbit antisera prepared against chick oviduct chromatin by comparison to NHP–DNA complexes of both liver and spleen. Moreover, these antisera markedly decreased transcription of chick chromatin by *E. coli* RNA polymerase. The specificity resides in the nonhistone proteins since these differences in complement fixation were found for both the intact and dehistonized chromatins. Differences were also found in chromatin from different stages of development of chick oviduct (Chytil and Spelsberg, 1971; Spelsberg *et al.*, 1971a). Recently, similar antigenic differences were reported for NHP of rat thymus and liver, Novikoff hepatoma, Walker tumor, and rat 30D hepatoma (Wakabayashi and Hnilica, 1972).

In summary, the NHP of chromatin contain a number of common pro-

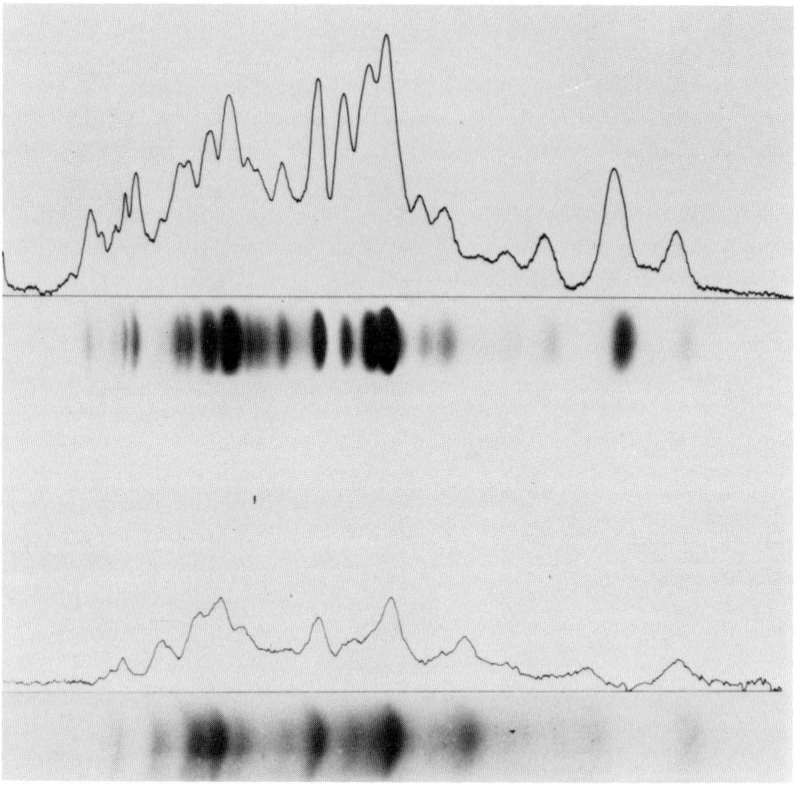

Fig. 15 Electrophoretic separation of phenol-soluble nuclear proteins. Top: rat liver; bottom: rat spleen. (From Teng *et al.*, 1971.)

teins, probably including a number of enzymes. In addition, there appear to be some organ- and species-specific NHP. Although the functions of these proteins are largely unknown, some of these apparently have gene regulatory capacities.

G. Effects of NHP on Template Activities

The capacity of chromatin as a template for synthesis of RNA by DNA-dependent RNA polymerase is much less than that of highly purified DNA. In early studies, histones were found to restrict RNA synthesis, particularly the amount of RNA synthesized (see p. 236). In more recent investigations, qualitatively different RNA's produced from chromatin of various sources were analyzed by RNA–DNA hybridization (Georgiev *et al.*, 1966; Paul and Gilmour, 1966, 1968). In addition, the amount of

RNA synthesized was also quantitated. The RNA transcribed from chromatin of mouse tissues competed with RNA extracted from the corresponding tissue but not with *E. coli* RNA (Paul and Gilmour, 1966b). Moreover, the RNA produced was reported to be specific for the tissue from which the chromatin was derived. Similar studies performed on dehistonized chromatin also indicated there was species specificity of the RNA produced (Spelsberg and Hnilica, 1971a, b). Although evidence for specificity was obtained in these studies, the hybridization techniques used in these experiments were only for RNA of relatively low Cot values, indicating that they originated from relatively highly reiterated DNA sequences.

The logical extension of this work was to reconstitute chromatin with various types of proteins. Chromatin was shown to have essentially native properties after dissociation and reconstitution (Bekhor *et al.*, 1969; Bonner *et al.*, 1968a; Huang and Huang, 1969). The reconstitution was achieved after the chromatin was solubilized in 2 *M* NaCl and 5 *M* urea in the initial solvent with subsequent removal of the urea and salt by dialysis. The reconstituted chromatin possessed the ability to synthesize RNA that was indistinguishable from RNA derived from normal chromatin (Gilmour and Paul, 1969). Using the nonhistone proteins from a number of rabbit tissues for reconstitution experiments, it was found that the type of RNA produced was dependent on the source of NHP (Gilmour and Paul, 1970). Thus, organ specificity seems to relate to the NHP of the chromatin.

Recently, similar studies have been applied to the comparison of tumor and normal tissues. When Walker tumor nonhistone proteins were used to activate the synthesis of RNA from normal rat liver chromatin, the RNA produced was similar to that produced by Walker tumor chromatin (Kostraba and Wang, 1972). In the converse experiment, activation of Walker tumor chromatin by rat liver nonhistone proteins resulted in production of RNA similar to rat liver RNA. Here, also, the nonhistone chromatin proteins seemed to dictate template specificity.

Nuclear phosphoproteins have been shown to stimulate transcription by RNA polymerase. The complex formed by addition of rat liver or kidney nuclear phosphoproteins to rat DNA had greater template activities than DNA alone (Teng *et al.*, 1971). The phosphoproteins may constitute a positive control mechanism for transcription (Stein *et al.*, 1974; Olson *et al.*, 1974b). As noted earlier, phosphorylation of the histones was also related to increased gene activity.

There is, then, considerable evidence that the nonhistone proteins may enhance transcription reactions. Although their role in cancer cells is not understood, the loss of growth control may occur at the transcriptional level. That specific nonhistone chromosomal proteins may be important in this process was suggested by the finding that in mouse fibroblasts DNA synthesis is correlated with the appearance of a specific DNA-binding protein (Salas and Green, 1971). In any event, the NHP of eukaryotic cells appear to function as gene derepressors or to be synthesized as a result of gene derepression as opposed to the gene repressors of bacterial systems.

H. NHP and Hormones

There is considerable evidence that protein receptors in the plasma membranes and in the cytoplasm (cytosol) of specific cells carry some hormones or activator substances into the nucleus where they localize at specific gene sites (Anderson *et al.*, 1972; Edelman *et al.*, 1970; Hardin and Cherry, 1972; Jensen and DeSombre, 1972; Marver *et al.*, 1972; Nunez *et al.*, 1971; O'Malley and Means, 1974; O'Malley *et al.*, 1971b, 1972a; Raspe, 1971; Schjeide, 1970; Smellie, 1971; Williams and Gorski, 1972). Furthermore, some steroid hormones increase the rates of synthesis of specific NHP fractions (Barnea and Gorski, 1970; Shelton and Allfrey, 1970; Teng and Hamilton, 1970). The presence and functional activity of such receptors in endocrine tissues is being used clinically as a means for evaluation of prognosis and therapy of human breast and uterine cancer (Wittliff *et al.*, 1971).

In chick oviducts, estrogen induces growth and differentiation as well as production of ovalbumin and other egg white proteins. Another protein, avidin, is synthesized in response to progesterone administration (O'Malley *et al.*, 1969). When progesterone enters the cell, it forms a complex with a cytoplasmic receptor protein (Sherman *et al.*, 1970). This complex in turn migrates into the nucleus where it binds to chromatin (O'Malley *et al.*, 1971b). The cytosol receptor contained two types of subunits of which only one bound specifically to chromatin. Both of these, however, possessed equal progesterone-binding abilities.

There is specificity in this binding in that the natural target chromatin binds the steroid–receptor complex much better than chromatins from other cell types (Spelsberg *et al.*, 1971a; Steggles *et al.*, 1971). The chromatin proteins associated with the steroid–receptor complex are acidic and exhibit tissue specificity (O'Malley *et al.*, 1972b). Studies with reconstituted chromatins indicate that receptor capacity is retained by the NHP and DNA from the oviduct. When reconstitution was carried out with histones from various sources, no evidence was found for tissue specificity of the histones. On the other hand, when the chromatin was reconstituted with NHP from sources other than the oviduct, the binding capacity was lost. The NHP fraction, then, contains the specific binding elements for the cytosol receptor–hormone complex.*

An analogous system is found in uteri of rats. Estradiol enters the cells of target tissue and forms a complex with a cytoplasmic receptor. This complex is transported into the nucleus where it associates with chromatin (Jensen *et al.*, 1971). The formation of the nuclear receptor is dependent on the prior presence of the cytosol receptor–estradiol complex in the cell (Jensen *et al.*, 1968; Shyamala and Gorski, 1969). The formation of the chromatin receptor for estradiol in the nonhistone fraction of calf endometrium does not require exposure of the cell to estradiol (Alberga *et al.*, 1971). A specific NHP is induced by estrogen in rat uterine chromatin (Barker, 1971); it is associated with the AR histone F3, III) by disulfide linkages. Its relationship to other estrogen-induced proteins has not been established.

The above examples indicate that the NHP have a role in the mechanism of action of some hormones. Certain NHP of target chromatins serve as acceptors for hormone–receptor complexes while others are synthesized in response to hormonal stimulation. Presumably, the NHP play a role in repression and derepression of the genome of hormonally sensitive tissues, but their exact functions remain to be determined.

I. NHP Biosynthesis and Turnover

Unlike the histones, which are primarily synthesized during S phase in the cell cycle, the nonhistone proteins appear to be turned over faster than the histones throughout the cell cycle (Busch, 1965; Littau *et al.*, 1964; Steele and Busch, 1963). Like the histones, the NHP of HeLa cells have been shown to be synthesized in the cytoplasm (Kawashima *et al.*, 1971; Stein and Baserga, 1971). In synchronized HeLa cells the incorporation of [^{14}C]leucine into tightly bound chromatin NHP increases after mitosis and reaches its maximum rate in late G_1 phase, just prior to the beginning of DNA synthesis (Phillips, 1968b; Stein and Baserga, 1970). During the S and G_2 phases there is a declining rate of NHP synthesis until the rate is the same as in G_1. By the use of inhibitors of DNA synthesis it was shown that unlike histone synthesis the NHP synthesis was not coupled to DNA replication. In addition, electrophoresis of the NHP produced at different stages of the cell cycle revealed the presence of phase-specific proteins. For example, one protein peak is present in much greater quantity in the G_2 phase than in the G_1 or S phase; another pro-

* The chromatin acceptor protein was dissociable from DNA by 2 *M* NaCl and 5 *M* urea (see Table IV).

tein is present in the S and G_2 phases but absent in the G_1 phase. The turnover rate of individual proteins also varied greatly. In the G_2 phase, relatively high-molecular-weight proteins turned over rapidly, while lower-molecular-weight proteins seemed to be more metabolically stable (Borun and Stein, 1972). In the G_1 phase the opposite was found to be the case; the high-molecular-weight proteins generally turned over more slowly.

Stein and Farber (1972) have measured template activities of native and reconstituted chromatin from HeLa cells in S phase and in mitosis. Template activity is much lower in mitosis than in S phase. Reconstituted chromatins using pooled histones from S phase and mitotic chromatin, DNA from exponentially growing cells, and NHP from mitotic chromatin had the same template activity as mitotic chromatin. If the reconstitution was performed with S phase NHP, the template activity was similar to that of native S phase chromatin. This gives support to the concept that NHP may be responsible for shutting off RNA synthesis during mitosis.

Increased rates of synthesis of NHP were observed in other systems where cell division is induced by a change in the environment of cells. Uptake of labeled amino acids into chromosomal NHP increased in mouse salivary glands stimulated with isoproterenol (Stein and Baserga, 1970), in proliferating rat mammary gland explants (Stellwagen and Cole, 1969b), and in fibroblast cultures stimulated to divide by a change of medium (Rovera and Baserga, 1971). Guinea pig lymphoid cells stimulated by phytohemagglutinin (PHA) show a rapid increase in the synthesis of cell nonhistone proteins but not of histones (Levy *et al.*, 1973). Several of the NHP were preferentially synthesized during PHA activation.

Such studies point to the complex chain of events involved in cell division. An initial stimulus causes activation or derepression of certain genes required for cell division (Stein and Baserga, 1972). The macromolecules synthesized then participate in the biochemical events of cell division and cause further modifications in genetic expression throughout the cell cycle.

The above events should be distinguished from the nonreplicative genetic regulation of stationary cells. Early studies indicated that there were higher rates of labeled amino-acid incorporation into the NHP of nondividing cells of kidney, pancreas, and liver than into their histones (Allfrey *et al.*, 1955). In addition, active (euchromatin) had twice the amount of NHP as did inactive chromatin (heterochromatin) (Frenster, 1965; Frenster *et al.*, 1963; Littau *et al.*, 1964). In active chicken tissues there is a greater rate of synthesis of NHP than in inactive tissues (Dingman and Sporn, 1964). Increased NHP synthesis is also observed when rat mammary glands begin to lactate (Stellwagen and Cole, 1969b).

It is likely that there are two classes of regulatory chromosomal NHP: those concerned with cell division and those concerned with nonreplicative events in the life of the cell. The former are likely to be important in neoplasia and cell proliferation in general.

J. Nuclear Proteins and Carcinogens

Some carcinogens bind to proteins which are similar in some cases to the steroid-binding proteins. The "h-protein" is a nonhistone saline-soluble protein that can be isolated from both cytoplasm and nuclei and may function in a transport system similar to that for steroid hormones. The h-protein (Bakey and Sorof, 1969; Bakey *et al.*, 1969; Ketterer, 1971; Libby, 1970; Morey and Litwack, 1969; Singer and Litwack, 1971; Sorof *et al.*, 1963, 1966) is known to bind dye carcinogens. It has been postulated (Abell and Heidelberger, 1962) that the h-protein carries hydrocarbons including carcinogens from the cytoplasm to the nucleus.

The protein of liver cytosol that is designated as steroid binder 1 (Singer and Litwack, 1971) also binds the carcinogen 3-methylcholanthrene. This protein has been isolated and characterized (Kuroki and Heidelberger, 1972; Morey and Litwack, 1969) as a particle with a sedimentation coefficient of approximately 3.58 S and a molecular weight of 22,000. Another protein, ligandin, not only binds some carcinogen metabolites but binds anionic steroid hormone metabolites with different degrees of affinity (Sherod *et al.*, 1970). Apparently, the receptors for 3-methylcholanthrene and estrogen-binding are distinctly different.

The chemical carcinogens *N*-hydroxy-*N*-2-fluorenyldicetamide-[^{14}C] *p*-dimethylaminoazobenzene, and 7,12-dimethylbenz-*d*-anthracene have been shown to bind *in vivo* to histones and nonhistone proteins of rat liver nuclei (Jungmann and Schweppe, 1972). The specificity of binding seemed to be greater for the histones than for the acidic proteins, although no purified nonhistone fractions were tested. The carcinogen 2-acetamidofluorene also binds to NHP when injected into rats (Lotlikar and Paik, 1971).

Although it can be demonstrated that carcinogens bind to certain macromolecules, a direct link between binding and the carcinogenesis has not been established. It is conceivable that carcinogen binding might alter the regulation of genes involved in growth control by nuclear proteins and thereby produce a loss of control of growth and cell division.

IX. The Nuclear Enzymes

Of the nuclear enzymes, most is known about the RNA polymerases, largely as the result of the extension of early findings of their presence (Ro and Busch, 1964) to their isolation (Austoker *et al.*, 1972; Chambon *et al.*, 1973; Gissinger and Chambon, 1972; Kedinger and Chambon, 1972; Kedinger *et al.*, 1972; Meilhac *et al.*, 1972; Roeder and Rutter, 1969; Seifart *et al.*, 1972; Weaver *et al.*, 1971) and separation into subunits (Table VI). Major differences have been found in the RNA polymerases that transcribe the nucleolar rDNA (and possibly other nucleolar genes) and those that transcribe the extranucleolar genes with respect to their physical characteristics, solubility and sensitivity to inhibitors. Virtually nothing is known with certainty regarding the initiation or termination factors in mammalian cells or plants (Dounce and Umana, 1962; Hardin and Cherry, 1972; Hardin *et al.*, 1970, 1971) that may be analogous to the sigma, rho and other factors involved in bacterial RNA synthesis (Stein, 1972; Stein and Baserga, 1970, 1971, 1972; Stein and Borun, 1972; Stein and

TABLE VI
Nuclear Enzymes

1. RNA synthesis and processing
   - RNA polymerases A, B, etc.
   - RNA modification enzymes; methylases, formation of modified bases.
   - RNA trimming or special cleavage enzymes
   - RNases: exo- and endonucleolytic
2. DNA synthesis
   - "True" synthetases
   - Ligases
   - Excision enzymes
   - Terminal addition enzymes
   - DNases
   - Modification enzymes: methylases, etc.
3. Other modification and synthetic enzymes
   - Histone phosphokinases, methylases, acetylases, deacetylases, proteases
   - NHP kinases and methylases
   - Nucleoside kinases
   - NAD pyrophosphorylase
4. Dehydrogenases
   - Steroid dehydrogenase
   - Cytochrome oxidase
   - Glycerol-3-phosphate dehydrogenases
   - Glyceraldehyde-3-phosphate dehydrogenase
   - Succinate, malate, isocitrate, lactate, NADH, NADPH, glucose-6-phosphate, phosphogluconate
5. Transferases: glycosyl for glycogen phosphorylases and branching enzymes
6. Enzymes of uncertain function
   - ATPases
   - Carboxylesterases
   - Phosphatases
   - 5′-Nucleotidases
   - Phosphodiesterase

Hausen, 1970; Teng *et al.*, 1971). It is possible that some ribonucleoprotein particles containing low-molecular-weight RNA may serve similar functions in mammalian cells (Chapter 5). Studies on these enzymes in cancer cells and other cells have thus far been only preliminary (Akao *et al.*, 1972; Meilhac *et al.*, 1972; Monon, 1972; Ro and Busch, 1964; Wilhelm *et al.*, 1972a, b). After RNA molecules have been synthesized, many supplementary modifications and additions must be made; for example, in the case of nuclear U-2 RNA (see Chapter 5), the 5′-portion of the molecule is subjected to 25 highly specific modifications of 69 nucleotides including formation of 11 pseudouridylic acid residues, 11 2′-*O*-methyl residues, and 3 modified bases including the unique base $N^{2,2}$-7-trimethylguanine. In addition, the tRNA molecules undergo a variety of other modifications on specific nucleotides that can only occur as a result of activities of special enzymes that have definitive cellular localization.

Additional modifications of newly synthesized RNA molecules include the processing and/or "trimming" of newly synthesized high-molecular-weight RNA that is either preribosomal or premessenger HnRNA. The enzymatic reactions involved in these cleavages are specific and not random digestions of the precursor molecules (Busch and Smetana, 1970; Prestayko and Busch, 1973; Prestayko *et al.*, 1972).

In addition to the enzymes of RNA synthesis, modification, trimming, and other metabolism, the nucleus must contain at some point all of the enzymatic factors involved in the synthesis, repair, and modification of DNA. It is obvious that much needs to be learned about the similarities and differences of these molecules in many cells, but thus far the problems involved in their isolation and characterization have been formidable.

The only other enzyme for which unequivocal demonstration of nucleolar localization has been effected is NAD synthetase (pyrophosphorylase), which is localized to the nucleolus (Busch and Smetana, 1970). There is no clear understanding at present for this localization and there has been some concern as to whether this enzyme may not be the "true" synthetic enzyme for NAD; needless to say, this question has been raised for many other enzymes, particularly for the cytoplasmic localization of enzymes that appear to be involved in DNA synthesis (Busch, 1965; Patel *et al.*, 1967).

Protein-Modifying Enzymes

Enzymes that modify proteins, including the histones and the nuclear phosphoproteins, are also found in the nucleus; although there is much interest in their functions, the factors controlling their activities are just now being studied.

Three enzymes (A, B_1, and B_2), which transfered acetate from acetyl coenzyme A to histones, were isolated from rat thymus nuclei or chromatin by Gallwitz and Sures (1972). Enzyme A preferentially acetylated the VLR (F1, I) histones but had activity toward other histones, whereas the GAR (F2a1, IV) histone was the preferred substrate for the B_1 and B_2 enzymes. Gallwitz (1971) has also demonstrated by chromatographic election profiles that histone acetyl transferases from various rat organs are tissue specific. Recently, a histone-specific deacetylase for the GAR (F2a1, IV) and AR (F3, III) histones was highly purified (Vidali *et al.*, 1972); this enzyme is a phosphorylated, acidic protein that binds to chromatin *in vitro*.

Similarly, enzymes participating in phosphorylation and dephosphorylation of nuclear proteins have been partially characterized. Langan (1968, 1969a) first demonstrated a cyclic AMP-dependent histone kinase specific for a certain serine residue in the VLR (F1, I) histones in rat liver. The degradative counterpart, a histone phosphatase, was also partially purified, although it lacked absolute specificity for the VLR (F1, I) histone (Meisler and Langan, 1969). Another cyclic AMP-independent kinase which acts on a second site of the VLR (F1, I) was later observed by Langan (1971). Further purification of both cyclic AMP-dependent and AMP-independent chromatin-associated kinases has been achieved by Yamamura *et al.* (1970). Phosphoprotein kinase activity which phosphorylates a number of nucleolar proteins has been found to be present in isolated nucleoli (Kang *et al.*, 1974).

Two nuclear protein methylases have been partially purified. Both require S-adenosylmethionine as a methyl donor. Protein methylase I is found both in nuclei and cytosol, and methylates guanidino groups of arginine particularly in histones (Paik and Kim, 1968). Protein methylase III seems to be found exclusively in the nuclei of species examined (rat and calf) and is specific for the methylation of ϵ-amino groups of lysine residues in histones (Paik and Kim, 1970b). The corresponding demethylases have not been found, which is perhaps indicative of the low turnover of methyl groups in histones.

A neutral protease is present in the deoxyribonucleoprotein complex of calf thymus and is capable of degrading the histones of that complex (Furlan and Jericijo, 1967). A similar enzyme has been isolated from rat liver chromatin (Garrels *et al.*, 1972). This enzyme seems to preferentially degrade the VLR (F1, I) histone. It is of interest that this histone also is reported to have the greatest turnover. Such examples of specific nuclear protein-modifying enzymes give support to the idea that protein modification may be important either in cell division or in specific gene regulation.

The nucleus has been studied extensively by cytochemists for many other enzymes including a variety of dehydrogenases, transferases, and hydrolyases (Table VI). Although there is convincing evidence for the presence of some of these enzymes in the nucleus, its outer membrane, and the intermembrane space (Berezney *et al.*, 1972; Vorbrodt, 1974), it is not clear what actual functional role these enzymes play. Within the limits of error in isolation of nuclei and nucleoli and the diffusion problems that attend cytochemical studies, it is somewhat uncertain whether these enzymes really have nuclear functions. In this connection, extensive discussions of enzyme and particle localization still absorb the attention of many cytologists and histochemists (Prestayko and Busch, 1973; Vorbrodt, 1974) as well as biochemists (Busch, 1962, 1966, 1965; Busch and Smetana, 1970; Siebert, 1967; Siebert *et al.*, 1971).

In summary, significant progress has been made in the isolation, purification, and enzymatic characterization of nuclear and nucleolar RNA polymerases. Less meaningful progress has emerged from studies on DNA synthetases and there is little clear understanding yet of the roles of the multiple modification enzymes of the nucleus that affect specific bases of the nucleic acids or particular amino acids of the nuclear proteins. Although a variety of other enzymes have been localized to the nucleus that carry out oxidative, transfer, and hydrolytic reactions, their roles in nuclear function are as yet largely unexplored. The relationship of all of these enzymes to the proteins involved in gene control and the many nuclear proteins now readily visualized in two-dimensional gel systems are topics that will require extensive study in the future, particularly with respect to the differences between cancer cells and other cells.

REFERENCES

Abell, C. W., and Heidelberger, L. (1962). *Cancer Res.* **22,** 931.

Adams, G. H. M., Vidali, G., and Neelin, J. M. (1970). *Can. J. Biochem.* **48,** 33.

Adler, K., Beyreuther, K., Fanning, E., Geisler, N., Gronenborn, B., Klemm, A., Müller-Hill, B., Pfahl, M., and Schmitz, A. (1972). *Nature (London)* **237,** 322.

Akao M., Kuroda, K., and Miyaki, K. (1972). *Gann* **63,** 1.

Alberga, A., Massol, N., Raynaud, J.-P., and Beulieu, E. (1971). *Biochemistry* **10,** 3835.

Albert, A. E. (1972). *Chem. Biol. Interactions* **4,** 287.

Alberts, B. M., Amodio, F. J., Jenkins, M., Gutman, E. D., and Ferris, F. L. (1968). *Cold Spring Harbor Symp. Quant. Biol.* **33,** 289.

Allfrey, V. G. (1954). *Proc. Nat. Acad. Sci. U.S.* **40,** 881.

Allfrey, V. G. (1969). *In* "Biochemistry of Cell Division" (R. Baserga, ed.), pp. 179–205. Thomas, Springfield, Illinois.

Allfrey, V. G. (1970). *In* "Aspects of Protein Synthesis" (C. B. Anfinsen, Jr., ed.), Part A, pp. 247–366. Academic Press, New York.

Allfrey, V. G. (1971). *In* "Histones and Nucleohistones" (D. M. P. Phillips, ed.), pp. 241–294. Plenum Press, New York.

Allfrey, V. G., Daly, M. M., and Mirsky, A. E. (1955). *J. Gen. Physiol.* **38,** 415.

Allfrey, V. G., Littau, V. C., and Mirsky, A. E. (193). *Proc. Nat. Acad. Sci. U.S.* **49,** 414.
Allfrey, V. G., Faulkner, R., and Mirsky, A. E. (1964). *Proc. Nat. Acad. Sci. U.S.* **51,** 786.
Anderson, J., Clark, J. H., and Peck, E. J., Jr. (1972). *Biochem J.* **126,** 561.
Ando, T., Iwai, K., Ishii, S. I., Azegami, M., and Nakahara, C. (1962). *Biochim. Biophys. Acta* **56,** 628.
Ando, T., and Suzuki, K. (1966). *Biochim. Biophys. Acta* **121,** 427.
Ando, T., and Suzuki, K. (1967). *Biochim. Biophys. Acta* **140,** 375.
Ando, T., and Watanabe, S. (1969). *Int. J. Prot. Res.* **1,** 221.
Ansevin, A. T., and Brown, B. W. (1971). *Biochemistry* **10,** 1133.
Arnold, E. A., and Young, K. E. (1972). *Biochim. Biophys. Acta* **257,** 482.
Austoker, J., Cox, D., and Mathias, A. P. (1972). *Biochem. J.* **129,** 1139.
Bakey, B., and Sorof, S. (1969). *Cancer Res.* **29,** 22.
Bakey, B., Sorof, S., and Siebert, G. (1969). *Cancer Res.* **29,** 28.
Balhorn, R., Rieke, W. O., and Chalkley, R. (1971). *Biochemistry* **10,** 3952.
Balhorn, R., Balhorn, M., Morris, H. P., and Chalkley, R. (1972a). *Cancer Res.* **32,** 1775.
Balhorn, R., Chalkley, R., and Granner, D. (1972b). *Biochemistry* **11,** 1094.
Balhorn, R., Oliver, D., Hohmann, P., Chalkley, R., and Granner, D. (1972c). *Biochemistry* **11,** 3915.
Barker, K. L. (1971). *Biochemistry* **10,** 284.
Barnea, A., and Gorski, J. (1970). *Biochemistry* **9,** 1899.
Barret, A., and Dingle, J. (eds.) (1971). "Tissue Proteinases." North-Holland Publ., Amsterdam.
Bartley, J. A., and Chalkley, R. (1968). *Biochim. Biophys. Acta* **160,** 224.
Bartley, J. A., and Chalkley, R. (1972). *J. Biol. Chem.* **247,** 3647.
Baserga, R., and Stein, G. (1971). *Fed. Proc.* **30,** 1752.
Bekhor, I., Kung, G. M., and Bonner, J. (1969). *J. Mol. Biol.* **39,** 351.
Berezney, R., Macauley, L. K., and Crane, F. L. (1972). *J. Biol. Chem.* **247,** 5549.
Birnie, G. D. (1972). "Subcellular Components: Preparation and Fractionation." Univ. Park Press, Baltimore, Maryland.
Bloch, D. P. (1969). *Genetics Suppl.* **61(1),** 93.
Bloch, D. P., and Brack, S. D. (1964). *J. Cell Biol.* **22,** 327.
Bloch, D. P., and Godman, G. C. (1955). *J. Biophys. Biochem. Cytol.* **1,** 17.
Bhorjee, J. S., and Pederson, T. (1972). *Proc. Nat. Acad. Sci. U.S.* **69,** 3345.
Bonner, J. (1965). "The Molecular Biology of Development." Oxford Univ. Press, London and New York.
Bonner, J., and Ts'o, P. O. P. (eds.) (1964). "The Nucleohistones." Holden-Day, San Francisco, California.
Bonner, J., Chalkley, G. R., Dahmur, M., Fambrough, D., Fujimura, F., Huang, R. C., Huberman, J., Jensen, R., Marushige, K., Ohlenbusch, H., Olivera, B., and Widholm, J. (1968a). *Methods Enzymol.* **12,** Part B, 3.
Bonner, J., Dahmus, M., Fambrough, D., Huang, R., Marushige, K., and Tuan, D. (1968b). *Science* **159,** 47.
Borun, T. W., and Stein, G. S. (1972). *J. Cell Biol.* **52,** 208.
Borun, T. W., Scharff, M., and Robbins, E. (1967). *Proc. Nat. Acad. Sci. U.S.* **58,** 1977.
Bradbury, E. M., Cary, P. D., Crane-Robinson, C., Riches, P. L., and Johns, E. W. (1971). *Nature (London) New Biol.* **233,** 265.
Brandt, W. F., and Von Holt, C. (1972). *FEBS Lett.* **23,** 357.
Buckingham, R. H., and Stocken, L. A. (1970). *Biochem. J.* **117,** 157.
Busch, H. (1962). "Biochemistry of the Cancer Cell." Academic Press, New York.

Busch, H. (1965). "Histones and Other Nuclear Proteins." Academic Press, New York.
Busch, H., and Davis, J. R. (1958). *Cancer Res.* **18,** 1241.
Busch, H., and Smetana, K. (1970). "The Nucleolus." Academic Press, New York.
Busch, H., and Steele, W. J. (1964). *Advan. Cancer Res.* **8,** 42.
Busch, H., Starbuck, W. C., Singh, E. J., and Ro, T. S. (1964). *Symp. Growth, Amherst* pp. 51–71.
Busch, H., Choi, Y. C., Starbuck, W. C., Olson, M. O. J., and Wikman, J. (1971). *In* "Drugs and Cell Regulation" (E. Mihich, ed.), pp. 51–62. Academic Press, New York.
Busch, H., Choi, Y. C., Daskal, I., Inagaki, A., Olson, M. O. J., Reddy, R., Ro Choi, T. S., Shibata, H., and Yeoman, L. C. (1972). *In Karolinska Symp. Res. Res. Methods Reproductive Tissue* (E. Dizfaluzy ed.). Stockholm.
Bustin, M., and Cole, R. D. (1968). *J. Biol. Chem.* **243,** 4500.
Bustin, M., and Stollar, B. D. (1972). *J. Biol. Chem.* **247,** 5716.
Bustin, M., and Stollar, B. D. (1973). *Biochemistry* **12,** 1124.
Byvoet, P. (1966). *J. Mol. Biol.* **17,** 311.
Byvoet, P., and Morris, H. P. (1971). *Cancer Res.* **31,** 468.
Byvoet, P., Shepherd, G. R., Hardin, J. M., and Noland, B. J. (1972). *Arch. Biochem. Biophys.* **148,** 558.
Candido, E. P. M., and Dixon, G. H. (1971). *J. Biol. Chem.* **246,** 3182.
Candido, E. P. M., and Dixon, G. H. (1972). *Proc. Nat. Acad. Sci. U.S.* **69,** 2015.
Chalkley, G. R., and Maurer, H. R. (1965). *Proc. Nat Acad. Sci. U.S.* **54,** 498.
Chambon, P., Gissinger, F., Kedinger, C., Mandel, J. L., and Meilhac, M. (1974). *In* "The Cell Nucleus" (H. Busch, ed.), Vol. III, p. 269. Academic Press, New York.
Chaudhuri, S., Stein, G., and Baserga, R. (1972). *Proc. Soc. Exp. Biol. Med.* **139,** 1363.
Chytil, F., and Spelsberg, T. C. (1971). *Nature (London) New Biol.* **233,** 215.
Clark, R. J., and Felsenfeld, G. (1971). *Nature (London) New Biol.* **229,** 101.
Clark, R. J., and Felsenfeld, G. (1972). *Nature (London) New Biol.* **240,** 226.
Coelingh, J. P., Monfoort, C. H., Rozijim, T. H., GeversLeuven, J. A., Schiphof, R., Steyn-Parve, E. P., Braunitzer, G., Schrenk, B., and Ruffus, A. (1972). *Biochim. Biophys. Acta* **285,** 1.
Comings, D. E. (1972). *Advan. Hum. Genet.* **3,** 237.
Crippa, M. (1970). *Nature (London)* **227,** 1138.
Cuatrecasas, P. (1971). *J. Agr. Food Chem.* **19,** 600.
Cuatrecasas, P. (1972). *Advan. Enzymol.* **36,** 29.
Davis, R. H., Copenhaver, J. H., and Carver, M. J. (1972). *J. Neurochem.* **19,** 473.
DeLange, R. J., Fambrough, D. M., Smith, E. L., and Bonner, J. (1969a). *J. Biol. Chem.* **244,** 319.
DeLange, R. J., Fambrough, D. M., Smith, E. L., and Bonner, J. (1969b). *J. Biol. Chem.* **244,** 5669.
DeLange, R. J., Hooper, J. A., and Smith, E. L. (1972). *Proc. Nat. Acad. Sci. U.S.* **69,** 882.
DeLange, R. J., Hooper, J. A., and Smith, E. L. (1973). *J. Biol. Chem.* **248,** 3261.
Dick, C., and Johns, E. W. (1969). *Biochim. Biophys. Acta* **175,** 414.
Diczfalusy, E. (1972). "Gene Transcription in Reproductive Tissue." Karolinska Inst. Stockholm.
Dijkstra, J., and Weide, S. S. (1972). *Exp. Cell Res.* **71,** 337.
Dingman, C. W., and Sporn, M. B. (1964). *J. Biol. Chem.* **229,** 3483.
Dounce, A. L. (1971). *Amer. Sci.* **59,** 74.
Dounce, A. L., and Umana, R. (1962). *Biochemistry* **1,** 811.

Dounce, A. L., Chanda, S. K., Ickowicz, R., Volkman, O., Palemiti, M., and Turk, R. (1972). *In* "Gene Transcription in Reproductive Tissue" (E. Diczfalusy, ed.). Karolinska Inst., Stockholm.
DuPraw, E. J. (1968). "DNA and Chromosomes." Holt, New York.
Dwivedi, R. S., Dutta, K. S., and Bloch, D. P. (1969). *J. Cell Biol.* **43,** 51.
Edelman, I. S., and Fanestil, D. D. (1970). *In* "Biochemical Actions of Hormones" (G. Litwack, ed.), pp. 324–331. Academic Press, New York.
Elgin, S. C. R., and Bonner, J. (1970). *Biochemistry* **9,** 4440.
Elgin, S. C. R., and Bonner, J. (1972). *Biochemistry* **11,** 772.
Elgin, S. C. R., Froehner, S. C., Smart, J. S., and Bonner, J. (1971). *Advan. Cell Mol. Biol.* **1,** 1.
Epifanova, O. L. (1971). *In* "The Cell Cycle and Cancer" (R. Baserga, ed.), pp. 143–190. Marcel and Decker, New York.
Estes, M. K., Huang, E.-S., and Pagano, J.-S. (1972). *Abstr. Amer. Soc. Microbiol.* 198.
Evans, J. H , Holbrook, D. J., and Irvin, J. L. (1962). *Exp. Cell Res.* **28,** 126.
Farina, F. A., Adelman, R. C., HoLo, C., Morris, H. P., and Weinhouse, S. (1968). *Cancer Res.* **28,** 1897.
Felix, K. (1960). *Adv. Protein Chem.* **15,** 1.
Flamm, W. G., and Birnstiel, M. L. (1964). *Biochim. Biophys. Acta* **87,** 101.
Frearson, P. M., and Crawford, L. V. (1972). *J. Gen. Virol.* **14,** 141.
Frearson, D. M., and Kirby, K. S. (1964). *Biochem. J.* **94,** 578.
Frenster, J. H. (1965). *Nature (London)* **206,** 680.
Frenster, J. H., Allfrey, V. G., and Mirsky, A. E. (1963). *Proc. Nat. Acad. Sci. U.S.* **50,** 1026.
Furlan, M., and Jericijo, M. (1967). *Biochim. Biophys. Acta* **147,** 145.
Gabel, N. W. (1972). *Perspect. Biol. Med.* **15,** 640.
Gallwitz, D. (1971). *FEBS Lett.* **13,** 306.
Gallwitz, D., and Mueller, G. C. (1969a). *J. Biol. Chem.* **244,** 5947.
Gallwitz, D., and Mueller, G. C. (1969b). *Science* **163,** 1351.
Gallwitz, D., and Mueller, G. C. (1970). *FEBS Lett.* **6,** 83.
Gallwitz, D., and Sekeris, C. E. (1969). *Hoppe-Seyler's Z. Physiol. Chem.* **350,** 150.
Gallwitz, D., and Sures, I. (1972). *Biochim. Biophys. Acta* **263,** 315.
Garrels, J. I., Elgin, S. C. R., and Bonner, J. (1972). *Biochem. Biophys. Res. Commun.* **46,** 545.
Georgiev, G. P. (1969). *Ann. Rev. Genet.* **3,** 155.
Georgiev, G. P., Anenieva, L. A., and Kozlov, Y. U. (1966). *J. Mol. Biol.* **22,** 365.
Gershey, E. L., Vidali, G., and Allfrey, V. G. (1968). *J. Biol. Chem.* **243,** 5018.
Gershey, E. L., Haslett, G. W., Vidali, G., and Allfrey, V. G. (1969). *J. Biol. Chem.* **244,** 4871.
Gilbert, W., and Müller-Hill, B. (1966). *Proc. Nat. Acad. Sci. U.S.* **56,** 1891.
Gilmour, R. S., and Paul, J. (1969). *J. Mol. Biol.* **40,** 137.
Gilmour, R. S., and Paul, J. (1970). *FEBS Lett.* **9,** 242.
Gissinger, F., and Chambon, P. (1972). *Eur. J. Biochem.* **28,** 277.
Goldstein, L. (ed.) (1967). "The Control of Nuclear Activity." Prentice-Hall, Englewood Cliffs, New Jersey.
Goldstein, L., and Prescott, D. M. (1967). *J. Cell Biol.* **33,** 637.
Goldstein, L., and Prescott, D. M. (1968). *J. Cell Biol.* **36,** 53.
Gronow, M., and Griffiths, G. (1971). *FEBS Lett.* **15,** 340.
Gross, P. R. (1968). *Ann. Rev. Biochem.* **37,** 631.
Gurley, L. R., and Hardin, J. M. (1968). *Arch. Biochem. Biophys.* **128,** 285.
Gurley, L. R., Enger, M. D., and Walters, R. A. (1973). *Biochemistry* **12,** 237.

Gutierrez-Cernosek, R. M., and Hnilica, L. S. (1971). *Biochim. Biophys. Acta* **247,** 348.
Hadler, S. C., Smart, J. E., and Bonner, J. (1971). *Biochim. Biophys. Acta* **236,** 253.
Hamilton, T. H., and Teng, C.-S. (1969). *Genetics Suppl.* **61,** 382.
Hancock, R. (1969). *J. Mol. Biol.* **40,** 457.
Hardin, J. W., and Cherry, J. H. (1972). *Biochem. Biophys. Res. Commun.* **48,** 299.
Hardin, J. M., Noland, B. J., and Shepherd, G. R. (1971). *Exp. Cell Res.* **68,** 459.
Hardin, J. W., O'Brien, T. J., and Cherry, J. H. (1970). *Biochim. Biophys. Acta* **224,** 667.
Hayashi, T., and Iwai, H. (1970). *J. Biochem.* **68,** 415.
Hearst, J. E., and Botchan, M. (1970). *Ann. Rev. Biochem.* **39,** 151.
Hempel, K., Lange, H. W., and Birkhofer, L. (1968a). *Naturwissenschaften* **55,** 37.
Hempel, K., Lange, H. W., and Birkhofer, L. (1968b). *H.S.Z. Physiol. Chem.* **349,** 603.
Hnilica, L. S. (1964). *Experientia* **20,** 13.
Hnilica, L. S. (1967). *Progr. Nucl. Acid Res. Mol. Biol.* **7,** 25.
Hnilica, L. S. (1972). "The Structure and Biological Functions of Histones." Chem. Rubber Press, Cleveland, Ohio.
Hohmann, P., and Cole, R. D. (1971). *J. Mol. Biol.* **58,** 533.
Holbriok, D. J., Irvin, J. L., Irvin, E. M., and Rotherham, J. (1960). *Cancer Res.* **20,** 1329.
Holbrook, D. J., Evans, J. H., and Irvin, J. L. (1962). *Exp. Cell Res.* **28,** 120.
Hooper, J. A., Smith, E. L., Sommer, K. R., and Chalkley, R. (1973). *J. Biol. Chem.* **248,** 3275.
Hsiang, M. W., and Cole, R. D. (1973). *J. Biol. Chem.* **248,** 2007.
Huang, R. C. C., and Bonner, J. (1962). *Proc. Nat. Acad. Sci. U.S.* **48,** 1216.
Huang, R. C., and Huang, P. C. (1969). *J. Mol. Biol.* **39,** 365.
Itzhaki, R. (1970). *Biochem. Biophys. Res. Commun.* **41,** 25.
Iwai, K., Ishikawa, K., and Hayashi, H. (1970). *Nature (London)* **226,** 1056.
Jensen, E. V., and DeSombre, E. R. (1972). *Ann. Rev. Biochem.* **41,** 203.
Jensen, E. V., Mumata, M., Brechex, P. I., and DeSombre, E. R. (1971). *In* "The Biochemistry of Steroid Hormone Action" (R. M. S. Smellie, ed.), p. 133. Academic Press, New York.
Jensen, E. V., Suzuki, Y., Haroashima, T., Stumpf, W., Jungblut, P., and DeSombre, E. R. (1968). *Proc. Nat. Acad. Sci. U.S.* **59,** 632.
Jobe, A., Riggs, A. D., and Bourgeois, S. (1972). *J. Mol. Biol.* **64,** 181.
Johns, E. W. (1971). *In* "Histones and Nucleohistones" (D. M. P. Phillips, ed.), pp. 1–83. Plenum Press, New York.
Johns, E. W., and Forrester, S. (1969). *Eur. J. Biochem.* **8,** 547.
Johns, E. W., and Hoare, T. A. (1970). *Nature (London)* **226,** 650.
Jones, O. W., and Berg, P. (1966). *J. Mol. Biol.* **22,** 199.
Jungmann, R. A., and Schweppe, J. S. (1972). *Cancer Res.* **32,** 952.
Jungmann, R. A., and Schweppe, J. S. (1972). *J. Biol. Chem.* **247,** 5535.
Kawashima, K., Izawa, M., and Sato, S. (1971). *Biochim. Biophys. Acta* **232,** 192.
Kang, Y. J., Olson, M. O. J., Jones, C., and Busch, H. (1974). Submitted for publication.
Kaye, A. M., and Sheratzky, D. (1969). *Biochim. Biophys. Acta* **190,** 527.
Kedinger, C., and Chambon, P. (1972). *Eur. J. Biochem.* **28,** 283.
Kedinger, C., Gissinger, F., Gniazdowski, M., Mandel, J.-L., and Chambon, P. (1972). *Eur. J. Biochem.* **28,** 269.
Keller, P. C. (1972). "The Role of Chromosomes in Cancer Biology." Springer-Verlag, Berlin and New York.
Kellermayer, M., and Busch, H. (1973). *Physiol. Chem. Phys.* **5,** 313.
Ketterer, B. (1971). *Biochem. J.* **126,** 3P.

Kinkade, J. M., and Cole, R. D. (1966a). *J. Biol. Chem.* **241,** 5790.
Kinkade, J. M., and Cole, R. D. (1966b). *J. Biol. Chem.* **241,** 5798.
Kistler, W. S., Geroch, M. E., and Williams-Ashman, H. G. (1973). *Biol. Chem.* **248,** 4532.
Kleiman, L., and Huang, R. C. C. (1971). *J. Mol. Biol.* **55,** 503.
Klein, A., and Bonhoeffer, F. (1972). *Ann. Rev. Biochem.* **41,** 301.
Kleinsmith, L. J., Allfrey, V. G., and Mirsky, A. E. (1966). *Proc. Nat. Acad. Sci. U.S.* **55,** 1182.
Kleinsmith, L. J., Heidema, J., and Carroll, A. (1970). *Nature (London)* **226,** 1025.
Klyszejko-Stefanowicz, L., and Polanowska, Z. (1971). *Post. Biochem.* **17,** 601.
Knecht, M. E., and Busch, H. (1971). *Life Sci.* **10,** 1297.
Kossel, A. (1884). *Z. Physiol. Chem.* **8,** 511.
Kossel, A. (1928). "The Protamines and Histones." Longmans and Green, London.
Kostraba, N. C., and Wang, T. Y. (1971). *Cancer Res.* **31,** 1663.
Kostraba, N. C., and Wang, T. Y. (1972). *Cancer Res.* **32,** 2348.
Kuroki, T., and Heidelberger, C. (1972). *Biochemistry* **11,** 2116.
Lam, D. M. K., and Bruce, W. R. (1971). *J. Cell. Physiol.* **78,** 13.
Langan, T. A. (1966). *In* "Regulation of Nuclei Acid and Protein Biosynthesis" (V. V. Konigskerger and L. Bosch, eds.), pp. 233. Elsevier, Amsterdam.
Langan, T. A. (1968). *Science* **162,** 579.
Langan, T. A. (1969a). *J. Biol. Chem.* **244,** 5763.
Langan, T. A. (1969b). *Proc. Nat. Acad. Sci. U.S.* **64,** 1276.
Langan, T. A. (1971). *Fed. Proc.* **30,** 1089A.
Langan, T. A., Rall, S. C., and Cole, R. D. (1971). *J. Biol. Chem.* **246,** 1942.
Leighton, T. J., Dill, B. C., Stock, J. J., and Philips, C. (1971). *Proc. Nat. Acad. Sci. U.S.* **68,** 677.
LeStourgeon, W. M., and Rusch, H. P. (1971). *Science* **174,** 1233.
Levy, R., Levy, S., Rosenberg, S. A., and Simpson, R. T. (1973). *Biochemistry* **12,** 224.
Levy, S., Simpson, R. T., and Sober, H. A. (1972). *Biochemistry* **11,** 1547.
Li, H.-J., and Bonner, J. (1971). *J. Biochem.* **10,** 1461.
Libby, P. R. (1970). *Biochim. Biophys. Acta* **213,** 234.
Liew, C. C., Haslett, C. W., and Allfrey, V. G. (1970). *Nature (London)* **226,** 414.
Littau, V. C., Allfrey, V. G., Frenster, J. H., and Mirsky, A. E. (1964). *Proc. Nat. Acad. Sci. U.S.* **52,** 93.
Loeb, J. (1968). *C. R. Acad. Sci. Paris* **262,** 2183.
Losick, R. (1972). *Ann. Rev. Biochem.* **41,** 409.
Lotlikar, P. D., and Paika, W. K. (1971). *Biochem. J.* **124,** 443.
Louie, A. J., and Dixon, G. H. (1972). *Proc. Nat. Acad. Sci. U.S.* **69,** 1975.
Lukanidin, E. M., Georgiev, G. P., and Williamson, R. (1971). *FEBS Lett.* **19,** 152.
Lukanidin, E. M., Zalmanzon, E! S., Komaromi, L., Samarina, O. P., and Georgiev, G. P. (1972). *Nature (London) New Biol.* **238,** 193.
MacGillivray, A. J., Cameron, R. J., Krauze, D., Rickwood, D., and Paul, J. (1972a). *Biochim. Biophys. Acta* **277,** 384.
MacGillivray, A. J., Paul, J., and Threlfall, G. (1972b). *Advan. Cancer Res.* **15,** 93.
Malamud, D. (1971). *In* "The Cell Cycle and Cancer" (R. Baserga, ed.), pp. 129–142. Dekker, New York.
Mandal, R. C. (1968). *Sci. Cult.* **34,** 78.
Marushige, K., and Dixon, G. H. (1969). *Develop. Biol.* **19,** 397.
Marushige, K., Ling, V., and Dixon, G. H. (1969). *J. Biol. Chem.* **244,** 5953.
Marver, D., Goodman, D., and Edelman, I. S. (1972). *Kidney Int.* **1,** 210.

Maskos, K. (1971). *Acta Biochim. Pol.* **18**, 57.
McClure, M. E., and Hnilica, L. S. (1970). *Int. Cancer Congr., 10th, Houston, 1970* Abstr. #441.
Meilhac, M., Tysper, Z., and Chambon, P. (1972). *Eur. J. Biochem.* **28**, 291.
Meisler, M. H., and Langan, T. A. (1969). *J. Biol. Chem.* **244**, 4961.
Menon, I. A. (1972). *Brain Res.* **42**, 529.
Miescher, F. (1897). *In* "Die Histochemischen und Physiologischen Arbeiten" (von F. Miescher, ed.), pp. 359. Vogel Vlg., Leipzig.
Mirsky, A. E., and Silverman, B. (1972). *Proc. Nat. Acad. Sci. U.S.* **69**, 2115.
Mohberg, J., and Rusch, H. P. (1969). *Arch. Biochem. Biophys.* **134**, 577.
Morey, K. S., and Litwack, G. (1969). *Biochemistry* **8**, 4813.
Murray, K. (1964). *Biochemistry* **3**, 10.
Murray, K., and Milstein, C. (1967). *Biochem. J.* **105**, 491.
Neelin, J. M., Callahan, P. R., Lamb, D. C., and Murray, K. (1964). *Can. J. Biochem.* **42**, 1743.
Nemer, M., and Lindsay, D. (1969). *Biochem. Biophys. Res. Commun.* **35**, 156.
Nohara, H., Takahashi, T., and Ogata, K. (1966). *Biochim. Biophys. Acta* **127**, 282.
Nohara, H., Takahashi, T., and Ogata, K. (1968). *Biochim. Biophys. Acta* **154**, 529.
Nunez, E., Engelmann, F., Benassayag, C., Savu, L., Crepy, O., and Jayle, M.-F. (1971). *C. R. Acad. Sci. Paris* **272**, 2396.
Ogawa, Y., Quagliarotti, G., Jordan, J., Taylor, C. W., Starbuck, W. C., and Busch, H. (1969). *J. Biol. Chem.* **244**, 4387.
Olins, D. E. (1969). *J. Mol. Biol.* **43**, 439.
Olson, M. O. J., Sugano, N., Yeoman, L. C., Johnson, B. R., Jordan, J., Taylor, C. W., Starbuck, W. C., and Busch, H. (1972). *Physiol. Chem. Phys.* **4**, 10.
Olson, M. O. J., Starbuck, W. C., and Busch, H. (1974a). *In* "The Molecular Biology of Cancer" (H. Busch, ed.), pp. 309–353. Academic, New York.
Olson, M. O. J., Orrick, L. R., Jones, C., and Busch, H. (1974b). *J. Biol. Chem.* **249**, April-May.
O'Malley, B. W., and Means, A. (1974). *In* "The Cell Nucleus" (H. Busch, ed.). Academic Press, New York, Vol. III.
O'Malley, B. W., Sherman, R. M., Toft, D. O., Spelsberg, W. T., and Steggles, A. W. (1971a). *Advan. Biosci.* **7**, 213.
O'Malley, B. W., Toft, D. O., and Sherman, M. R. (1971b). *J. Biol. Chem.* **246**, 117.
O'Malley, B. W., Rosenfeld, G., Comstock, J. P., and Means, A. R. (1972a). *In* "Gene Transcription in Reproductive Tissue" (E. Diczfalusy, ed.), pp. 281–395. Karolinska Inst., Stockholm.
O'Malley, B. W., Spelsberg, T. C., Schrader, W. T., Chytil, F., and Steggles, A. W. (1972b). *Nature (London)* **235**, 141.
Ord, M. G., and Stocken, L. A. (1966). *Biochem. J.* **98**, 5P.
Ord, M. G., and Stocken, L. A. (1968). *Biochem J.* **107**, 403.
Orrick, L. R., Olson, M. O. J., and Busch, H. (1973). *Proc. Nat. Acad. Sci. U.S.* **70**, 1316.
Paik, W. K., and Kim, S. (1968). *J. Biol. Chem.* **243**, 2108.
Paik, W. K., and Kim, S. (1970a). *Biochem. Biophys. Res. Commun.* **40**, 224.
Paik, W. K., and Kim, S. (1970b). *J. Biol. Chem.* **245**, 88.
Paik, W. K., and Kim, S. (1970c). *J. Biol. Chem.* **245**, 6010.
Paine, P. L., and Feldherr, C. M. (1972). *Exp. Cell Res.* **74**, 84.
Palau, J., Pardon, J. F., and Richards, B. M. (1967). *Biochim. Biophys. Acta* **138**, 633.
Panyim, S., Bilek, D., and Chalkley, R. (1971). *J. Biol. Chem.* **246**, 4206.

Panyim, S., and Chalkley, R. (1969a). *Arch. Biochem. Biophys.* **130,** 337.
Panyim, S., and Chalkley, R. (1969b). *Biochemistry* **8,** 3972.
Panyim, S., and Chalkley, R. (1969c). *Biochem. Biophys. Res. Commun.* **37,** 1042.
Pardon, J. F., and Wilkins, M. H. F. (1972). *J. Mol. Biol.* **68,** 115.
Patel, G. L. (1972). *Life Sci.* **11,** 1135.
Patel, G., and Wang, T.-Y. (1965). *Biochim. Biophys. Acta* **95,** 314.
Patel, G., Howk, R., and Wang, T.-Y. (1967). *Nature (London)* **215,** 1488.
Paul, J., and Gilmour, R. S. (1966a). *J. Mol. Biol.* **16,** 242.
Paul, J., and Gilmour, R. S. (1966b). *Nature (London)* **210,** 992.
Paul, J., and Gilmour, R. S. (1968). *J. Mol. Biol.* **34,** 305.
Phillips, D. M. P. (1963). *Biochem. J.* **87,** 258.
Phillips, D. M. P. (1968a). *Biochem. J.* **107,** 135.
Phillips, D. M. P. (1968b). *Experientia* **24,** 668.
Phillips, D. M. P. (1971). *In* "Histones and Nucleohistones" (D. M. P. Phillips, ed.), p. 47. Plenum, New York.
Phytila, M. J., and Sherman, F. G. (1968). *Biochem. Biophys. Res. Commun.* **31,** 340.
Pietsch, P. (1969). *Cytobios* **1,** 375.
Pitot, H. C. (1968). *Cancer Res.* **28,** 1880.
Platz, R. D., Kishi, M., and Kleinsmith, L. J. (1970). *FEBS Lett.* **12,** 38.
Pogo, B. G. T. (1966). *Proc. Nat. Acad. Sci. U.S.* **55,** 805.
Pogo, B. G. T. (1968). *Proc. Nat. Acad. Sci. U.S.* **59,** 1337.
Pogo, B. G. T., Pogo, A. O., and Allfrey, V. G. (1969). *Genetics. Suppl.* **61,** 373.
Potter, M. (1968). *Cancer Res.* **28,** 1891.
Potter, V. R. (1968). *Cancer Res.* **28,** 1901.
Prescott, D. (1966). *J. Cell Biol.* **31,** 1.
Prestayko, A. W., and Busch, H. (1973). *Methods Cancer Res.* **9,** 155.
Prestayko, A. W., Lewis, B. C., and Busch, H. (1972). *Biochim. Biophys. Acta* **269,** 90
Rall, S. C., and Cole, R. D. (1971). *J. Biol. Chem.* **246,** 7175.
Raspe, G. (1971). *Advan. Biosci.* **7,** 165.
Reid, B. R., and Cole, R. D. (1964). *Proc. Nat. Acad. Sci. U.S.* **51,** 1044.
Reid, B. R., Stellwagen, R. H., and Cole, R. D. (1968). *Biochim. Biophys. Acta* **155,** 593.
Riggs, A. D., Bourgeois, S., Newby, R. F., and Cohn, M. (1968). *J. Mol. Biol.* **34,** 365.
Ro, T. S., and Busch, H. (1964). *Cancer Res.* **24,** 1630.
Robbins, E., and Borun, T. W. (1967). *Proc. Nat. Acad. Sci. U.S.* **57,** 409.
Roeder, R. G., and Rutter, W. J. (1969). *Nature (London)* **224,** 234.
Rovera, G., and Baserga, R. (1971). *J. Cell. Physiol.* **77,** 201.
Ruddon, R. W., and Rainey, C. H. (1971). *FEBS Lett.* **14,** 170.
Sadgopal, A., and Bonner, J. (1969). *Biochim. Biophys. Acta* **186,** 349.
Salas, T., and Green, H. (1971). *Nature (London)* **229,** 165.
Schjeide, O. A. (1970). *In* "Cellular Differentiation (O. A. Schjeide and J. de Vellis, ed.), pp. 169–200. Van Nostrand Reinhold, Princeton, New Jersey.
Sedwick, W. D., Wang, T. S.-F., and Korn, D. (1972). *J. Biol. Chem.* **247,** 5026.
Seifart, K. H., Benecke, B. J., and Juhasz, P. P. (1972). *Arch. Biochem. Biophys.* **151,** 519.
Sekeris, C. E., Sekeris, K. E., and Gallwitz, D. (1967). *H.S.Z. Physiol. Chem.* **348,** 1660.
Sharma, A. K., and Sharma, A. (1972). "Chromosomal Techniques." Univ. Park Press, Baltimore, Maryland.
Shaw, L. M. J., and Huang, R. C. C. (1970). *Biochemistry* **9,** 4530.
Shelton, K. R., and Neelin, J. M. (1971). *Biochemistry* **10,** 2342.
Shelton, K. R., and Allfrey, V. G. (1970). *Nature (London)* **228,** 132.

Shepherd, G. R., Noland, B. J., and Hardin, J. M. (1971a). *Arch. Biochem. Biophys.* **142,** 299.

Shepherd, G. R., Hardin, J. M., and Noland, B. J. (1971b). *Arch. Biochem. Biophys.* **143,** 1.

Sherman, M. R., Corvol, P. L., and O'Malley, B. W. (1970). *J. Biol. Chem.* **245,** 6085.

Sherod, D., Johnson, G., and Chalkley, R. (1970). *Biochemistry* **9,** 4611.

Sherton, C. C., and Wool, I. G. (1972). *J. Biol. Chem.* **247,** 4460.

Shih, T. Y., and Bonner, J. (1970a). *J. Mol. Biol.* **48,** 469.

Shih, T. Y., and Bonner, J. (1970b). *J. Mol. Biol.* **50,** 333.

Shih, T. Y., and Fasman, G. D. (1971). *Biochemistry* **10,** 1675.

Shih, T. Y., and Fasman, G. D. (1972). *Biochemistry* **11,** 398.

Shyamala, G., and Gorski, J. (1969). *J. Biol. Chem.* **244,** 1097.

Siebert, G. (1967). *Methods Cancer Res.* **3,** 47–59.

Siebert, G., Ord, M. G., and Stocken, L. A. (1971). *Biochem. J.* **122,** 721.

Siebert, G., Villalobos, J., Jr., Ro, T. S., Steele, W. J., Lindenmayer, G., Adams, H., and Busch, H. (1966). *J. Biol. Chem.* **241,** 71.

Simpson, R. T. (1972). *Biochemistry* **11,** 2003.

Singer, S., and Litwack, G. (1971). *Cancer Res.* **31,** 1364.

Smellie, R. M. S. (1971). "The Biochemistry of Steroid Hormone Action." Academic Press, New York.

Sonnenberg, B. P., and Zubay, G. (1965). *Proc. Nat. Acad. Sci. U.S.* **54,** 415.

Sorof, S., Young, E. M., Coffey, C. B., and Morris, H. P. (1966). *Cancer Res.* **26,** 81.

Sorof, S., Young, E. M., McCue, M. M., and Fetterman, P. L. (1963). *Cancer Res.* **23,** 864.

Spelsberg, T. C., and Hnilica, L. S. (1971a). *Biochim. Biophys. Acta* **228,** 202.

Spelsberg, T. C., and Hnilica, L. S. (1971b). *Biochim. Biophys. Acta* **228,** 212.

Spelsberg, T. C., Steggles, A. W., and O'Malley, B. W. (1971a). *J. Biol. Chem.* **246,** 4188.

Spelsberg, T. C., Wilhelm, J. A., and Hnilica, L. S. (1971b). *In* "Sub-Cellular Biochemistry" (B. D. Roodyn, ed.), pp. 1–107. Plenum Press, New York.

Stedman, E., and Stedman, E. (1950). *Nature (London)* **166,** 780.

Steele, W. J., and Busch, H. (1963). *Cancer Res.* **23,** 1153.

Steele, W. J., and Busch, H. (1964). *Exp. Cell Res.* **33,** 68.

Steggles, A. W., Spelsberg, T. C., and O'Malley, B. W. (1971). *Biochem. Biophys. Res. Commun.* **43,** 20.

Stein, G. (1972). *In* "The Pathology of Transcription and Translation" (E. Farber, ed.), pp. 21–35. Dekker, New York.

Stein, G., and Baserga, R. (1970). *J. Biol. Chem.* **245,** 6097.

Stein, G., and Baserga, R. (1971). *Biochem. Biophys. Res. Commun.* **44,** 218.

Stein, G., and Farber, J. (1972). *Proc. Nat. Acad. Sci. U.S.* **69,** 2918.

Stein, G. S., and Baserga, R. (1972). *Advan. Cancer Res.* **15,** 287.

Stein, G. S., and Borun, T. W. (1972). *J. Cell Biol.* **52,** 292.

Stein, G. S., Spelsberg, T. C., and Kleinsmith, L. J. (1974). *Science* **183,** 817.

Stein, H., and Hausen, P. (1970). *Cold Spring Harbor Symp. Quant. Biol.* **35,** 709.

Stellwagen, R. H., and Cole, R. D. (1969a). *Ann. Rev. Biochem.* **38,** 951.

Stellwagen, R. H., and Cole, R. D. (1969b). *J. Biol. Chem.* **244,** 4878.

Stevely, W. S., and Stocken, L. A. (1968). *Biochem. J.* **110,** 187.

Stollar, B. D., and Ward, M. (1970). *J. Biol. Chem.* **245,** 1261.

Sugano, N., Olson, M. O. J., Yeoman, L. C., Johnson, B. R., Taylor, C. W., Starbuck, W. C., and Busch, H. (1972). *J. Biol. Chem.* **247,** 3589.

Sung, M. T., and Dixon, G. H. (1970). *Proc. Nat. Acad. Sci. U.S.* **67,** 1616.

Sung, M. T., Dixon, G. H., and Smithies, O. (1971). *J. Biol. Chem.* **246,** 1358.
Teng, C.-S., and Hamilton, T. H. (1970). *Biochem. Biophys. Res. Commun.* **40,** 1231.
Teng, C.-S., Teng, C. T., and Allfrey, V. G. (1971). *J. Biol. Chem.* **246,** 3597.
Tidwell, T., Allfrey, V. G., and Mirsky, A. E. (1968). *J. Biol. Chem.* **243,** 707.
Tobey, R. A., Peterson, D. F., and Anderson, E. C. (1971). *In* "The Cell Cycle and Cancer" (R. Baserga, ed.), pp. 309–353. Dekker, New York.
Ursprung, H., Smith, K. D., Sofer, W. H., and Sullivan, P. T. (1968). *Science* **160,** 1075.
Vendrely, R., Knobloch-Mazen, A., and Vendrely, C. (1960). *Biochem. Pharm.* **4,** 1928.
Venis, M. A. (1972). *Biochem. J.* **127,** 29.
Vidali, G., Boffa, L. C., and Allfrey, V. G. (1972). *J. Biol. Chem.* **247,** 7365.
Vidali, G., Gershey, E. L., and Allfrey, V. G. (1968). *J. Biol. Chem.* **243,** 6361.
Von Hahn, H. P. (1971). *Beitr. Pathol. Bd.* **144,** 327.
Von Hippel, P. H., and McGhee, J. D. (1972). *Ann. Rev. Biochem.* **41,** 231.
Vorbrodt, A. (1974). *In* "The Cell Nucleus" (H. Busch, ed.). Academic Press, New York.
Wagner, T., and Spelsberg, T. C. (1971). *Biochemistry* **10,** 2599.
Wakabayashi, K., and Hnilica, L. S. (1972). *J. Cell Biol.* **55,** 271a.
Wang, T. Y. (1966). *J. Biol. Chem.* **241,** 2913.
Wang, T. Y. (1972). *Exp. Cell Res.* **69,** 217.
Weaver, R. F., Blatti, S. P., and Rutter, W. J. (1971). *Proc. Nat. Acad. Sci. U.S.* **68,** 2294.
Weisenthal, L. M., and Ruddon, R. W. (1972). *Cancer Res.* **32,** 1009.
Weisenthal, L. M., and Ruddon, R. W. (1973). *Cancer Res.* **33,** 2923.
Wigle, D. T., and Dixon, G. H. (1971). *J. Biol. Chem.* **246,** 5636.
Wilhelm, J. A., Ansevin, A. T., Johnson, A. W., and Hnilica, L. S. (1972a). *Biochim. Biophys. Acta* **272,** 220.
Wilhelm, J. A., Groves, C. M., and Hnilica, L. S. (1972b). *Experientia* **28,** 514.
Wilhelm, J. A., and McCarty, K. S. (1970). *Cancer Res.* **30,** 418.
Williams, D. L., and Gorski, J. (1972). *In* "Gene Transcription in Reproductive Tissue" (E. Diczfalusy, ed.), p. 420. Karolinska Inst., Stockholm.
Wittliff, J. L., Hilf, R., Brooks, W. F., Jr., Savlov, E. D., Hall, T. C., and Orlando, R. A. (1971). *Cancer Res.* **32,** 1983.
Wong, K.-Y., Patel, J., and Krause, M. O. (1972). *Exp. Cell Res.* **69,** 456.
Yamamura, H., Takeda, M., Kumor, A., and Mishizuka, Y. (1970). *Biochem. Biophys. Res. Commun.* **40,** 675.
Yeoman, L. C., Olson, M. O. J., Sugano, N., Jordan, J., Taylor, C. W., Starbuck, W. C., and Busch, H. (1972). *J. Biol. Chem.* **247,** 6018.
Yeoman, L. C., Taylor, C. W., and Busch, H. (1973a). *Biochem. Biophys. Res. Commun.* **51,** 1956.
Yeoman, L. C., Taylor, C. W., Jordan, J. J., and Busch, H. (1973b). *Biochem. Biophys. Res. Commun.* **53,** 1067.
Yoshida, M., and Shimura, K. (1972). *Biochim. Biophys. Acta* **263,** 690.
Zbarsky, I. B., and Georgiev, G. P. (1959). *Biochim. Biophys. Acta* **32,** 301.
Zhelabovskaya, S. M., and Berdyshev, G. D. (1972). *Exp. Geront.* **7,** 313.
Zubay, G., Morse, D. E., Schrenk, W. J., and Miller, J. H. M. (1972). *Proc. Nat. Acad. Sci. U.S.* **69,** 1100.

7

Animal Nuclear DNA-Dependent RNA Polymerases

P. Chambon, F. Gissinger, C. Kedinger, J. L. Mandel, and M. Meilhac

I. Introduction

Although DNA-dependent RNA polymerase (E.C. 2.7.7.6) was first identified in rat liver nuclei (Weiss and Gladstone, 1959; Weiss, 1960), until recently most of the studies dealing with animal enzymes have not been carried out on purified enzyme, but on isolated nuclei or unpurified chromatin ("aggregate enzyme").

Depending mainly on the ionic strength of the incubation medium, isolated nuclei synthesized predominantly a GC-rich ribosomal-like RNA at low ionic strength or a more DNA-like RNA at high ionic strength (Widnell and Tata, 1964, 1966; Blackburn and Klemperer, 1967; Chambon *et al.*, 1968; Pegg and Korner, 1967; Johnson *et al.*, 1969). At low ionic strength the polymerase activity was localized principally in the nucleolus, whereas at high ionic strength extranucleolar (nucleoplasmic) synthesis was predominant (Maul and Hamilton, 1967; Pogo *et al.*, 1967; Jacob *et al.*, 1968; Ro and Busch, 1967; Siebert *et al.*, 1966; Johnson *et al.*, 1971). These studies indicated also that nucleolar RNA synthesis was preferentially stimulated by Mg^{2+}, whereas the nucleoplasmic RNA synthesis was better stimulated by Mn^{2+}. These results suggested that animal nuclei could contain two distinct RNA polymerase species (Widnell and Tata, 1966), but other explanations, implicating an effect of the salt concentration on the deoxyribonucleoprotein template, were also proposed (Chambon *et al.*, 1968).

The observation of Stirpe and Fiume (1967), showing that α-amanitin, a toxin of the toadstool *Amanita phalloides* (Wieland, 1968; Wieland and Wieland, 1972), specifically inhibited the RNA synthesis catalyzed at high ionic strength by isolated mouse liver nuclei, was the first observation strongly supporting the hypothesis of multiple RNA polymerases in animal nuclei. Subsequently several groups have resolved and purified multiple RNA polymerase activities from a variety of animal cells (Cold Spring Harbor Symposium on Quantitative Biology, 1970; Goldberg and

Moon, 1970). This achievement opened a new era in the field of animal RNA polymerases, since studies on the possible regulatory role of RNA polymerase can be adequately performed only with purified enzymes transcribing exogenous templates, as has been previously brilliantly demonstrated for prokaryotic RNA polymerases (Burgess, 1971; Bautz ,1972). RNA polymerase A activity is not inhibited by α-amanitin but RNA polymerase B activity is inhibited at very low concentrations of α-amanatin (Kedinger *et al.*, 1970).

Since an exhaustive article on mammalian nuclear RNA polymerases has been recently published (Jacob, 1973) this present review, although general, reflects the views and interests of the authors.

II. Purification of Mammalian Nuclear RNA Polymerases

A. *Solubilization and Recovery*

Early attempts to solubilize and to purify RNA polymerase from a variety of tissues (for references, see Jacob, 1973) failed because the recovery of soluble enzyme was low, which led to a great instability of the enzyme on further purification. The reason for this failure is related to the intranuclear state of RNA polymerase, which is mostly bound within the chromatin in a tight transcription complex (DNA-RNA polymerase–RNA complex) insoluble in buffers of low ionic strength (Chambon *et al.*, 1968). Recently three types of methods have been devised which result in higher yields of solubilized enzyme from a variety of tissues: (1) homogenization of whole tissue, homogenization or incubation of isolated cells, nuclei, or nucleoli in adequate buffers of various ionic strengths for various lengths of times at either 4° or 37°C (Seifart and Sekeris, 1969; Stein and Hausen, 1970a; Voigt *et al.*, 1970; Bagshaw and Malt, 1971; Chesterton and Butterworth, 1971a, b; Furth *et al.*, 1970; Mertelsmann, 1969; Goldberg and Moon, 1970; Sugden and Keller, 1973; for other references, see Cold Spring Harbor Symposium, 1970); (2) sonication in a medium of low ionic strength (Goldberg *et al.*, 1969; Jacob *et al.*, 1970a, b); (3) sonication in a medium of high ionic strength (KCl or ammonium sulfate) (Roeder and Rutter, 1969; Kedinger *et al.*, 1970; Roeder and Rutter, 1970a; Kedinger *et al.*, 1972; Mandel and Chambon, 1971a; Weaver *et al.*, 1971; Mainwaring *et al.*, 1971; for other references, see Cold Spring Harbor Symposium, 1970). An excellent summary of the methods used by the various investigators is in the recent review of Jacob (1973).

It is difficult to assess the amounts of enzyme activity actually solubilized using the three types of methods. Many factors could affect the activity measured in crude homogenates of tissues or nuclei, since the various enzyme activities (see below) have different requirements for optimal activity when bound to the chromatin (Widnell and Tata, 1966; Stirpe and Fiume, 1967; Chambon *et al.*, 1968). Furthermore, a substantial part of the total activity is soluble and not bound to chromatin, even before the solubilizing treatment (Kedinger *et al.*, 1972). This soluble activity is not measured unless exogenous DNA is added to the incubation mixture for RNA synthesis, and has a lower ionic strength optimum (below 0.1 *M* ammonium sulfate) than the "aggregate" chromatin-bound enzyme (above 0.2 *M* ammonium sulfate). Finally, possible stimulating factors present in the initial homogenate could be lost during the solubilizing step (Stein and Hausen, 1970a). Another method, not based on the determination of the enzyme activity, is required to estimate the actual amount of solubilized enzyme activity. Such a determination was recently performed in our laboratory for the B RNA polymerases by comparing the amounts of labeled amanitin which could be bound to the enzyme preparation before and after solubilization by sonication in a medium of high ionic strength (Kedinger *et al.*, 1972; Chambon *et al.*, 1972). Using this assay, more than 90% of the B enzyme molecules were found to be solubilized. At the present time, it is not possible to determine the amount of solubilization for the A type of RNA polymerase actvity but this should become feasible in the near future using an immunological assay (see below).

Sonication in a medium of high ionic strength is the best method to solubilize quantitatively the RNA polymerase activity. The authors have used sonication successfully for quantitative solubilization of RNA polymerase from either whole tissues (calf thymus, cells in culture, rat testis, rat liver, rat uterus) or purified nuclei (rat brain, rat liver, calf thymus). Moreover, this method should prevent proteolysis which could occur when the solubilization process involves a prolonged incubation of the lysate. In some instances enzyme A activity can be selectively solubilized by incubating rat liver nuclei in a low salt, nonlytic medium (Chesterton and Butterworth, 1971b). Such selectivity was not observed with calf thymus (Gissinger, unpublished results).

B. *Isolation and Purification of Multiple Forms of RNA Polymerase*

The existence of multiple RNA polymerase species is supported by three lines of evidence obtained independently in several laboratories. Multiple peaks of enzyme activity were obtained by chromatography on substituted cellulose or Sephadex columns (Roeder and Rutter, 1969, 1970a, b; Kedinger *et al.*, 1970, 1971; Chesterton and Butterworth, 1971a, b, c; Jacob *et al.*, 1971; Mandel and Chambon, 1971a; Mainwaring *et al.*, 1971; Tocchini-Valentini and Crippa, 1970a, b; Smuckler and Tata, 1971, 1972; Doenecke *et al.*, 1972; for other references, see Cold Spring Harbor Symposium, 1970) and these activities appeared to have distinctive intranuclear localizations (see below). Furthermore, two classes of enzyme (A and B) were distinguished, according to the inhibitory effect of α-amanitin (Kedinger *et al.*, 1970). Subsequently the complete purification and structural analysis of multiple RNA polymerase activities from calf thymus and rat liver has firmly established the multiplicity of RNA polymerases (see below).

Purification to homogeneity was achieved for the mixture of the rat liver B enzymes (Mandel and Chambon, 1971a; Chesterton and Butterworth, 1971c; Weaver *et al.*, 1971) and for calf thymus AI, BI, and BII enzymes (Gissinger and Chambon, 1972; Kedinger and Chambon, 1972) using various combinations of ammonium sulfate fractionation, centrifugation through glycerol gradients, and chromatography on DEAE–cellulose, DEAE–Sephadex, Sepharose, hydroxyapatite, and phosphocellulose. Milligram quantities of pure enzymes were obtained from calf thymus, which is the tissue having the highest RNA polymerase content. Addition of glycerol to all buffers increases the stability of the enzyme activities throughout the purification, but the final yield of activity was nevertheless low (Kedinger and Chambon, 1972, Gissinger and Chambon, 1972). This does not seem to be due to the loss of stimulating factors during the purification.

Recent studies have shown that up to eight identifiable protein bands having polymerase activity could be resolved from crude extracts of mouse liver nuclei by polyacrylamide gel electrofocusing (Bagshaw and Drysdale, 1973). Whether these multiple bands correspond to eight different RNA polymerase species or are aggregation artifacts remains to be seen.

Multiple RNA polymerases have also been observed in lower eukaryotes such as yeast (Ponta *et al.*, 1972; Dezelée *et al.*, 1972; Adman *et al.*, 1972; Brogt and Planta, 1972), an aquatic fungus (Horgen and Griffin, 1971), and maize (Strain *et al.*, 1971).

C. *Nomenclature of the Various Animal RNA Polymerases*

The terminology of the animal RNA polymerases is complex, because the various investigators classified the enzymes according to different criteria. Roeder and Rutter (1969) called I, II, and III, the three enzyme activities separated on DEAE–Sephadex in the order of their elution from column. The difficulty with this kind of nomenclature became apparent when additional enzyme activities were found which were eluted at lower salt concentrations than enzyme III. Furthermore, such a terminology could be misleading, since the position of a given enzyme activity could change from one tissue to another; many factors could affect the elution profile when a given protein present in a crude extract is chromatographed on a substituted cellulose column. We therefore proposed (Kedinger *et al.*, 1971) a terminology based on a criterion that is firm and easy to determine, the sensitivity to α-amanitin. The subunit composition of the enzymes was also considered in their characterization, since ultimately the nomenclature should be based on the subunit structure. Enzymes of class A are not inhibited by α-amanitin at any concentration, whereas enzymes of class B are inhibited at very low concentration of amanitin (10^{-9}–10^{-8} *M*). The recently discovered "cytoplasmic" RNA polymerase activity, which is inhibited at much higher α-amanitin concentration (10^{-5}–10^{-4}) *M*), would then belong to a new class C (Seifart *et al.*, 1972; Amalric *et al.*, 1972).

It has also been suggested that the intranuclear localization and the distinctive ionic requirements (see below) of the various enzyme activities could be incorporated in the nomenclature. This would not be very helpful because (*a*) the isolation of active nucleoli from many tissues is difficult; and (*b*) ionic requirements, to be distinctive, should be determined under very stringent conditions (see below). Table I lists the various RNA polymerase activities that have been isolated. An additional peak of activity after DEAE–cellulose or Sephadex chromatography does not necessarily mean that a new enzyme is present. It could correspond either to an *in vitro* modification of a previously known enzyme by using a modified solubilizing procedure, or to a chromatography artifact, as recently shown by Smuckler and Tata (1972) for enzyme pre-A, which on rechromatography was resolved into the two classic AI and B enzyme activities.

III. Intracellular Localization and Relative Concentrations

Enzymes AI, AII, and AIV are of nucleolar origin, whereas enzymes AIII and B are found in the nucleoplasm (see Table I for references). Of course, this does not prove that in the intact nucleus some of the B activity is not localized in the nucleolus and that some of the enzyme AI + AII activity is not localized in the nucleoplasm. In fact, an anucleolate mutant of *Xenopus laevis*, which makes no rRNA and which lacks definitive nucleoli, has a normal amount of RNA polymerase AI (Roeder *et al.*, 1970). The use of fluorescent antibodies should help to clarify this point. Similarly, enzyme C, which is found in the cytoplasm after cell homogenization could nevertheless be of nuclear origin.

TABLE I
Nomenclature and Localization of Animal DNA-Dependent RNA Polymerases

Our terminology			
Class of enzyme	Enzymes	Other terminology	Localization
A (insensitive to amanitin)	Enzyme AI[a,b,c]	Enzyme I[d], enzyme IA[f]	Nucleolar[b,e,g]
	Enzyme AII[b]	Enzyme I[d], enzyme IB[f]	Nucleolar[h]
	Enzyme AIII[a]	Enzyme III[d]	Nucleoplasmic[e]
	Enzyme AIV	Enzyme IV[f], pre-A[j], IA[n]	Nucleolar[i,n]
B (sensitive to low concentrations of amanitin, 10^{-9}–10^{-8} *M*)	Enzyme BI[a,l]	Enzyme II[d,k], IIA[f]	Nucleoplasmic[e,g]
	Enzyme BII[a,l]	Enzyme II[d,k], IIB[f]	Nucleoplasmic[e,g]
C (sensitive to high concentrations of amanitin, 10^{-5}–10^{-4} *M*)	Enzyme C[m]	—	Cytoplasmic[m]

[a] Kedinger *et al.* (1971).
[b] Butterworth *et al.* (1971).
[c] Gissinger and Chambon (1972).
[d] Roeder and Rutter (1969).
[e] Roeder and Rutter (1970a).
[f] Jacob (1973).
[g] Jacob *et al.* (1970a).
[h] Chesterton and Butterworth (1971a).
[i] Jacob *et al.* (1971).
[j] Smuckler and Tata (1972).
[k] Weaver *et al.* (1971).
[l] Kedinger and Chambon (1972).
[m] Seifart *et al.* (1972).
[n] Sajdel and Jacob (1971).

It is presently impossible to assess the *in vivo* amounts of the various enzymes for reasons already discussed in Section II, A. Enzymes AI and B have been found in various ratios in all extracts of tissues or cells so far examined. Enzyme AIII, which seems to be very unstable, was only found as a minor component in rat liver (Blatti *et al.*, 1970; Jacob *et al.*, 1971; Sajdel and Jacob, 1971), sea urchin (Roeder and Rutter, 1969; 1970b), and *Xenopus laevis* embryos (Roeder *et al.*, 1970). We did not detect it in calf thymus, perhaps because its level is very low compared with the levels of enzymes AI and B. Enzyme AII was found in rat liver nuclei only when using a low salt extraction method (Chesterton and Butterworth, 1971a). We did not find any AII enzyme in calf thymus, using either high salt-sonication (Gissinger and Chambon, 1972) or low salt-extraction methods (Gissinger, unpublished results). This observation suggests that the rat liver enzyme AII, which is progressively released during a long period of incubation at 37°C (Chesterton and Butterworth, 1971a), could be derived *in vitro* from enzyme AI. Comparison of the subunit patterns of purified enzymes AI and AII and immunological studies are necessary to assess this possibility.

A tissue specificity of at least some of the RNA polymerase species is an interesting possibility, providing one of the mechanisms which could regulate the transcription of the genes specifically expressed in a given tissue. The ratio of enzyme BI to enzyme BII is much higher in rat liver than in calf thymus. Whether this difference is a characteristic of the animals or organs, or results from selective loss of enzymes during the purification is presently unknown (Mandel and Chambon, 1971a). There is some evidence that rat liver nuclei could contain an additional B enzyme (B0) (Mandel and Chambon, 1971a).

IV. Structure of the Multiple RNA Polymerases

A. *Molecular Weight*

Sedimentation studies of a mixture of B RNA polymerases (calf thymus, rat liver, or KB cells: Chambon *et al.*, 1970; Weaver *et al.*, 1971; Chesterton *et al.*, 1972; Sugden and Keller, 1973), of calf thymus (CT) AI RNA polymerase (Chambon *et al.*, 1970), and of rat liver (RL) AI and AII RNA polymerases (Chesterton *et al.*, 1972) through glycerol gradients have shown that both classes of enzyme sediment faster than the *Escherichia coli* core enzyme (MW 380,000–400,000: Burgess, 1971) at about 14–15 S. B enzymes sediment slightly faster than A enzymes. These observations suggest a molecular weight (MW) of about 500,000. In contrast to *E. coli* RNA polymerase (Berg and Chamberlin, 1970), there is no drastic modification of the sedimentation rate of animal enzymes when the ionic strength increased from 0.15 to 1.5 (Kedinger *et al.*, 1974; Sugden and Keller, 1973).

The MW of the enzymes was also estimated by electrophoresis in polyacrylamide gels of increasing porosity (Hedrick and Smith, 1968). Values of 550,000 ± 10%, 600,000 ± 10%, and 570,000 ± 10% were found for calf thymus AI, BI, and BII enzymes, respectively (Chambon *et al.*, 1972; Kedinger *et al.*, 1974).

B. *Subunit Composition of Calf Thymus RNA Polymerase AI, BI, and BII*

Purified calf thymus AI, BI, and BII enzymes move essentially as single bands with distinctive R_f values when electrophoresed on polyacrylamide gels under nondenaturating conditions (Chambon *et al.*, 1972, 1973; Gissinger and Chambon, 1972; Kedinger and Chambon, 1972). Since the MW's are very similar, these differences are mainly related to charge differences. The subunit composition of the enzymes was investigated by polyacrylamide gel electrophoresis in the presence of sodium dodecylsulfate (SDS) and the MW of the various subunits was estimated according to Weber and Osborn (1969) (Chambon *et al.*, 1972, 1973). Molar ratios of the various subunits were obtained by densitometry of the various bands after staining of the SDS gels. Although similar values were obtained using either Coomassie Blue or Amido Black, which suggests that the proposed molar ratios do not reflect a preferential binding of the dye to some of the subunits, the values for the molar ratio of the small subunits should nevertheless be considered as tentative, since the error is very large when measuring by densitometry the stoichiometry of polypeptide chains whose MW's differ by an order of magnitude. The subunit patterns of the three calf thymus enzymes are shown in Fig. 1 (Chambon *et al.*, 1972, 1973; Gissinger and Chambon, 1972; Kedinger and Chambon, 1972) and the results are summarized in Table II. The overall structures of AI, BI, and BII are $[(197{,}000)_1\ (126{,}000)_1\ (51{,}000)_1\ (44{,}000)_1\ (25{,}000)_2\ (16{,}500)_2]$, $[(214{,}000)_1\ (140{,}000)_1\ (34{,}000)_{1\text{-}2}\ (25{,}000)_2\ (16{,}500)_{3\text{-}4}$ and $[(180{,}000)_1\ (140{,}000)_1\ (34{,}000)_{1\text{-}2}\ (25{,}000)_2\ (16{,}500)_{3\text{-}4}]$, respectively. There is some evidence that the enzymes could contain some additional components of even lower MW's (Kedinger *et al.*, 1974). The best proof that all of these polypeptide chains belong to the RNA polymerase enzymes and are not contaminants is that all of these subunits were also found when we analyzed the material eluted from the polyacrylamide gels after nondenaturating electrophoresis of the purified enzymes (Kedinger *et al.*, 1974).

These results indicate that the basic structure of the mammalian enzymes is similar to that of the prokaryotic enzyme in that they consist of two high MW subunits accompanied by several small subunits (Burgess, 1971). These studies firmly establish, on structural grounds, the multiplicity of nuclear RNA polymerases and make unlikely an interconversion between A and B enzymes, as proposed by Chesterton and Butterworth (1971a). Significant structural differences exist between A and B enzymes, whereas the difference between the B species appears to lie in only one subunit, that is, the heavier of the two large subunits. The structural differences between AI and B enzymes are confirmed by immunological studies (Kedinger *et al.*, 1974) which show that there is no cross-reaction between an antibody against pure calf thymus AI enzyme and a purified mixture of B enzymes (Fig. 2). Possible charge differences between subunits of identical MW are not revealed by polyacrylamide gel

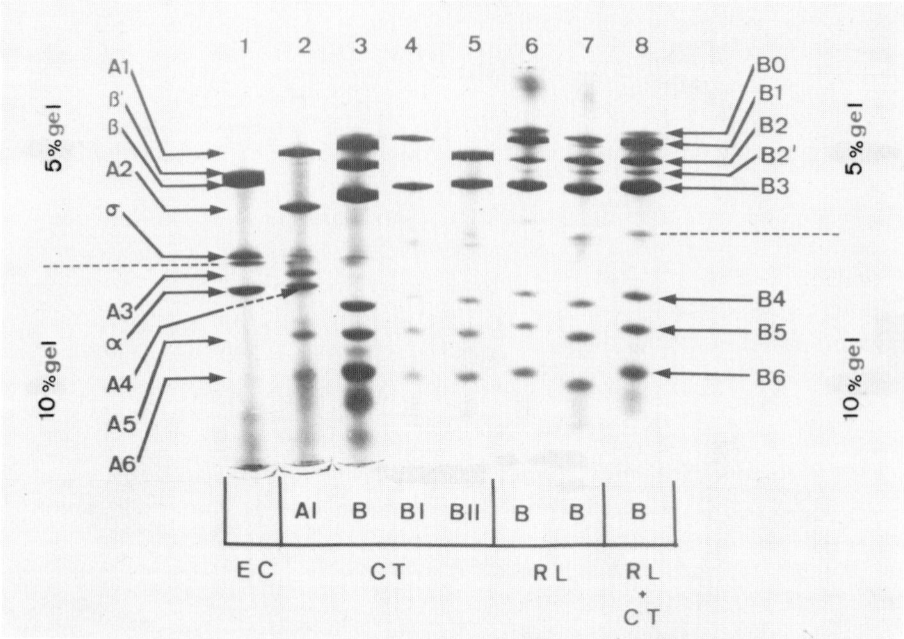

Fig. 1 Subunit pattern of calf thymus RNA polymerases AI, BI, BII and a mixture of purified rat liver B enzymes. Comparison with *E. coli* holoenzyme. About 1.5 activity unit of each enzyme was denatured and SDS polyacrylamide gel electrophoresis was carried out using mixed gels (5 and 10% in the upper and the lower parts of the gel, respectively) in order to resolve both the high- and the low-molecular-weight subunits (Chambon *et al.*, 1972). The gels belong to two series run independently (series A : gels 1, 2, and 3; series B : gels 4, 5, 6, 7, and 8). Gel 1, *E. coli* holoenzyme; gel 2, CT RNA polymerase AI (Gissinger and Chambon, 1972); gel 3, mixture of purified CT B enzymes (Kedinger and Chambon, 1972); gel 4, CT RNA polymerase BI (Kedinger and Chambon, 1972); gel 5; CT RNA polymerase BII (Kedinger and Chambon, 1972); gel 6, mixture of purified rat liver B enzymes (Chambon *et al.*, 1972); gel 7, mixture of purified rat liver B enzymes (Chesterton and Butterworth, unpublished observation); gel 8; coelectrophoresis of B enzymes from calf thymus and rat liver. EC, *E. coli* holoenzyme; CT, calf thymus; RL, rat liver.

electrophoresis in the presence of SDS. Comparison of the charges of the various subunits is currently under investigation in our laboratory, since the finding that the MW's of the two smallest subunits of both AI and B enzymes are identical raises the interesting possibility that there could be a common pool of low MW subunits for A and B enzymes.

Weaver *et al.* (1971) have found only one predominant nucleoplasmic B enzyme in calf thymus with the substructure $(190{,}000)_1$ $(150{,}000)_1$ $(35{,}000)_1$ $(25{,}000)_1$. The possible relationship between this enzyme and the BII enzyme will be discussed in Section IV, D.

TABLE II
Current Knowledge Concerning the Subunit Structure of Mammalian RNA Polymerases

	Subunit	MW	Molar ratio
CT[a] form BI[b,c]	—	—	—
	B1	214,000	1
	—	—	—
	—	—	—
	B3	140,000	1
	B4	34,000	1–2
	B5	25,000	2
	B6	16,500	3–4
CT form BII[b,c]	—	—	—
	—	—	—
	B2	180,000	1
	—	—	—
	B3	140,000	1
	B4	34,000	1–2
	B5	25,000	2
	B6	16,500	3–4
RL[d] B mixture[b]	B0	220,000	0.2
	B1	214,000	0.4
	B2	180,000	0.3
	B2′	165,000	0.1
	B3	140,000	1
	B4	34,000	1–2
	B5	25,000	2
	B6	16,500	3–4
RL B mixture[e]	—	—	—
	B1	215,000	0.2
	B2	180,000	0.5
	B2′	165,000	0.1
	B3	140,000	1
	B4	34,000	1–2
	B5	25,000	1–3
	B6	16,500	1–3
CT form AI[b,f]	A1	197,000	1
	A2	126,000	1
	A3	51,000	1
	A4	44,000	1
	A5	25,000	2
	A6	16,500	2
RL form AII[e]	A1	170,000	1
	A2	126,000	1
	A3	?	?
	A4	40,000	?
	A5	?	?
	A6	?	?

[a] CT, calf thymus.
[b] Chambon *et al.* (1972, 1973).
[c] Kedinger and Chambon (1972).
[d] RL, rat liver.
[e] Chesterton and Butterworth, unpublished results.
[f] Gissinger and Chambon (1972).

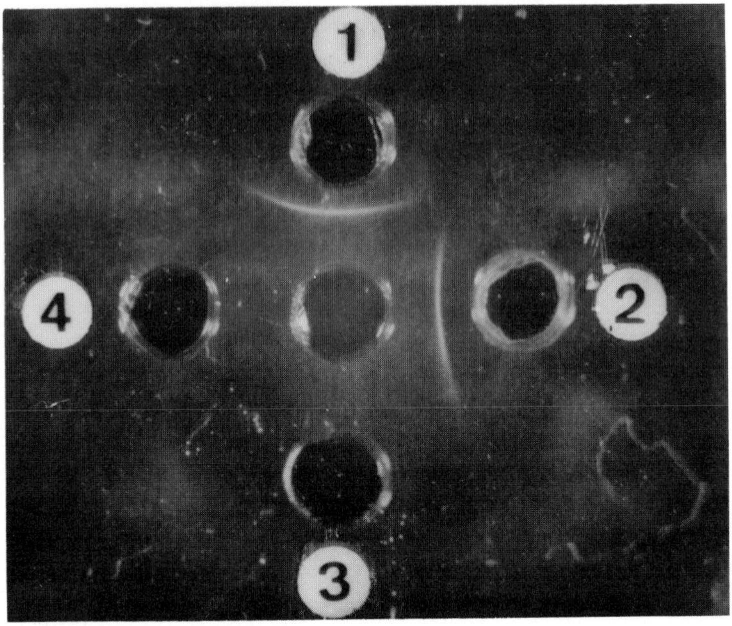

Fig. 2 Double immunodiffusion reaction. The center well contained rabbit antiserum to purified calf thymus RNA polymerase AI (Gissinger and Chambon, 1972). Peripheral wells were filled as follows: 1, purified calf thymus RNA polymerase AI (fraction GG, Gissinger and Chambon, 1972); 2, partially purified calf thymus RNA polymerase AI (fraction SE, Gissinger and Chambon, 1972); 3, mixture of partially purified calf thymus RNA polymerases B (fraction PC1, Kedinger and Chambon, 1972); 4, mixture of purified calf thymus RNA polymerases B (fraction PC2, Kedinger and Chambon, 1972).

C. Subunit Composition of Rat Liver B Polymerases

Analysis of the purified B activity from rat liver by polyacrylamide gel electrophoresis under nondenaturating conditions revealed the presence of two components similar to calf thymus BI and BII enzymes (Mandel and Chambon, 1971a). The identity of these two rat liver components with calf thymus BI and BII enzymes was also supported by the migration of labeled amanitin with the two protein bands (Mandel and Chambon, 1971a). This similarity was confirmed by subunit analysis of a mixture of purified rat liver B enzymes (Chambon *et al.*, 1972, 1973; Chesterton and Butterworth, unpublished results) (Fig. 1 and Table II), where polypeptide chains corresponding to calf thymus B1, B2, B3, B4, B5, and B6 were observed. In addition two minor components, B0 (220,000 daltons) and B2′ (165,000 daltons) were detected. As already discussed (Section III) this B0 component could be one of the subunits of a rat liver-specific B0 enzyme with a structure $(220,000)_1$ $(140,000)_1$ $(34,000)_{1\text{-}2}$ $(25,000)_2$ $(16,500)_{3\text{-}4}$. The subunit analysis of the purified rat liver B enzyme of Weaver *et al.* (1971) also suggests the presence of two forms of B activity with subunit structures [$(190,000)_1$ $(150,000)_1$ $(35,000)_1$ $(25,000)_1$] and [$(170,000)_1$ $(150,000)_1$ $(35,000)_1$ $(25,000)_1$], whereas Seifart *et al.* (1972) found for the subunits of their purified rat liver B enzyme activity the same MW's as Chambon *et al.*, (1972, 1973) and Chesterton and Butterworth (unpublished results).

Very little is known about the structure of A enzymes from rat liver. Unpublished results of Chesterton and Butterworth (Table II) suggest that rat liver enzyme AII also contains two large subunits, which are similar to the calf thymus A1 and A2 subunits, except that the MW of the rat liver A1 subunit would be 170,000 instead of 197,000. At the present time, the structural difference between rat liver enzyme AI and AII is not known. Structural similarities between calf thymus AI enzyme and rat liver nucleolar A enzymes are also indicated by immunological studies, which show that an antibody against calf thymus AI enzyme cross-reacts with a partially purified rat liver AI enzyme (Kedinger *et al.*, 1974). These results and the striking structural resemblance between the calf thymus, KB cell, and rat liver B enzymes suggest that no drastic change has occurred in the structure of RNA polymerases during the evolution of the mammals.

D. *Origin of the Various B RNA Polymerases*

It has been suggested that BII enzyme could be derived from BI enzyme by proteolysis during the first stages of purification (Weaver *et al.*, 1971; Chambon *et al.*, 1973). This possibility is very unlikely, since Kedinger and Chambon (1972) could not demonstrate any significant conversion of form BI to form BII by aging a crude extract of calf thymus for 2 hr at room temperature. In contrast to the result of Weaver *et al.* (1971), addition of a proteolytic inhibitor during the purification did not change the BI to BII ratio in a calf thymus purified B enzyme preparation (Kedinger *et al.*, 1974). It remains to be investigated whether enzyme BII is formed *in vivo* by a specific proteolytic conversion of B1 to B2 subunit [as in the case of the β-subunit of *Bacillus subtilis* polymerase, which is modified during sporulation (Leighton *et al.*, 1972; Millet *et al.*, 1972)], or is the product of a gene different from that coding for the B1 subunit.

The differences between the stoichiometry of the various subunits as determined by Weaver *et al.* (1971) and Chambon *et al.*, (1972, 1973) are probably due to the inaccuracy of the method discussed in Section IV, B.

The apparent absence of the B6 subunit (16,500 daltons) in the B enzyme preparations of Weaver *et al.* (1971) is also noteworthy and may be due to the use of 5% polyacrylamide gels. A significant discrepancy exists between the MW of the large subunits of the B enzymes as determined by the group of Rutter (190,000, 170,000, 150,000: Weaver *et al.*, 1971) and the other investigators. Chesterton and Butterworth (Table II and Fig. 1; unpublished results) and Chambon *et al.* (1972; Fig. 1) have found an additional minor component (B2′) in their rat liver enzyme B preparations with a MW of 165,000. This component could belong to an additional B polymerase form, which cannot be separated from BII on nondenaturating polyacrylamide gel (Chesterton and Butterworth, unpublished results). This suggests that the B activity purified by Weaver *et al.* (1971) from calf thymus or rat liver could represent a mixture of form BII and of this putative additional B form. This possibility is strongly supported by the inability of these authors to resolve their B preparation into two enzymes, corresponding to BI and BII, using disc polyacrylamide gel electrophoresis under nondenaturating conditions (Weaver *et al.*, 1971), whereas authentic BI and BII enzymes are readily separated under these conditions (Kedinger *et al.*, 1971; Mandel and Chambon, 1971a; Kedinger and Chambon, 1972). Coelectrophoresis of the enzyme preparations of the two groups should clarify this point.

The purified KB cell polymerase B activity contains, like the rat liver B enzymes, two major components of MW 220,000 and 140,000, and a minor component of MW 170,000 (Sudgen and Keller, 1973). Yeast A and B enzymes also possess two large subunits. The MW's of the two large subunits of yeast enzymes A (Ponta *et al.*, 1972) and B (Ponta *et al.*, 1972; Dezelée *et al.*, 1972) are very similar to those of the animal AI and BII enzymes, respectively.

V. General Properties of Animal RNA Polymerases

A. *Basic Requirements for RNA Synthesis*

The requirements for RNA synthesis are the same as for the bacterial enzyme: the four nucleoside triphosphates, DNA, and Mg^{2+} or Mn^{2+} (Goldberg and Moon, 1970; Chesterton and Butterworth, 1971b; for other references, see Cold Spring Harbor Symposium, 1970). A sulfhydryl reagent is not mandatory for enzymatic activity. Indirect evidence for the dependence of enzyme activity on sulfhydryl groups is derived from the inactivation of enzyme with p-chloromercuribenzoate (Gissinger *et al.*, 1974). Thioglycerol and dithiothreitol are the two most effective compounds for stabilizing the enzymes and for stimulating RNA synthesis (Goldberg and Moon, 1970).

The K_s values for UTP and CTP are around 0.02 m*M* for both A and B calf thymus enzymes. The K_s values for ATP and GTP are 0.04 and 0.03 m*M* for calf thymus AI RNA polymerase, and 0.08 and 0.07 m*M* for calf thymus B enzymes, respectively (Gissinger *et al.*, 1974). These numbers are close to the corresponding K_s values for the bacterial enzymes. The higher K_s values for ATP and GTP, especially in the case of the B enzymes, could be related to the preferential initiation of RNA chains by these nucleotides (see below), since it was found for the bacterial enzyme that the K_s values for initiation are higher than those for elongation of RNA chains (Wu and Goldthwait, 1969a, b). Although there is no synthesis in the presence of GTP, UTP, or CTP alone, some poly(A) synthesis is observed in the presence of ATP (Chambon *et al.*, 1970).

A DNA template is an absolute requirement in the reaction, but the pyrimidine strand of synthetic deoxypolynucleotides, poly(dT) or poly(dC), is also transcribed (Blatti *et al.*, 1970; Lentfer and Lezius, 1972), whereas poly(dG) and poly(dA) are very poorly transcribed. These results are also in keeping with preferential chain initiation with GTP and ATP. In contrast to the bacterial enzyme, enzymes A and B are unable to use poly(A) and poly(U) as templates, at least under the incubation conditions of Chambon *et al.* (1970). It is not known whether the animal enzymes are able to catalyze an unprimed synthesis of poly(rIC) or poly(rAU) like the bacterial enzyme (Krakow and Karstadt, 1967; Smith *et al.*, 1966).

Both calf thymus A and B enzymes catalyze the incorporation of inorganic pyrophosphate into ribonucleoside triphosphates (Meilhac *et al.*, 1972). As with the bacterial enzyme (Dunn and Bautz, 1969), pyrophosphate exchange can occur in the presence of only two ribonucleoside triphosphates under conditions where no RNA synthesis can be measured (Meilhac *et al.*, 1972). Both AI and B enzymes are acidic proteins which migrate toward the anode at pH 8.5 (Mandel and Chambon, 1971a; Gissinger and Chambon, 1972; Kedinger and Chambon, 1972). The optimal pH is 8 for both A and B enzymes (Chesterton and Butterworth, 1971b; Gissinger *et al.*, 1974).

B. Divalent Cation and Ionic Strength Optima

The nucleolar enzymes (AI, AII) have optimal activities at low ionic strength (below 40 m*M* ammonium sulfate) and are equally stimulated by Mn^{2+} or Mg^{2+}. In contrast, higher ionic strengths (around 100–120 m*M* ammonium sulfate) and Mn^{2+} rather than Mg^{2+} are required for optimal activity of the B enzymes (Cold Spring Harbor Symposium, 1970; Chesterton and Butterworth, 1971b; Jacob, 1973). Enzyme AIII resembles the B enzymes in its divalent cation requirement, but its activity is unaffected by the ionic strength between 0 and 200 m*M* ammonium sulfate (Roeder and Rutter, 1969).

In fact, we found (Gissinger *et al.*, 1974) that these optima values are dependent on both the nature of the template and its concentration in the incubation medium. For instance, both calf thymus enzymes AI and B show their optimal activities at 40–50 m*M* ammonium sulfate and in the presence of Mn^{2+} when transcribing poly(dAT). At low calf thymus DNA concentration (0.5–1 μg/0.25 ml incubation) both calf thymus AI and B enzymes utilize Mn^{2+} and Mg^{2+} equally well and the optimal values for ionic strength decrease to 5 m*M* ammonium sulfate for enzyme AI and to 50 m*M* ammonium sulfate for the B enzymes (Gissinger *et al.*, 1974). Finally, calf thymus enzyme AI cannot transcribe the twisted circular SV40 DNA (form I) in the presence of Mg^{2+}; with this template the optimal ionic strengths are 16 m*M* for calf thymus enzyme AI and 16–25 m*M* for calf thymus and rat liver B enzymes (Mandel and Chambon, 1971b, 1974a). These results cast some doubt on the validity of the relationships that can be suggested between the soluble purified enzymes and the activities that can be detected in crude nuclear preparations by comparing their divalent cation and ionic strength requirements.

C. Initiation, Elongation, and Termination of RNA Chains

As with the bacterial enzyme (Burgess, 1971), initiation (i.e., formation of the first phosphodiester bond) of RNA synthesis by animal enzymes is preceded by a preinitiation process which, in the case of the B enzymes, involves at least two events: primary binding of enzyme to DNA, which occurs readily at low temperature, and a step that is time- and temperature-dependent, suggesting that it is related to the opening of the two DNA strands over a short local region, possibly at the initiation site (Meilhac *et al.*, 1972).

Studies of the incorporation of [γ–^{32}P]-labeled nucleoside triphosphates have shown that most, if not all, of the RNA chains are initiated by GTP or ATP. The ATP to GTP ratio is different for calf thymus AI and B enzymes and varies according to the template and the divalent cation present in the incubation mixture (native calf thymus DNA versus native SV40 DNA form I, native calf thymus DNA versus denatured calf thymus DNA, Mn^{2+} versus Mg^{2+} (Chambon *et al.*, 1970; Mandel and Chambon, 1974b; Gissinger *et al.*, 1974).

The rate of elongation of RNA chains also varies according to the origin of the template and the nature of the stimulating divalent cation. Rates of 5–10 nucleotides/sec are observed when AI or B enzymes transcribe SV40 DNA form I, whereas rates of about 30 nucleotides/sec are found when calf thymus DNA is transcribed by B enzymes in the presence of Mn^{2+} (Mandel and Chambon, 1974b; Meilhac and Chambon, 1973). This latter value is similar to that measured *in vitro* for the bacterial enzyme (Richardson, 1969) and probably close to the *in vivo* rate of RNA synthesis in mammals (Greenberg and Penman, 1966). The size of RNA chains synthesized by the B enzymes on native calf thymus DNA is increased when the stimulating factors of Stein and Hausen or Seifart are added (see below). Whether these factors stimulate only the extent of elongation or also the rate of elongation is unknown.

Very little is known about chain termination, RNA release, and reinitiation (see Sections VI, C and D). It has been reported that RNA synthesis catalyzed by rat liver polymerase B on native DNA continues to proceed for a longer time in the presence of spermine (Stirpe and Novello, 1970a). The mechanism of this stimulation is unknown. The role of the various subunits in the different steps of the transcription process remains to be investigated.

VI. Template Specificity of Animal RNA Polymerases

Various deproteinized "native" animal DNA's were used as templates for the multiple RNA polymerases of a variety of tissues. It was generally found (Roeder and Rutter, 1969; Gniazdowski *et al.*, 1970; Chesterton and Butterworth, 1971b; for other references, see Cold Spring Harbor Symposium, 1970) that the AI enzyme prefers native DNA, whereas the B and AIII enzymes transcribe native and denatured DNA's more or less equally efficiently. In some cases the base composition of the RNA synthesized by the B enzymes is not complementary to the template DNA (Chambon *et al.*, 1970; Smuckler and Tata, 1972), suggesting a selective transcription of some parts of the DNA. None of these studies demonstrated that gene expression in animal cells is regulated by distinct RNA polymerases with different template specificities, as suggested by their multiplicity and their different intranuclear localization. Several approaches have been used to try to demonstrate such template specificities.

A. *Transcription of Cellular DNA*

The most direct way to support the above hypothesis would be to demonstrate that the various enzymes specifically transcribe different parts of the deproteinized chromosomal DNA. Up to now, a direct experimental approach to answer this question failed for one of the following reasons. First, all of the available preparations of animal DNA's contain a relatively high number of single-stranded nicks ("high" compared to the number of true promoter sites) where RNA synthesis can be initiated nonspecifically (Vogt, 1969; Gniazdowski *et al.*, 1970). This situation undoubtedly results in masking any specific transcription of unique genes. Second, the comparison of the RNA transcribed *in vitro* from unique genes with RNA synthesized *in vivo* cannot be performed using the conventional DNA–RNA hybridization–competition method owing to the complexity of the animal genome. In the only case where this would be feasible, since the *in vivo* product is available in large quantities, i.e., ribosomal RNA, the corresponding genes cannot be readily isolated from chromosomal DNA. Where they have been isolated, their transcription did not exhibit any specificity with respect to the intranuclear origin of the enzyme: Neither *Xenopus laevis* polymerase AI nor polymerase B shows any striking preference for purified rDNA versus bulk DNA, and neither enzyme transcribes specifically the regions of the rDNA which are transcribed *in vivo* (Roeder *et al.*, 1970), possibly because initiation at single-stranded nicks or at the ends of the molecules was obscuring the pattern of transcription.

A less direct approach consists of looking for a possible specificity of the sites where RNA synthesis is initiated by the various RNA polymerases. Using this approach, aided by the inhibitor AF/013 (see below), we have demonstrated that the sites on calf thymus DNA, which confer resistance to the inhibitor upon calf thymus AI and B, and *E. coli* holoenzyme RNA polymerases, are different (Chambon *et al.*, 1972, 1973). There is therefore very little hope of transcribing accurately animal DNA's or chromatin with a bacterial RNA polymerase. These studies also suggest that the purified calf thymus B enzymes can initiate on "true" double-stranded regions of the DNA and are not lacking an initiation factor, whereas enzyme AI, like the *E. coli* core enzyme, initiates at single-stranded nicks or at the ends of the DNA molecules (Meilhac and Chambon, 1973). This could mean either that enzyme AI lacks an initiation factor required for initiation on double-stranded calf thymus DNA, or that the sites at which enzyme AI can initiate specifically are very rare, which could be the case if enzyme AI were involved specifically in the transcription of ribosomal cistrons, as suggested by its intranuclear localization.

The number of sites that can protect B enzymes against the inhibitory effect of AF/013 was determined by measuring the incorporation of [γ-^{32}P]-ATP or GTP. The number of such sites is low, in the order of 40,000–60,000 (depending on the presence of Mn^{2+} or Mg^{2+} during the incubation) per calf thymus haploid genome, which corresponds to an average of one initiation site for a DNA weight of 27–43 $\times 10^6$ daltons (Chambon *et al.*, 1972, 1973; Meilhac and Chambon, 1973). Whether these sites correspond to physiological initiation sites is unknown.

B. Transcription of Deoxyribonucleoprotein

Obviously the *in vivo* template for animal RNA polymerases is not a "naked" DNA; it has been shown that histones, acidic proteins, and possibly specific chromosomal RNA's possess some role in the regulation of transcription (for an excellent review, see MacGillivray *et al.*, 1972). Using rat liver chromatin, Butterworth *et al.* (1971) have shown that rat liver AI polymerase is virtually inactive in the transcription of chromatin, whereas form B polymerase actively transcribes chromatin. A comparison of the activity of rat liver form B polymerase and *Micrococcus lysodeikticus* RNA polymerase demonstrates that chromatin is a more efficient template for the rat liver enzyme. There is also some evidence that the animal and the bacterial enzymes bind to and transcribe from different sites on the chromatin DNA. With respect to AI enzyme, there are two interpretations of these results: Either enzyme AI cannot initiate on chromatin because its DNA does not contain any nicks (or because the nicks are hidden), or the amount of accessible ribosomal DNA is too low to detect its transcription. For B enzymes, these observations fit with the conclusion that they can initiate on intact double-stranded DNA from sites that are different from those used by a bacterial enzyme. Calf thymus chromatin is also a better template for a calf thymus RNA polymerase than for *E. coli* RNA polymerase (Keshgegian and Furth, 1972). For reasons already discussed (Section VI,A) it was impossible, in these studies, to investigate the fidelity of the transcription.

C. Transcription of Viral DNA's

It is clear from the above considerations that one of the prerequisites to demonstrate unambiguously the template specificity of purified RNA polymerases is to use templates of defined structure for which the *in vivo* products of transcription are known and are easy to compare with the RNA's made *in vitro*. In this respect and for obvious reasons, the studies performed so far have used either bacteriophage DNA's or animal viral DNA's.

1. TRANSCRIPTION OF BACTERIOPHAGE DNA'S

Although phage DNAs are not physiological templates for the animal RNA polymerases, they were used in early attempts to demonstrate the template specificity of the animal enzymes, because they are relatively easy to prepare in a native and intact state and because their *in vivo* and *in vitro* products of transcription by the *E. coli* enzyme are well characterized (for references, see Bautz, 1972).

Calf thymus RNA polymerases AI and B can bind to native T4 or λ phage DNA's, but are unable to use efficiently the initiation sites which are present in these DNA's and which are readily used by the *E. coli* holoenzyme (Gniazdowski *et al.*, 1970; Chambon *et al.*, 1970; Blatti *et al.*, 1970; Burgess, 1971, Bautz, 1972). The basis for this specificity lies in the native double-stranded structure of the DNA, since denatured or double-stranded nicked phage DNA's are efficiently transcribed. Addition of the *E. coli* σ-factor, which powerfully stimulates the activity of *E. coli* core enzyme on intact native DNA's (Burgess *et al.*, 1969; Burgess, 1971; Bautz, 1972), does not stimulate transcription of native T4 or T2 DNA's by the animal enzymes (Gniazdowski *et al.*, 1970; Blatti *et al.*, 1970; Furth *et al.*, 1972). This result is not surprising in view of the structural differences that exist between the animal enzymes and the bacterial enzyme (Section IV,B), but it is in contrast with the *in vivo* reported effect of the σ-factor, which stimulates RNA synthesis when injected in amphibian oocytes (Crippa and Tocchini-Valentini, 1970; Tocchini-Valentini and Crippa, 1970a). The reason for this puzzling discrepancy remains to be established. The RNA synthesized by a calf thymus RNA polymerase on T2 DNA is very similar to the RNA transcribed by the *E. coli* core enzyme (Furth *et al.*, 1972). This result suggests that the very limited transcription of phage DNA's by animal enzymes is initiated more or less at random at single-stranded nicks (Bautz, 1972).

In many respects the calf thymus enzymes behave like the bacterial core enzyme, which can also bind to T4 phage DNA (Darlix *et al.*, 1969), but does not initiate RNA synthesis unless σ-factor is added (Travers and Burgess, 1969). Whether the inability of the animal enzymes to transcribe phage DNA's reflects an intrinsic property of these enzymes, an inability to recognize the initiation signals present on these DNA's, or the loss of a σ-like initiation factor during the course of the purification, is still unknown.

2. TRANSCRIPTION OF ANIMAL VIRAL DNA's

Studies on the transcription *in vitro* of phage DNA's by purified *E. coli* RNA polymerase have been extremely useful in elucidating some of the mechanisms involved in the regulation of transcription in prokaryotes (Bautz, 1972). It can be expected that similar studies using the purified animal RNA polymerases, and DNA's from animal viruses, which do not contain endogenous polymerases, could throw some light upon the role of the various animal RNA polymerases in the regulation of transcription in eukaryotes.

Twisted, circular SV40 DNA (form I) and adenovirus 2 DNA are efficiently transcribed *in vitro* by rather crude RNA polymerase preparations of KB cells which contain both forms of enzymes (Keller and Goor, 1970; Green *et al.*, 1970). The *in vitro* transcription of these two DNA's by mammalian cell RNA polymerase preparations is asymmetrical and primarily from the DNA strand that is also transcribed *in vitro* by *E. coli* enzyme (Westphal, 1970; Green and Hodap, 1971) and *in vivo* "early" during infection (Sambrook *et al.*, 1972; Lindstrom and Dulbecco, 1972; Khoury *et al.*, 1972; Green *et al.*, 1970). SV40 DNA form I is also transcribed asymmetrically by a partially purified rat liver B polymerase (Herzberg and Winocour, 1970). These results suggest that the mammalian enzymes exhibit DNA strand selection and asymmetrical transcription, which are two important criteria for a faithful transcription by the *E. coli* holoenzyme (Geiduschek and Haselkorn, 1969; Burgess, 1971; Bautz, 1972).

In contrast with these results, SV40 form I is transcribed symmetrically by pure calf thymus AI and B and rat liver B RNA polymerases (Mandel and Chambon, 1971b, 1974b). On the other hand, calf thymus RNA polymerase AI, like the *E. coli* holoenzyme (Westphal and Kiehn, 1970), can pass over its own initiation site on SV40 DNA form I and synthesize RNA which has at least twice the length of a DNA strand, whereas the RNA's made by the calf thymus or rat liver B enzymes have a definite maximum size which corresponds roughly to 18 S RNA (Mandel *et al.*, 1973). These results indicate that, besides the difference already described at the level of chain initiation, there are also differences between AI, B, and *E. coli* enzymes at the level of chain termination.

The reason for the discrepancy between our results and those of Keller and Goor (1970) and of Herzberg and Winocour (1970) is unknown, but could lie either in the origin of the enzyme or in its degree of purity. Whether the 18 S RNA, which is synthesized *in vitro* by the B enzymes, is related to the mRNA's of similar sizes that are present in SV40-infected cells (Martin and Byrne, 1970; Weinberg *et al.*, 1972) is not known, but it is unlikely. It was indeed recently shown for both SV40 and adenovirus that the mRNA's found in the cytoplasm are probably derived from much larger nuclear precursors, which in the case of SV40 virus could possible contain some cellular RNA sequences (Parsons *et al.*, 1971; Wall *et al.*, 1972; Jaenisch, 1972; MacGuire *et al.*, 1972; Martin and Byrne, 1970; Sokol and Carp, 1971; Hirai and Defendi, 1972; Weinberg *et al.*, 1972). *In vivo* most of the transcription of adenovirus and SV40 DNA seems to be catalyzed by a form B enzyme (see Section VIII,D).

The symmetry of the *in vitro* transcription catalyzed by the pure animal enzymes is noteworthy. On the basis of the results obtained with the bacteria, it was believed that to be faithful transcription should be asymmetrical. In fact, recent results suggest that *in vivo* transcription can be at least in part symmetrical for some animal viral DNA's (SV40 and adenovirus: Aloni, 1972; Lucas and Ginsberg, 1972), mitochondrial DNA (Aloni and Attardi, 1971), and even of cellular DNA (Stampfer *et al.*, 1972).

The ability of the animal RNA polymerases to transcribe a twisted, circular DNA seems to prove that these enzymes can initiate on an intact native DNA and that no additional initiation factors are required. It was recently shown (Maestre and Wang, 1971; Dean and Lebowitz, 1971; Delius *et al.*, 1972; Hossenlopp *et al.*, 1974) that twisted, circular DNA's are not perfectly double-stranded and contain, owing to conformational constraints, small denaturedlike regions. Therefore, the ability of the animal enzymes to transcribe a twisted double-stranded DNA does not prove that they can initiate on a regular double-stranded DNA.

In conclusion, although the studies of the transcription of animal viral DNAs have revealed some distinct properties of the various animal enzymes, they have not led, up to now, to a clear-cut answer to the basic question of the role of animal RNA polymerases in the control of transcription. There are several reasons for this failure: (*a*) The *in vivo* pattern of transcriptional regulation is in fact unknown, the situation being in any case much more complicated than was initially believed owing to (1) the possible integration of the viral genome, even during the lytic infection (Hirai and Defendi, 1972), (2) the formation of giant nuclear RNA precursors, which could contain some cellular RNA sequences, consequently increasing the importance of the posttranscriptional events, and (3) the discovery of symmetrical transcription; (*b*) in the case of SV40, the twisted, circular DNA form I contains some denatured regions which can possibly act *in vitro* as unspecific initiation sites.

D. Factors

Studies on bacterial RNA polymerases have shown that protein factors which are either found free or bound to the enzyme play a major role in the positive control of gene transcription in prokaryotes by acting on initiation or termination of RNA synthesis (for reviews, see Burgess, 1971 and Bautz, 1972). If such control mechanisms exist in eukaryotes, one should find similar factors in animal cells.

Protein factors, which stimulate the transcription of native animal DNA's by the B enzymes, were found by two groups of investigators in rat liver (Seifart and Sekeris, 1969; Seifart, 1970; Seifart *et al.*, 1973) and in calf thymus (Stein and Hausen, 1970a, b; Hameister *et al.*, 1972). These proteins are rather specific for B enzymes and act only when these enzymes transcribe a native DNA at low ionic strength, in the presence of Mn^{2+}. Both rat liver and calf thymus factors have been extensively purified and are free of nucleases. They appear to consist of a family of closely related basic proteins (pK 8.5–9.6) of low MW (25,000–35,000 daltons). Recent studies (Hameister *et al.*, 1972; Seifart *et al.*, 1973; Lentfer and Lezius, 1972) show that these proteins are not initiation factors like the bacterial σ-factor, but act mainly by stimulating the chain-elongation step and also by promoting some release of the synthesized RNA chains and reinitiation.

It is not known whether the factors act nonspecifically by stimulating the elongation of all the RNA molecules or specifically by promoting the elongation of only some RNA transcripts. The effect of the factors on the transcription of either rat liver or calf thymus chromatin has not been investigated.

A rat liver nucleolar protein factor that appears to specifically stimulate the transcription of native DNA by enzyme AI was recently isolated (Higashinakagawa and Muramatsu, 1972). The function and mechanism of action of this factor is presently unknown, as well as its relationship with the short-lived protein which might regulate the transcription of the nucleolar ribosomal genes *in vivo* (Muramatsu *et al.*, 1970).

It should be pointed out that the term "factor" for all these proteins is particularly misleading, since it suggests a functional similarity with the bacterial transcriptional factors, which specifically affect the synthesis of well-defined classes of RNA on well-defined DNA (Burgess, 1971; Bautz, 1972). As already discussed, this is clearly not the case for the animal factors. In this respect, it is interesting to recall that polyamines stimulate RNA synthesis catalyzed by either the bacterial (Fox and Weiss, 1964) or the animal (Stirpe and Novello, 1970a) B RNA polymerases, and that even histones can stimulate RNA synthesis when added at low concentrations (Konishi and Koide, 1971).

VII. Inhibitors of Animal RNA Polymerases

A. *Amanitins*

α-Amanitin, a toxic isolated from the mushroom *Amanita phalloides* (Wieland, 1964, 1968; Wieland and Wieland, 1972; Fiume and Wieland, 1970), specifically inhibits RNA synthesis catalyzed at high ionic strength by isolated nuclei (Stirpe and Fiume, 1967; Stirpe and Novello, 1970a, b;

Novello and Stirpe, 1969, 1970). The RNA polymerase activity measured at low ionic strength is resistant to the poison. This suggested that there could be two types of RNA polymerase in animal cells and that α-amanitin might act specifically on one of them. This is indeed the case, as demonstrated simultaneously by several groups of investigators, using solubilized enzymes (Jacob *et al.*, 1970a; Kedinger *et al.*, 1970; Lindell *et al.*, 1970; Chesterton and Butterworth, 1971b). Neither A enzymes, nor *E. coli* polymerases are inhibited by α-amanitin. This specificity is the basis for our nomenclature of the animal RNA polymerases (Kedinger *et al.*, 1971). Some other naturally occurring and chemically modified amatoxins also inhibit B polymerases. There is a striking parallel between *in vivo* toxicity of these compounds and either their *in vitro* inhibitory action on solubilized B RNA polymerases (Buku *et al.*, 1971) or the values of the dissociation constants of the pure enzyme–amanitin complexes (Meilhac and Chambon, 1974). These results indicate that the main *in vivo* targets of amanitins are the B RNA polymerases.

The mechanism of action of amanitins was extensively studied in our laboratory (Chambon *et al.*, 1970, 1972; Meilhac *et al.*, 1970; Meilhac and Chambon, 1974). Amanitins act by binding to free or DNA-bound B enzymes; the stoichiometry of the reaction is probably 1:1. Amanitins do not bind to A enzymes or to *E. coli* polymerase. The affinity of the amanitins for the B enzymes is very high ($K_D \sim 3 \times 10^{-9}$ *M* for *O*-methyl-γ-amanitin). The half-life of the complex at 4°C is at least 30 hr. This allows the determination of the amount of amanitin bound to the enzyme by using an assay where the enzyme is retained on a nitrocellulose filter (Chambon *et al.*, 1972). There is no difference between the affinity of amanitins for calf thymus BI and BII enzymes or B enzymes from other rat tissues. B enzymes are the only substances to which amanitins bind with such an affinity in crude tissue homogenates, which permits a direct determination of the number of B polymerase molecules in such homogenates (Chambon *et al.*, 1972).

Amanitins inhibit all the steps of the RNA polymerase reaction except binding of the enzyme to DNA. Elongation of RNA chains is immediately stopped when amanitins are added (Jacob *et al.*, 1970a; Kedinger *et al.*, 1970; Lindell *et al.*, 1970). Neither RNA nor DNA are released from the RNA–enzyme–DNA transcription complex (Meilhac and Chambon, 1974). It is unknown to which subunit of B polymerases amanitins bind. Contrary to a previous report (Chambon *et al.*, 1970), it is also very likely that chain initiation is inhibited, since pyrophosphate exchange is inhibited even in the presence of only two nucleoside triphosphates under conditions where chain elongation cannot occur (Meilhac and Chambon, 1974). The mechanism of action of amanitins resembles that of steptolydigin, which inhibits the bacterial enzyme by blocking chain elongation as well as chain initiation (Cassani *et al.*, 1971). At very high concentrations, streptolydigin also inhibits both A and B calf thymus polymerases (Chambon *et al.*, 1970). It is not known if the inhibition mechanism is the same as for *E. coli* RNA polymerase.

Chan *et al.* (1972) have recently obtained mutants of hamster ovary cells which are resistant to α-amanitin. There is some evidence that the B polymerases of such mutants are selectively modified, suggesting that the mutation affects a structural gene of one of the subunits of the B polymerases.

B. *Rifamycin Derivatives*

Nuclear animal RNA polymerases are resistant to rifampicin (for references, see Cold Spring Harbor Symposium, 1970). In contrast, some semisynthetic derivatives of rifamycin (like AF/05 and AF/013) inhibit both classes of animal RNA polymerases (Meilhac *et al.*, 1972), but at much higher concentrations than those of rifampicin which inhibit the bacterial enzymes. Unlike rifampicin these derivatives are not specific for DNA-dependent RNA polymerases, since RNA-dependent DNA polymerases of tumor viruses (Gurgo *et al.*, 1971) and even other unrelated enzymes like hexokinase (Riva *et al.*, 1972) are inhibited over the same concentration range.

Despite this lack of specificity, these derivatives can be used for investigating the mechanism of *in vitro* transcription catalyzed by animal RNA polymerases. It was shown that they selectively inhibit chain initiation on native DNA (Meilhac *et al.*, 1972; Onishi and Muramatsu, 1972). The inhibition of elongation, which was noticed at high concentration of AF/013 by Juhasz *et al.* (1972), is very probably due to the presence of wide denatured regions in their "native" DNA (Meilhac and Chambon, 1973).

Studies on the mode of action of these inhibitors on B enzymes show that inhibition takes place at two steps of the preinitiation process: Both the primary binding of enzyme to DNA and a temperature- and time-dependent step are inhibited (Meilhac *et al.*, 1972). In this respect, the rifamycin derivatives differ from rifampicin, which does not inhibit the binding of the bacterial enzyme to DNA (Burgess, 1971), and act more like heparin sulfate on *E. coli* RNA polymerase (Zillig *et al.*, 1970). After preincubation of B enzymes and native DNA at 37°C, RNA synthesis becomes almost completely resistant to AF/013, suggesting that the formation of the resistant complex is related to the opening of the two DNA strands over a short local region at the initiation site. As already discussed

(Section VI,A), these derivatives have been very useful in assessing the mechanism of initiation of RNA synthesis by AI and B enzymes and in demonstrating that the two forms of animal enzymes initiate at different sites, which furthermore differ from those used by *E. coli* holoenzyme.

C. Other Inhibitors

Cycloheximide rapidly inhibits the synthesis of ribosomal RNA when administered *in vivo,* suggesting either that the presence of some short-lived protein(s) is required for the normal level of transcription of the ribosomal genes or that the nucleolar polymerase itself is sensitive to cycloheximide (Muramatsu *et al.,* 1970). *In vitro* studies revealed the inhibition of DNA-dependent RNA polymerase activity in the nucleolar preparation isolated from rats treated with cycloheximide, but little inhibition was noted for the extranucleolar nuclear fraction from the same sources (Muramatsu *et al.,* 1970; Yu and Feigelson, 1972). However, Higashinakagawa and Muramatsu (1972) have shown that neither the solubilized nucleolar nor the extranucleolar rat liver RNA polymerases are inhibited *in vitro* by addition of cycloheximide. Similar results were obtained in our laboratory (Gissinger, unpublished results) using either crude or highly purified calf thymus RNA polymerases. Moreover, the relative activities of solubilized nucleolar and extranucleolar RNA polymerase do not seem to be affected by *in vivo* cycloheximide treatment (Higashinakagawa *et al.,* 1972), although somewhat different results were reported by Yu and Feigelson (1972). On the other hand, Timberlake *et al.* (1972a) found a marked inhibition of their preparation of rat liver RNA polymerase AI at a very low concentration of cycloheximide. The reason for these discrepancies is unknown, but it seems related to the preparation of the enzyme, since the enzyme prepared by the latter group exhibits a very strange and puzzling time course for RNA synthesis. Cycloheximide also inhibits RNA synthesis catalyzed by AI-type enzymes of lower eukaryotes, such as an aquatic fungus and a water mold (Horgen and Griffin, 1971; Timberlake *et al.,* 1972b).

The exotoxin *Bacillus thuringiensis,* when given *in vivo* to mice, preferentially depresses synthesis of nuclear rRNA in the liver (Mackedowski *et al.,* 1972). *In vitro,* it inhibits the animal RNA polymerases as well as *E. coli* RNA polymerase by competition with ATP (Smuckler and Hadjiolov, 1972; Beebee *et al.* 1972). The two groups disagree as to whether the toxin preferentially affects the nucleoplasmic (Smuckler and Hadjiolov) or the nucleolar (Beebee *et al.*) enzymes. Studies in our laboratory (Meilhac, unpublished results), conducted with either partially or highly purified preparations of calf thymus and rat liver AI and B polymerases, do not reveal any significant difference in the sensitivity of the two types of enzyme. Further investigations are required to elucidate the reasons for these contradictory results.

VIII. Physiological Role of the Various Animal Nuclear RNA Polymerases

If the multiplicity of animal polymerases has, as expected, a role in the regulation of transcription by providing specific means of controlling independently the synthesis of the various classes of cellular RNA's, it should be possible to correlate the synthesis of the different RNA's with the various nuclear polymerase activities that have been so far identified. Specific inhibitors are the tools of choice for such studies, and it was expected that amanitins, which specifically inhibit *in vitro* the B enzymes, would be very useful for elucidating the *in vivo* role of the multiple enzymes. As usual with animal cells, reality turned out to be more complicated than expected! Our ideas concerning the *in vivo* involvement of the various nuclear RNA polymerases in RNA synthesis came essentially from three types of experiments: (1) studies on the effect of α-amanitin on the synthesis of the various classes of RNA's, when nuclei are incubated *in vitro* under conditions (salt and divalent cations) optimal for the solubilized activities; (2) studies on the alterations of RNA synthesis when α-amanitin is administered *in vivo;* (3) studies on the effect of α-amanitin on viral RNA synthesis in cells infected with DNA viruses which do not contain an endogenous RNA polymerase.

A. *Role of the Nucleolar AI RNA Polymerase*

The nucleolar localization of enzyme AI and the observation that, under conditions favoring polymerase AI activity (low salt, amanitin present), the RNA synthesized by isolated rat liver nuclei has a GC content similar to that of ribosomal RNA, suggest that this enzyme activity is involved in rRNA synthesis (Blatti *et al.*, 1970). DNA–RNA hybridization studies also indicated that rRNA synthesis is at least in part carried out by an α-amanitin-resistant enzyme (Blatti *et al.*, 1970). More recently Reeder and Roeder (1972), using isolated *Xenopus laevis* nuclei, have unequivocally demonstrated the total insensitivity of rRNA synthesis to α-amanitin, which rules out B RNA polymerases as the direct agents of rRNA synthesis. In addition, their data demonstrate that rRNA is also actively synthesized in high salt and remains the main RNA product in the presence of α-amanitin, whereas it is over-shadowed by synthesis of other

RNA's in the absence of the drug. Their results do not rule out the possible involvement of polymerase form AIII, which is also resistant to α-amanitin, in rRNA synthesis. This possibility is not very likely, since (1) nuclei of *Xenopus laevis* which make predominantly rRNA in the presence of α-amanitin appear to contain five- to tenfold more of form AI than form AIII (Reeder and Roeder, 1972); of course, this does not exclude that the *in vivo* situation is very different, due either to a selective *in vivo* stimulation of AIII activity or to selective losses of this activity during solubilization of the enzymes; (2) an antibody against pure calf thymus AI polymerase inhibits all of the amanitin-resistant RNA polymerase activity in crude homogenates of calf thymus (Kedinger and Gissinger, unpublished results).

Since enzyme AIII was not found in this tissue, our results mean that either enzyme AI or an unknown and very closely related enzyme is in fact responsible for rRNA synthesis. In addition, Zylber and Penman (1971) have shown that synthesis of high-molecular-weight nucleolar RNA in isolated HeLa nuclei is not inhibited by amanitin. Taken together, these results suggest very strongly that RNA polymerase AI is responsible for rRNA synthesis. If this is the case, then the results of Reeder and Roeder (1972) also demonstrate that enzyme AI is still very active when nuclei are incubated at high ionic strength in the presence of amanitin. Therefore, the activity measured at high ionic strength in the presence of amanitin may not necessarily be enzyme AIII activity (Blatti *et al.*, 1970), even if actinomycin is added at low concentration to specifically inhibit rRNA synthesis (Zylber and Penman, 1971).

Contrary to expectations, the synthesis of ribosomal RNA precursors and of rRNA is rapidly and significantly inhibited in rat liver *in vivo* after administration of α-amanitin (Jacob *et al.* 1970b, c; Niessing *et al.*, 1970; Tata *et al.*, 1972; Sekeris and Schmid, 1972). This unexpected finding agrees with results obtained by electron microscopy, showing that treatment with amanitin *in vivo* causes a fragmentation of liver nucleoli (Fiume and Laschi, 1965; Fiume *et al.*, 1969; Meyer-Schultz and Porte, 1971; Marinozzi and Fiume, 1971; Petrov and Sekeris, 1971). On the other hand, treatment with α-amanitin does not result in any significant inhibition of the amanitin-resistant RNA polymerase activity of isolated nuclei (Stirpe and Fiume, 1967; Tata *et al.*, 1972; Sekeris and Schmid, 1972), although Jacob *et al.* (1970c) found some inhibition in isolated nucleoli.

Several explanations could account for the inhibition of rRNA synthesis *in vivo* despite the lack of inhibition of the purified AI enzyme and of the amanitin-resistant activity measured in isolated liver nuclei of amanitin-treated rats: (1) Amantin could be converted in rat liver to an inhibitory metabolite which might inhibit the nucleolar enzyme and could be lost during the preparation of the nuclei; (2) the synthesis of a mRNA by the amanitin-sensitive B enzymes is continuously required for the synthesis of a rapidly renewed protein which stimulates *in vivo* the nucleolar polymerase (Muramatsu *et al.*, 1970; Higashinakagawa and Muramatsu, 1972); this protein could be a factor stimulating initiation on the ribosomal genes; (3) the activity of the nucleolar RNA polymerase is regulated *in vivo* by a factor that is directly sensitive to α-amanitin or one of its metabolites; (4) the inhibition of rRNA synthesis is mechanical and due to a condensation of the chromosomes containing the nucleolar organizers consecutive to an arrest of RNA synthesis catalyzed by the B enzymes (Marinozzi and Fiume, 1971).

The first and the fourth hypotheses are supported by autoradiographic studies showing that RNA synthesis is not inhibited in nucleoli of *Triturus* oocytes (Bucci *et al.*, 1971) or in nucleoli of *Chironomus* salivary glands (Wobus *et al.*, 1971; Beerman, 1971; Egyhazi *et al.*, 1972). The results on *Triturus* oocytes are in agreement with those of Tocchini-Valentini and Crippa (1970a, b) showing that α-amanitin does not affect RNA synthesis in growing oocytes of *Xenopus* at stage 4 which synthesize predominantly (more than 97%) rRNA. Analysis of the RNA's made in salivary glands incubated in the presence of α-amanitin also confirms that the synthesis of the rRNA precursors is not inhibited (Serfling *et al.*, 1972; Egyhazi *et al.*, 1972). On the contrary, when α-amanitin was injected into ligated *Calliphora* larvae, the synthesis of rRNA was also inhibited, although less than that of giant heterogeneous RNA (Shaaya and Clever, 1972).

B. Role of the Nucleoplasmic B RNA Polymerases

Several lines of evidence demonstrate that the α-amanitin-sensitive nucleoplasmic B enzymes synthesize the giant nuclear heterogeneous RNA, which is presumably the precursor of mRNA. Zylber and Penman (1971) showed that synthesis of the nucleoplasmic giant heterogeneous RNA is severely depressed when HeLa nuclei are incubated under conditions favoring B polymerases (high ionic strength, low concentration of actinomycin in order to inhibit the nucleolar enzyme). In agreement with these results, most of the non-rRNA synthesis is inhibited by α-amanitin when isolated *Xenopus laevis* nuclei are incubated *in vitro* (Reeder and Roeder, 1972). The synthesis of rat liver nonribosomal RNA's is rapidly inhibited after *in vivo* administration of α-amanitin (Jacob *et al.*, 1970c; Niessing *et al.*, 1970; Tata *et al.*, 1972). Similar results were obtained after administration of α-amanitin to *Calliphora* larvae (Shaaya and Clever, 1972). Autoradiographic studies reveal an almost complete shut-off of [^{3}H]uridine incorporation in the extranucleolar regions of polytene chromosomes after incubation of *Chironomus* salivary glands in the presence of α-amanitin (Beermann, 1971; Wobus *et al.*, 1971; Egyhazi *et al.*, 1972). Similarly, when *Triturus* oocytes are incubated *in vitro*, α-amanitin causes an early and striking retraction of the loops of the lampbrush chromosomes, accompanied by a cessation of [^{3}H]uridine incorporation along the chromosome axis (Bucci *et al.*, 1971). Finally, *in vivo* administration of α-amanitin to rats results in a very rapid chromosome condensation (Marinozzi and Fiume, 1971) accompanied by a decrease in the number of ribonucleoprotein perichromatin fibrils (Petrov and Sekeris, 1971) which are most probably the morphological counterparts of giant heterogeneous RNA (Monneron and Bernhard, 1969; Petrov and Bernhard, 1971).

Smuckler and Tata (1972) observed that *in vivo* liver RNA synthesis remains inhibited for much longer than the nuclear B RNA polymerase activity assayed *in vitro* after *in vivo* α-amanitin administration to rats. This puzzling difference, which was not found by Sekeris and Schmid (1972), could possibly be explained by a release of enzyme-bound amanitin during the isolation of nuclei.

C. Role of the Nucleoplasmic AIII Polymerase

All of the evidence concerning the possible *in vivo* role of this enzyme activity is circumstantial. It indicates that a nucleoplasmic α-amanitin-resistant activity might catalyze the synthesis of a fraction of the giant heterogeneous RNA as well as 5 and 4 S RNA's. Zylber and Penman (1971) and Price and Penman (1972b) have indeed observed that some giant heterogeneous and 4–5 S RNA's are synthesized in the nucleoplasm when HeLa nuclei are incubated under conditions favoring soluble polymerase AIII (high salt concentration in the presence of α-amanitin). When dipteran salivary glands are incubated in the presence of α-amanitin, some residual chromosomal band labeling persists (Beerman, 1971; Serfling *et al.*, 1972; Egyhazi *et al.*, 1972) and biochemical analysis of the synthesized RNA's shows that they consist both of giant heterogeneous RNA's and 8 S, 5 S, and 4 S RNA's (Serfling *et al.*, 1972; Egyhazi *et al.*, 1972). Similar results were obtained by Shaaya and Clever (1972), who injected α-amanitin into *Calliphora* larvae. Montecuccoli *et al.* (1972) reported that liver 4 S RNA synthesis is only partially inhibited after administration of α-amanitin to rats. It should be stressed that none of these results exclude that polymerase AI could be involved in these nucleoplasmic amanitin-resistant syntheses, since (1) the data related to the intranuclear localization of polymerase AI do not exclude that some of this enzyme activity could be extranucleolar (Roeder and Rutter, 1970b); and (2) Reeder and Roeder have clearly shown that polymerase AI is functional at high salt concentrations (Reeder and Roeder, 1972).

D. *Involvement of Nuclear RNA Polymerases in Transcription of Viral DNA in Vivo*

As already discussed (Section VI,C,2), some animal DNA viruses offer a model system in which a small number of genes are expressed through mechanisms reflecting those of the host cell. With respect to the *in vivo* role of the multiple nuclear RNA polymerases, DNA viruses could be useful in demonstrating that the soluble enzyme activities that have been characterized *in vitro* are actually operating *in vivo.*

The properties of the RNA polymerase activities that transcribe the adenovirus genome have been studied recently. In isolated nuclei from infected cells the major portion of the adenovirus DNA is transcribed by a nucleoplasmic α-amanitin-sensitive activity which closely resembles polymerase B activity, no matter whether the nuclei were isolated from cells early or late in infection (Price and Penman, 1972a; Wallace and Kates, 1972; Chardonnet *et al.*, 1972). Ledinko (1971) and Chardonnet *et al.* (1972) also showed that adenovirus multiplication is inhibited when the cells are exposed to α-amanitin. Since cycloheximide fails to inhibit the synthesis of early virus-specified RNA (Parsons and Green, 1971; Chardonnet *et al.*, 1972), the most likely explanation of these results is that one of the B RNA polymerases is utilized, at least early in infection, for transcribing the adenovirus DNA. Recent results indicate that, in isolated nuclei of infected cells, SV40 DNA is also transcribed by an α-amanitin-sensitive activity (Jackson and Sugden, 1972).

In addition, nuclei from adenovirus-infected HeLa cells contain an activity which is resistant to α-amanitin and which synthesizes a discrete 5.5 S viral-specific RNA (Price and Penman, 1972a, b). Whether this activity is the AI or AIII enzyme is presently unknown, although circumstantial evidences suggest that polymerase AIII could be responsible for the synthesis of this low-molecular-weight virally specified RNA (Price and Penman, 1972b).

IX. *In Vivo* Regulation of Animal RNA Polymerase Activities

Qualitative as well as quantitative changes in the transcription pattern are observed in a great variety of physiological or nonphysiological conditions [(for references, see, for instance, Tata (1970); MacGillivray *et al.*

(1972); Jacob (1973); a complete bibliography of this subject is obviously beyond the scope of this review)]. Basically these changes can be accounted for by several mechanisms acting separately or simultaneously: (1) modification in the availability of the chromatin template involving either negative or positive control mechanisms; (2) changes affecting RNA polymerases themselves by one of the following mechanisms: (*a*) quantitative variation in the amount of active polymerase molecules, either by activation or inhibition of preexisting molecules or by regulation of their biosynthesis; (*b*) qualitative modifications leading to new template specificities and which can be due either to the synthesis of factors that bind to preexisting enzyme molecules, structural modification of preexisting RNA polymerases (by proteolysis, for instance), or synthesis of entirely new species of RNA polymerase.

Changes in RNA polymerase activity were observed in a variety of conditions, but the underlying mechanisms remain hypothetical, since the investigations were carried out with "aggregate" polymerase preparations in which the enzyme was dependent on the endogenous chromatin template. The discovery of multiple RNA polymerases and their solubilization has led to more meaningful studies aimed at distinguishing between the various possible mechanisms described above. Up to now, only changes in the quantities of soluble active enzymes have been observed (Sajdel and Jacob, 1971; Smuckler and Tata, 1971; Mainwaring *et al.*, 1971, Roeder and Rutter, 1970b; Chesterton *et al.*, 1972). The mechanism of these variations in the amounts of active polymerases is unknown. The determination of the number of polymerase molecules in crude tissue extracts using either labeled amanitin in the case of the B enzymes (see Section VII,A) or immunological assays should distinguish between variations in the rate of synthesis of the enzymes and activation of preexisting enzyme molecules. It is not known if these quantitative variations are accompanied by qualitative changes, since the only way to detect such changes would be either to analyze the subunit structure of the enzymes or to demonstrate a modification in their template specificities. The first type of study is technically very difficult, whereas the other alternative requires the characterization of the synthesized RNA (for discussion of this problem, see Section VI).

Finally it is worthwhile to recall the results of Roeder *et al.* (1970) showing that anucleolate *Xenopus laevis* embryos contain a normal amount of polymerase AI, as well as those of Roeder and Rutter (1970b) showing that polymerase AI is present in the nuclei of sea urchin blastulae in the absence of rRNA synthesis. These observations and similar results of Tocchini-Valentini and Crippa (1970 a, b) appear to rule out any regulatory mechanism in which the basic components of RNA polymerase AI are limiting within the cell. A similar conclusion is reached if one considers the number of B polymerases molecules per cell as measured by binding of labeled amanitin (Chambon *et al.*, 1972). This number ranges from 1×10^4 to 6×10^4 per haploid genome in cells exhibiting very different metabolic activities. As pointed out by Kafatos (see the discussion of the paper of Chambon *et al.*, 1972), it is very likely that these numbers exceed the number of polymerase molecules that are required to catalyze the *in vivo* RNA synthesis of mRNA, indicating that polymerases B molecules are present in saturating amount. These observations do not exclude that an essential activating factor is limiting, as suggested by Tocchini-Valentini and Crippa (1970a, b).

X. Conclusion and Prospects

Although spectacular advances have been made in the knowledge of mammalian DNA-dependent RNA polymerases during the last three years, the involvement of the multiple enzymes in the control of transcription is far from being understood. The only hard facts concern the demonstration that it is possible to isolate from animal cells several distinct forms of RNA polymerase which, at best, are characterized by (*a*) their intranuclear localization; (*b*) their sensitivity to amanitins; (*c*) their chromatographic properties; (*d*) their structure and immunological properties. Only AI, AII, BI, and BII RNA polymerase fulfill almost all of these criteria. Structural analysis of AIII activity is still required to prove unequivocally that it differs from the AI enzyme. Divalent cation and salt requirements, as well as a preference for native versus denatured DNA, appear to be looser criteria to distinguish among the various activities.

All of the other aspects of the problem remain open to question, particularly those concerning the possible role of the multiple RNA polymerases in the regulation of transcription:

1. Although it is certain that there are at least two forms of RNA polymerase operating *in vivo,* either sensitive or resistant to amanitins, it is unknown whether the multiple amanitin-sensitive and amanitin-resistant enzymes *in vitro* have their *in vivo* equivalents or correspond to artifacts formed during the solubilization and purification steps. Furthermore, although it is certain that the ribosomal, 4 S, and 5 S RNA's are synthesized by amanitin-resistant activities whereas the giant heterogeneous RNA's are made by an amanitin-sensitive activity, it is not firmly known which of the multiple solubilized amanitin-sensitive or resistant enzymes are involved *in vivo* in the synthesis of the various cellular RNA's (Section VIII).

2. *In vitro* studies have shown that amanitins are highly specific for

B polymerases *in vitro;* the mechanism of their inhibitory effect has been elucidated. It remains to be understood why α-amanitin is less specific, when administered *in vivo,* and also inhibits the synthesis of rRNA's. At least four hypotheses could account for this observation (Section VII,A).

3. Despite numerous studies, it is still unknown whether the purified enzymes can initiate on a "true" native double-stranded DNA because all of the available animal DNA's prepared up to now contain a substantial amount of single-stranded nicks, where initiation can occur nonspecifically (Section VI,C,1). That enzyme AI cannot transcribe a chromatin template (Section VI,B) does not obligatorily mean that enzyme AI is not able to initiate on this template, but could mean that the number of available AI-specific initiation sites is very low, resulting in a barely detectable RNA synthesis: Contrary to expectations, transcription of SV40 DNA by purified enzymes led to an ambiguous answer owing to the peculiar structure of SV40 DNA form I (Section VI,C,2). With respect to this problem, it is important to remember that *in vivo* eukaryotic DNA is largely associated with proteins, which might influence its template activity. Therefore, *in vitro* use of a "naked" DNA may not adequately reflect physiological conditions.

4. For these and additional reasons, it is not known whether the purified enzymes exhibit any template specificity and could specifically transcribe some regions of a cellular or viral genome, because up to now studies with isolated ribosomal genes have failed to reveal a specificity of the nucleolar AI polymerase for this template, possibly because too many single-stranded nicks were present (Section VI,A). For the other genes the comparison of the *in vitro* synthesized RNA with *in vivo* mRNA was impossible, using conventional DNA–RNA hybridization, due to the complexity of animal genomes (Section VI,A). Except for the rRNA's, very little is known on the nature of the cellular or viral RNA's which are synthesized in the cell nucleus; for instance, it is not really known whether *in vivo* transcription is completely asymmetrical or partially symmetrical (Section VI,C,2).

5. Although an elongation factor has been characterized, the search for specific initiation or termination factors is impeded by the impossibility of assaying for specific transcription products (Section VI,D); the same difficulty arises when one wants to look for possible qualitative changes in RNA polymerases, which could occur when the pattern of transcription is modified either in physiological or nonphysiological conditions (Section IX).

Obviously the proper way to answer all of these and related questions is to find systems in which the template is well defined and the transcription products easy to compare with their *in vivo* counterparts. In this respect, provided significant progress can be made in the knowledge of its *in vivo* pattern of transcription, study of the *in vitro* transcription of the adenovirus genome could throw some light on the possible involvement of the various RNA polymerases in the regulation of transcription in mammalian cells. The other alternatives are (*a*) to study the transcription of very high-molecular-weight "unnicked" cellular DNA, as recently obtained in our laboratory (Gross-Bellard *et al.*, 1973); (*b*) to take advantage of the possibility of synthesizing the complementary DNA of an available mRNA with a viral reverse transcriptase, in order to study by hybridization the specific transcription of this RNA, using as template the chromatin of the cells which synthesize specifically this mRNA (hemoglobin, ovalbumin or histone–mRNA's, for instance); (*c*) to study the transcription of high-molecular-weight DNA or chromatin prepared from virally transformed cells, since the assay for viral transcripts can easily be performed using conventional DNA–RNA hybridization. In addition, the possibility of determining quantitative and qualitative variations of the RNA polymerases by binding of labeled amanitin, immunological assays, and measurements of the solubilized activities should be very helpful in the above studies, as well as for investigating the *in vivo* regulation of RNA polymerase activities. Unfortunately, rapid, spectacular advances cannot be expected, since, in contrast with prokaryotes (Burgess, 1971; Bautz, 1972), animal studies not only cannot make use of adequate genetic tools, but are also made more difficult by the limited amount of available material.

ACKNOWLEDGEMENTS

We are grateful to all those who have provided us with unpublished information. Special thanks are due to Mrs. Brigitte Chambon for secretarial assistance and to Dr. P. Fellner for reviewing the final manuscript. The investigations by the authors were supported by grants from the CNRS, the INSERM, the Délégation à la Recherche Scientifique et Technique, the Commissariat à l'Energie Atomique, and La Fondation pour la Recherche Médicale Française.

REFERENCES

Adman, R., Schultz, L. D., and Hall, B. D. (1972). *Proc. Nat. Acad. Sci. U.S.* **69,** 1702.
Aloni, Y. (1972). *Proc. Nat. Acad. Sci. U.S.* **69,** 2404.
Aloni, Y., and Attardi, G. (1971). *Proc. Nat. Acad. Sci. U.S.* **68,** 1757.
Amalric, F., Nicoloso, M., and Zalta, J. P. (1972). *FEBS Lett.* **22,** 67.
Bagshaw, J. C., and Drysdale, J. W. (1973). *Ann. N.Y. Acad. Sci.* **209,** 363.
Bagshaw, J. C., and Malt, R. A. (1971). *Biochem. Biophys. Res. Commun.* **42,** 1207.
Bautz, E. K. F. (1972). *Progr. Nucl. Acid Res. Mol. Biol.* **12,** 129–160.
Beebee, T., Korner, A., and Bond, R. P. M. (1972). *Biochem. J.* **127,** 619.
Beerman, W. (1971). *Chromosoma* **34,** 152.

Berg, D., and Chamberlin, M. (1970). *Biochemistry* **9**, 5055.
Blackburn, K. J., and Klemperer, H. G. (1967). *Biochem. J.* **102**, 168.
Blatti, S. P., Ingles, C. J., Lindell, T. J., Morris, P. W., Weaver, R. F., Weinberg, F., and Rutter, W. J. (1970). *Cold Spring Harbor Symp. Quant. Biol.* **35**, 649.
Brogt, T. M., and Planta, R. J. (1972). *FEBS Lett.* **20**, 47.
Bucci, S., Nordi, I., Mancino, G., and Fiume, L. (1971). *Exp. Cell Res.* **69**, 462.
Buku, A., Campadelli-Fiume, G., Fiume, L., and Wieland, T. (1971). *FEBS Lett.* **14**, 42.
Burgess, R. R. (1971). *Ann. Rev. Biochem.* **40**, 711.
Burgess, R. R., Travers, A. A., Dunn, J. J., and Bautz, E. K. F. (1969) *Nature (London)* **221**, 43.
Butterworth, P. H. W., Cox, R. F., and Chesterton, C. J. (1971). *Eur. J. Biochem.* **23**, 229.
Cassani, G., Burgess, R. R., Goodman, H. M., and Gold, L. (1971). *Nature (London)* **230**, 197.
Chambon, P., Ramuz, M., Mandel, P., and Doly, J. (1968). *Biochim. Biophys. Acta* **157**, 504.
Chambon, P., Gissinger, F., Mandel, J. L., Kedinger, C., Gniazdowski, M., and Meilhac, M. (1970). *Cold Spring Harbor Symp. Quant. Biol.* **35**, 693.
Chambon, P., Gissinger, F., Kedinger, C., Mandel, J. L., Meilhac, M., and Nuret, P. (1972). *Acta Endocrinol.* **168**, 222.
Chambon, P., Meilhac, M., Walter, S., Kedinger, C., Mandel, J. L., and Gissinger, F. (1973). *In* "Gene Expression and Its Regulation" (A. Hollender, ed.), p. 75. Plenum Press, New York.
Chan, V. L., Whitmore, G. F., and Siminovitch, L. (1972). *Proc. Nat. Acad. Sci. U.S.* **69**, 3119.
Chardonnet, Y., Gazzolo, L., and Pogo, B. G. T. (1972). *Virology* **48**, 305.
Chesterton, C. J., and Butterworth, P. H. W. (1971a). *FEBS Lett.* **12**, 301.
Chesterton, C. J., and Butterworth, P. H. W. (1971b). *Eur. J. Biochem.* **19**, 232.
Chesterton, C. J., and Butterworth, P. H. W. (1971c). *FEBS Lett.* **15**, 181.
Chesterton, C. J., Humphrey, S. M., and Butterworth, P. H. W. (1972). *Biochem. J.* **126**, 675.
Crippa, M., and Tocchini-Valentini, G. P. (1970). *Nature (London)* **226**, 1243.
Cold Spring Harbor Symp. Quant. Biol., Transcription Genet. Mater. (1970). **35**, 641–737.
Darlix, J. L., Sentenac, A., Ruet, A., and Fromageot, P. (1969) *Eur. J. Biochem.* **11**, 43.
Dean, W. W., and Lebowitz, J. (1971). *Nature (London)* **231**, 5.
Delius, H., Mantell, N. J., and Alberts, B. (1972). *J. Mol. Biol.* **67**, 341.
Dezelée, S., Sentenac, A., and Fromageot, P. (1972). *FEBS Lett.* **21**, 1.
Doenecke, D., Pfeiffer, C., and Sekeris, C. E. (1972). *FEBS Lett.* **21**, 237.
Dunn, J. J., and Bautz, E. K. F. (1969). *Biochem. Biophys. Res. Commun.* **36**, 925.
Egyhazi, E., D'Monte, B., and Edström, J. E. (1972). *J. Cell Biol.* **53**, 523.
Fiume, L., and Laschi, L. (1965). *Sperimentale* **115**, 228.
Fiume, L., and Wieland, T. (1970). *FEBS Lett.* **8**, 1.
Fiume, L., Marinozzi, V., and Nordi, I. (1969). *Brit. J. Exp. Pathol.* **50**, 270.
Fox, C. F., and Weiss, S. B. (1964). *J. Biol. Chem.* **239**, 175.
Furth, J. J., Nicholson, A., and Austin, G. E. (1970). *Biochim. Biophys. Acta* **213**, 124.
Furth, J. J., Pizer, L. I., Austin, G. E., and Fujii, K. (1972). *Life Sci.* **11**, 1001.
Geiduschek, E. P., and Haselkorn, R. (1969). *Ann. Rev. Biochem.* **38**, 647.

Gissinger, F., and Chambon, P. (1972). *Eur. J. Biochem.* **28**, 277.

Gissinger, F., Kedinger, C., and Chambon, P. (1974). *Biochim.*, in press.

Gniazdowski, M., Mandel, J. L., Gissinger, F., Kedinger, C., and Chambon, P. (1970). *Biochem. Biophys. Res. Commun.* **38**, 1033.

Goldberg, M. L., and Moon, H. D. (1970). *Arch. Biochem. Biophys.* **141**, 258.

Goldberg, M. L., Moon, H. D., and Roseman, W. (1969). *Biochim. Biophys. Acta* **171**, 192.

Green, M., and Hodap, M. (1971). *J. Mol. Biol.* **64**, 305.

Green, M., Parsons, J. T., Piña, M., Fujinaga, K., Calfier, H., and Landgraf-Leurs, I. (1970) *Cold Spring Harbor Symp. Quant. Biol.* **35**, 803.

Greenberg, H., and Penman, S. (1966). *J. Mol. Biol.* **21**, 527.

Gross-Bellard, M., Oudet, P., and Chambon, P. (1973). *Europ. J. Biochem.* **36**, 32.

Gurgo, C., Ray, R. K., Thiry, L., and Green, M. (1971). *Nature (London)* **229**, 111.

Hameister, H., Wilson, M., and Stein, H. (1972). *Studia Biophys. (Berlin)* **31/32**, 33.

Hedrick, J. L.; and Smith, A. J. (1968). *Arch. Biochem. Biophys.* **126**, 155.

Herzberg, M., and Winocour, E. (1970). *J. Virol.* **6**, 667.

Higashinakagawa, T., and Muramatsu, M. (1972). *Biochem. Biophys. Res. Commun.* **47**, 1.

Hirai, K., and Defendi, V. (1972). *J. Virol.* **9**, 705.

Horgen, P. A., and Griffin, D. H. (1971). *Proc. Nat. Acad. Sci. U.S.* **68**, 338.

Hossenlopp, P., Oudet, P., and Chambon, P. (1974). *Europ. J. Biochem.* **41**, 397.

Jackson, A., and Sugden, W. (1972). *J. Virol.* **10**, 1086.

Jacob, S. T. (1973). *Progr. Nucl. Acid Res. Mol. Biol.* **13**, 93–136.

Jacob, S. T., Sajdel, E. M., and Munro, H. N. (1968). *Biochim. Biophys. Acta* **157**, 421.

Jacob, S. T., Sajdel, E. M., and Munro, H. N. (1970a). *Biochem. Biophys. Res. Commun.* **38**, 765.

Jacob, S. T., Sajdel, E. M., Muecke, W., and Munro, H. N. (1970b). *Cold Spring Harbor Symp. Quant. Biol.* **35**, 681.

Jacob, S. T., Muecke, W., Sajdel, E. M., and Munro, H. N. (1970c). *Biochem. Biophys. Res. Commun.* **40**, 334.

Jacob, S. T., Sajdel, E. M., and Munro, H. N. (1971). *Advan. Enzyme Regulat.* **9**, 169.

Jaenisch, R. (1972). *Nature (London)* **235**, 46.

Johnson, J. D., Jant, B. A., Sokoloff, L., and Kaufman, S. (1969). *Biochim. Biophys. Acta* **179**, 526.

Johnson, J. D., Jant, B. A., Kaufman, S., and Sokoloff, L. (1971). *Arch. Biochem. Biophys.* **142**, 489.

Juhasz, P. P., Benecke, B. J., and Seifart, K. H. (1972). *FEBS Lett.* **27**, 30.

Kedinger, C., Gniazdowski, M., Mandel, J. L., Gissinger, F., and Chambon, P. (1970). *Biochem. Biophys. Res. Commun.* **38**, 165.

Kedinger, C., and Chambon, P. (1972). *Eur. J. Biochem.* **28**, 283.

Kedinger, C., Nuret, P., and Chambon, P. (1971). *FEBS Lett.* **15**, 169.

Kedinger, C., Gissinger, F., Gniazdowski, M., Mandel, J. L., and Chambon, P. (1972). *Eur. J. Biochem.* **28**, 269.

Kedinger, C., Gissinger, F., and Chambon, P. (1974). *Europ. J. Biochem.* (in press).

Keller, W., and Goor, R. (1970). *Cold Spring Harbor Symp. Quant. Biol.* **35**, 671.

Keshgegian, A. A., and Furth, J. J. (1972). *Biochem. Biophys. Res. Commun.* **48**, 757.

Khoury, G., Byrne, J. C., and Martin, M. A. (1972). *Proc. Nat. Acad. Sci. U.S.* **69**, 1925.

Konishi, G., and Koide, S. S. (1971). *Experientia* **27**, 262.

Krakow, J. S., and Karstadt, M. (1967). *Proc. Nat. Acad. Sci. U.S.* **58,** 2094.
Ledinko, N. (1971). *Nature* (*London*) **233,** 247.
Leighton, T. J., Freese, P. K., Doi, R. H., Waren, R. A. J., and Kelln, R. A. (1972). *In* "Spores V," pp. 238–246. Amer. Soc. for Microbiol.
Lentfer, D., and Lezius, A. G. (1972). *Eur. J. Biochem.* **30,** 278.
Lindell, T. J., Weinberg, F., Morris, P. W., Roeder, R. G., and Rutter, W. J. (1970). *Science* **170,** 447.
Lindstrom, D. M., and Dulbecco, R. (1972). *Proc. Nat. Acad. Sci. U.S.* **69,** 1517.
Lucas, J. J., and Ginsberg, H. S. (1972). *Biochem. Biophys. Res. Commun.* **49,** 39.
MacGillivray, A. J., Paul, J., and Threlfall, G. (1972). *Advan. Cancer Res.* **15,** 93.
MacGuire, P. M., Swart, C., and Hodge, L. D. (1972). *Proc. Nat. Acad. Sci. U.S.* **69,** 1578.
Mackedonski, V. V., Nikolaev, N., Sebesta, K., and Hadjiolov, A. A. (1972). *Biochem. Biophys. Acta* **272,** 56.
Maestre, M. F., and Wang, J. C. (1971). *Biopolymers* **10,** 1021.
Mainwaring, W. I. P., Mangan, F. R., and Peterken, B. M. (1971). *Biochem. J.* **123,** 619.
Mandel, J. L., and Chambon, P. (1971a). *FEBS Lett.* **15,** 175.
Mandel, J. L., and Chambon, P. (1971b). *C. R. Soc. Biol.* **165,** 509.
Mandel, J. L., and Chambon, P. (1974a). *Europ. J. Biochem.* **41,** 367.
Mandel, J. L., and Chambon, P. (1974b). *Europ. J. Biochem.* **41,** 379.
Mandel, J. L., Kedinger, C., Gissinger, F., Chambon, P., and Fried, A. H. (1973). *FEBS Lett.* **29,** 109.
Marinozzi, V., and Fiume, L. (1971). *Exp. Cell Res.* **67,** 311.
Martin, M. A., and Byrne, J. C. (1970). *J. Virol.* **6,** 463.
Maul, G. G., and Hamilton, T. H. (1967). *Proc. Nat. Acad. Sci. U.S.* **57,** 1371.
Meilhac, M., and Chambon, P. (1973). *Europ. J. Biochem.* **35,** 454.
Meilhac, M., and Chambon, P. (1974) *Biochem. Biophys. Acta* (in press).
Meilhac, M., Kedinger, C., Chambon, P., Faulstich, H., Govidan, M. V., and Wieland, T. (1970). *FEBS Lett.* **9,** 258.
Meilhac, M., Tysper, S., and Chambon, P. (1972). *Eur. J. Biochem.* **28,** 297.
Mertelsmann, R. (1969). *Eur. J. Biochem.* **9,** 311.
Meyer-Schultz, F., and Porte, A. (1971). *Cytobiologie* **3,** 387.
Millet, J., Kerjan, P., Aubert, J. P., and Szulmajster, J. (1972). *FEBS Lett.* **23,** 47.
Monneron, A., and Bernhard, W. (1969). *J. Ultrastruct. Res.* **27,** 266.
Montecuccoli, G., Novello, F., and Stirpe, F. (1972). *FEBS Lett.* **25,** 305.
Muramatsu, M., Shimada, N., and Higashinakagawa, T. (1970). *J. Mol. Biol.* **53,** 91.
Niessing, J., Schneiders, B., Kunz, W., Seifart, K. H., and Sekeris, C. E. (1970). *Z. Naturforsch.* **25b,** 119.
Novello, F., and Stirpe, F. (1969). *Biochem. J.* **112,** 721.
Novello, F., and Stirpe, F. (1970). *FEBS Lett.* **8,** 57.
Onishi, T., and Muramatsu, M. (1972). *Biochem. J.* **128,** 1361.
Parsons, J. T., and Green, M. (1971). *Virology* **45,** 154.
Parsons, J. T., Gardner, J., and Green, M. (1971). *Proc. Nat. Acad. Sci. U.S.* **68,** 557.
Pegg, A. E., and Korner, A. (1967). *Arch. Biochem. Biophys.* **118,** 362.
Petrov, P., and Bernhard, W. (1971). *J. Ultrastruct. Res.* **35,** 386.
Petrov, P., and Sekeris, C. E. (1971). *Exp. Cell Res.* **69,** 393.
Pogo, A. O., Littau, V. C., Allfrey, V. G., and Mirsky, A. E. (1967). *Proc. Nat. Acad. Sci. U.S.* **57,** 743.
Ponta, H., Ponta, U., and Wintersberger, E. (1972). *Eur. J. Biochem.* **29,** 110.

Price, R., and Penman, S. (1972a). *J. Virol.* **9,** 621.
Price, R., and Penman, S. (1972b). *J. Mol. Biol.* **70,** 435.
Reeder, R. H., and Roeder, R. G. (1972). *J. Mol. Biol.* **67,** 433.
Richardson, J. P. (1969). *Progr. Nucl. Acid Res. Mol. Biol.* **9,** 75.
Riva, S., Fietta, A., and Silvestri, L. G. (1972). *Biochem. Biophys. Res. Commun.* **49,** 1263.
Ro, T. S., and Busch, H. (1967). *Biochim. Biophys. Acta* **134,** 184.
Roeder, R. G., and Rutter, W. J. (1969). *Nature (London)* **224,** 234.
Roeder, R. G., and Rutter, W. J. (1970a). *Proc. Nat. Acad. Sci. U.S.* **65,** 675.
Roeder, R. G., and Rutter, W. J. (1970b). *Biochemistry* **9,** 2543.
Roeder, R. G., Reeder, R. H., and Brown, D. D. (1970). *Cold Spring Harbor Symp. Quant. Biol.* **35,** 727.
Sajdel, E. M., and Jacob, S. T. (1971). *Biochem. Biophys. Res. Commun.* **45,** 707.
Sambrook, J., Sharp, P. A., and Keller, W. (1972). *J. Mol. Biol.* **70,** 57.
Seifart, K. H. (1970). *Cold Spring Harbor Symp. Quant. Biol.* **35,** 719.
Seifart, K. H., and Sekeris, C. E. (1969). *Eur. J. Biochem.* **7,** 408.
Seifart, K. H., Benecke, B. J., and Juhasz, P. P. (1972). *Arch. Biochem. Biophys.* **151,** 519.
Seifart, K. H., Juhasz, P. P., and Benecke, B. J. (1973). *Eur. J. Biochem.* **33,** 181.
Sekeris, C. E., and Schmid, W. (1972). *Febs Lett.* **27,** 41.
Serfling, E., Wobus, U., and Panitz, R. (1972). *FEBS Lett.* **20,** 148.
Shaaya, E., and Clever, U. (1972). *Biochim. Biophys. Acta* **272,** 373.
Siebert, G., Villalobos, J. Jr., Ro, T. S., Steele, W. J., Lindenmayer, G., Adams, H., and Busch, H. (1966). *J. Biol. Chem.* **241,** 71.
Smith, D. A., Ratliff, R. L., Trujillo, T. T., Williams, D. L., and Hayes, F. N. (1966). *J. Biol. Chem.* **241,** 1915.
Smuckler, E. A., and Hadjiolov, A. A. (1972). *Biochem. J.* **129,** 153.
Smuckler, E. A., and Tata, J. R. (1971). *Nature (London)* **234,** 37.
Smuckler, E. A., and Tata, J. R. (1972). *Biochem. Biophys. Res. Commun.* **49,** 16.
Sokol, F., and Carp, R. I. (1971). *J. Gen. Virol.* **11,** 177.
Stampfer, M., Rosbash, M., Huang, A. S., and Baltimore, D. (1972). *Biochem. Biophys. Res. Commun.* **49,** 217.
Stein, H., and Hausen, P. (1970a). *Eur. J. Biochem.* **14,** 270.
Stein, H., and Hausen, P. (1970b). *Cold Spring Harbor Symp. Quant. Biol.* **35,** 709.
Stirpe F., and Fiume, L. (1967). *Biochem. J.* **105,** 779.
Stirpe, F., and Novello, F. (1970a). *Eur. J. Biochem.* **15,** 505.
Stirpe, F., and Novello, F. (1970b). *In Symp. Mech. Toxicity* (W. N. Aldridge, ed.), pp. 147–161. Macmillan, New York.
Strain, E. C., Mullinix, K. P., and Bogorad, L. (1971). *Proc. Nat. Acad. Sci. U.S.* **68,** 2647.
Sudgen, B., and Keller, W. (1973). *J. Biol. Chem.* **248,** 3777.
Tata, J. R. (1970). *In* "Biochemical Actions of Hormones" (G. Litwack, ed.), Vol. 1, p. 89. Academic Press, New York.
Tata, J. R., Hamilton, M. J., and Shilds, D. (1972). *Nature (London)* **238,** 161.
Timberlake, W. E., Hagen, G., and Griffin, D. H. (1972a). *Biochem. Biophys. Res. Commun.* **48,** 823.
Timberlake, W. E., Mc Dowell, L., and Griffin, D. H. (1972b). *Biochem. Biophys. Res. Commun.* **46,** 942.
Tocchini-Valentini, G. P., and Crippa, M. (1970a). *Cold Spring Harbor Symp. Quant. Biol.* **35,** 737.
Tocchini-Valentini, G. P., and Crippa, M. (1970b). *Nature (London)* **228,** 993.

Travers, A. A., and Burgess, R. R. (1969). *Nature (London)* **222,** 537.
Vogt, V. (1969). *Nature (London)* **223,** 854.
Voigt, H. P., Kaufmann, R., and Matthei, H. (1970). *FEBS Lett.* **10,** 257.
Wall, R., Philipson, L., and Darnell, J. E. (1972). *Virology* **50,** 27.
Wallace, R. D., and Kates, J. (1972). *J. Virol.* **9,** 627.
Weaver, R. F., Blatti, S. P., and Rutter, W. J. (1971). *Proc. Nat. Acad. Sci. U.S.* **68,** 2994.
Weber, K., and Osborn, M. (1969). *J. Biol. Chem.* **244,** 4406.
Weinberg, R. A., Warnaar, S. O., and Winocour, E. (1972). *J. Virol.* **10,** 193.
Weiss, S. B. (1960). *Proc. Nat. Acad. Sci. U.S.* **46,** 1020.
Weiss, S. B., and Gladstone, L. (1959). *J. Amer. Chem. Soc.* **81,** 4118.
Westphal, H. (1970). *J. Mol. Biol.* **50,** 407.
Westphal, H., and Kiehn, E. D. (1970). *Cold Spring Harbor Symp. Quant. Biol.* **35,** 819.
Widnell, C. C., and Tata, J. R. (1964). *Biochim. Biophys. Acta* **87,** 531.
Widnell, C. C., and Tata, J. R. (1966). *Biochim. Biophys. Acta* **123,** 478.
Wieland, T. (1964). *Pure Appl. Chem.* **9,** 145.
Wieland, T. (1968). *Science* **159,** 946.
Wieland, T., and Wieland, O. (1972). *In* "Microbiological Toxins," Vol. 8, pp. 249–280. Academic Press, New York.
Wobus, U., Panitz, R., and Serfling, E. (1971). *Experientia* **27,** 1202.
Wu, C. W., and Goldthwait, D. A. (1969a). *Biochemistry* **8,** 4450.
Wu, C. W., and Goldthwait, D. A. (1969b). *Biochemistry* **8,** 4458.
Yu, F. L., and Feigelson, P. (1972). *Proc. Nat. Acad. Sci. U.S.* **69,** 2833.
Zillig, W., Zechel, K., Rabussay, D., Schechner, M., Sethi, V. S., Palm, P., Heil, A., and Seifert, W. (1970). *Cold Spring Harbor Symp. Quant. Biol.* **35,** 47.
Zylber, E., and Penman, S. (1971). *Proc. Nat. Acad. Sci. U.S.* **68,** 2861.

8

Cytochemistry of Nuclear Enzymes

Andrzej Vorbrodt

I. Introduction

The total number of enzymes listed and classified in the book "Enzyme Nomenclature" (1965) was 847. Six years later, in "Enzyme Nomenclature" (1971) this number had increased to 1805. Most of the enzymes listed, however, are of bacterial origin, and probably only about one-fourth of the total number of presently known enzymes can be found in animal tissues. Pearse (1968), emphasized that from about 400 enzymes found in vertebrate tissues approximately 20% can be detected histochemically. This quantity corresponds to approximately 80 enzymes that can be demonstrated in tissue sections. Another 20 or 30 enzyme activities can now be studied (employing existing principles) to give a total of about 100 enzymes accessible for cytochemical detection.

The progress of ultrastructural cytochemistry has made it possible to introduce the concept of enzymatic markers of cell organelles (Novikoff and Essner, 1962; Novikoff and Novikoff, 1972). However, the nucleus is the only cell constituent which does not have its "own" enzymatic marker. This results from a lack of reliable cytochemical methods for the detection of at least a typical nuclear enzyme, e.g., RNA polymerase or NAD pyrophosphorylase. The concentration of almost all cytochemically demonstrable enzymes is much higher in cytoplasmic organelles than in the cell nucleus. Thus the activity of cytoplasmic enzymes masks and efficiently competes with the nuclear enzymes for the substrates present in the incubation medium. Many of the problems related to impediments in cytochemical detection of nuclear enzymes will be presented and discussed here.

By common usage, the term "histochemistry" refers to the localization of chemical constituents at the level of cells and tissues. The term "cytochemistry" refers to localization of chemical constituents at the level of subcellular structures like the nucleus, nucleolus, and cytoplasmic organelles. Our considerations will refer solely to quantitative cytochemistry.

II. Technical Problems

A. *Preparation of Biological Material*

Although methodological and technical problems are not the main purpose of this chapter, it seems desirable to mention briefly some problems of the preparation of biological material for cytochemical studies of nuclear enzymes. The choice of the proper method for examination of tissue samples is of great importance for the final results. The advantages and disadvantages of the utilizable techniques are collated in Table I.

The use of paraffin-embedded material for cytochemical detection of nuclear enzymes should be regarded as unsatisfactory. This technique was widely used in the 1940–1950 period that produced artifactual results related to nuclear localization of phosphatase activity (we shall return to this topic in succeeding sections.)

For *in situ* studies it is necessary to obtain properly prepared tissue sections. However, the main disadvantage of this method is a considerable thickness of the sections obtained from the cryostat and freezing microtome (about 10–15 μm). The sections commonly used for cytochemical studies in the electron microscope (two-step technique) obtained from a

TABLE I
Methods of Preparation of Biological Material for Cytochemical Examination of Cell Nuclei

Methods	Advantages	Disadvantages
I. Nuclei *in situ*		
A. Light microscopy		
1. Acetone fixation / Freeze substitution / Freeze-drying } paraffin embedding	Not recommended	Inhibition of enzymes activity, favorable for diffusion artifacts
2. Unfixed or fixed cryostat sections	Good preservation of enzymatic activity	Thickness of sections favors the interference of cytoplasmic enzymes
3. Formol–calcium- or glutaraldehyde-fixed frozen sections	Good preservation of the structure, fairly well-preserved enzymatic activity	Thickness of sections favors the interference of cytoplasmic enzymes
4. Unsectioned cells, cultivated *in vitro*, tissue prints, dried, fixed or unfixed	Good preservation of the structure, fairly well-preserved enzymatic activity	Possible interaction of cytoplasmic enzymes; problem of substrate penetration
B. Electron microscopy		
1. Usual two-step technique	Good ultrastructure preservation; fair preservation of enzymatic activity	Interference of cytoplasmic enzymes
2. Ultrathin frozen sections, fixed or unfixed	Fair preservation of enzymatic activity and nuclear ultrastructure, avoidance of interaction of cytoplasmic enzymes Recommended for cytochemical studies of nuclei *in situ*	Possibility of release of soluble enzymes
II. Isolated nuclei		
A. Unfixed or fixed nuclei isolated in aqueous media	Good preservation of insoluble enzymes activity; the interference of cytoplasmic enzymes is omitted	Release of soluble enzymes; problem of substrate penetration
B. Nuclei isolated in nonaqueous media	Good preservation of soluble enzymes	Poor preservation of ultrastructural details; loss of lipids.

"chopper" or from freezing microtome are even thicker (30–50 μm). As the bulk of these sections consists of cytoplasmic components with numerous organelles rich in most of cytochemically demonstrable enzymes, the detection of nuclear enzymes is confronted with many difficulties. This problem is illustrated in Fig. 1, which represents the possible events occurring in commonly used sections (40 μm thick) and in ultrathin frozen sections (about 1000 Å thick) during incubation for detection of nuclear enzymes. The components of the incubation medium (thick arrows) penetrate into the thick section (A) along the intracellular spaces or directly into cytoplasm or nuclei on the section surface. The concentration of substrate will be unchanged in the sectioned nuclei (3 and 4). The composition of the incubation medium penetrating into nuclei lying inside the section (1 and 2) is probably changed (small arrows) because the substrate was already in contact with enzyme molecules (small circles) in the cell membranes and in the cytoplasmic organelles. The reaction products (small black dots) can move together with penetrating medium and adsorb at the nuclear structures. These events are frequently observed when the lead precipitation technique is used for visualization of several hydrolytic enzymes, especially phosphatases. The use of ultrathin frozen sections offers quite different possibilities. In Fig. 1 the line B is drawn on the same scale as the thick section (A) and corresponds to the thickness of 1000 Å. The line C represents an ultrathin section of the same thickness as B, but it is shown under higher magnification to illustrate the enzyme molecules and the reaction product. The possibilities of the formation of artifacts because of eventual movement of the reaction products and substrates inside the section are drastically diminished or almost completely avoided. Contact between enzyme molecules and constituents of the incubation medium (substrates) is extremely easy, and the reaction products can be formed at the sites of enzymatic activity without any tendency to translocation and movement. One of the main disadvantages of this technique is the formation of small amounts of the reaction product resulting from relatively low concentration of the enzyme molecules present in the ultrathin section. Therefore, the time of incubation should be more or less prolonged. Another disadvantage seems to be the release of soluble enzymes which are not tightly bound to the structures. This possibility is of lesser importance, however, and in some instances can be even advisable, because of avoidance of interference of lysoenzymes such as lysosomal hydrolases. The technique of cryoultramicrotomy seems to be the most convenient method for preparation of tissue material for cytochemical examination of enzyme activity in nuclei *in situ*.

In some cytochemical studies, the fraction of isolated nuclei can be used; the advantages and disadvantages of this method are also listed in

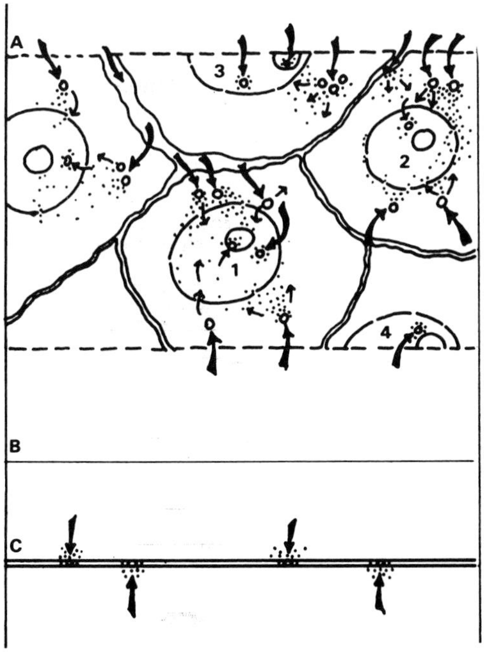

Fig. 1 Diagram illustrating the possible events occurring in thick (40 μm) and in ultrathin (1000 Å) frozen sections during incubation (for detailed explanation see text).

Table I. During the isolation procedure, some soluble enzymes can leak out of the nuclei. This phenomenon probably does not take place if the isolation procedure is performed in nonaqueous media (Behrens, 1932; Siebert, 1967). Another drawback to the use of isolated nuclei are the difficulties in penetration of some components of the incubation medium across the nuclear envelope.

B. Techniques of Enzymatic Activity Detection in Cell Nuclei

Most of the methods used for detection of oxidoreductases are essentially based on the reduction of tetrazolium salts to formazans.

In spite of relatively successful trials (Barnett *et al.*, 1959; Rosa and Tsou, 1964) the tetrazolium salts are not commonly used for the visualization of dehydrogenases activity under the electron microscope. When tetrazolium methods are used for the detection of dehydrogenases in cell nuclei some difficulties can arise. One problem (Novikoff and Essner, 1960) is the inability of nuclear structures to bind tetrazolium salts. This feature of nuclear components can be responsible for lack of positive reactions in cell nuclei studied *in situ,* i.e., in tissue sections incubated in media containing tetrazolium salts and examined both under light and electron microscope. Another important factor is the poor ability of tetrazolium salts to cross the nuclear envelope. The insufficient permeability of nuclear envelope for tetrazolium salts and for other substrate components like sodium succinate can be responsible for a negative reaction obtained for succinate and some other dehydrogenases in isolated nuclei (Vorbrodt *et al.,* 1965; Anders and Bukhvalov, 1972).

Thus, a negative reaction for some dehydrogenases in isolated nuclei is not necessarily an indication of the lack of the particular enzyme activity, but rather of the low sensitivity of the tetrazolium method. Hopefully, the introduction of potassium ferricyanide as hydrogen acceptor will improve the techniques of detecting dehydrogenase activity. In this method the final reaction product appears in the form of copper ferrocyanide which is easily visible under the electron microscope (Ogawa *et al.,* 1968). Recently, a new method was developed for the ultrastructural localization of cytochrome oxidase; it is based on the oxidative polymerization of diaminobenzidine (Seligman *et al.,* 1968, 1970).

From the large group of the transferases, only a few enzymes can be detected cytochemically. Most interesting is the localization of RNase activity, which unequivocally is present in the cell nucleus. Cytochemical methods of detection of a majority of the hydrolytic enzymes are based on metal ion capturing reactions and will be mentioned in the sections concerning the localization of particular enzymes.

Besides the above mentioned classic cytochemical techniques, there are some methods that can be applied for localization of nuclear enzymes. One method is histoautoradiography. At present this technique is utilized only on a limited scale, especially for indirect visualization of DNA or RNA polymerases. Autoradiography can be also utilized for detection of some enzymes by means of radioactive isotope-labeled specific inhibitors (Ostrowski and Barnard, 1961; Darzynkiewicz *et al.,* 1967). The great progress of ultrastructural autoradiography, which consists mainly of production of new ultrafine grain emulsion and decreasing sizes of silver grains, offers new possibilities for more effective application of this technique in studies on nuclear enzymes localization and activity (Lettré and Paweletz, 1966; Wisse and Tates, 1968; Kuroiwa and Tanaka, 1971; Fakan and Bernhard, 1971).

Another technique that should offer great progress in enzymatic studies in the near future is immunohistochemistry (Coons and Kaplan, 1950; Coons, 1956). At present this technique is being systematically improved and adapted for electron microscopy (Nakane and Pierce, 1966; Leduc *et al.,* 1955, 1968, 1969; Kuhlmann, 1970).

III. Cytochemically Detected Enzymes in the Cell Nucleus

A. *Oxidoreductases*

The results of quantitative biochemical studies indicate that at least 13 oxidoreductases can be found in isolated cell nuclei (Betel, 1967; Siebert, 1968; Zbarsky and Perevoschchikova, 1970; Pokrovsky *et al.*, 1970).

Some enzymes found in nuclear fractions by quantitative biochemical methods have not yet been demonstrated cytochemically. They are glycerol-3-phosphate dehydrogenase (EC 1.1.1.8), 3-α-hydroxysteroid dehydrogenase (EC 1.1.1.50), β-hydroxysteroid dehydrogenase (EC 1.1.1.51) and glyceraldehyde phosphate dehydrogenase (EC 1.2.1.12). Enzymes that are detected biochemically in cell nuclei, such as malate (EC 1.1.1.38) and isocitrate (EC 1.1.1.42) dehydrogenases, are not visualized with cytochemical methods. On the contrary, succinate dehydrogenase (EC 1.3.99.1) activity, considered a typical mitochondrial enzyme and not found biochemically in cell nuclei, was observed in the nucleoli of some normal and neoplastic epithelial cells (Chatterjee and Mitra, 1961, 1962), as well as in neurons (Tewari and Bourne, 1962). In the nucleoli of fibroblasts (L line) and human amnion FL line cells cultivated *in vitro*, Zawistowski *et al.* (1965) found positive reaction for isocitrate, alcohol, glutamate, lactate, and DL-α-glycerophosphate dehydrogenases, but never for succinate dehydrogenase.

However, in the experiments performed on nuclei isolated from normal and thioacetamide-treated rat liver, we have never observed even traces of positive reactions when succinate dehydrogenase activity was cytochemically detected (Vorbrodt *et al.*, 1965). The penetration of sodium succinate across the nuclear envelope may be insufficient for obtaining a positive reaction inside the nucleus. This fact may be responsible for the low sensitivity of cytochemical methods of detection of dehydrogenases inside isolated cell nuclei. The concentration of the two above-mentioned enzymes (malate and isocitrate dehydrogenases) in cell nuclei may be too low for cytochemical detection. On the other hand, the very delicate positive reaction for succinate dehydrogenase observed in the nucleoli of some cells may result from nonenzymatic reduction of tetrazolium salts by still unknown chemical agents present in the particular cell components. These positive reactions were obtained only in tissue sections, where direct contact between nucleolar components and substrate constituents was achieved.

The careful biochemical analyses performed on cell nuclei isolated both in aqueous and nonaqueous media, as well as the results of hundreds of cytochemical studies performed on tissue sections, do not show the presence of succinate dehydrogenase activity in cell nuclei. Chemical analysis of isolated nucleoli also indicates the lack of succinate dehydrogenase activity in these nuclear components (Siebert, 1966; Busch, 1968). In pathologically altered cells, the nucleoli may have positive cytochemical reactions for some enzymes, but they should be considered as exceptions to the rule.

The intranuclear localization of other enzymes, such as lactate dehydrogenase (EC 1.1.1.27), reduced NAD dehydrogenase (EC 1.6.99.2), and reduced NADP (EC 1.6.99.1) dehydrogenase (tetrazolium reductase), was studied solely in the light microscope; the precise association with particular nuclear components is still unknown. In nuclei obtained from thioacetamide-treated rat liver cells the reduced NAD–tetrazolium reductase was absent from greatly enlarged nucleoli (Vorbrodt *et al.*, 1965), and gave a diffuse reaction in the nucleoplasm. In nuclei obtained from normal rat liver the reactions for reduced NAD–tetrazolium reductase, reduced NADP–tetrazolium reductase, lactate dehydrogenase, and glucose-6-phosphate dehydrogenase (EC 1.1.1.49) were uniformly distributed in the whole nucleoplasm and slightly concentrated in the region of nuclear envelope.

NADP-linked glucose-6-phosphate dehydrogenase activity, visualized by using potassium ferricyanide as the acceptor of hydrogen and examined under the electron microscope, was found in the region of the nuclear envelope (Nagasaka, 1970). In sections of mouse liver cell the reaction product was present in the chromatin components adjacent to the inner membrane of the nuclear envelope. In the testis of the pond snail the NADP-linked phosphogluconate dehydrogenase (EC 1.1.1.44) activity was also found in the inner membrane of the nuclear envelope of some spermatids.

Information on the presence of cytochrome oxidase activity inside the cell nucleus was very scarce until Tewari and Bourne (1962) showed positive reactions for cytochrome oxidase inside the nucleoli and in the region of the nuclear envelope of Purkinje's cells. Recently, biochemical studies showed cytochrome oxidase activity inside the cell nuclei and in the inner membrane of the nuclear envelope (Betel, 1967; Zbarsky and Perevoschchikova, 1969; Pokrovsky *et al.*, 1970). Using the diaminobenzidine (DAB) method for cytochemical detection of cytochrome oxidase, Rupec *et al.* (1971) observed a positive reaction in the inner membrane of the nuclear envelope and in the euchromatin regions of isolated rat thymus nuclei. The electron-opaque bodies that showed enzymatic activity and localized preferentially in the euchromatinic spaces most probably were identical to the RNA-containing coiled bodies (Monneron and Bernhard, 1969). Similar results were obtained recently (Bukhvalov *et al.*,

1972) on isolated rat liver cell nuclei, isolated membranes of these nuclei, and in isolated rat thymus nuclei. The highest cytochrome oxidase activity is confined to the inner membrane of nuclear envelope, and lower activity to the outer membrane and to chromatin components. These results have been obtained only in isolated nuclei, and further experiments on tissue sections are needed before making final estimation of the specificity and validity of the method.

Probably, the introduction of copper ferrocyanide and diaminobenzidine techniques for detection of some oxidoreductases at the ultrastructural level will enable further progress in our knowledge on enzymology of cell nucleus.

B. Transferases

Many enzymes belonging to this group are of great biological importance and could be considered as typical enzymatic markers of the nucleus. Among them the RNA nucleotidyl transferases and DNA nucleotidyl transferases (polymerases) are of special interest for the role of the cell nucleus as a cellular center of genetic information (see Chapter 7). Unfortunately, cytochemical methods for the detection of these enzymatic activities have not been developed. From the large group of transferases only a few enzymes can be detected cytochemically.

1. GLYCOSYLTRANSFERASES

Glycosyltransferases are commonly known as typical cytoplasmic enzymes and had not been biochemically detected in cell nuclei. Nevertheless, in a few cases these enzymes were claimed to be present in cell nuclei. Godlewski (1961) observed a positive reaction for glycogen phosphorylase (EC 2.4.1.1) and the branching factor (EC 2.4.1.1.18) in the nuclei of some Novikoff ascites hepatoma cells; the reaction was not present in the nucleoli. A positive reaction for glycogen phosphorylase was also observed in the nuclei of some epithelial cells of precancerous lesions of the human uterine cervix (Godlewski, 1964). In the nuclei of desquamated corneal cells in rabbits and monkeys a positive reaction for glycogen phosphorylase was noted by Shantaveerappa *et al.* (1966).

At present, however, a positive cytochemical reaction for glycogen synthetase (EC 2.4.1.11) in cell nuclei has not been reported.

2. RNA NUCLEOTIDYLTRANSFERASES (EC 2.7.7.6)

The localization of RNA nucleotidyltransferases, commonly called RNA polymerases, was studied in frozen section of plant tissues using three nonlabeled and one tritium-labeled nucleoside triphosphates. The sites of the enzyme activity were visualized by light microscopic autoradiography (Fisher, 1968). Using similar techniques applied to electron microscopy, Maul and Hamilton (1967) observed the sites of two DNA-dependent RNA polymerase activities in isolated rat liver cell nuclei. One enzyme, activated with Mg^{2+} and responsible for the synthesis of ribosomal RNA, was found in the nucleolus. Another, activated by Mn^{2+} and $(NH_4)_2SO_4$, was localized mainly in the chromatin. The enzyme synthesized mainly DNA-like RNA. The resolution of the autoradiograms presented in this paper was not sufficient for precise indications of the exact sites of the enzyme activity at the ultrastructural level. Recent progress in ultrastructural autoradiography makes possible more precise studies in this field. A good example of this progress is that Fakan and Bernhard (1971) where the resolution of autoradiograms was adequate for the localization of labeling sites at the level of nucleolar granular, fibrillar, or amorphous components.

Another technique that could be applied for the detection of the sites of activity of RNA or DNA polymerases is immunocytochemistry. The use of this method is limited by difficulties associated with isolation of sufficient amounts of pure enzyme and with the production of specific antibodies.

3. RIBONUCLEASES (EC 2.7.7.16, EC 2.7.7.17)

In the animal tissues exist at least several enzymes depolymerizing RNA as exonucleolytic (phosphodiesterases I and II), and endonucleolytic (RNase I: EC 2.7.7.16, acid RNase: EC 2.7.7.17, 5′-RNases). The classification of these enzymes is complicated by their modes of action, formation of the reaction products, substrate specificity, and their existence inside or outside the cell (Egami and Nakamura, 1969).

The cytochemical methods used for the detection of RNA depolymerizing activity are based on three different principles.

a. Substrate Film Method. This technique was elaborated and introduced by Daoust (1957) and progressively improved (Amano and Daoust, 1961; Daoust and Amano, 1960; Daoust, 1965). Unfortunately, with this method it is impossible to differentiate the activity of different types of ribonucleases and phosphodiesterases.

Daoust and Amano (1964) reported that it was possible to recognize the intracellular sites of RNase activity. This activity was localized in the perinuclear zones of cells of the Malpighian layer of stratified squamous epithelia. Weak resolution, however, makes it impossible to get reliable information concerning the intranuclear localization of the enzymatic activity, although such activity could not be excluded.

b. Immunohistochemical Technique. The localization of nucleolytic enzymes by means of immunofluorescence techniques was studied by Marshall (1954). A modified technique was used by Ehinger (1965), who found localization of RNase in the nucleoli of rat exocrine pancreatic cells. More recently the localization of alkaline and acid ribonucleases was studied by Morikawa (1967). Alkaline RNase was found in the nuclei or in the nuclear envelope of bovine liver and kidney cells as well as in the nuclei of mononuclear cells, probably macrophages. Acid RNase was present only in the nuclei of leukocytes fixed in acetone or formalin, and a very weak reaction was observed in the nuclei of kidney cells fixed in formalin. The microphotographs showing the distribution of the enzymes in the pancreas, spleen, and in leukocytes proved the specificity of the reaction. The immunological purity of the antigens and specificity of the antibodies were studied carefully and included immunodiffusion and immunoelectrophoresis.

c. Azo Dye-Coupling Techniques. The most specific substrates for detection of RNase activity seems to be at present α-naphthyluridine-3′-phosphate which in the incubation medium is hydrolyzed by RNase to 2′,3′-UMP and free naphthol. Enzymatically liberated α-naphthol is coupled with suitable diazotate by standard azo dye-coupling techniques, giving an insoluble dye (Zan-Kowalczewska *et al.*, 1966; Shugar and Sierakowska, 1967). Unfortunately, despite the high specificity of this method, the pictures thus obtained show that only alkaline (pancreatic type) RNase localized in cell membrane can be visualized with this substrate.

The same is true not only for formol–calcium-fixed frozen sections but also for smears of normal and leukemic blood cells, where a positive reaction was observed solely in the cytoplasmic components (Gluzman *et al.*, 1971). Further studies along this line may lead to the synthesis of proper substrates for the detection of some other types of ribonucleases. Both the lysosomal enzyme (acid RNase) as well as 5′-RNases, which are present in cell nuclei (Heppel, 1966; Lazarus and Sporn, 1967; de Lamirande, 1967), cannot be detected by means of presently known substrates.

d. Lead Phosphate Precipitation Technique. The lead phosphate precipitation technique for the detection of RNase activity is essentially based on Gomori's acid phosphatase procedure (Gomori, 1953), which was adapted by Aronson *et al.* (1958) for the visualization of DNase II activity. A method for RNase detection has been described by Taper (1968), using RNA as substrate, with exogenous phosphatase (acid or alkaline) in the incubation medium for the liberation of terminal phosphate groups exposed by tissue ribonucleases. The liberated phosphate is trapped by lead ions and finally visualized for light microscopy with yellow ammonium sulfide. This is a typical two-step method: (*a*) liberation of mono- or oligonucleotides with exposed terminal phosphates; (*b*) hydrolysis of these phosphate esters by exogenous phosphatase. In such procedures, the possibility exists of appearance of diffusion artifacts and adsorption of exogenous phosphatase on the nuclear structures. Another disadvantage in the use of RNA as a substrate is that it can be attacked by several different types of RNases and phosphodiesterases (Sierakowska and Shugar, 1960; Shugar and Sierakowska, 1967).

As described by Taper *et al.* (1971a, b), the reaction for both alkaline and acid RNases was localized mainly in the cell nuclei. Unfortunately, the authors did not describe an intranuclear localization of enzymatic activity. Zotikov and Bernhard (1970) used glutaraldehyde-fixed ultrathin frozen sections of rat liver, kidney, pancreas, and hepatoma. RNase activity was detected with a lead phosphate precipitation technique very similar to that described by Taper ((1968). The reaction product for all nuclease (alkaline and acid RNase and acid DNase) activity was invariably present in the condensed chromatin, while the nucleoli were unstained. This finding raises questions about the specificity of the reaction, because the presence of ribonuclease activity in the nucleoli is well established (Baltus, 1954, 1962; Chakravorty and Busch, 1967; Siebert, 1966, 1968; Busch, 1968).

In summary, the data indicate that a specific substrate for RNase is α-naphthyluridine-3′-phosphate. Although it cannot be used yet for visualization of nuclear RNase activity, further work along this line may lead to elaboration of new substrates specific for different types of ribonucleases. Basic differences exist in the pattern of intracellular localization of RNase activity when the results of immunohistochemical and lead precipitation techniques are compared. The former gives a positive reaction in nucleoli (but not in chromatin and nucleoplasm) and in cytoplasmic granules, while with the latter heavy precipitates of the reaction products on the condensed chromatin are observed both in the light and electron microscope. These facts impair the validity of the lead techniques for visualization of RNase activity; they need further improvement and reexamination.

C. *Hydrolases*

1. CARBOXYLESTERASE (EC 3.1.1.1)

Among numerous publications concerning the cytochemical visualization of esterase activity, information on the possible intranuclear localization of the enzyme was presented in only one paper. The enzyme, organophosphate-sensitive esterase (probably nonspecific carboxyl esterase), was detected using tritium-labeled inhibitor and ultrastructural autoradiography (Barnard *et al.*, 1970). The obtained electronogram showed, in addition to the cytoplasmic labeling, relatively high concentration of radioactive material in the nuclei of epithelial cells of convoluted tubules (mainly proximal) of mouse kidney. Parallel examination of mouse liver revealed labeling only over the cytoplasm. Although some translocation of the enzyme from cytoplasm to the nucleus cannot be excluded, careful analysis of the autoradiograms and grain counts attest to the reliability of this finding. The presence of carboxylesterase in the nuclei of kidney cells and absence in the liver cell nuclei may be considered as a reflection of biochemical specialization of different organs at the level of cellular structure.

2. ALKALINE PHOSPHATASE (EC 3.1.3.1)

Since the introduction of histochemical methods for detection of alkaline phosphatases (Takamatsu, 1939; Gomori, 1939) hundreds of papers concerning this enzyme have been published. In a paper on enzymatic histochemistry of human embryo development (Rossi *et al.*, 1954) the number of references related to histochemistry of phosphatases with special regard to problem of nuclear localization of alkaline phosphatase was 141. This short chapter cannot include all the papers dealing with nuclear localization of alkaline phosphatases.

In tissue sections incubated in Gomori's medium the nuclei in the vicinity of highly active sites have positive reactions for alkaline phosphatase in both the nuclei and nucleoli. The very first experiments of Gomori (1951) and Novikoff (1951) showed that the appearance of positive reactions in cell nuclei is the effect of a secondary adsorption of calcium phosphate only; the nuclei do not contain any enzyme either originally or by secondary adsorption. According to Gomori (1953), staining of the nuclei occurs in the Ca^{2+}–Co^{2+} method when the concentration of Ca^{2+} is low, i.e., about 0.01–0.03 *M*. If the concentration is increased to 0.1 *M*, all the diffusion artifacts are completely eliminated. These statements now have the same validity as twenty years ago.

The artifactual staining of cell nuclei is probably caused both by the secondary adsorption of the reaction product, and displacement of the enzyme and secondary adsorption on the nuclear structure which occur during penetration of the fixative. The incomplete penetration of the fixative into the tissue block offers favorable conditions for translocation and secondary adsorption of the enzyme. The problem of nuclear staining is discussed in a manual of histochemistry (Pearse, 1968) and concludes that nuclear staining observed with Gomori's technique for alkaline phosphatase activity is due mainly to artifacts.

The use of ultrathin frozen sections efficiently prevents the appearance of diffusion artifacts even with prolonged incubation of highly active sections. The electronmicrograph of Fig. 2 shows a glutaraldehyde-fixed, ultrathin frozen section of a rat kidney proximal convoluted tubule. This section was incubated in Gomori's medium for alkaline phosphatase activ-

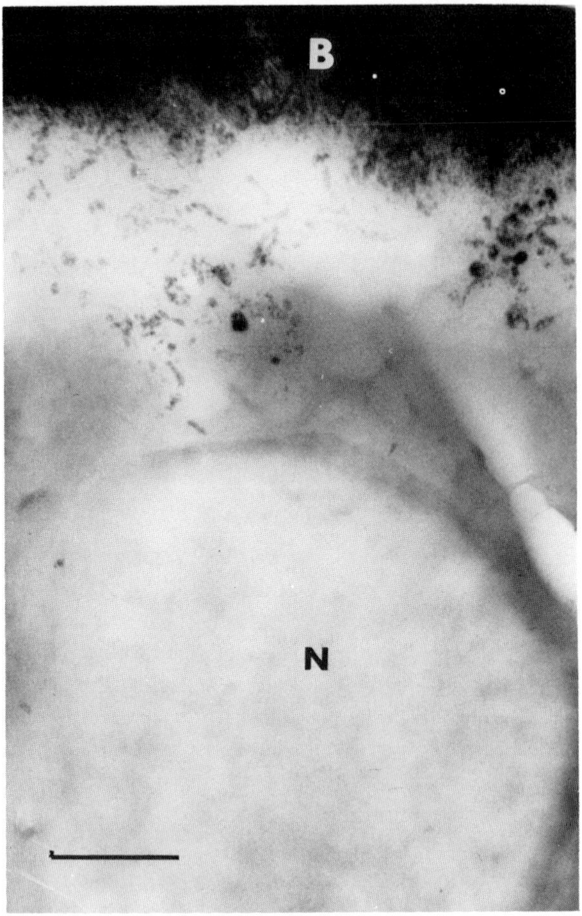

Fig. 2 Glutaraldehyde-fixed, ultrathin frozen section of the proximal convoluted tubule of rat kidney, incubated for visualization of alkaline phosphatase activity. The reaction product is highly concentrated in the brush border (B); the cell nucleus (N) remains completely unstained (× 17,500; rule indicates 1 μm).

ity detection for 1 hr at 37°C. Despite the high concentration of the reaction product in the brush border (B) only traces of artifactual staining of cell nuclei (N) can be noted.

It is impossible here to discuss all the details related to the technical problems of cytochemical visualization of alkaline phosphatase in cell nuclei. Biochemical studies (Siebert, 1966, 1968) on cell nuclei isolated from different tissues and organs do not show the presence of detectable quantities of alkaline phosphatase. These findings agree with the results obtained with modern methods of preparation of tissue samples and detection of enzymatic activity including Gomori's method, azo dye techniques, and immunohistochemical techniques (Yasuda, 1968).

In accordance with the principle that "the exception proves the rule," there are some marginal cases that should be mentioned. For example, Chevremont and Firket (1953) observed a positive reaction for alkaline phosphatase in the nuclei of actively growing fibroblasts and myoblasts *in vitro*. A positive reaction in the nuclei was obtained both with the Gomori's and azo dye techniques.

The general conclusion can be drawn that, using properly fixed and prepared tissue sections, cytochemical reactions for alkaline phosphatase activity are not obtained in the nuclei of the majority of cells, including highly active tissues like absorptive epithelia of small intestine or kidney proximal convoluted tubules.

3. ACID PHOSPHATASE (EC 3.1.3.2)

This enzyme, which is a typical marker of the lysosomes (Novikoff and Essner, 1962; Novikoff and Novikoff, 1972) is not detectable cytochemically in the nuclei of most animal and plant cells. However, in tissue samples prepared for electron microscopy (two-step technique) frequently there are characteristic artifacts which can be interpreted as positive reactions in cell nuclei. Two examples of these artifacts are presented in Figs. 3 and 4. In Fig. 3 the glutaraldehyde-fixed frozen section (40 μm thick) of rat thyroid gland was incubated for 15 min at 37°C in Gomori's acid phosphatase medium. In the electron micrograph a "stream" of reaction products is clearly visible that was translocated from active sites and scattered throughout cytoplasm as well as accidentally adsorbed on the nuclear components. This kind of artifact is easily recognizable and does not lead to misinterpretation. More controversial is the pattern of distribution of lead phosphate in cell nuclei (Fig. 4). In the electron micrograph a sample is shown of regenerating rat liver fixed in glutaraldehyde, cut with freezing microtome into 40 μm-thick sections, and incubated in Gomori's medium with CMP as a substrate. In this ultrathin section the

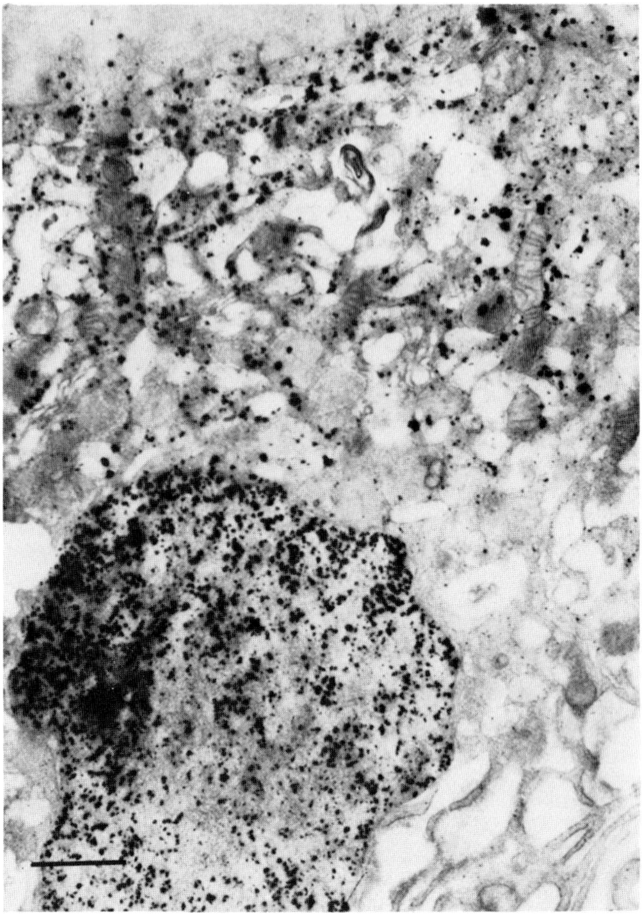

Fig. 3 Typical artifactual staining of the cell nucleus after incubation of glutaraldehyde-fixed, frozen, thick (40 μm) section in Gomori's medium for acid phosphatase detection; after incubation the section was postfixed in OsO_4 and embedded in Epon (for explanation see text) (× 12,000; rule indicates 1 μm).

lead phosphate precipitates are localized almost solely in the condensed chromatin of cell nucleus. The interpretation of the origin of this type of artifact is essentially the same as previously described (Fig. 1). The substrate molecules present in the penetrating incubation medium were probably hydrolyzed by cytoplasmic enzymes and part of the reaction product was translocated and finally adsorbed on the chromatin.

The lack of positive reaction for acid phosphatase in cell nuclei of almost all plants and animal tissues (even those with high activity in nu-

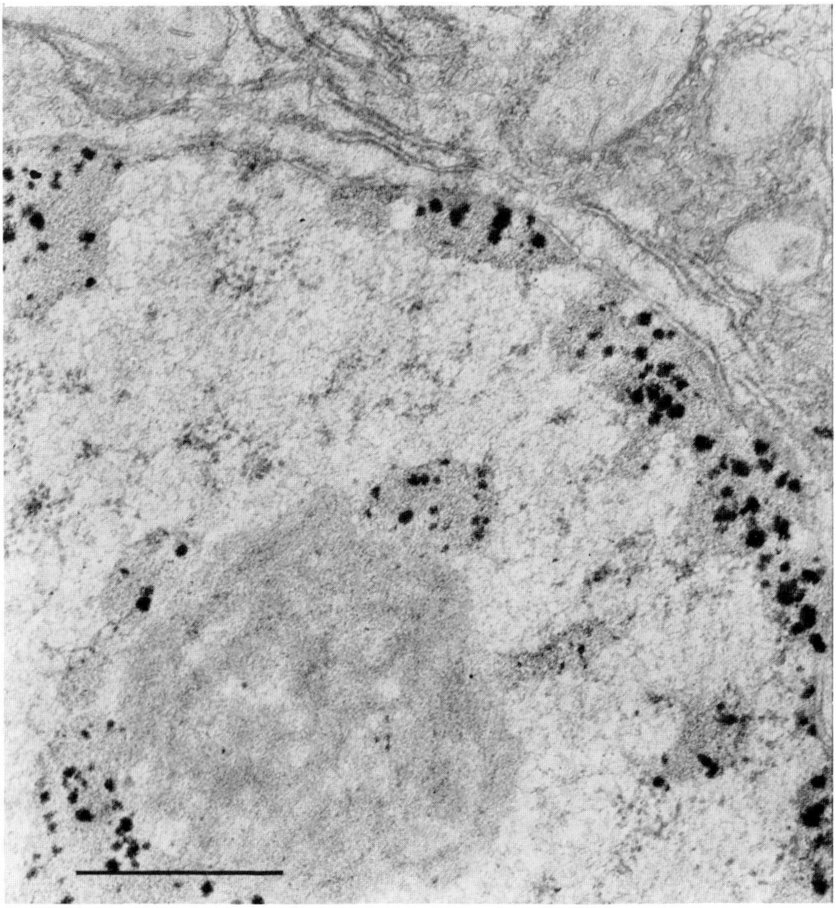

Fig. 4 Another example of artifactual adsorption of the reaction product on condensed chromatin, after incubation of glutaraldehyde-fixed thick (40 μm) section in Gomori's acid phosphatase medium containing cytidine-5′-phosphate as a substrate (× 28,000; rule indicates 1 μm).

merous lysosomes) is in agreement with recent data on isolated cell nuclei and nucleoli (Siebert, 1966, 1968; Busch, 1968). As to alkaline phosphatase, however, there are also some exceptions. In nucleoli isolated from starfish oocytes, Vincent (1952) found relatively high activity of acid phosphatase. Love *et al.* (1969) observed positive reactions for acid phosphatases in the nuclei and nucleolini of four types of HeLa cells and three lines of human cells cultivated *in vitro*. The nuclear enzyme differs in pH optimum and in sensitivity to inhibitors from the cytoplasmic enzyme. Some evidence excludes the possibility that nuclear and nucleolinar staining was due to nonenzymatic hydrolysis of the substrate, diffusion from the cytoplasm, or selective affinity of this structure to lead. In a more recent paper Soriano and Love (1971) examined the localization of the reaction product by electron microscopy. The localization of acid phosphatase in the nucleolus was similar in the light and electron microscopes and corresponded to the light fibrillar areas or "fibrillar centers" (Recher *et al.*, 1969). A similar distribution of acid phosphatase activity was also observed recently in our laboratory (Fig. 5). The *in vitro* cultivated neoplastic cells derived from SV40-induced tumor of hamster liver showed an invariably strong positive reaction for acid phosphatase in small intranucleolar granules corresponding to the nucleolini. A positive reaction appeared after a 30-min incubation period in Gomori's medium. Bankowski and Vorbrodt (1962) and Bankowski (1963) observed in the nucleoli of isolated rat liver nuclei a positive reaction after incubation in the medium containing β-glycerophosphate or 5′-AMP (pH 6–6.5). The reaction product was frequently observed in small nucleolar granules resembling the nucleolini.

4. 5′-NUCLEOTIDASE (EC 3.1.3.5)

The activity of 5′-nucleotidase is highly concentrated in the cell membrane and for this reason is considered as a cytochemical marker of this structure (Novikoff and Essner, 1962; Novikoff and Novikoff, 1972). Recent biochemical data indicate that this enzyme is not present in the nuclei and nucleoli isolated from some representative tissues (Siebert,

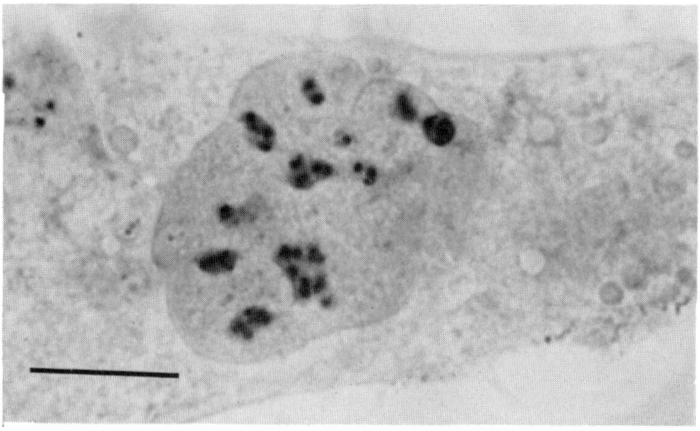

Fig. 5 The reaction for acid phosphatase activity in the nucleoli (nucleolini) of neoplastic cells cultivated *in vitro* (see text) ($\times$ 1800; rule indicates 10 μm).

1966, 1968; Busch, 1968). In glutaraldehyde fixed sections incubated in Wachstein-Meisel's medium (1957), containing nucleoside 5′-monophosphate (usually AMP) and prepared for electron microscopy, nuclear staining generally is not observed. In isolated nuclei incubated at pH 6.5 with AMP as substrate, Bankowski (1963) observed a positive reaction in the nucleoli, whereas in the same rat liver, fixed in formol–calcium, the positive reaction appeared solely in the cell membrane and in bile canaliculi. In alkaline medium containing 5′-AMP and calcium salts, Novikoff (1952) found a positive reaction in cell nuclei. This finding was confirmed on isolated rat liver nuclei showing a strong positive reaction even after 15 min. of incubation (Novikoff *et al.*, 1958). As calcium phosphate has a great tendency to adsorb onto nuclear and nucleolar structures, these findings were probably artifacts.

Several 5′-nucleotidase isoenzymes found in rat and mouse tissues differ in pH optima and intracellular localization. One group of isoenzymes with greatest activity at pH 5.0 dephosphorylates deoxyribonucleotides faster than ribonucleotides. Another group of isoenzymes with optimal pH between 7 and 7.5 hydrolyzes predominantly ribonucleotides (Hardonk *et al.*, 1968). These findings suggest the presence of one of the 5′-nucleotidase isoenzymes in the nucleoli.

5. PHOSPHODIESTERASES (EC 3.1.4.1)

Biochemical analysis of intracellular distribution of phosphodiesterases I and II show their cytoplasmic localization (Razzell, 1961; van Dyck and Wattiaux, 1968; Erecinska *et al.*, 1969). Cytochemical investigation on phosphodiesterase I localization in rat liver with the aid of α-naphthylthymidine 5′-phosphate as a substrate revealed most of the enzyme activity in cellular membranes delimiting bile canaliculi (Sierakowska *et al.*, 1963). Different results were obtained (Wolf *et al.*, 1968a, b) by new indigogenic substrates enabling cytochemical visualization of phosphodiesterases I and II activities in tissue sections. The localization of phosphodiesterase I (pH 9) was essentially similar to that described by Sierakowska *et al.* (1963). When 5-bromo-4-chloro-3-indolylthymidine 3′-phosphate was incubated at pH 4.8–5.2 with fresh frozen sections of spleen, kidney, liver, and intestine of rat and pig, the positive reaction appeared mainly in cell nuclei. At pH 5.9 the staining was reduced in the nuclei and almost exclusively confined to the lysosomes.

Gluzman and Petrenko (1972) used an incubation medium containing NAD, 5′-nucleotidase of snake venom, and lead nitrate for cytochemical detection of phosphodiesterase I. In normal and leukemic bone marrow cells the reaction product was localized only in cytoplasmic components both in normal and leukemic cells. With the method of Shanta *et al.* (1966) for detection of cyclic 3′, 5′-nucleotide phosphodiesterase activity, Coulson and Kennedy (1972) observed a positive reaction in the nuclear membrane of human small lymphocytes. These results should be taken cautiously because the localization of the reaction product was examined only under the light microscope and applied controls were insufficient (lack of the control medium containing 5′-AMP instead of cyclic AMP).

In conclusion, only in the experimental work of Wolf *et al.* (1968a, b) was a nuclear localization of phosphodiesterase reported. These results still need reexamination.

6. DEOXYRIBONUCLEASE I (EC 3.1.4.5) AND DEOXYRIBONUCLEASE II (EC 3.1.4.6)

The following techniques can be applied for cytochemical detection of deoxyribonucleases: (*a*) the substrate film method of Daoust and Amano (1960) and Daoust (1965); (*b*) the immunohistochemical technique as described by Marshall (1954); (*c*) the indigogenic method described by Tsou *et al.* (1968); and (*d*) the lead phosphate precipitation method introduced by Aronson *et al.* (1958).

Biochemical analysis of nuclei and nucleoli isolated both in aqueous and nonaqueous media indicates that DNase I and DNase II are present in cell nuclei, and in contrast to RNases are absent from the nucleoli (Allfrey and Mirsky, 1953; Swingle and Cole, 1964; Siebert, 1966, 1968; Busch, 1968). These basic differences in intranuclear distribution of DNase and RNase activities should be considered when cytochemical studies of these enzymes are performed.

The substrate film of Daoust and Amano (1960, 1964) does not permit the exact localization of DNase activity at the intracellular level, and consequently cannot give valid informations concerning the presence or absence of these enzymatic activities in the cell nucleus.

The immunohistochemical method was used by Marshall (1954) for demonstration of intracellular localization of DNase I activity. He found that the enzyme is mainly concentrated in the apical portion of the acinar cell cytoplasm of the pancreas. These findings do not provide information on the nuclear localization of the enzyme.

For detection of DNase activity two indigogenic substrates were synthesized: 5-iodoindoxylthymidine 5′-phosphate (I) and 5-iodoindoxylthymidine 3′-phosphate (II). According to the authors (Tsou *et al.*, 1968), these substrates can be used for the cytochemical study of phosphodiesterasic hydrolysis of the 3′- or 5′-bonds in DNA and thus are potentially useful for the study of deoxyribonucleases activity. The main technical advantage of these substrates is a low diffusion of the 5-iodoindoxyl and a high electron density of this indigo dye enabling studies under the electron microscope. Cytochemical studies show some differences between these two substrates, especially in the appearance of the reaction in cell nuclei. In the alkaline medium (pH 9) the positive reaction in the nuclei of endothelial cells was observed only with substrate I. Kidney and liver cells also hydrolyzed only substrate I. A preliminary study performed on neoplastic cells showed a lack of the enzyme activity, which might be explained by the presence of a specific inhibitor of DNase in these cells. DNase II could be also responsible for hydrolysis of 5-bromo-4-chloro-3-indolylthymidine 3′-phosphate at pH 4.8–5.2 in cell nuclei (Wolf *et al.*, 1968a, b). The results obtained with indigogenic substrates still have to be confirmed in different laboratories.

The method of detection of DNase II activity using lead phosphate precipitation technique has been described by Aronson *et al.* (1958). In cryostat-cut frozen sections postfixed in acetone–formalin–water mixture, a strong positive reaction was obtained in cell nuclei of different tissues and organs. This method was modified by the author (1961), who recommended the use of formol—calcium–fixed frozen sections and incubation in a medium differing from the original in its higher pH (5.9), a lower concentration of lead nitrate (final concentration 0.002 *M*) and a lower concentration of DNA. These modifications resulted in the shortening of the incubation time and a demonstration of both lysosomal and nuclear localization of the reaction product. In almost all animal and plant tissues that were examined, the positive reaction of cell nuclei was localized in the region of the nuclear envelope and in the condensed chromatin including the nucleolus-associated chromatin. The nucleoli were almost completely unstained. These observations were confirmed (Zotikov and Bernhard, 1970) by ultrastructural studies using glutaraldehyde-fixed ultrathin frozen sections and an incubation medium in which the lead nitrate concentration was twice as high as in Vorbrodt's medium. The heavy precipitates of the reaction products were invariably localized in the condensed chromatin while the nucleolus remained unstained.

Taper *et al.* (1971a) observed a strong positive reaction for alkaline and acid DNases in the cell nuclei of normal rat liver and a gradual weakening of the intensity of the reaction in the course of carcinogenesis. These authors also observed (Taper *et al.*, 1971b) the weakening or disappearance of the nuclear staining in malignant tumors of the human central nervous system. In benign tumors, the activity was similar to that of normal tissues from which tumors originated. The reappearance of high activity of DNases in necrotic areas of malignant tumors was considered as the result of splitting of the DNase–inhibitor complex.

However, all these findings should be taken with reservations. One of the main imperfections of Aronson's method is the use of exogenous acid phosphatase, a possible source of artifacts. Rosenthal *et al.* (1969) and Berg *et al.* (1972) showed that in ATPase medium the reaction product contains not only a "simple" lead phosphate but also nucleotides, lead, and phosphate. The precipitation of the nucleotides increases with increasing lead nitrate concentration in the incubation medium. Little or no ATP is precipitated in the absence of inorganic phosphate, but precipitation occurs after addition of inorganic phosphate. This phenomenon indicates that liberation of orthophosphate and formation of lead phosphate is accompanied by precipitation of nucleotides. Thus, the relatively high concentration of acid phosphatase in the incubation medium is favorable for liberation of orthophosphate ions, which in the presence of lead promote the formation of precipitates composed of mono- and oligonucleotides. These precipitates (Rosenthal *et al.*, 1969; Moses *et al.*, 1966) have a great affinity for the heterochromatinic regions of cell nucleus but not for the nucleolus (Fig. 3).

Probably, the increase of lead nitrate concentration in Zotikov's incubation medium not only inhibits DNase activity but also is favorable for formation of the precipitates in excess.

The second source of artifacts is the affinity of acid phosphatase for nuclear structures. As shown in the author's laboratory, acid phosphatase is easily adsorbed on isolated or sectioned nuclei. In Fig. 6, the localization of the reaction product is shown in isolated rat liver nuclei exposed to different experimental conditions. Nuclei incubated for visualization of DNase II for 1 hr (Fig. 6 A) have relatively high concentrations of the reaction product in the region of the nucleolus-associated chromatin. Nuclei incubated for 3 hr in the same medium (Fig. 6 B) had a strong positive reaction in the chromatin; the nuclear envelope is almost completely unstained. In Fig. 6 C, the nuclei were incubated in a medium containing 5′-UMP instead of DNA. Acid phosphatase was not present in this medium. The lack of positive reaction in the nuclei indicates that there is no phosphatase activity. Only in some nucleoli can very delicate positive reactions be noted in the form of small dots. Figure 6 D shows cell nuclei treated with 0.01% solution of acid phosphatase for 30 min at 4°C, briefly washed, and incubated in the medium containing 5′-UMP. The positive reaction in such nuclei indicates that acid phosphatase was adsorbed on the nuclear structures and then dephosphorylated the nucleotides present in the incubation medium.

Bearing all these reservations in mind, the precipitates observed in cell nuclei after incubation in original or modified Aronson's medium are probably mixtures liberated by true enzymatic action of DNase II oligo- and mononucleotides, the orthophosphate liberated from these nucleotides by

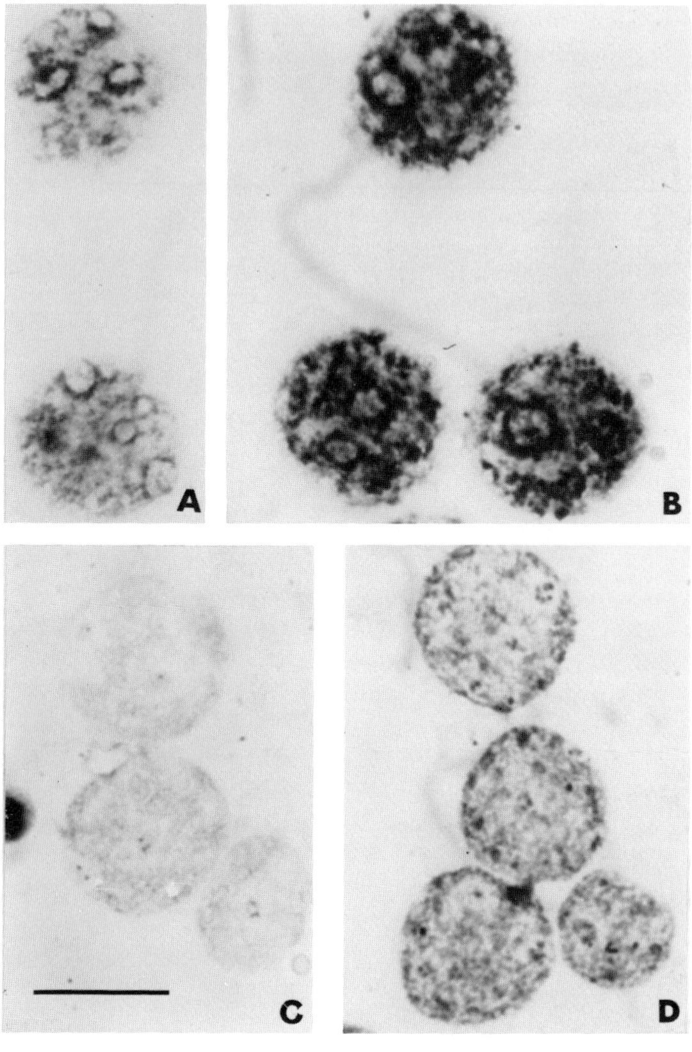

Fig. 6 Microphotographs illustrating the localization of the reaction product of DNase II activity in isolated rat liver cell nuclei (A and B); the results of control experiments related to the adsorption of acid phosphatase on nuclear structures (C and D) (for detailed explanation see text) (× 1800; rule indicates 10 μm).

acid phosphatase, lead, and lead phosphate. The final localization of the reaction product in the cell nucleus depends on the sites of adsorption of acid phosphatase, on the sites of adsorption of lead containing precipitates, and, probably, on the sites of the true enzymatic action of DNase II. In contradiction to biochemical data, the identical localization of the reaction product for DNase II and RNases in the condensed chromatin (Zotikov and Bernhard, 1970) suggest that the pictures obtained illustrate sites of adsorption rather than the sites of specific activity of two different enzymes.

The disappearance or weakening of the reaction of nucleases in malignant tumor nuclei (Taper *et al.*, 1971a, b) may be explained by affinity changes of nuclear components to lead-containing precipitates as well as to a more dispersed state of the chromatin.

Generally, the lead phosphate precipitation technique for detection of DNase II activity requires more scrupulous reexamination.

7. ATP PHOSPHOHYDROLASES (EC 3.6.1.3)

The intracellular localization of ATP phosphohydrolases (ATPases) is one of the most controversial problems in enzymatic cytochemistry, especially if methodical problems are considered. The series of articles dealing with the problem of nonenzymatic hydrolysis of ATP by lead ions and other possible sources of artifacts occurring in ATPase cytochemistry have impaired confidence in these methods (Rosenthal *et al.*, 1966, 1969; Moses *et al.*, 1966). It is not possible to consider all problems concerning the detection of ATPases in cell components. According to biochemical data (Siebert, 1968) in cell nuclei isolated from different animal organs, two types of ATP phosphohydrolases can be found. One, ATPase A, located in the nucleolus, is activated by Mg^{2+} and has its highest activity at pH 5.9. A very similar enzyme with a high substrate specificity was isolated from pig kidney nuclei and called GTPase A (Siebert, 1960, 1963). The second, ATPase B is activated by Ca^{2+} and has its highest activity at pH 8.4. This latter enzyme is mainly localized in the extranucleolar part of the nucleus (Fischer *et al.*, 1959; Siebert, 1966). The localization of two very similar types of ATPases were studied with independent cytochemical methods (Vorbrodt, 1957; Bankowski and Vorbrodt, 1957). One type of enzyme, corresponding to ATPase A, was detected in slightly acid medium (pH 6–6.5) containing sodium succinate as a buffer, lead acetate, and $MgCl_2$. The reaction product was mainly concentrated in the nucleoli while the nucleoplasm stained relatively weakly. The second type, corresponding to ATPase B, was detected at pH 8.3 in the presence of $CaCl_2$ and $MgCl_2$. The reaction product was homogeneously distributed in the nucleoplasm (Vorbrodt, 1960). In further studies on these enzymes (ATPase B) on smears and cryostat sections or on isolated nuclei (Bankowski and Vorbrodt, 1962; Bankowski, 1963) some technical problems have arisen because of diffusion artifacts and inconsistency in the localization of the reaction products. Consequently, further cytochemical work was not performed on the localization of ATPase B.

Further experiments evaluated the results of biochemical data (Siebert, 1963, 1967) and cytochemical studies (Coleman, 1965). The localization of the nuclear enzyme(s) splitting ATP and GTP at pH 5.9, corresponding to ATPase A, was carefully studied in tissue sections and in isolated cell nuclei by light (Vorbrodt *et al.*, 1964) and electron microscopy (Vorbrodt, 1967). Because of the possible interference (interaction) of cytoplasmic enzymes during incubation of thick (40 μm) sections for electron microscopy (see comments to Fig. 1), further studies were performed using glutaraldehyde-fixed, ultrathin frozen sections prepared according to Bernhard (1965) and Bernhard and Leduc (1967). The reaction product for ATPase A activity both with ATP (Fig. 7) or GTP (Fig. 8) as substrate was present in the nucleolus and in the interchromatinic granules (Vorbrodt and Bernhard, 1968). The appearance of the reaction products occurring in the form of very homogeneous, minute, and dense precipitates precisely localized in particular parts of the nuclear structure strongly suggested that it does not result from adsorption of the product but from enzymatic action. In contrast, the reaction product appearing in thick sections (40 μm), incubated with ATP or GTP as a substrate, was localized in the nucleolus and scattered all over the chromatin, nucleoplasm, and cytoplasm as dustlike, minute precipitates (Fig. 9); the mitochondria were not covered with the precipitates. Comparison of the pictures obtained in ultrathin frozen sections (Figs. 7 and 8) with those obtained in thick sections (40 μm) used in two-step preparation for electron microscopy (Figs. 3, 4, and 9) indicates unequivocally the validity of ultrathin frozen sectioning in nuclear cytochemistry. The presence of nuclear ATPase showing optimal activity in the slightly acid pH range (pH 6) and the alkaline range (pH 8–9) corresponding to ATPases A and B was also confirmed by Klein (1966) in isolated nuclei from embryonic chick myocardium. The localization of ATPase A activity corresponds strictly to the localization of nuclear ribonucleoprotein components.

In some experimental work where Washstein-Meisel's medium was applied for cytochemical detection of nuclear ATPase activity (Raikhlin and Shubin, 1970), the cytoplasmic ATPase-rich structures could interfere with nuclear enzymes, especially if thick sections (30–40 μm) are incubated before dehydration and embedding in Epon. The lead precipitates resulting from the action of cytoplasmic enzymes, and possibly from nonenzymatic hydrolysis of ATP by lead ions, could be adsorbed at the chromatin components of cell nucleus.

Actually we have no alternative methods for visualization of ATP phosphohydrolases in cell nuclei. The present methods of detection of nuclear ATPase activity are still far from faultless and one should try to avoid all supposed sources of errors. Consequently, it is advisable to use ultrathin frozen sections or isolated nuclei for maximal reduction of the

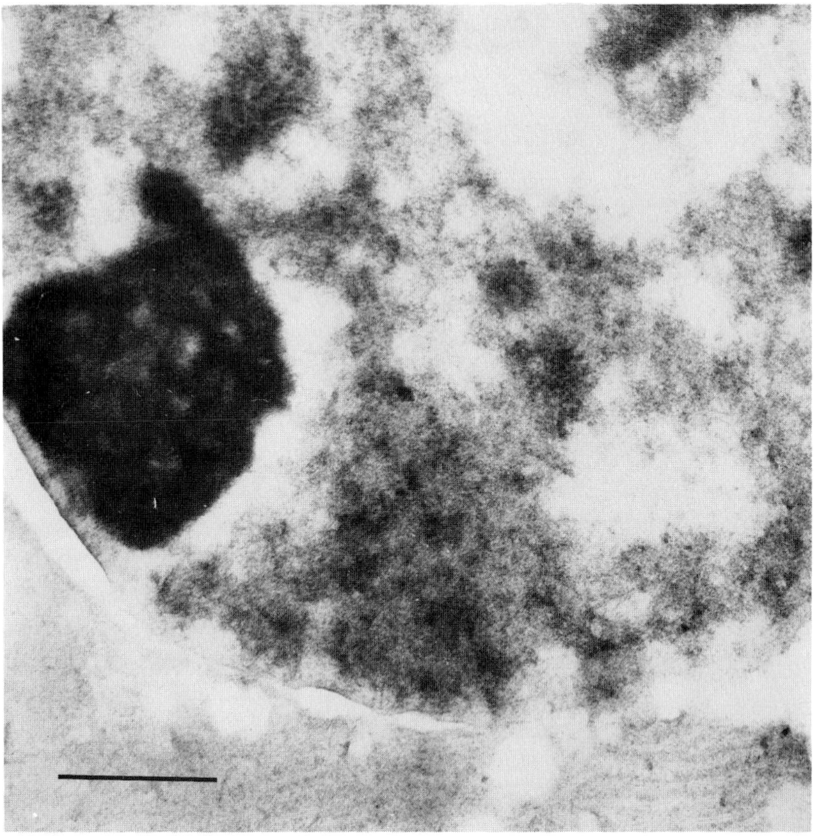

Fig. 7 Glutaraldehyde-fixed, ultrathin frozen section of rat liver incubated in the medium for visualization of ATPase A activity for 60 min at 37°C. The reaction product is highly concentrated in the nucleolus, in the interchromatinic granules, and in the intranuclear substance; chromatin remains completely unstained (× 21,000; rule indicates 1 μm.)

interaction of cytoplasmic enzymes. It is also possible to diminish artifacts and nonenzymatic hydrolysis of ATP by the introduction into the incubation medium of some lead-complexing factors like Tiron, succinate, or cysteine (Deimling, 1964; Malendowicz, 1972).

8. NUCLEOSIDE DIPHOSPHATASE (EC 3.6.1.6)

Nucleosidediphosphatase (ADPase) is considered an enzymatic marker of the endoplasmic reticulum hydrolyzing IDP, UDP, and GDP, but not CDP or ADP (Novikoff and Heus, 1963). In some cells, this enzyme is highly concentrated in the inner elements of the Golgi apparatus, where

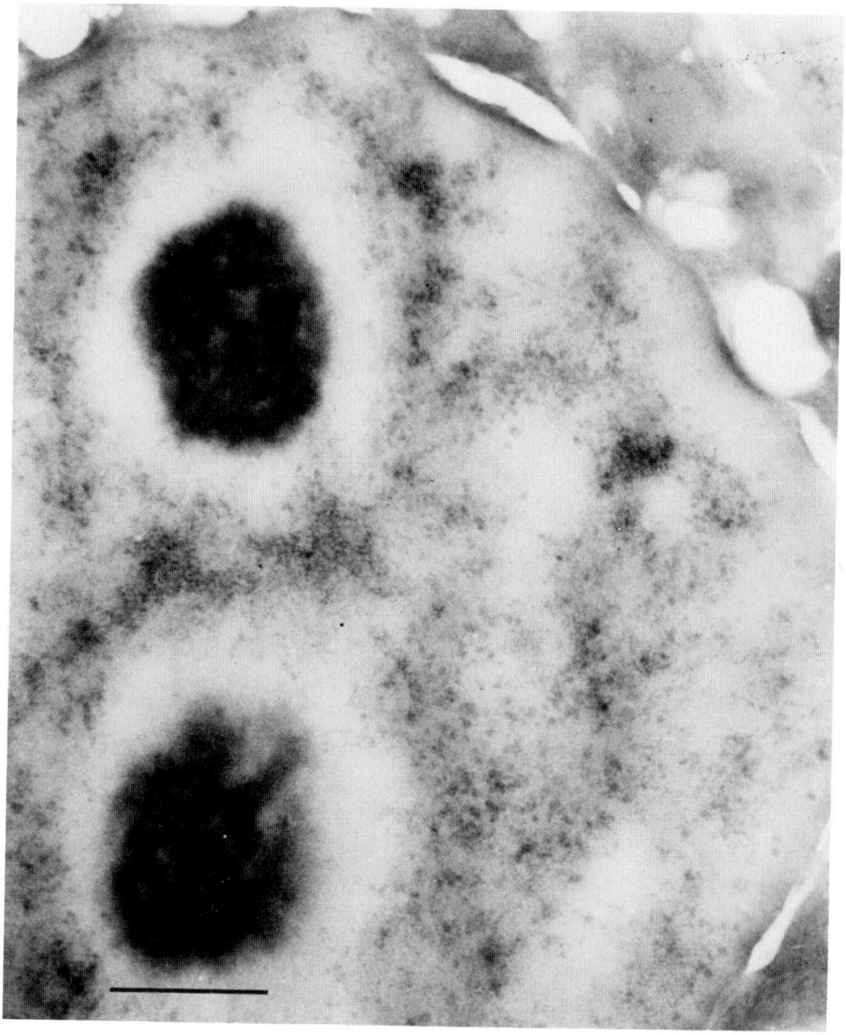

Fig. 8 The section was prepared and processed the same way as in Fig. 7, except that in the incubation medium GTP was used instead of ATP. The reaction product is concentrated in two nucleoli and in the interchromatinic granules (× 21,000; rule indicates 1 μm).

another enzyme that splits thiamine pyrophosphate is found. By light and electron microscopy the cytochemical reaction for thiamine pyrophosphatase (TPPase) activity is a reliable enzymatic marker of the Golgi apparatus (Novikoff and Goldfischer, 1961; Novikoff and Novikoff, 1972). In numerous neurons, especially in the dorsal root ganglia, a positive reaction for TPPase and NDPase appears not only in the Golgi apparatus

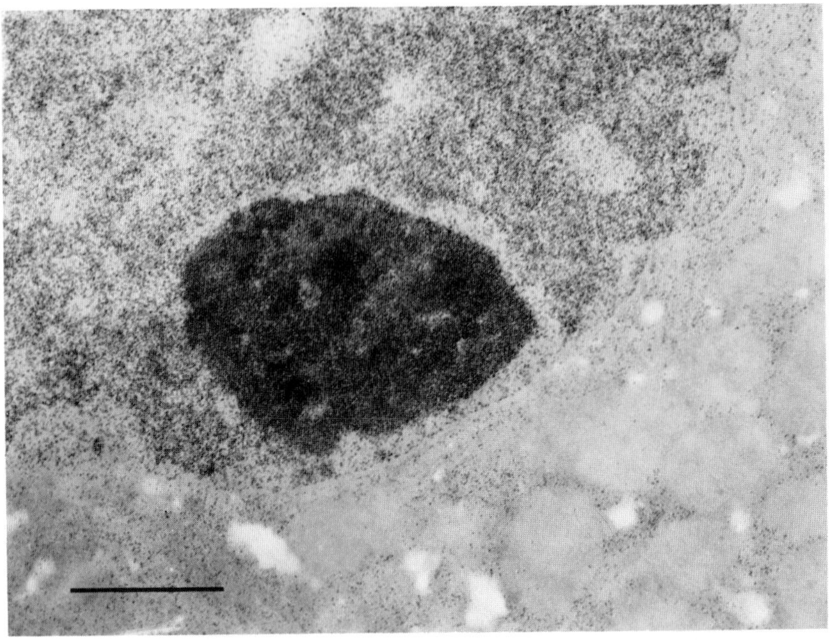

Fig. 9 Glutaraldehyde-fixed thick (40 μm) frozen section of rat liver incubated 30 min at 37°C for visualization of GTPase A activity, and embedded in Epon. The reaction product is highly concentrated in the nucleolus and in the intranuclear substance. The condensed chromatin, including nucleolus-associated chromatin, is almost unstained, although dustlike, minute precipitates are scattered throughout the entire nucleus and cytoplasm, except the mitochondria (× 20,000; rule indicates 1 μm).

but also inside the nucleoli (Novikoff, 1967). A positive reaction for TPPase in cell nuclei of supporting tissue cells in such organs as human stomach, mouse stomach and testis, and rat lung was reported (Yasuzumi *et al.*, 1965). These authors did not observe any reaction product in cell nuclei of the epithelia of the same organs.

IV. Cytochemically Demonstrable Enzymes in the Nuclear Envelope

In tissue sections incubated for cytochemical visualization of several enzymes and examined by electron microscopy, there are reaction products in the space between the two leaflets (membranes) of the nuclear envelope. The results of several cytochemical studies indicate the presence of the following enzymatic activities in the perinuclear space:

1. NADP-linked glucose-6-phosphate dehydrogenase (Nagasaka, 1970).
2. Plant oxidase (BED oxidase) in corn root tips (Nir and Seligman, 1971).
3. Peroxidase (endogenous) in various animal cells (Bainton and Farquhar, 1970; Novikoff *et al.*, 1971; Herzog and Miller, 1972).
4. Acetylcholinesterase in some neurons (Novikoff *et al.*, 1966).
5. Acid phosphatase in embryonic blood-forming cells (Vorbrodt, 1967), in eosinophilic leukocytes (Bainton and Farquhar, 1970), and probably in other cells.
6. Glucose-6-phosphatase in various animal and plant cells (Sabatini *et al.*, 1963; Lazarus and Barden, 1964; Goldfischer *et al.*, 1964).
7. Aryl sulfatase in plant cells (Poux, 1965), in eosinophilic leukocytes (Bainton and Farquhar, 1970).
8. Nucleoside diphosphatase in various animal cells (Novikoff and Heus, 1963; Pelletier and Novikoff, 1972).
9. Thiamine pyrophosphatase in some animal cells (Goldfischer *et al.*, 1964; Pelletier and Novikoff, 1972).
10. Acetyl-CoA carboxylase (EC 6.4.1.2) in rat liver cells (Yates *et al.*, 1969).

All these enzymes were detected in the nuclear envelope and in the endoplasmic reticulum, but they were not found inside the cell nucleus. Therefore they cannot be regarded as nuclear but rather as belonging to the endoplasmic reticulum system. The structural continuity of the perinuclear space with that of the endoplasmic reticulum is well established, and the presence of the same enzymatic activities in these spaces is not surprising.

The nuclear envelope is not a continuous structure but is interrupted at intervals by the pores. It is believed that these structures facilitate the interchange of materials between the nucleus and the cytoplasm by energy-requiring activities. Yasuzumi and Tsubo (1966) performed cytochemical studies on the localization of ATPase activity in the epithelial cells of the mouse choroid plexus and found strongly positive reactions in the form of fine globular elements showing high electron density, which were situated in the region of the nuclear pores. These findings would be of great interest and importance except for the unusual procedure employed, which impairs the validity of the results obtained.

The problem of ATPase activity in the nuclear envelope was studied cytochemically several years ago. Biochemical analysis of isolated nuclear membrane revealed the presence of glucose-6-phosphatase, Mg-activated ATPase (pH 7.2), and DPNH–cytochrome c reductase (Kashing and Kasper, 1969). In properly prepared tissue samples incubated for a short time (10–20 min) in Wachstein-Meisel's medium for ATPase detection, a positive reaction is not present inside the cell nucleus or in the perinuclear space. However, after prolonged incubation of the glutaraldehyde-fixed or unfixed section, a positive reaction for ATPase appears in the chromatin and in cell envelope (Raikhlin and Shubin, 1970). The interpretation of this observation relates to the possibility of the artifactual adsorption of lead phosphate on these structures (Moses *et al.*, 1966) as noted in previous sections (Figs. 3 and 4). In contrast to the lack of the reaction product for ATPase in the perinuclear space in tissue sections, isolated nuclei show the presence of such reaction products in the nuclear envelope. Bankowski (1963) in scrupulous studies showed a positive reaction for ATPase A in the envelope of unfixed nuclei isolated from rat liver. After fixation or aging, the reaction disappeared from the envelope and appeared in the nucleoli. The presence of ATPase A in the envelope of rat liver nuclei (unfixed) was confirmed by electron microscopy (Vorbrodt, 1967). Recently, Bukhvalov *et al.* (1971) demonstrated a positive reaction for ATPase A (pH 6) on the surface of nuclear envelope in isolated mouse liver nuclei (unfixed). In the same nuclear preparation, the reaction for ATPase B (pH 8.3) was mainly concentrated in the nucleoli and chromatin. The authors' opinion was that the activity of these enzymes was completely (ATPase A) or almost completely (ATPase B) inhibited by short glutaraldehyde fixation. In these studies, the localization of ATPase B is doubtful, because, according to our experience, the detection of this enzyme activity is especially capricious due to the great affinity of calcium phosphate for the nucleolus, chromatin, and nuclear membrane structures.

There are similar disadvantages when detecting cytochrome oxidase activity in cell nuclei. This enzyme was recently demonstrated cytochemically in the envelope of isolated rat thymus nuclei (Rupec *et al.*, 1971). The localization of this enzyme was also observed in isolated rat liver nuclei, as well as in isolated membranes of these nuclei and in isolated rat thymus nuclei (Bukhvalov *et al.*, 1972). These findings indicate that some enzymes which are probably present in the nuclear envelope (inner or outer membrane) are very sensitive to the action of the fixatives, and probably could be studied only in unfixed, ultrathin, frozen sections.

V. Concluding Remarks

The nucleus is the center of genetic information of the cell; consequently one can assume that its structure and chemical composition are the reflection of highly specialized functions. From the structural and functional point of view one can recognize the following four nuclear systems or compartments: (*a*) chromosomal complex (chromatin), (*b*) nucleolar complex, (*c*) nuclear envelope, and (*d*) nuclear sap (soluble space).

The majority of enzymes visualized by cytochemical procedures in cell nucleus are also present in cytoplasmic organelles, and for this reason it is difficult to consider any one of them as typical nuclear marker. From six groups of known enzymes only about twenty representatives of oxidoreductases, transferases, and hydrolases can be actually visualized in the nucleus; lyases, isomerases, and ligases are still not demonstrable by means of cytochemical procedure. The diagram presented in Fig. 10 illustrates the present status of our knowledge of the distribution of cytochemically detectable enzymes in nuclear components, including perinuclear space.

Several enzymes, even those essential for nuclear function, are soluble and this fact delimits the potentialities of cytochemistry applied to nuclear enzymology.

The intranuclear localization of enzymes engaged in the metabolism of nucleic acids is of special interest. Hitherto existing information indicates that the enzymes related to RNA metabolism are mainly concentrated in the nucleolus, although they can be also found in the nuclear

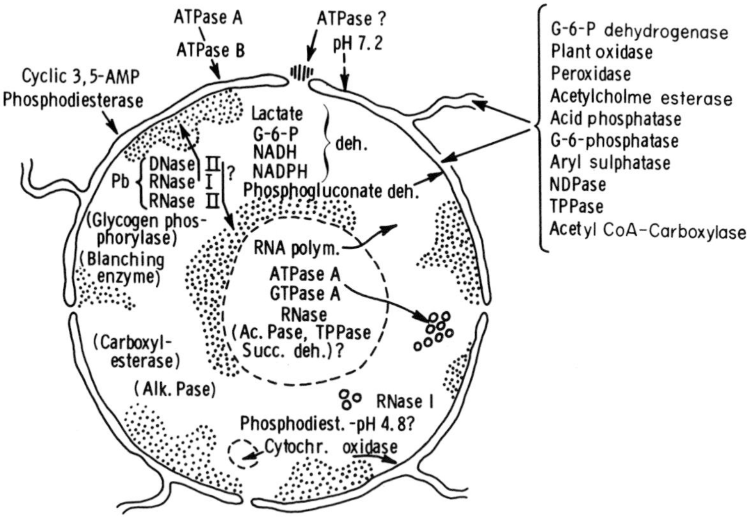

Fig. 10 Diagram illustrating localization of cytochemically detectable enzymes in different components of cell nucleus. The dotted areas represent the condensed chromatin; the interchromatinic granules are presented as small circles; the nucleolus is encircled by interrupted line. The enzymes in parentheses were observed only in some isolated cases. The question mark indicates that the presence of the enzyme is still questionable. Alk. Pase, alkaline phosphatase; Ac. Pase, acid phosphatase; TPPase, thiamine pyrophosphatase; Succ. deh., succinic dehydrogenase.

body, probably in the interchromatinic and perichromatinic regions. These enzymes include RNA polymerases, ribonucleases, and ATPase A.

Phosphodiesterase II and DNase II, which are related to DNA metabolism, were found in the nucleus by means of indigogenic methods. The activities of RNase I, RNase II, and DNase II, visualized by lead phosphate precipitation technique, were found solely in the condensed chromatin. As the validity of the results obtained with both indigogenic and lead precipitation techniques is questionable, this information should be confirmed by more specific and reliable methods.

The progress in cytochemistry of cytoplasmic organelles is much more advanced than in the field of nuclear cytochemistry. This fact is a reflection of a much higher concentration of a majority of cytochemically detectable enzymes in cytoplasmic structures than in the nucleus. This high concentration of cytoplasmic enzymes renders difficult the studies of enzymology of the cell nucleus *in situ*, i.e., in tissue sections. The maximal reduction of the interference of cytoplasmic enzymes can be achieved by the use of ultrathin frozen sections or isolated nuclei or nucleoli.

The progress in cryoultramicrotomy, the elaboration of new staining reactions—especially azo-dye and indigogenic techniques—as well as advances in immunohistochemistry will undoubtedly offer new possibilities for cytochemical studies of nuclear enzymology.

REFERENCES

Allfrey, V. G., and Mirsky, A. E. (1953). *J. Gen. Physiol.* **36,** 227.
Amano, H., and Daoust, R. (1961). *J. Histochem. Cytochem.* **9,** 161.
Anders, V. N., and Bukhvalov, I. B. (1972). *Bull. Exp. Biol.* **2,** 109 (Russian text).
Aronson, J., Hempelmann, I. H., and Okada, S (1958). *J. Histochem. Cytochem.* **6,** 255.
Bainton, D. F., and Farquhar, M. G. (1970). *J. Cell. Biol.* **45,** 54.
Baltus, E. (1954). *Biochim. Biophys. Acta* **15,** 263.
Baltus, E. (1962) *Biochim. Biophys. Acta* **55,** 82.
Bańkowski, Z. (1963). *Folia Histochem. Cytochem.* **1,** 17.
Bańkowski, Z., and Vorbrodt, A. (1957). *Bull. Acad. Pol. Sci. Cl. II B,* **5,** 245.
Bańkowski, Z., and Vorbrodt, A. (1962). *Ann. Histochim.* **7,** 31.
Barnard, E. A., Budd, G. C., and Ostrowski, K. (1970). *Exp. Cell Res.* **60,** 405.
Barrnett, R. J., Karmaker, S. S., Seligman, A. M. (1959). *J. Histochem. Cytochem.* **7,** 300.
Behrens, M. (1932). *Z. Physiol. Chem.* **209,** 59.
Berg, G. G., Lyon, D., and Campbell, M. (1972). *J. Histochem. Cytochem.* **20,** 39.
Bernhard, W. (1965). *Ann. Biol.* **4,** 5.
Bernhard, W., and Leduc, E. H. (1967). *J. Cell Biol.* **34,** 757.
Betel, J. (1967). *Biochim. Biophys. Acta* **143,** 62.
Bukhvalov, I. B., Zbarsky, I. B., Raikhlin, N. T., Troitzkaya, L. P., Filippova, N. A. (1971). *Cytologia* **13,** 654 (Russian text).

Bukhvalov, I. B., Dmitrjev, H. A., Troitzkaya, L. P., Zbarsky, I. B., Perevoschchikova, K. A., Raikhlin, N. T., and Filippova, N. A. (1972). *Cytologia* (in press) (Russian text).
Busch, H. (1968). *In* "Comprehensive Biochemistry" (M. Florkin and E. H. Stotz, eds.) Vol. 23, pp. 39–76. Elsevier, New York.
Chakravorty, A. K., and Busch, H. (1967). *Cancer Res.* **27**, 789.
Chatterjee, P. R., and Mitra, S. (1961). *Nature (London)* **192**, 285.
Chatterjee, P. R., and Mitra, S. (1962). *J. Histochem. Cytochem.* **10**, 6.
Chèvremont, M., and Firket, H. (1953). *Int. Rev. Cytol.* **2**, 261.
Coleman, J. R. (1965). *J. Cell Biol.* **27**, 20.
Coons, A. H. (1956). *Int. Rev. Cytol.* **5**, 1.
Coons, A. H., and Kaplan, M. H. (1950). *J. Exp. Med.* **91**, 1.
Coulson, A. S., and Kennedy, L. (1972). *Proc. Leucocyte Culture Conf. 6th* 91–96.
Daoust, R. (1957). *Exp. Cell Res.* **12**, 203.
Daoust, R. (1965). *Int. Rev. Cytol.* **18**, 191.
Daoust, R., and Amano, H. (1960). *J. Histochem. Cytochem.* **8**, 131.
Daoust, R., and Amano, H. (1964). *J. Histochem. Cytochem.* **12**, 429.
Darżynkiewicz, Z., Rogers, A. W., and Barnard, E. A. (1967). *J. Histochem. Cytochem.* **14**, 915.
Deimling, O. H. (1964). *Histochemie* **4**, 48.
de Lamirande, G. (1967). *Cancer Res.* **27**, 1722.
Egami, F., and Nakamura, K. (1969). "Microbial Ribonucleases." Springer-Verlag, Berlin and New York.
Ehinger, B. (1965). *Histochemie* **5**, 145.
"Enzyme Nomenclature" (1965). Elsevier, New York.
"Enzyme Nomenclature-Final Draft" (1971). Brisbane.
Erecińska, M., Sierakowska, H., and Shugar, D. (1969). *Eur. J. Biochem.* **11**, 465.
Fakan, S., and Bernhard, W. (1971). *Exp. Cell Res.* **67**, 129.
Fischer, R., Siebert, G., and Adloff, E. (1959). *Biochem. Zeitschr.* **332**, 131.
Fisher, D. B. (1968). *Int. Congr. Histochem. Cytochem., 3rd* p. 60. Springer-Verlag, New York.
Gluzman, D. F., and Petrenko, M. H. (1972). *Ukr. Biochim. J.* **44**, 259 (Russian text).
Gluzman, D. F., Novikova, V. F., and Stotskaya, L. N. (1971). *Bull. Exp. Biol. Med.* **9**, 90.
Godlewski, H. G. (1961). *Acta Histochem.* **11**, 58.
Godlewski, H. G. (1964). *Acta Un. Int. Cancer.* **20**, 706.
Goldfischer, S., Essner, E., and Novikoff, A. B. (1964). *J. Histochem. Cytochem.* **12**, 72.
Gomori, G. (1939). *Proc. Soc. Exp. Biol. Med.* **42**, 23.
Gomori, G. (1951). *J. Lab. Clin. Med.* **37**, 526.
Gomori, G. (1953). "Microscopic Histochemistry." Chicago Press, Chicago, Illinois.
Hardonk, M. J., Kondstaal, J., and Hoedemaeker, P. J. (1968). *In Int. Congr. Histochem. Cytochem. 3rd* p. 94. Springer-Verlag, New York.
Heppel, L. A. (1966). *In* "Procedures in Nucleic Acid Research" (G. L. Cantoni and D. R. Davies eds.), pp. 31–36. Harper, New York.
Herzog, V., and Miller, F. (1972). *J. Cell. Biol.* **53**, 662.
Kashing, D. M., and Kasper, C. B. (1969). *J. Biol. Chem.* **244**, 3786.
Klein, R. L. (1966). *J. Histochem. Cytochem.* **14**, 669.
Kuhlmann, W. D. (1970). *In* "Electron Microscopy" (P. Favard, ed.), Vol. I, pp. 535–536. Soc. Fr. Microsc. Electron., Paris.
Kuroiwa, T., and Tanaka, N. (1971). *J. Cell Biol.* **49**, 939.

Lazarus, S. S., and Barden, H. (1964). *J. Histochem. Cytochem.* **12,** 792.

Lazarus, H. M., and Sporn, M. B. (1967). *Proc. Nat. Acad. Sci. U.S.* **57,** 1386.

Leduc, E. H., Coons, A. H., and Conolly, J. M. (1955). *J. Exp. Med.* **102,** 61.

Leduc, E. H., Avrameas, S., and Bouteille, M. (1968). *J. Exp. Med.* **127,** 109.

Leduc, E. H., Scott, G. B., and Avrameas, S., (1969). *J. Histochem. Cytochem.* **17,** 211.

Lettré, H., and Paweletz, N. (1966). *Naturwissenschaften* **53,** 268.

Love, R., Studziński, G. P., and Walsh, R. J. (1969). *Exp. Cell Res.* **58,** 62.

Malendowicz, L. (1972). *Folia Histochem. Cytochem.* **10,** 237.

Marshall, J. M. (1954). *Exp. Cell Res.* **6,** 240.

Maul, G. G., and Hamilton, T. H. (1967). *Proc. Nat. Acad. Sci. U.S.* **57,** 1371.

Monneron, A., and Bernhard, W. (1969). *J. Ultrastruct. Res.* **27,** 266.

Morikawa, S. (1967). *J. Histochem. Cytochem.* **15,** 662.

Moses, H. L., Rosenthal, A. S., Beaver, D. L., and Schuffman, S. S. (1966). *J. Histochem. Cytochem.* **14,** 702.

Nagasaka, M. (1970). *In* "Electron Microscopy" (P. Favard, ed.) Vol. I, p. 555. Soc. Fr. Microsc. Electron, Paris.

Nakane, P. K., and Pierce, B. G. (1966). *J. Histochem. Cytochem.* **14,** 920.

Nir, I., and Seligman, A. M. (1971). *J. Histochem. Cytochem.* **19,** 611.

Novikoff, A. B. (1951). *Science* **113,** 320.

Novikoff, A. B. (1952). *Exp. Cell Res. Suppl.* **2,** 123.

Novikoff, A. B. (1967). *In* "The Neuron" (H. Hyden, ed.), pp. 255–318. Elsevier, New York.

Novikoff, A. B., and Essner, E. (1960). *Amer. J. Med.* **29,** 102.

Novikoff, A. B., and Essner, E. (1962). *Fed. Proc.* **21,** 1130.

Novikoff, A. B., and Goldfischer, S. (1961). *Proc. Nat. Acad. Sci. U.S.* **47,** 802.

Novikoff, A. B., and Heus, M. (1963). *J. Biol. Chem.* **238,** 710.

Novikoff, A. B., and Novikoff, P. N. (1972). *In* "Biomembranes" (L. A. Manson, ed.), Vol. II, pp. 33–39. Plenum Press, New York.

Novikoff, A. B., Hausman, D. H., and Podber, E. (1958). *J. Histochem. Cytochem.* **6,** 61.

Novikoff, A. B., Quintana, N., Villaverde, H., and Forschirm, R. (1966). *J. Cell. Biol.* **29,** 525.

Novikoff, A. B., Beard, M. E., Albala, A., Sheid, B., Quintana, N., and Biempica, L. (1971). *J. Microsc.* **12,** 381.

Ogawa, K., Saito, T., and Mayahara, H. (1968). *J. Histochem. Cytochem.* **16,** 49.

Ostrowski, K., and Barnard, E. A. (1961). *Exp. Cell Res.* **25,** 465.

Pearse, A. G. E. (1968). "Histochemistry," Vol. I, 3rd ed. Churchill, London.

Pelletier, G., and Novikoff, A. B. (1972). *J. Histochem. Cytochem.* **20,** 1.

Pokrovsky, A. A., Zbarsky, I. B., Gapparov, M. M., Perevoschchikova, K. A., Tutelian, W. A., Dielektorskaya, L. N., Laschnieva, N. W., and Ponomareva, L. G. (1970). *In* "The Cell Nucleus and its Ultrastructures," pp. 156–160. Moskva (Russian text).

Poux, N. (1965). *J. Histochem. Cytochem.* **13,** 520.

Raikhlin, N. T., and Shubin, A. S. (1970). *Folia Histochem. Cytochem.* **8,** 121.

Razzell, W. E. (1961). *J. Biol. Chem.* **236,** 3028.

Recher, L., Whitescarver, J., and Briggs, L. (1969). *J. Ultrastruct. Res.* **29,** 1.

Rosa, C. G., and Tsou, K. C. (1964). *J. Histochem. Cytochem.* **12,** 23.

Rosenthal, A. S., Moses, H. L., Beaver, D. L., and Schuffman, S. S. (1966). *J. Histochem. Cytochem.* **14,** 698.

Rosenthal, A. S., Moses, H. L., Ganote, C. E., and Tice, L. (1969). *J. Histochem. Cytochem.* **17,** 839.

Rossi, F., Pescetto, G., and Reale, E. (1954). *C. R. Ass. Anat.* **I,** 1–100.
Rupec, M., Brühl, R., and Sekeris, C. E. (1971). *Exp. Cell Res.* **66,** 157.
Sabatini, D., Bensch, K., and Barrnett, R. J. (1963). *J. Cell Biol.* **17,** 19.
Seligman, A. M., Karnovsky, M. J., Wasserkrug, H. L., and Hanker, J. S. (1968). *J. Cell Biol.* **38,** 1.
Seligman, A. M., Wasserkrug, L. H., and Plapinger, R. E. (1970). *Histochemie* **23,** 43.
Shanta, T. R., Woods, W. D., Waitzman, M. B., and Bourne, G. H. (1966). *Histochemie,* **7,** 177.
Shantaveerappa, T. R., Waitzman, H. R., and Bourne, G. H. (1966). *Histochemie* **7,** 81.
Shugar, D., and Sierakowska, H. (1967). *Progr. Nucl. Acid Res.* **7,** 369–429.
Siebert, G. (1960). *In* "The Cell Nucleus" (J. S. Mitchell, ed.), pp. 176–184. Butterworths, London and Washington, D.C.
Siebert, G. (1963). *Exp. Cell Res. Suppl.* **9,** 389.
Siebert, G. (1966). *Nat. Cancer Inst. Monogr.* **23,** 285.
Siebert, G. (1967). *Methods Cancer Res.* **2,** 287–301.
Siebert, G. (1968). *In* "Comprehensive Biochemistry" (M. Florkin and E. H. Stotz, eds.), Vol. XXIII, pp. 1–37. Elsevier, New York.
Sierakowska, H., and Shugar, D. (1960). *Acta Biochim. Pol.* **7,** 475.
Sierakowska, H., Szemplińska, H., and Shugar, D. (1963). *Acta Biochim. Pol.* **10,** 399.
Soriano, R. Z., and Love, R. (1971). *Exp. Cell Res.* **65,** 467.
Swingle, K. F., and Cole, L. (1964). *J. Histochem. Cytochem.* **12,** 442.
Takamatsu, H. (1939). *Tr. Soc. Pathol. Jap.* **29,** 492.
Taper, H. S. (1968). *Ann. Histochim.* **14,** 301.
Taper, H. S., Brucher, J. M., and Fort, L. (1971a). *Cancer Res.* **28,** 482.
Taper, H. S., Fort, L., and Brucher, J. M. (1971b). *Cancer Res.* **31,** 913.
Tewari, H. B., and Bourne, G. H. (1962). *J. Histochem. Cytochem.* **10,** 619.
Tsou, K. C., Chang, M. Y., and Matsukawa, S. (1968). *Int. Congr. Histochem. Cytochem., 3rd* pp. 273–274. Springer-Verlag, New York.
van Dyck, J. M., and Wattiaux, R. (1968). *Eur. J. Biochem.* **7,** 15.
Vincent, W. S. (1952). *Proc. Nat. Acad. Sci. U.S.* **38,** 139.
Vorbrodt, A. (1957). *Exp. Cell Res.* **12,** 154.
Vorbrodt, A. (1960). *Bull. Acad. Pol. Sci. Cl. II B* **8,** 89.
Vorbrodt, A. (1961). *J. Histochem. Cytochem.* **9,** 647.
Vorbrodt, A. (1967). *Folia Histochem. Cytochem.* **5,** 239.
Vorbrodt, A. (1967). *Acta Histochem.* **28,** 215.
Vorbrodt, A., and Bernhard, W. (1968). *J. Microsc.* **7,** 195.
Vorbrodt, A., Krzyzowska-Gruca, St., and Steplewski, Z. (1964). *Bull. Acad. Pol. Sci.* **12,** 337.
Vorbrodt, A., Steplewski, Z., and Zolnierczyk, Z. (1965). *Folia Histochem. Cytochem.* **3,** 137.
Wachstein, M., and Meisel, E. (1957). *Amer. J. Clin. Pathol.* **27,** 13.
Wisse, E., and Tates, A. D. (1968). *Electron Microsc.* **2,** 465.
Wolf, P. L., Horwitz, J. P., Freisler, J. V., von der Muehll, E., and Vasquez, J. (1968a). *Int. Congr. Histochem. Cytochem., 3rd* p. 295. Springer-Verlag, New York.
Wolf, P. L., Horwitz, J. P., Freisler, J. V., Vasquez, J., and von der Muehll, E. (1968b). *Biochim. Biophys. Acta* **159,** 212.
Yasuda, K. (1968). *Acta Histochem. Cytochem.* **1,** 219.
Yasuzumi, G., and Tsubo, J. (1966). *Exp. Cell Res.* **43,** 281.
Yasuzumi, G., Suhigara, R., Nakai, Y., Sugioka, T., and Enomoto, K. (1965). *J. Electron Microsc.* **14,** 346.

Yates, R. D., Higgins, J. A., and Barrnett, R. J. (1969). *J. Histochem. Cytochem.* **17,** 379.

Zan-Kowalczewska, H., Sierakowska, H., and Shugar, D. (1966). *Acta Biochim. Pol.* **13,** 237.

Zawistowski, S., Bławat, F., Kowalska, Z., and Towiańska, A. (1965). *Folia Histochem. Cytochem.* **3,** 283.

Zbarsky, I. B., and Perevoschchikova, K. A. (1969). *Nature* (*London*) **221,** 257.

Zbarsky, I. B., and Perevoschchikova, K. A. (1970). *In* "The Cell Nucleus and its Ultrastructures," pp. 144–155. Moskva (Russian text).

Zotikov, L., and Bernhard, W. (1970). *J. Ultrastruct. Res.* **30,** 642.

9

Nuclear Protein Synthesis

LeRoy Kuehl

I. Introduction*

Until about ten years ago it was frequently assumed that nuclear proteins were synthesized in the nucleus. Isolated nuclei had been shown to incorporate radioactive amino acids into protein, and labeled amino acids

* In writing this chapter, no attempt was made to provide a complete survey of the literature. Instead, papers were selected for inclusion in the bibliography because they were considered representative or because they illustrated particular points. Abbreviations used without definition are those recommended by the *Journal of Biological Chemistry*.

administered to intact animals often appeared rapidly in the proteins of the cell nuclei; yet to generalize from these observations that the nucleus was the sole site of synthesis of nuclear proteins was clearly unwarranted.

During the past ten years it has been shown in several experimental systems that a substantial portion of the protein in the cell nucleus is of cytoplasmic origin. Furthermore, 18 S ribosomal RNA has been shown to be absent in some preparations of nuclei, implying that the small ribosomal subunit and consequently the ability to synthesize protein are also absent. In the light of these observations, some molecular biologists have stated categorically that protein synthesis is restricted to the cytoplasm, a conclusion which is equally unwarranted.

Our present knowledge concerning nuclear protein synthesis may be summarized as follows: First, it has been shown convincingly that, in some cells, nuclear proteins are produced in the cytoplasm. For the few cases that have been studied, the evidence favors the viewpoint that a substantial proportion of the nuclear protein is made outside the nucleus. Second, there is good evidence that in thymocytes, which are unusual in that the nucleus comprises a very large proportion of the total cell volume, the nucleus is a site of protein synthesis. In these cells, a considerable proportion of the nuclear protein seems to be made in the nucleus. Finally, there is a mass of data, often contradictory and never compelling, suggesting that the nuclei of other cell types also synthesize protein, and a much smaller body of evidence suggesting that they do not. The reader is referred to the excellent review of Goldstein (1970) for a treatment of the subject matter of this chapter from a different perspective.

II. Extranuclear Synthesis of Nuclear Proteins

A. *Evidence from Labeling Kinetics*

Probably the earliest study suggesting that nuclear proteins were made in the cytoplasm was that of Zalokar (1960), who injected [^{3}H]leucine into *Drosophila* and observed by autoradiographic techniques that, during the first minute after injection, nearly all incorporation occurred in the cytoplasm; only at longer times did labeled protein appear in the nucleus. These experiments suggested but did not prove that nuclear proteins were produced in the cytoplasm. The delay in nuclear labeling might, for example, have reflected the time required for the radioactive amino acid to cross the nuclear envelope to reach a nuclear site of protein synthesis. Bloch and Brack (1964), using grasshopper testis, and Kedes *et al.* (1969), using sea urchin embryos, obtained similar results.

In recent years more sophisticated studies have provided better evidence for the cytoplasmic synthesis of nuclear proteins. Wu and Warner (1971) derived equations describing the appearance of label in nascent proteins, completed proteins, and total proteins (sum of nascent and completed proteins) following a pulse of radioactive amino acid. Using HeLa cells, it was shown that the appearance of label in the nuclear fraction followed the equation for completed proteins, indicating that, within the limits of sensitivity of the method, all nuclear proteins were made in the cytoplasm. It should be noted, however, that preparation of the nuclei involved a detergent wash. This wash may have extracted nascent proteins from the nuclei since it was found to contain both nascent and completed proteins.

Some of the most convincing evidence for cytoplasmic synthesis of nuclear proteins has been provided by experiments in which proteins labeled in the cytoplasm during a brief exposure to a radioactive amino acid are observed to move into the nucleus after addition of a high concentration of unlabeled amino acid or an inhibitor of protein synthesis to prevent further incorporation. Such experiments have been performed with mouse fibroblasts by Zetterberg (1966a) and with HeLa cells by Speer and Zimmerman (1968) and Stein and Baserga (1971). The latter investigators observed that during a 5-min incubation in the presence of either excess unlabeled amino acid or cycloheximide, radioactivity incorporated during a brief labeling period decreased in cytoplasmic proteins and concomitantly increased in three nuclear protein fractions, which together accounted for all the proteins in the isolated nuclei. A quantitative evaluation of the data led to the conclusion that more than 90% of each of the three nuclear protein fractions is synthesized in the cytoplasm.

Ling *et al.* (1969) have provided an equally convincing demonstration of cytoplasmic synthesis of a nuclear protein. They found that after a 30-sec pulse of radioactive arginine, labeled protamine could be identified in the cytoplasmic fraction of trout testis, but that the nuclear protamines remained unlabeled. Upon chasing with unlabeled arginine, there was a decrease in the labeled protamine in the cytoplasm and a concomitant increase in the nucleus. Furthermore, a cytoplasmic fraction incorporated labeled amino acid into protamine *in vitro,* whereas a nuclear fraction did not.

Hemoglobin is present in the nucleus as well as the cytoplasm of avian erythrocytes. Kabat (1968) presented autoradiographic evidence that the nuclear hemoglobin originates in the cytoplasm. Chicken erythrocytes were labeled for 12 min at which time about 93% of the incorporated radioactivity was in the cytoplasm; the cells were then chased with unlabeled amino acid. During the chase, no additional incorporation occurred; there was a redistribution of the previously incorporated radioisotope, however, so that after 2 hr only about 85% of the label was in the cytoplasm, the rest having migrated into the nucleus. Since the principal protein synthesized by erythrocytes is hemoglobin, the data suggested that this was the protein which had migrated into the nucleus. In Section III,A evidence is presented for the opposite conclusion, namely, that both the nuclear and the cytoplasmic hemoglobin are produced in the nucleus in avian erythrocytes.

In the studies cited above, the specific activities of cytoplasmic proteins have been higher than those in the nucleus at short labeling times. Usually this is not the case. In Section III,A a number of studies are presented in which it has been observed that nuclear proteins become labeled as rapidly as or even more rapidly than cytoplasmic proteins.

B. *Identification of Nascent Nuclear Proteins on Cytoplasmic Polysomes*

Robbins and Borun (1967) and Borun *et al.* (1967) identified a small class of polysomes in HeLa cells which is present only during histone synthesis and which incorporates labeled amino acids into a product resembling histones. These polysomes were isolated from a cell homogenate which had been centrifuged to remove nuclei and were, presumably, of cytoplasmic origin. The authors concluded that histones were synthesized in the cytoplasm, then transferred to the nucleus. Similar results were obtained by Gallwitz and Mueller (1969a) using HeLa cells and by Kedes *et al.* (1969) with sea urchin embryos.

The studies of Ling *et al.* (1969) suggest that protamines are also synthesized on cytoplasmic polysomes. These investigators prepared polysomes from the cytoplasm of trout testis cells actively engaged in protamine biosynthesis. Sedimentation analysis of the resulting polysomes revealed a prominent diribosome peak from which nascent protamine molecules could be isolated and identified. These experiments, together with the pulse-labeling experiments described earlier, provide good evidence that the cytoplasm is the principal site of protamine biosynthesis in trout testis.

All the studies described in this section can be criticized on the grounds that the polysomes were, in no case, unequivocally shown to be solely of cytoplasmic origin. Loss of materials from nuclei during isolation in aqueous media has frequently been observed, and, although it seems unlikely, it is possible that the polysomes in the above studies were of nuclear origin, particularly since tissues were usually homogenized in detergent-containing solutions.

C. Other Studies

Using the technique of nuclear transplantation in amoeba, Byers *et al.* (1963) performed some of the early experiments demonstrating cytoplasmic synthesis of nuclear proteins. Surgically enucleated amoebae were allowed to incorporate labeled amino acid, then, following a chase with unlabeled amino acid, were implanted with unlabeled nuclei. The latter became labeled, even though control experiments demonstrated that the enucleate cells no longer contained significant amounts of free radioactive amino acid at the time of implantation with the unlabeled nucleus. Nuclear proteins in amoeba can be divided into two classes. One class migrates rapidly between the nucleus and the cytoplasm; the other is held tightly in the nucleus. Representatives of both classes of proteins migrated into the unlabeled nucleus from the labeled cytoplasm.

Using autoradiographic and microinterferometric techniques, Zetterberg (1966b) provided evidence for cytoplasmic synthesis of nuclear proteins in exponentially growing mouse fibroblasts. He showed by autoradiography that the rate of incorporation of [^{3}H]leucine per unit volume of cytoplasm was constant during interphase. By interferometry, it was established that an increase in dry mass occurred in the cytoplasm during the first half of interphase but that during the latter half of interphase the increase in dry mass was limited mainly to the nucleus, suggesting that during this part of the cell cycle the protein synthesized in the cytoplasm was being transferred to the nucleus. Quantitative evaluation of the data suggested that about 65% of the mass increase in the nucleus resulted from the transport into the nucleus of material synthesized in the cytoplasm.

D. Site of Synthesis

Four distinct classes of ribosomes can be distinguished in the cytoplasm of most animal cells: (1) free ribosomes, (2) those bound to the endoplasmic reticulum, (3) those bound to the outer nuclear membrane, and (4) mitochondrial ribosomes. Plant cells have, in addition, ribosomes associated with the chloroplasts. Each of these classes appears to synthesize particular kinds of proteins. Thus, Siekevitz and Palade (1960), Redman (1969), and Sherr and Uhr (1970) have furnished evidence that proteins produced for export from the cell are made on polysomes bound to the endoplasmic reticulum. Free polysomes, on the other hand, seem to produce intracellular proteins (Redman, 1969; Ganoza and Williams, 1969), and mitochondrial ribosomes probably synthesize some, but not all, mitochondrial proteins (Beattie *et al.*, 1967; Lizardi and Luck, 1972).

The demonstration that some nuclear proteins are made in the cytoplasm raises the question: On which class, or classes, of cytoplasmic ribosomes are these proteins synthesized? Kuehl and Sumsion (1971) examined the labeling kinetics of nuclear and cytoplasmic aldolase, glyceraldehyde-3-phosphate dehydrogenase, and lactic dehydrogenase in rats treated with puromycin, an antibiotic that has been shown to inhibit the synthesis of cytoplasmic proteins to a greater extent than those of the nucleus. At short labeling times, the specific activities of the nuclear enzymes were several times as high as those of the corresponding enzymes in the cytoplasm, demonstrating that the labeled nuclear enzymes had not arisen from the cytoplasmic enzyme pool. Since free polysomes are thought to release finished polypeptide chains directly into the soluble compartment of the cytoplasm, the labeled nuclear enzymes had presumably not been synthesized on free cytoplasmic ribosomes. If they were of cytoplasmic origin, they must have been made by one of the membrane-bound classes of ribosomes.

Gorovsky (1969) presented evidence that the ribosomes associated with the outer nuclear membrane are involved in the synthesis of nuclear proteins in *Tetrahymena. In vitro* labeling studies suggested a presursor–product relationship between the nascent proteins on the ribosomes attached to the outer nuclear membrane and the proteins inside the nucleus. Furthermore, puromycin caused a premature release of labeled proteins from the outer membrane ribosomes which was accompanied by a parallel increase in the labeled proteins inside the nucleus.

E. *Migration of Proteins from the Cytoplasmic Site of Synthesis into the Nucleus*

Since the nuclear envelope is interrupted by numerous pores, and since these are known to permit passage of macromolecules, it is possible that proteins produced in the cytoplasm might move into the nucleus by simple diffusion through the nuclear pores. Studies on the transport of secretory proteins, however, suggest that other alternatives also exist. Thus, Siekevitz and Palade (1960), Caro and Palade (1964), Redman and Sabatini (1966), and Jamieson and Palade (1968) have shown that digestive enzymes produced by the acinar cells of the guinea pig pancreas are synthesized by membrane-bound ribosomes, move vectorially into the lumen of the endoplasmic reticulum, then pass to the Golgi apparatus, where they are packaged into secretory granules, in which form they are transported to the margin of the cell. Investigations by Ashley and Peters (1969) and Zagury *et al.* (1970) indicate that serum albumin synthesized by rat hepatocytes and immunoglobulins produced by mouse plasma cells are transported in a similar fashion.

The lumen of the endoplasmic reticulum is contiguous with the perinuclear space. Therefore, proteins synthesized by the ribosomes attached to the endoplasmic reticulum as well as those bound to the outer nuclear membrane could, if discharged vectorially through the membrane, enter the nucleus without passing through the soluble part of the cytoplasm. The results of Kuehl and Sumsion (1971) and of Gorovsky (1969) presented in the previous section suggest that this might be the case.

III. Intranuclear Protein Synthesis

A. *Evidence from Labeling Kinetics*

A number of investigators have injected intact organisms with radioactive amino acids, subsequently detected labeled proteins in the cell nuclei, and concluded that these proteins must have been synthesized in the nuclei. Such a conclusion is justified, however, only if it can be demonstrated that the labeled material detected in the nucleus is actually protein and that this protein did not migrate into the nucleus from a cytoplasmic site of synthesis. Unfortunately, most investigators have failed to recognize these possibilities or have not used an experimental approach which would circumvent them. As a result, protein synthesis by the cell nucleus has never been demonstrated unequivocally by pulse-labeling studies, although some such studies do suggest that the nucleus is a site of protein synthesis.

Since information concerning the rate of migration of proteins between the cytoplasm and nucleus is essential for the interpretation of pulse-labeling experiments, a brief summary of our knowledge in this area will be presented at this point. The problem has been studied by a number of investigators, most of whom have found that the nuclear envelope serves as an effective barrier to the free diffusion of macromolecules. Thus, Feldherr and Feldherr (1960) reported that fluorescein-labeled rabbit γ-globulin introduced into the cytoplasm of moth oocytes by microinjection had, after 10 min, diffused from the site of injection to fill the entire cytoplasm, but had not entered the nucleus. In a similar type of experiment, Feldherr (1969) injected polyvinylpyrrolidone-stabilized colloidal gold particles, 25–55 Å in diameter (the size of small globular proteins), into the cytoplasm of amoebae, frog oocytes, and cockroach oocytes. After 50 min, the concentration of gold particles in the nucleus was only about half that in the cytoplasm for amoebae, and only 1–2% that in the cytoplasm for oocytes.

Labeled nuclei were transplanted from donor amoebae into unlabeled animals by Byers *et al.* (1963), who then followed the migration of protein from the labeled into the unlabeled nuclei by electron microscopic autoradiography. Although radioactivity could be detected in the host nuclei a few minutes after transplantation, it required about an hour for them to attain half of the radioactivity present at equilibrium and 4–5 hr to reach equilibrium. Kedes *et al.* (1969) exposed sea urchin embryos to radioactive leucine, then observed the distribution of label at various times after chasing with a large excess of unlabeled leucine. At short times, nearly all the radioactivity was localized in the cytoplasm; after 15 min significant label appeared in the nucleus and after about 4 hr equilibrium was attained, at which time 40–60% of the label was nuclear.

Kuehl and Sumsion (1971) reported that in rats treated with a low level of puromycin the specific radioactivities of nuclear aldolase, glyceraldehyde-3-phosphate dehydrogenase, and lactic dehydrogenase were several times as high as the corresponding cytoplasmic enzymes following a short period of incorporation of labeled amino acid. In preliminary studies, Kuehl (1972) found that when further protein synthesis was inhibited, the specific radioactivities of the corresponding nuclear and cytoplasmic enzymes approached one another, presumably due to movement of the enzymes between the two cell compartments, but that equilibrium had still not been established after 90 min.

These and other studies suggest that proteins do migrate across the nuclear envelope, but at a rate considerably slower than would be expected if the proteins could diffuse freely into the nucleus. They suggest that, in many systems, 10 minutes is too short a time to permit significant migration, but that equilibrium may be approached after several hours. This rule of thumb must be used cautiously, however, since Loewenstein *et al.* (1966) and Feldherr (1968, 1969, 1970) have shown that the permeability of the nuclear envelope may vary greatly from one type of cell to another and for a single type of cell under different physiological conditions. In some early kinetic studies purporting to demonstrate protein synthesis by the cell nucleus, very long labeling times were employed (24 hr in one investigation) and appreciable migration of newly synthesized proteins between cytoplasm and nucleus may have occurred. In more recent studies, shorter labeling times have been used and some of the results do suggest that the nucleus is a site of protein synthesis.

Schultze and Maurer (1967) found, using an autoradiographic technique, that 5 min after injection of a labeled amino acid into mice, the concentration of protein-bound radioactivity in the nuclei of a number of cell types was the same as, or higher than, that in the cytoplasm. Furthermore, the ratio of nuclear to cytoplasmic activity remained constant for labeling times up to 5 hr. Hnilica *et al.* (1966) prepared histones from isolated heptoma nuclei and nucleoli at several times after injection of radioactive lysine into tumor-bearing rats and observed substantial labeling of both nuclear and nucleolar histones at the shortest labeling times employed (about 15 min). Interestingly, the specific activity of the nucleolar histones was nearly twice as great as those from the nucleus, an observation interpreted by the authors as indicating histone synthesis by the nucleolus.

The nucleoli of onion root cells have become labeled 5 min after immersion of the roots in a solution containing radioactive arginine (Chouinard and LeBlond, 1967), and appreciable incorporation into the proteins of uterine nuclei has been found 10 min after injection of radioactive methionine into ovariectomized rats (Hamilton *et al.*, 1968). Finally, Kuehl (1967) and Kuehl and Sumsion (1971) have found, in rat liver, that the specific activity of total nuclear protein is approximately equal to that of total cytoplasmic protein, that the specific radioactivities of nuclear aldolase, glyceraldehyde-3-phosphate dehydrogenase, and lactic dehydrogenase are about equal to the specific activities of the corresponding cytoplasmic enzymes 30 sec after injection of labeled amino acid into the animal, and that the nuclear to cytoplasmic specific activity ratios are not significantly altered in animals labeled for longer times.

The results of these and other investigations demonstrate that in a wide variety of experimental situations nuclear proteins become labeled very quickly, more quickly, in fact, than might be predicted if the proteins were synthesized in the cytoplasm and transported into the nucleus; even for very short incorporation times the specific activities of the nuclear proteins may be as high as or higher than those in the cytoplasm. Such observations suggest that the nucleus is a site of protein synthesis. However, other investigators using similar techniques have described systems in which cytoplasmic proteins become labeled, but little or no radioactivity is found in the nucleus at short times after administration of a radioactive amino acid. In interpreting the results of *in vivo* labeling experiments, the tacit assumption is that a nuclear protein produced in the cytoplasm would, following synthesis, first enter the soluble space of the cytoplasm then migrate into the nucleus. An alternate possibility, which has been discussed on pp. 350–351, is that proteins produced on cytoplasmic ribosomes might move into the nucleus without passing through the soluble part of the cytoplasm. Such a mechanism might permit proteins to pass from a cytoplasmic site of synthesis into the nucleus more rapidly than if they were required to enter the nucleus via the soluble space of the cytoplasm.

Several labeling studies have yielded very unusual results. Monesi (1964) examined mouse spermatids autoradiographically at various times after administration of labeled amino acids. Incorporation into the cytoplasm was high during early stages of development, then decreased steadily, becoming zero during the last developmental stage. The extent of incorporation into the nucleus was highly dependent on the amino acid used for labeling. For six amino acids, incorporation was low during the early stages of development, then decreased to zero. Arginine incorporation, however, was very high during the later stages of spermatid development, when other amino acids were not being incorporated. Monesi suggests that this is due to the synthesis of arginine-rich histones in the nucleus during the later stages of spermatid development. This explanation, however, overlooks the fact that arginine-rich histones contain high amounts of leucine and lysine, two amino acids which were not incorporated during late spermatid development. An alternative explanation of the data is that arginine was not being incorporated into newly synthesized polypeptide chains, but, rather, was being attached to existing proteins. Enzymes which catalyze such a reaction have been described by Kaji (1968) and Soffer (1968).

Another interesting labeling study which has led to a rather startling conclusion is that of Hammel and Bessman (1964). These investigators incubated pigeon erythrocytes with radioactive amino acids, then separated the cells into a cytoplasmic and a nuclear fraction. Total radioactivity in the nuclear fraction rose rapidly, then leveled off, whereas the activity in the cytoplasm rose more slowly and continued to rise after net incorporation into the nuclear fraction had ceased. These observations suggested a transfer of proteins synthesized in the nuclei into the cytoplasm. This interpretation was supported by the results of a pulse-chase experiment, in which proteins labeled during the pulse were shown to migrate from the nuclear fraction to the cytoplasm during the chase. Much of the rapidly labeled protein in the nuclear fraction could be washed from the nuclei with sucrose and sodium chloride solutions. The labeled protein in such washes was shown by chromatographic analysis and fingerprinting to consist predominantly of globin. From these results Hammel and Bessman concluded that, in the avian erythrocyte, most, if not all, of the hemoglobin is synthesized in the nucleus and transferred to the cytoplasm. Such a conclusion may not be warranted. The nuclei that were used in most experiments were, as indicated by the published electron micrographs, heavily contaminated with endoplasmic reticulum, and the labeled "nuclear" proteins may have been nascent polypeptide chains associated with the membrane-bound ribosomes. Some control experiments were done with nuclei which had been isolated in nonaqueous media and which were, therefore, free from contaminating endoplasmic reticulum. The radioactive product in these experiments was never shown to be globin, however.

Kabat (1968), from the results of a radioautographic study, concluded that, in avian erythrocytes, the hemoglobin is synthesized exclusively in the cytoplasm. However, the labeled protein in Kabat's study was never unequivocally shown to be hemoglobin. The most reasonable conclusion concerning the site of hemoglobin synthesis in the avian erythrocyte is that the definitive study remains to be done.

B. *Identification of Components Required for Protein Synthesis in the Nucleus*

The demonstration that all the components required for protein synthesis (activating enzymes, transfer RNA's, polysomes, etc.) are present in the nucleus would constitute indirect evidence that the nucleus is a site of protein synthesis. In fact, many of these components have been identified in isolated nuclei. Several complications, however, make interpretation of the results difficult. First, any component, especially if present in low concentrations, may be the result of cytoplasmic contamination. Many of the constituents of the protein-synthesizing system of mitochondria and chloroplasts are different from those of the cytoplasm. It has, therefore, been relatively easy to establish that each of these organelles possesses a unique system for protein synthesis. Protein-synthesizing components that have been identified in nuclei, on the other hand, have usually proved to be identical with those in the cytoplasm and the possibility of cytoplasmic contamination is not so easily dismissed. A second problem is that some components of the protein-synthesizing system may be present in the nucleus in an inactive form. Thus, most of the steps in the assembly of a ribosome are believed to take place in the nucleus. The data of Warner (1971), however, suggests that the completion of the ribosome may occur in the cytoplasm, and nuclear structures presumed to be ribosomes may, in fact, be inactive ribosomal precursor particles. Messenger RNA and transfer RNA are, likewise, produced in the nucleus as precursors which undergo further processing prior to assuming their usual function in the cell.

1. RIBOSOMES

Preparation of ribosomes from isolated nuclei was reported a number of years ago by Frenster *et al.* (1960, 1961), Wang (1961, 1964), Georgiev *et al.* (1961), and Pogo *et al.* (1962). Frenster *et al.* (1960) and Wang (1961) presented data indicating that the isolated particles actively incorporated amino acids into polypeptide chains. At the time of these studies, it was not generally appreciated that the endoplasmic reticulum is contiguous with the nuclear envelope and that the outer surface of the nuclear envelope is, itself, studded with ribosomes. As a consequence, these studies were not designed in such a way as to eliminate the possibility that the "nuclear" ribosomes were really perinuclear contaminants, although yields of ribosomes were in some instances so high as to make this seem unlikely. More recently, this difficulty has been appreciated and steps have been taken to circumvent it. Thus, Bach and Johnson (1966, 1967) isolated HeLa cell nuclei by a procedure involving several washes with Triton X-100, a nonionic detergent which has been shown by Blobel and Potter (1966), Sadowski and Steiner (1968), and Barton *et al.*, (1971), to remove the outer nuclear membrane. Extraction of these nuclei with a DNA-containing buffer yielded polysomes which, when properly supplemented, incorporated labeled amino acids into proteins. These proteins were shown to be different from those produced by cytoplasmic polysomes. Sadowski and Howden (1968) reported the isolation of polysomes from Triton-washed nuclei from rat liver. In this case liberation of the polysomes was accomplished by treating the nuclei with deoxycholate and DNase.

McCarty *et al.* (1966) purified rat liver nuclei by sedimentation through a sucrose gradient containing EDTA and RNase to remove the ribosomes adhering to the nuclear envelope. Ribosomes isolated from these nuclei were similar to cytoplasmic ribosomes by several criteria including their ability to function in incorporation of amino acids into proteins.

Thus it is possible to isolate functional ribosomes from nuclei which have been treated to remove adhering fragments of endoplasmic reticulum and outer membrane ribosomes. In some instances, yields have been comparable to those obtained from cytoplasm. These observations strongly suggest that nuclei do contain functional ribosomes.

By electron microscopy, three classes of intranuclear particles which superficially resemble ribosomes have been identified: perichromatin granules, interchromatin granules, and nucleolar granules. For a more detailed discussion of these particles, the reader is referred to articles by Monneron and Bernhard (1969), Busch and Smetana (1970) and Simard (1970).

The perichromatin granules are almost certainly not ribosomes. They usually have a diameter of approximately 400 Å, about twice that of cytoplasmic ribosomes. Furthermore, Watson (1962) has estimated that there are 500–2000 perichromatin granules per liver cell nucleus. This is about two orders of magnitude fewer than would be required to account for the number of ribosomes which can be extracted from isolated liver cell nuclei.

About ten years ago it was thought that interchromatin granules or nucleolar granules might be ribosomes; in recent years most investigators have become more cautious about making such an identification. Although both types of granules have been demonstrated to be comprised, like ribosomes, of ribonucleoprotein and although both are approximately the same size as cytoplasmic ribosomes, there are differences which suggest that neither of these particles may be identical with cytoplasmic ribosomes. Thus, Monneron and Bernhard (1969) have described morphological differences between interchromatin granules and ribosomes, and Bernhard (1966) has reported that nucleolar granules are smaller and more irregular in shape than ribosomes. Furthermore, the subunit structure characteristic of cytoplasmic ribosomes and association into polysomelike complexes has never been observed for either of these intranuclear particles.

Therefore, the ribosomes which can be extracted from isolated nuclei cannot be identified *in situ* with the electron microscope and the question as to whether nuclei possess functional ribosomes must still be regarded as unanswered.

2. ACTIVATING ENZYMES AND TRANSFER RNA'S

Hopkins (1959) prepared an enzyme fraction from isolated calf thymus and chicken kidney nuclei which catalyzed an amino acid-dependent exchange between pyrophosphate and ATP. Twelve different L-amino acids stimulated the exchange; D-amino acids were inactive. These observations strongly suggested that activating enzymes (aminoacyl–tRNA synthetases) were present in nuclei. In thymocytes about 70% of the total activating enzyme activity of the cell was confined to the nucleus; the corresponding figure for chicken kidney cells was 7.6%. Hopkins also obtained evidence that transfer RNA is present in isolated thymus nuclei. RNA isolated from nuclei which had been incubated with a radioactive amino acid contained significant amounts of label which could be released by dilute alkali or RNase, but not by acid. In a later paper (Hopkins *et al.*, 1961), it was shown that the radioactive amino acid was bound to a terminal adenosine residue of the RNA.

Several investigators have published data indicating that isolated nucleoli contain activating enzymes and transfer RNA's. Frog egg nucleoli were reported by Brandt and Finamore (1963) to catalyze an exchange reaction between pyrophosphate and ATP. The exchange was stimulated by amino acids, suggesting that the nucleoli contained activating enzymes. Lamkin and Hurlbert (1972), using a different assay, found activating enzymes in the isolated nucleoli of Novikoff hepatoma cells. Evidence that nucleoli from these cells contain transfer RNA has also been published (Nakamura *et al.*, 1968). The isolated nucleoli were found to contain 4S RNA which, when isolated and incubated with radioactive amino acids, ATP, and activating enzymes, became labeled.

A comparison of the cytoplasmic, nuclear, and nucleolar transfer RNA's of Novikoff hepatoma cells has been made by Ritter and Busch (1971). No significant differences were found between the valyl and leucyl transfer RNA's from the three cell fractions upon co-chromatography on hydroxylapatite, BD–cellulose, and DEAE–Sephadex. The authors concluded that there was little, if any, nucleolus or nucleus specific valyl or leucyl transfer RNA in Novikoff cells.

C. *Protein Synthesis by Isolated Nuclei and Nuclear Subfractions*

The most direct way to establish that protein synthesis takes place within the nucleus would be to demonstrate that isolated nuclei are, themselves, capable of protein synthesis. A considerable number of investigators have employed this approach and most have found that isolated nuclei do, indeed, incorporate labeled amino acids into proteins. A number of such studies are summarized in Table I. As will be discussed in a subsequent section, interpretation of the results of these investigations is fraught with difficulty and, although it seems highly probable that the nuclei of some kinds of cells do produce proteins, this cannot be regarded as conclusively proved.

In discussing nuclear protein synthesis, it is convenient to consider first the results obtained with thymus and spleen nuclei. Extensive and carefully controlled investigations have been made with such nuclei, and results obtained by different investigators have, in most particulars, agreed very well. For other types of nuclei, the findings of different laboratories have often been at variance. Furthermore, data obtained using thymus and spleen nuclei suggest that they possess a protein-synthesizing system similar to that found in the cytoplasm, whereas some of the results which have been obtained with other kinds of nuclei are not easily reconciled with current concepts of protein biosynthesis.

1. THYMUS AND SPLEEN NUCLEI

The most extensive series of investigations on protein synthesis in isolated nuclei was done by a group of workers at Rockefeller University and reported in papers by Allfrey (1954), Allfrey *et al.* (1955, 1957, 1964), Hopkins (1959), Hopkins *et al.* (1961), and Frenster *et al.* (1960, 1961). Most of these studies were made with calf thymus nuclei prepared by differential centrifugation from dilute sucrose solutions. Sometimes the nuclei were further purified by centrifugation through gradients of Ficoll (a high polymer of sucrose), although this treatment resulted in a partial loss of activity. In most experiments, a suspension of nuclei was incubated with a radioactive amino acid, and the label incorporated into material insoluble in hot trichloroacetic acid was measured. The isolated nuclei contained all the components required for protein synthesis. Incorporation was not stimulated by addition of nucleoside triphosphates or factors from the soluble portion of the cell.

A number of observations suggested that the radioactive amino acids had been incorporated in internal positions in the polypeptide chains and were not simply adsorbed or otherwise attached to preexisting protein molecules. Thus, radioactivity was not released when the labeled proteins were treated with sodium hydroxide or ninhydrin, D-amino acids were not incorporated, and a radioactive amino acid, once incorporated, did not exchange with unlabeled amino acid from the medium. Strong support for the view that incorporation had occurred in internal positions in the polypeptide chain was subsequently provided by Reid and Cole (1964), who observed that all the tryptic peptides of the lysine-rich histone fraction became labeled upon incubation of the isolated nuclei with a radioactive amino acid.

The amino-acid incorporation catalyzed by thymus nuclei proceeds by a mechanism similar in many respects to that operating in the microsomal system. As discussed in an earlier section, activating enzymes, transfer RNA, and ribosomes were shown, by the Rockefeller group, to be present in the isolated nuclei. These were presumed to be involved in the incorporation process, in which they appeared to function in the same manner as their cytoplasmic counterparts. Also, incorporation was blocked by puromycin, an inhibitor of protein synthesis in microsomal systems.

Incorporation by the isolated nuclei was substantial; on a weight basis their activity was about one-third that of whole thymus cells and comparable to that reported by Sachs (1957), Hoagland *et al.* (1964), and others for liver microsomes. This is significant since it renders less likely the possibility that the incorporation was due to contamination by cytoplasmic microsomes or whole cells or that it was the result of nonspecific adsorption of the labeled amino acid to components of the nuclei. The activity of the isolated nuclei was quite labile. It was lost upon storage, upon freezing or thawing, or upon treating the nuclei with dense sucrose, Ficoll, or detergents.

The energy required for peptide-bond formation was generated in the isolated nuclei by an oxidative process employing endogenous substrates. Incorporation did not occur in a nitrogen atmosphere and was blocked by a number of inhibitors of oxidative phosphorylation including cyanide, antimycin A, and dinitrophenol. Addition of glucose, fructose, and α-

TABLE I
Amino Acid Incorporation by Isolated Nuclei

	Purification steps employed		Requirements for incorporation			Percent inhibition by						
Source of nuclei	Centrif. through dense sucrose[a]	Detergent washes	Nucleoside triphosphates[b]	Soluble factors from supn't.	Relative incorporation rate[c]	CN^-	2,4-Dinitrophenol	Puromycin	Cycloheximide	Chloramphenicol	RNase	Reference
Thymus, calf	0	0	0	0	50	76 (1.0 m*M*)	84 (0.20 m*M*)			0 (0.32 m*M*)	0	Allfrey *et al.* (1957)
Thymus, calf	+	0	0	0	7		89 (0.20 m*M*)	90 (0.21 m*M*)			5 (1 mg/ml)	Allfrey *et al.* (1964)
Thymus, calf	0	0	0	0	90			70–95 (0.10 m*M*)			8 (0.1 mg/ml)	Reid *et al.* (1968)
Thymus, rat	+	0	0	0	8						2 (0.01 mg/ml)	Uete (1967)
Spleen, rat	0	0	0	0	900	70 (1.0 m*M*)	70 (0.20 m*M*)	93 (0.08 m*M*)			22 (1 mg/ml)	Tsuzuki and Naora (1968)
Spleen, rat	0	0	0	0	100		66 (0.20 m*M*)		85 (0.08 m*M*)		17 (1 mg/ml)	Tsuzuki (1972)
Liver, rat	+	0	0	0			72 (2.0 m*M*)			40 (0.50 m*M*)	8 (0.1 mg/ml)	Rendi (1960)
Liver, rat	+	0	±	0	10						74 (0.01 mg/ml)	Uete (1967)
Liver, rat	+	0	0	0	30	83 (0.20 m*M*)	72 (0.10 m*M*)	0 (0.21 m*M*)	7 (1.8 m*M*)	95 (0.50 m*M*)	0 (0.1 mg/ml)	Ono and Terayama (1968)
Liver, rat	+	0	0	0	100	36 (0.20 m*M*)		3 (0.21 m*M*)		0 (0.50 m*M*)		Weser and Koolman (1969)
Liver, rat	+	+	0	0	40	85 (8.0 m*M*)	84 (8.0 m*M*)	68 (4.8 m*M*)	25 (3.6 m*M*)	72 (7.4 m*M*)	0 (1 mg/ml)	Anderson *et al.* (1972)

Liver, mouse	+	0	0	0	1							Arnold *et al.* (1968)
Brain, rat	0	0	0	0	7			87 (0.10 mM)	76 (0.18 mM)		0 (0.1 mg/ml)	Burdman and Journey (1969)
Brain, rat	+	0	0	0	100							Løvtrup-Rein (1970)
Brain, rat	+	+	+	+	2			84 (0.21 mM)	57 (0.36 mM)	6.5 (0.62 mM)	85 (0.1 mg/ml)	Fleischer-Lambropoulos and Reinsch (1971)
Prostate, rat	+	+	0	0	30	40 (1.0 mM)	30 (1.0 mM)	26 (2.1 mM)	34 (3.6 mM)	27 (3.1 mM)	0 (1.0 mg/ml)	Anderson *et al.* (1971)
Kidney, rat	0	0	0	0			78 (0.8 mM)					Rees *et al.* (1962)
HeLa cells	+	±	+	+	4			90 (0.15 mM)	51 (1.0 mM)	2 (0.15 mM)	96 (0.05 mg/ml)	Gallwitz and Mueller (1969b)
HeLa cells	0	+	0	0	50	54 (0.10 mM)	68 (0.20 mM)	94 (1.0 mM)	0 (1.1 mM)	0 (0.10 mM)	0 (0.2 mg/ml)	Zimmerman *et al.* (1969)
Sea urchin embryos	+	0	?	0	1							Løvtrup-Rein (1972)
Drosophila salivary glands	0	+	0	0	3500							Helmsing (1971)
Pea seedling stems	0	0	±	0	50							Birnstiel *et al.* (1962)
Tobacco-cell culture	+	0	0	0	70					0 (6.2 mM)	0 (0.2 mg/ml)	Flamm *et al.* (1963)

[a] Centrifugation through Ficoll (a sucrose polymer) was considered equivalent to centrifugation through sucrose.

[b] A ± indicates that incorporation was stimulated by addition of the indicated component, but was not totally dependent upon it.

[c] Values given are expressed as picocuries of labeled amino acid incorporated per min per mg of protein from an incubation mixture containing 1 μCi per ml. of radioactive amino acid. Whenever possible, initial rates of incorporation were used in making the calculations. It was sometimes necessary to assume a value for the counting efficiency in the radioactivity measurements or for the amount of protein in the isolated nuclei. Comparison of the incorporation rates obtained by different investigators requires that initial rates are proportional to the number of nuclei in the incubation mixture and that the isolated nuclei do not contain appreciable pools of unlabeled amino acids. However, as shown by Reid *et al.* (1968) and Anderson *et al.* (1972), these requirements may not always be met Comparison of the incorporation rates reported by different investigators using different materials must, therefore, be made with caution.

ketoglutarate to the incubation mixture resulted in little or no stimulation of incorporation. These observations are consistent with the results of McEwen *et al.* (1963) and Conover (1970a, b) which demonstrated that thymus nuclei can perform an oxygen-dependent synthesis of ATP using endogenous substrates.

Incorporation was dependent on sodium ions, and potassium would not substitute for sodium. Since incorporation by cytoplasmic ribosomes requires potassium (or ammonium) ions rather than sodium, the sodium ion dependence of the nuclear incorporation indicates that the observed incorporation was not due to contaminating cytoplasmic microsomes. The sodium dependence of the nuclear system is now believed to be explained by the observation of Allfrey *et al.* (1961) and Karjalainen (1966) that thymus nuclei possess a sodium-dependent system for active transport of amino acids. Some investigators have suggested that sodium dependence is characteristic of nuclear protein synthesis. This viewpoint is not justified, however, since, even for thymus nuclei, not all investigators have found a sodium dependence. Thus, Uete (1967) reported that sodium was inhibitory, and Herranen and Brunkhorst (1962) found that sodium could be either stimulatory or inhibitory, depending on the pH of the incubation mixture and the time which elapsed between sacrifice of the animal and preparation of the nuclei.

The Rockefeller group reported that pretreatment of the isolated nuclei with DNase substantially reduced their ability to incorporate labeled amino acids. Activity was restored by incubating the treated nuclei with DNA or RNA. It seems likely that loss of activity upon DNase treatment was a secondary effect resulting from inhibition of nuclear oxidative phosphorylation by histones released by digestion of the DNA (Allfrey and Mirsky, 1958; McEwen *et al.*, 1963) and that reversal of the inhibition by exogenous polyanions was due to their ability to form complexes with these histones. DNase inhibition is by no means characteristic of nuclear protein synthesis and even incorporation by thymus nuclei was found by Uete (1967) to be unaffected by DNase treatment.

RNase, in contrast to DNase, had little effect on protein synthesis by intact nuclei even at very high concentrations, a surprising observation in view of the high sensitivity of protein synthesis in microsomal systems to RNase. Failure of RNase to inhibit nuclear protein synthesis was not due to an inability of the enzyme to penetrate the nuclei or to digest the nuclear RNA once inside the nucleus, since about half of the nuclear RNA was released by RNase treatment.

In an attempt to find which nuclear proteins had become labeled, the Rockefeller group developed a procedure for separating the proteins of the isolated thymus nuclei into several fractions. Upon incubation of the nuclei with a radioactive amino acid, each of these fractions became labeled. The nonhistone chromosomal proteins became most highly labeled; the histones had the lowest specific activity. More recently Reid and Cole (1964) and Reid *et al.* (1968) reported that four lysine-rich histones, a fraction containing arginine-rich and slightly lysine-rich histones, and a residual fraction all became labeled upon incubation of isolated thymus nuclei with radioactive amino acids.

A number of observations suggested that incorporation was due to isolated nuclei rather than to whole cells or cytoplasmic contaminants: (1) Electron microscopic autoradiographs showed extensive incorporation over the isolated nuclei; most of this label was internal rather than being peripherally located, suggesting that incorporation was not due to cytoplasmic contaminants adhering to the nuclei. (2) Comparison of incorporation by nuclei-rich and cell-rich fractions indicated that incorporation by contaminating whole cells could account for only a small portion of the incorporation observed in the nuclear preparations. (3) Addition of cytoplasmic fractions to the nuclear preparations did not augment incorporation. (4) Incorporation by isolated nuclei differed from protein synthesis in whole cells by virtue of its DNase sensitivity and requirement for sodium ions and from protein synthesis by microsomal systems by virtue of its DNase sensitivity, RNase resistance, sodium-ion requirement, failure to respond to added nucleoside triphosphates or supernatant factors, and sensitivity to inhibitors of oxidative phosphorylation. (5) The rate of incorporation by the isolated thymus nuclei was very high.

The results of other investigators who have studied protein synthesis in isolated thymus nuclei (Breitman and Webster, 1958; Herranen and Brunkhorst, 1962; Reid and Cole, 1964; Reid *et al.*, 1968; and Uete, 1966, 1967) have agreed quite well with those of the Rockefeller group. In view of this and of the extensive work which has been done with isolated thymus nuclei, it is difficult to avoid the conclusion that thymus nuclei do synthesize proteins and that the amount of protein produced in the nucleus must be a significant portion of the protein synthesized by the thymus cell.

In the thymocyte, the nucleus accounts for about 60% of the total cell volume. This is quite unusual; in most cells the nucleus amounts to no more than 20% of the cell volume (the figure for the hepatocyte, for example, is 7%). This has led some investigators to suggest that the metabolism of the thymocyte nucleus may also be unusual and that, in this cell, functions normally performed in the cytoplasm, such as protein synthesis, may have been taken over by the nucleus. The studies of Tsuzuki and Naora (1968) and Tsuzuki (1972) on protein synthesis in isolated rat spleen nuclei are interesting in this connection, since spleen, like thymus, contains a high proportion of cells with unusually large nuclei. Results with isolated spleen nuclei have generally been quite similar to those with isolated thymus nuclei (see Table I). The nuclei do not require added nucleoside triphosphates or supernatant factors for activity; incorporation is depressed by inhibitors of oxidative phosphorylation, by puromycin, and by DNase, but is relatively insensitive to RNase; and the amount of incorporation per unit weight of nuclei is high. Tsuzuki (1972) has, in a series of carefully controlled experiments, identified immunoglobulin G as one of the products of synthesis of the isolated spleen nuclei.

In summary, the following features appear to be characteristic of amino-acid incorporation by thymus and spleen nuclei: (1) The isolated nuclei contain all components necessary for protein synthesis; addition of nucleoside triphosphates, factors from the soluble portion of the cell, etc. does not stimulate incorporation. (2) Activity of the nuclei is labile and is lost upon freezing and thawing, exposure to concentrated sucrose solutions, and other equally mild treatments. (3) Incorporation is reduced in the presence of inhibitors of oxidative phosphorylation. (4) The ability to synthesize proteins is not destroyed by RNase as it is in microsomal systems. (5) The common inhibitors of microsomal protein synthesis also reduce incorporation by isolated nuclei. (6) Amino acids are incorporated in internal positions of the polypeptide chains of a variety of different nuclear proteins.

2. NUCLEI OF NONLYMPHOID TISSUES

Amino-acid incorporation by nuclei from nonlymphoid tissues has been studied by a number of investigators. The results of some of these studies are summarized in Table I. An inspection of this table reveals considerable diversity in the results obtained by different workers even when comparisons are restricted to studies on nuclei from a particular tissue. In the preceding section six features which are characteristic of amino-acid incorporation by isolated thymus and spleen nuclei were listed. Incorporation by nuclei from other tissues often displays most of these features. The correspondence is rarely complete, however, and nearly every investigator who has studied protein synthesis in the nuclei of nonlymphoid tissues has obtained results which differed in one or more important aspects from those reported for thymus and spleen nuclei. In view of the wide diversity of results which have been obtained for nuclei from nonlymphoid tissues, no attempt will be made to review these studies in detail. Instead, discussion will be limited to a few general observations.

As indicated in Table I, incorporation usually proceeds in the absence of added nucleoside triphosphates or factors from the soluble portion of the cell. With few exceptions, RNase, even in high concentrations, has little effect on activity. In these respects, the results are similar to those that have been obtained with thymus and spleen nuclei.

Results with inhibitors of protein synthesis have been variable (see Table I) and frequently difficult to understand. Zimmerman *et al.* (1969), for example, reported that incorporation of HeLa cell nuclei was sensitive to one common inhibitor of protein synthesis (puromycin), but not to another (cycloheximide). The explanation for such an observation is not apparent. The effects of various cations have also been quite variable. Rendi (1960), using liver nuclei, and Løvtrup-Rein (1970), using brain nuclei, reported a stimulation by sodium and an inhibition by potassium, whereas Fleischer-Lambropoulos and Reinsch (1971) reported that the activity of brain nuclei was stimulated slightly by either sodium or potassium. Uete (1967) and Fleischer-Lambropoulos and Reinsch (1971) reported for liver and brain nuclei, respectively, that incorporation was stimulated by magnesium, whereas Rendi (1960) reported no effect of this cation on liver nuclei. As discussed previously, thymus nuclei, too, have shown a variable response to cations.

Many investigators have employed centrifugation through dense sucrose and detergent washes to remove cytoplasmic contaminants from the nuclei used to study amino-acid incorporation. Nuclei have sometimes been quite active following these treatments. Allfrey *et al.* (1964), however, demonstrated that, for thymus nuclei, such treatments destroy the ability of the nucleus to synthesize proteins. This poses several questions: Does resistance to dense sucrose and detergents imply that the mechanism of protein synthesis is different from that in thymus nuclei? Do such treatments alter the properties of the protein-synthesizing system of the nucleus? Is the incorporation catalyzed by the purified nuclei an artifact remaining after the biological activity of the nucleus has been destroyed?

Most nuclei, like those from the thymus, incorporate amino acids into a rather large and heterogeneous group of proteins. In some instances, however, incorporation is predominantly into nucleolar proteins. Thus, Birnstiel and Hyde (1962) found that when pea nuclei were incubated with radioactive amino acids for short times, the nucleoli contained nearly all the label, and Zimmerman *et al.* (1969) found that in HeLa cell nuclei more than 95% of the bound radioactivity was associated with the nucleoli. The latter study should be contrasted with that of Gallwitz and Mueller (1969b), who determined autoradiographically that, upon incubation of HeLa cell nuclei with a radioactive amino acid, label was incorporated throughout the nucleus.

3. NUCLEOLI

As discussed earlier, nucleoli have been shown to contain activating enzymes and transfer RNA, and several groups of investigators have found the nucleolus to be the predominant site of labeling when whole cells or nuclei were incubated with radioactive amino acids. One of these groups, that of Zimmerman *et al.* (1969), reported that the isolated nucleoli were, themselves, capable of incorporating labeled amino acids. In this study, most of the label was associated with 78S ribosome-like particles, particularly at short labeling times. Amino-acid incorporation by isolated nucleoli has also been described for pea nucleoli (Birnstiel and Hyde, 1963), frog egg nucleoli (Brandt and Finamore, 1963), ascites cell nucleoli (Izawa and Kawashima, 1969), and for Novikoff hepatoma cell nucleoli (Lamkin and Hurlbert, 1972). Results of these investigators were similar in that incorporation was not dependent on addition of soluble components of the cell or ATP and that incorporation was not inhibited by RNase. Both Birnstiel and Hyde (1963) and Lamkin and Hurlbert (1972) reported that incorporation was sensitive to puromycin; Izawa and Kawashima (1969) found incorporation to be insensitive to cycloheximide. Lamkin and Hurlbert made the interesting observations that tyrosine was incorporated about 400 times as well as valine and that label from cytoplasmic aminoacyl–tRNA was not incorporated.

4. CHROMATIN

Wang (1965, 1967), Wang and Patel (1967), Hu and Wang (1971), and Sekeris *et al.* (1966a, b) have reported that chromatin, or fractions derived from chromatin, incorporate labeled amino acids. The chromatin preparations employed may well have been contaminated with components of the nuclear membrane and nucleoli, a fact appreciated by the investigators. The reaction was not dependent on the addition of soluble components of the cytoplasm and proceeded in the absence of an added energy source, although Wang (1967) reported that incorporation was twice as great in the presence of an ATP-regenerating system. Incorporation was insensitive to RNase and, as shown by Sekeris *et al.* (1966a), responded differently to inhibitors of protein synthesis than does the cytoplasmic ribosomal system.

The investigation of Wang (1967) has provided interesting insights into the nature of chromatin-catalyzed, amino-acid incorporation. Wang prepared four protein fractions from rat liver chromatin. By several criteria, these fractions were different from one another. Every one of these fractions incorporated radioactive amino acids, although none of them contained much nucleic acid (the fraction with the highest nucleic acid content had 3.6% RNA and 0.69% DNA).

5. DIFFICULTIES IN INTERPRETATION OF RESULTS

Interpretation of the results which have been obtained with isolated nuclei and nuclear subfractions presents a number of difficulties. Some of these are as follows:

1. Synthesis of a protein from its constituent amino acids is a highly endergonic process, yet amino-acid incorporation by isolated nuclei and nuclear subfractions usually proceeds in the absence of an added energy source. This presents no difficulty in the case of thymus nuclei, which are known to have an endogenous nucleotide pool and a mechanism for oxidative phosphorylation; it is less easily explained for some other nuclei which do not have a significant nucleotide pool or a mechanism for regenerating nucleoside triphosphates; and is still more difficult to explain in the case of purified nuclear subfractions.

2. Although protein synthesis in cytoplasmic systems requires a considerable variety of soluble protein and RNA molecules, incorporation by isolated nuclei usually proceeds in the absence of added supernatant factors. This can be explained in the case of thymus nuclei isolated in isotonic sucrose solutions, for such nuclei are known to retain soluble proteins. It is well established that many other types of nuclei lose considerable amounts of soluble protein during isolation, particularly if the isolation procedure has involved detergent washes. The observation that amino-acid incorporation by these nuclei is independent of added supernatant factors is somewhat puzzling. Finally, some of the nuclear subfractions which have been shown to incorporate labeled amino acids [e.g., those studied by Wang (1967)] must lack many, if not all, of the soluble protein and RNA components involved in protein synthesis in cytoplasmic systems. Any protein synthesis catalyzed by these fractions must take place by a mechanism quite different and considerably less complex than that operating in the cytoplasm.

3. Protein synthesis by cytoplasmic systems involves messenger RNA, which is highly sensitive to pancreatic RNase, yet amino-acid incorporation by isolated nuclei and nuclear subfractions is usually unaffected by RNase, even at high concentrations. This cannot be due to an inability of the RNase to enter the nuclei or to function once inside, since intranuclear RNA species have frequently proven susceptible to attack by exogenous RNase. It is possible, of course, that intranuclear messenger RNA is inaccessible to RNase. Presumptive messenger RNA from the nucleus has frequently been found in association with nonribosomal proteins,

and it is possible that these have some protective function. It is interesting that Ling and Dixon (1970) have reported that the small cytoplasmic polysomes responsible for protamine biosynthesis are resistant to RNase. The nuclear subfractions studied by Wang (1967) contain very little RNA, yet still actively incorporate amino acids. In this case, it seems probable that RNA is not involved in the amino-acid incorporation. If this incorporation represents true protein synthesis, the synthetic mechanism must be quite different from that in the cytoplasm.

4. It has been observed frequently that the rate of incorporation of amino acids by isolated nuclei or nuclear subfractions varies greatly from one amino acid to another. This observation can be readily explained for thymus nuclei by assuming that the isolated nuclei have an endogenous pool of amino acids and that the concentrations of the different amino acids in this pool vary widely. The observation is more difficult to explain for some other kinds of nuclei which are believed to lose most of their soluble components during isolation and is quite inexplicable for isolated nuclear subfractions, which would not be expected to have significant amino acid pools. It has been mentioned previously that Lamkin and Hurlbert (1972) found that tyrosine was incorporated about 400 times as well as valine in isolated nucleoli, yet no nucleolar protein has been found to contain 400 times as much of the one amino acid as the other, nor is there reason to believe that isolated nucleoli contain a significant pool of free amino acids.

5. The capacity of such a wide variety of nuclear preparations and nuclear subfractions to actively catalyze amino acid incorporation is, itself difficult to explain. Nuclei can be sedimented through sucrose or Ficoll, washed with various detergents, or extracted with various salt solutions, and usually they still retain the ability to incorporate amino acids; they may be treated to yield nucleolar or chromatin fractions and the chromatin may be further separated into various protein components and each of the resulting nuclear subfractions actively promotes incorporation of labeled amino acids. This is certainly not the type of behavior one would predict for a system carrying out *de novo* synthesis of discrete proteins.

If one considers all the results that have been obtained with isolated nuclei and nuclear subfractions, two conclusions are almost inescapable: (1) Not all investigators have been studying the same phenomenon. The wide diversity of results which have been obtained, sometimes with the same experimental material, make this conclusion difficult to avoid. (2) Some investigators have not been studying nuclear protein synthesis at all. There are several ways by which incorporation by isolated nuclei might be explained without invoking nuclear protein synthesis.

One possibility is that the observed incorporation is due to nonspecific adsorption of the labeled amino acid to an existing macromolecule. An early but still very timely study by Brunish and Luck (1952) illustrates the potential severity of this problem. These investigators incubated purified deoxynucleoprotein or histone with various radioactive amino acids and determined the amount of radioactivity converted to a trichloroacetic acid-insoluble form during the incubation. A time- and temperature-dependent incorporation of radioactivity was observed for both the nucleoprotein and histone fractions. For the histone fraction, incorporation was pH dependent, the optimum being about pH 8.8. The incorporated radioactivity was not removed by repeated washings nor by dialysis against unlabeled amino acid. Only about 30% of the activity was removed from the histone fraction when it was dissolved in 0.5 *M* sodium hydroxide and reprecipitated with trichloroacetic acid. The amino-acid incorporation observed by Brunish and Luck might easily have passed for *in vitro* protein synthesis were it not for their observation that the extent of incorporation was considerably greater at 100° than at 37°C! The conclusion of the authors bears repeating: "It is difficult to believe that a system consisting of a protein and an amino acid heated at 100° in water represents protein synthesis as it occurs *in vivo*. Yet the tests which have been employed in more elaborate systems to demonstrate 'protein synthesis' have been carried out, and, if this were a more elaborate system incubated at 37°, one would be inclined to say that 'protein synthesis' had indeed been achieved."

Another possibility is that the observed incorporation may be the result of enzymatic addition of labeled amino acids to existing proteins. Enzymes which catalyze such reactions have been described by Hird *et al.* (1964), Kaji (1968), and Soffer (1968). The results of Monesi (1964) discussed on pp. 353–354 suggest that nuclei may catalyze such a reaction *in vivo*.

Finally it is possible that contaminating whole cells, microorganisms, or cytoplasmic ribosomes are responsible for the observed incorporation. As has been discussed previously, nuclei isolated by centrifugation from sucrose solutions are commonly contaminated with cytoplasmic ribosomes, attached either to the outer nuclear membrane or to fragments of the endoplasmic reticulum adhering to the nuclear envelope.

D. *Evidence against Intranuclear Protein Synthesis*

It has occasionally been suggested that the previously cited studies of Robbins and Borun (1967), Ling *et al.* (1969), and others, which demonstrate the synthesis of nuclear proteins on cytoplasmic ribosomes, indicate that the nucleus is not a site of protein synthesis. They, of course, do not so indicate, nor do they eliminate the possibility that nuclear proteins whose synthesis on cytoplasmic ribosomes has been demonstrated might also be made in the nucleus.

Penman (1966) demonstrated that 16S ribosomal RNA was absent from isolated HeLa cell nuclei. This implies that the small ribosomal subunit is also absent from the nuclei and that they are, therefore, unable to synthesize proteins, at least by the usual mechanism of protein synthesis. It should be noted, however, that isolation of the nuclei used by Penman involved a detergent wash and that, as a consequence, some nuclear components were probably lost during isolation. Similar findings were reported for amphibian oocyte nuclei by Rogers (1968). In these experiments, however, the nuclei were isolated by micro-manipulation, thus considerably reducing the danger that nuclear components were lost during isolation.

IV. Summary and Conclusions

It has been shown that in some cells nuclear proteins are produced in the cytoplasm. In those cells which have been studied, the evidence favors the view that the proportion of nuclear protein made outside the nucleus is substantial. There is also rather convincing evidence that in thymocytes, which are unusual in that the nucleus comprises a very large proportion of the total cell volume, the nucleus is a site of protein synthesis. In this case the evidence suggests that a considerable proportion of the nuclear protein is made in the nucleus. Although the nuclei of other cell types may also synthesize proteins, the evidence for this is contradictory and not compelling. For other cell types, therefore, the role of the nucleus in the synthesis of proteins is unclear. It seems possible that in all cells some nuclear proteins are synthesized in the cytoplasm and some in the nucleus, but that the proportion of nuclear protein made at the two loci varies greatly from one cell type to another.

Although a few of the questions which might be asked about nuclear protein synthesis have been answered, at least in a tentative and incomplete fashion, most remain as a subject for future research. Among those questions that remain are the following: (1) Are nuclei generally able to synthesize proteins or is this ability limited to a few unusual types such as those of the thymus? (2) Which proteins do nuclei make? (3) How much protein do they make? (4) Does protein synthesis in the nucleus proceed by the same mechanism as that in the cytoplasm? If not, what is the nature of the nuclear mechanism? (5) What is the physiological significance of nuclear protein synthesis? (6) Are nuclear proteins which are made in the cytoplasm synthesized on a particular class of ribosomes?

If so, which class? (7) How do proteins move from a cytoplasmic site of synthesis to their ultimate location in the nucleus?

NOTE ADDED IN PROOF

Since the original manuscript for this chapter was written, a review of protein synthesis in isolated nuclei (Anderson, 1972) and several research papers deserving of mention have appeared.

Additional information on the rate of migration of proteins between the nucleus and cytoplasm (discussed on pp. 351–352) has been provided by Paine and Feldherr (1972). These investigators injected proteins of different molecular weights into the cytoplasm of cockroach oocytes and followed the migration of the injected molecules into the nucleus. Proteins of low molecular weight (<20,000) moved rapidly into the nucleus. Those of moderate or high molecular weight (>40,000) moved into the nucleus slowly and even after 5 hr the nuclear concentrations of these proteins were far below the cytoplasmic levels.

An explanation for the observation that during the late stages of spermatid development arginine, but not other amino acids, is incorporated into mouse spermatid nuclei (Monesi, 1964; discussed on pp. 353–354), has been provided by the finding of Kistler *et al.* (1973) that rat testes and epididymal sperm contain protamine-like proteins with exceptionally high contents of arginine and low contents of other amino acids.

A number of investigations on protein synthesis in isolated nuclei have appeared recently. Unfortunately, most of these have done little to clear the confusion which exists in this field, although a study by Dravid and Wong (1972) does suggest that some of the results obtained with isolated nuclei may have been due to contaminating microorganisms. Burdman (1972) reported that cycloheximide and puromycin strongly depressed amino acid incorporation by isolated rat brain nuclei. Dravid and Wong (1972), on the other hand, found that incorporation by rat brain nuclei was unaffected by cycloheximide and only slightly inhibited by puromycin. The procedure for isolating the nuclei and the incubation conditions employed to obtain amino acid incorporation were quite similar in the two studies. Haglid (1972) made the interesting observation that astrocytoma nuclei were about 65 times as active as normal brain nuclei in incorporating labeled amino acids into protein. Of the many types of nuclei which have been investigated, only the highly polyploid nuclei of *Drosophila* salivary glands (Helmsing, 1971) have shown a comparably high rate of amino acid incorporation. The unusually high activity of astrocytoma nuclei cannot be considered characteristic of nuclei from malignant tissues, since nuclei from glioblastoma multiforme incorporated labeled amino acids only moderately more rapidly than those from normal brain and since, as shown by various investigators, HeLa cell nuclei are not unusually active in amino acid incorporation (see Table I). As discussed on p. 365, several investigators have reported that incorporation of labeled amino acids by isolated nuclei has been predominantly into nucleolar proteins. Another such instance has been described by Laval and Bouteille (1973). These investigators determined by autoradiography that there was about three times as much incorporation into the nucleoli as into the rest of the nucleus after incubation of Triton X-100 washed rat liver nuclei with radioactive amino acids.

References to Note Added in Proof

Anderson, K. M. (1972). *Int. J. Biochem.* **3,** 449.
Burdman, J. A. (1972). *Brain Res.* **41,** 413.
Dravid, A. R., and Wong, E. (1972). *J. Neurochem.* **19,** 2709.
Haglid, K. G. (1972). *J. Neurochem.* **19,** 19.
Kistler, W. S., Geroch, M. E., and Williams-Ashman, H. G. (1973). *J. Biol. Chem.* **248,** 4532.
Laval, M., and Bouteille, M. (1973). *Exp. Cell Res.* **79,** 391.
Paine, P. L., and Feldherr, C. M. (1972). *Exp. Cell Res.* **74,** 81.

REFERENCES

Allfrey, V. G. (1954). *Proc. Nat. Acad. Sci. U.S.* **40,** 881.
Allfrey, V. G., and Mirsky, A. E. (1958). *Proc. Nat. Acad. Sci. U.S.* **44,** 981.
Allfrey, V. G., Mirsky, A. E., and Osawa, S. (1955). *Nature (London)* **176,** 1042.
Allfrey, V. G., Mirsky, A. E., and Osawa, S. (1957). *J. Gen. Physiol.* **40,** 451.
Allfrey, V. G., Meudt, R., Hopkins, J. W., and Mirsky, A. E. (1961). *Proc. Nat. Acad. Sci. U.S.* **47,** 907.
Allfrey, V. G., Littau, V. C., and Mirsky, A. E. (1964). *J. Cell Biol.* **21,** 213.
Anderson, K. M., Crosthwait, H. C., and Slavik, M. (1971). *Exp. Cell Res.* **66,** 273.
Anderson, K. M., Slavik, M., and Elebute, O. P. (1972). *Can. J. Biochem.* **50,** 190.
Arnold, E. A., Young, D. E., and Stowell, R. E. (1968). *Exp. Cell Res.* **52,** 1.
Ashley, C. A., and Peters, T. (1969). *J. Cell Biol.* **43,** 237.
Bach, M. K., and Johnson, H. G. (1966). *Nature (London)* **209,** 893.
Bach, M. K., and Johnson, H. G. (1967). *Biochemistry* **6,** 1916.
Barton, A. D., Kisieleski, W. E., Wassermann, F., and Mackevicius, F. (1971). *Z. Zellforsch. Mikrosk. Anat.* **115,** 299.
Beattie, D. S., Basford, R. E., and Koritz, S. B. (1967). *Biochemistry* **6,** 3099.
Bernhard, W. (1966). *In Int. Symp. Nucleolus—Its Struct. Function.* Nat. Cancer Inst. Monogr. 23, pp. 13–38.
Birnstiel, M. L., and Hyde, B. B. (1962). *Fed. Proc. Fed Amer. Soc. Exp. Biol.* **21,** 156.
Birnstiel, M. L., and Hyde, B. B. (1963). *J. Cell Biol.* **18,** 41.

Birnstiel, M. L., Chipchase, M. I. H., and Hayes, R. J. (1962). *Biochim. Biophys. Acta* **55**, 728.

Blobel, G., and Potter, V. R. (1966). *Science* **154**, 1662.

Bloch, D. P., and Brack, S. D. (1964). *J. Cell Biol.* **22**, 327.

Borun, T. W., Scharff, M. D., and Robbins, E. (1967). *Proc. Nat. Acad. Sci. U.S.* **58**, 1977.

Brandt, E. E., and Finamore, F. J. (1963). *Biochim. Biophys. Acta* **68**, 618.

Breitman, T. R., and Webster, G. C. (1958). *Biochim. Biophys. Acta* **27**, 408.

Brunish, R., and Luck, J. M. (1952). *J. Biol. Chem.* **197**, 869.

Burdman, J. A., and Journey, L. I. (1969). *J. Neurochem.* **16**, 493.

Busch, H., and Smetana, K. (1970). "The Nucleolus," Chapter 2. Academic Press, New York.

Byers, T. J., Platt, D. B., and Goldstein, L. (1963). *J. Cell Biol.* **19**, 467.

Caro, L. G., and Palade, G. E. (1964). *J. Cell Biol.* **20**, 473.

Chouinard, L. A., and LeBlond, C. P. (1967). *J. Cell Sci.* **2**, 473.

Conover, T. E. (1970a). *Arch. Biochem. Biophys.* **136**, 541.

Conover, T. E. (1970b). *Arch. Biochem. Biophys.* **136**, 551.

Feldherr, C. M. (1968). *J. Cell Biol.* **39**, 49.

Feldherr, C. M. (1969). *J. Cell Biol.* **42**, 841.

Feldherr, C. M. (1970). *J. Cell Biol.* **47**, 60a.

Feldherr, C. M., and Feldherr, A. B. (1960). *Nature (London)* **185**, 250.

Flamm, W. G., Birnstiel, M. L., and Filner, P. (1963). *Biochim. Biophys. Acta* **76**, 110.

Fleischer-Lambropoulos, H., and Reinsch, I. (1971). *Hoppe-Seyler's Z. Physiol. Chem.* **352**, 593.

Frenster, J. H., Allfrey, V. G., and Mirsky, A. E. (1960). *Proc. Nat. Acad. Sci. U.S* **46**, 432.

Frenster, J. H., Allfrey, V. G., and Mirsky, A. E. (1961). *Biochim. Biophys. Acta* **47**, 130.

Gallwitz, D., and Mueller, G. C. (1969a). *Science* **163**, 1351.

Gallwitz, D., and Mueller, G. C. (1969b). *Eur. J. Biochem.* **9**, 431.

Ganoza, M. C., and Williams, C. A. (1969). *Proc. Nat. Acad. Sci. U.S.* **63**, 1370.

Georgiev, G. P., Samarina, O. P., Mantieva, V. L., and Zbarsky, I. B. (1961). *Biochim. Biophys. Acta* **46**, 399.

Goldstein, L. (1970). *Advan. Cell Biol.* **1**, 187.

Gorovsky, M. A. (1969). *J. Cell Biol.* **43**, 46a.

Hamilton, T. H., Teng, C.-S., and Means, A. R. (1968). *Proc. Nat. Acad. Sci. U.S.* **59**, 1265.

Hammel, C. L., and Bessman, S. P. (1964). *J. Biol. Chem.* **239**, 2228.

Helmsing, P. J. (1971). *Biochim. Biophys. Acta* **232**, 733.

Herranen, A., and Brunkhorst, W. (1962). *Biochim. Biophys. Acta* **65**, 523.

Hird, H. G., McLean, E. J. T., and Munro, H. N. (1964). *Biochim. Biophys. Acta* **87**, 219.

Hnilica, L. S., Liau, M. C., and Hurlbert, R. B. (1966). *Science* **152**, 521.

Hoagland, M. B., Scornik, O. A., and Pfefferkorn, L. C. (1964). *Proc. Nat. Acad. Sci. U.S.* **51**, 1184.

Hopkins, J. W. (1959). *Proc. Nat. Acad. Sci. U.S.* **45**, 1461.

Hopkins, J. W., Allfrey, V. G., and Mirsky, A. E. (1961). *Biochim. Biophys. Acta* **47**, 194.

Hu, A. L., and Wang, T. Y. (1971). *Arch. Biochem. Biophys.* **144**, 549.

Izawa, M., and Kawashima, K. (1969). *Biochim. Biophys. Acta* **190**, 139.

Jamieson, J. D., and Palade, G. E. (1968). *J. Cell Biol.* **39**, 589.
Kabat, D. (1968). *J. Biol. Chem.* **243**, 2597.
Kaji, H. (1968). *Biochemistry* **7**, 3844.
Karjalainen, E. (1966). *Acta Chem. Scand.* **20**, 586.
Kedes, L. H., Gross, P. R., Cognetti, G., and Hunter, A. L. (1969). *J. Mol. Biol.* **45**, 337.
Kuehl, L. (1967). *J. Biol. Chem.* **242**, 2199.
Kuehl, L. (1972). Unpublished observations.
Kuehl, L., and Sumsion, E. N. (1971). *J. Cell Biol.* **50**, 1.
Lamkin, A. F., and Hurlbert, R. B. (1972). *Biochim. Biophys. Acta* **272**, 321.
Ling, V., and Dixon, G. H. (1970). *J. Biol. Chem.* **245**, 3035.
Ling, V., Trevithick, J. R., and Dixon, G. H. (1969). *Can. J. Biochem.* **47**, 51.
Lizardi, P. M., and Luck, D. G. L. (1972). *J. Cell Biol.* **54**, 56.
Loewenstein, W. R., Kanno, Y., and Ito, S. (1966). *Ann. N. Y. Acad. Sci.* **137**, 708.
Løvtrup-Rein, H. (1970). *Brain Res.* **19**, 433.
Løvtrup-Rein, H. (1972). *Exp. Cell Res.* **72**, 188.
McCarty, K. S., Parsons, J. T., Carter, W. A., and Laszlo, J. (1966). *J. Biol. Chem.* **241**, 5489.
McEwen, B. S., Allfrey, V. G., and Mirsky, A. E. (1963). *J. Biol. Chem.* **238**, 758.
Monesi, V. (1964). *Exp. Cell Res.* **36**, 683.
Monneron, A., and Bernhard, W. (1969). *J. Ultrastruct. Res.* **27**, 266.
Nakamura, T., Prestayko, A. W., and Busch, H. (1968). *J. Biol. Chem.* **243**, 1368.
Ono, H., and Terayama, H. (1968). *Biochim. Biophys. Acta* **166**, 175.
Penman, S. (1966). *J. Mol. Biol.* **17**, 117.
Pogo, A. O., Pogo, B. G. T., Littau, V. C., Allfrey, V. G., Mirsky, A. E., and Hamilton, M. G. (1962). *Biochim. Biophys. Acta* **55**, 849.
Redman, C. M. (1969). *J. Biol. Chem.* **244**, 4308.
Redman, C. M., and Sabatini, D. D. (1966). *Proc. Nat. Acad. Sci. U.S.* **56**, 608.
Rees, K. R., Ross, H. F., and Rowland, G. F. (1962). *Biochem. J.* **83**, 523.
Reid, B. R., and Cole, R. D. (1964). *Proc. Nat. Acad. Sci. U.S.* **51**, 1044.
Reid, B. R., Stellwagen, R. H., and Cole, R. D. (1968). *Biochim. Biophys. Acta* **155**, 593.
Rendi, R. (1960). *Exp. Cell Res.* **19**, 489.
Ritter, P. O., and Busch, H. (1971). *Physiol. Chem. Phys.* **3**, 411.
Robbins, E., and Borun, T. W. (1967). *Proc. Nat. Acad. Sci. U.S.* **57**, 409.
Rogers, M. E. (1968). *J. Cell Biol.* **36**, 421.
Sachs, H. (1957). *J. Biol. Chem.* **228**, 23.
Sadowski, P. D., and Howden, J. A. (1968). *J. Cell Biol.* **37**, 163.
Sadowski, P. D., and Steiner, J. W. (1968). *J. Cell Biol.* **37**, 147.
Schultze, B., and Maurer, W. (1967). *In* "The Control of Nuclear Activity" (L. Goldstein, ed.), pp. 319–349. Prentice-Hall, Englewood Cliffs, New Jersey.
Sekeris, C. E., Schmid, W., Gallwitz, D., and Lukacs, I. (1966a). *Life Sci.* **5**, 969.
Sekeris, C. E., Schmid, W., Gallwitz, D., and Lukacs, I. (1966b). *Angew. Chem. Int. Ed. Engl.* **5**, 592.
Sherr, C. J., and Uhr, J. W. (1970). *Proc. Nat. Acad. Sci. U.S.* **66**, 1183.
Siekevitz, P., and Palade, G. E. (1960). *J. Biophys. Biochem. Cytol.* **7**, 619.
Simard, R. (1970). *Int. Rev. Cytol.* **28**, 169.
Soffer, R. L. (1968). *Biochim. Biophys. Acta* **155**, 228.
Speer, H. L., and Zimmerman, E. F. (1968). *Biochem. Biophys. Res. Commun.* **32**, 60.
Stein, G., and Baserga, R. (1971). *Biochem. Biophys. Res. Commun.* **44**, 218.

Tsuzuki, J. (1972). *Exp. Cell Res.* **72,** 453.
Tsuzuki, J., and Naora, H. (1968). *Biochim. Biophys. Acta* **169,** 550.
Uete, T. (1966). *Biochim. Biophys. Acta* **121,** 395.
Uete, T. (1967). *J. Biochem. (Tokyo)* **61,** 251.
Wang, T.-Y. (1961). *Biochim. Biophys. Acta* **51,** 180.
Wang, T.-Y. (1964). *Biochim. Biophys. Acta* **87,** 141.
Wang, T.-Y. (1965). *Proc. Nat. Acad. Sci. U.S.* **54,** 800.
Wang, T.-Y. (1967). *J. Biol. Chem.* **242,** 1220.
Wang, T.-Y., and Patel, G. (1967). *Life Sci.* **6,** 413.
Warner, J. R. (1971). *J. Biol. Chem.* **246,** 447.
Watson, M. L. (1962). *J. Cell Biol.* **13,** 162.
Weser, V. U., and Koolman, J. (1969). *Hoppe-Seyler's Z. Physiol. Chem.* **350,** 1273.
Wu, R. S., and Warner, J. R. (1971). *J. Cell Biol.* **51,** 643.
Zagury, D., Uhr, J. W., Jamieson, J. D., and Palade, G. E. (1970). *J. Cell Biol.* **46,** 52.
Zalokar, M. (1960). *Exp. Cell Res.* **19,** 184.
Zetterberg, A. (1966a). *Exp. Cell Res.* **43,** 526.
Zetterberg, A. (1966b). *Exp. Cell Res.* **42,** 500.
Zimmerman, E. F., Hackney, J., Nelson, P., and Arias, I. M. (1969). *Biochemistry* **8,** 2636.

PART III

Special Aspects of Nuclear Function

10

Effects of Female Steroid Hormones on Target Cell Nuclei

Bert W. O'Malley and Anthony R. Means

I. Introduction

In the course of experiments designed to determine the sequence of events responsible for steroid hormone effects in endocrine target cells, investigators have been repeatedly led to consider the nucleus as the primary site of action. Numerous experiments have supported the suggestion that steroid hormones regulate cell function by influencing target tissue protein synthesis (Gorski *et al.*, 1968; O'Malley *et al.*, 1969; Mueller, 1965; Hamilton, 1968; Segal and Scher, 1967; Williams-Ashman and Reddi, 1971). In most instances steroid hormone stimulation of target tissue protein synthesis is preceded by quantitative and often qualitative changes in cell RNA synthesis. The most obvious effect on nonmalignant cell RNA metabolism is a stimulation of nuclear, rapidly labeled heterogeneous RNA followed by increased production of ribosomal RNA and often transfer RNA (Greenman *et al.*, 1965; Hamilton, 1968; O'Malley *et al.*, 1968, 1969). It is probable but unproved that animal cell messenger RNA is a component of the nuclear, rapidly labeled heterogeneous RNA. Additional support for a primary effect of steroids on nuclear gene transcription is provided by the ability of actinomycin D to block most steroid hormone-mediated cell responses. General hypotheses involving a simple stimulation of messenger RNA must not be oversimplified because certain evidence is not entirely compatible with such a theory (Tomkins *et al.*, 1969). However, recent experiments have rather conclusively demonstrated that steroid hormones are capable of inducing a net increase in specific mRNA molecules in target cells (Chan *et al.*, 1973; Means *et al.*, 1972; Rosenfeld *et al.*, 1972a; O'Malley *et al.*, 1973).

If steroids do in fact regulate nuclear gene transcription, certain considerations should be compatible with such a theory. There must be a mechanism for limiting the steroid-induced response to target tissues. There should be a defined sequence of events which eventuates in the transport of a steroid molecule to its presumed nuclear site of action following penetration of the target cell membrane. The existence of mediators or "second messengers" must be delineated. The steroid hormone itself or an intracellular mediator should be capable of interacting at certain predetermined sites in the nucleus prior to alterations in DNA transcription. Changes in nuclear RNA synthesis should finally result in a net increase in the levels of specific mRNA's, which are in turn limited to steroid hormone target tissue and inducible by only the steroid in question. Increases in the intracellular levels of these mRNA molecules should precede fluctuations in the rate of synthesis of the corresponding specific proteins. Over the past decade a considerable body of experimental data relating to these theoretical considerations has accumulated. The remainder of this chapter will summarize present evidence favoring our prejudice that the target cell nucleus is a major determinant in steroid hormone induction of new cell functions. Discussions will be limited primarily to effects of estrogen in the rat uterus and progesterone in the chick oviduct since the bulk of the experimental data on mechanisms of female sex steroid action emanate from these two-model systems. However, the generality of these observations as applied to mechanisms of action of all steroid hormones has been recently reviewed (Means and O'Malley, 1972; Jensen and DeSombre, 1972; Edelman and Fanestil, 1970; Liao and Fang, 1969; Tomkins *et al.*, 1966).

II. Steroid Hormone Receptors

A. *Estrogen*

The concept of steroid hormone "receptors" initially resulted from studies which involved injections of physiological amounts of radioactive estradiol to immature rats (Jensen and Jacobson, 1962). It was noted that target tissue alone was capable of retaining [^{3}H]estradiol-17β against a marked concentration gradient with blood. These observations were confirmed by both biochemical and autoradiographic methods (Noteboom and Gorski, 1965; Jensen *et al.*, 1969), and eventually extended to numerous vertebrate species (Jensen and DeSombre, 1972). Upon homogenization of uterine target tissue, a soluble protein capable of binding [^{3}H]-estradiol was shown to exist in the cytoplasm (105,000 *g* supernatant fraction) (Toft and Gorski, 1966). This binding protein was considered to be a "receptor" because of its limitation to estrogen target tissue, its high binding affinity for estrogens ($K_d = 10^{-10}$), and its specific attraction for only biologically active estrogens, either naturally occurring or synthetic. Upon ultracentrifugation in a sucrose gradient, this cytoplasmic hormone–protein complex sedimented as a discrete band at about 8 S (Toft and Gorski, 1966; Rochefort and Baulieu, 1968). Addition of 0.3 *M* KCl to these gradients resulted in the reversible transformation of this 8 S complex to a more slowly sedimenting 4 S form (Korenman and Rao, 1968). Thus the cytoplasmic receptor appeared to contain a 4 S estradiol-binding unit which under conditions of low ionic strength underwent association with other binding entities to form an 8 S complex. Recent experiments have indicated that, in the presence of ionic conditions in the presumed physiological range, an intermediate 6 S form may predominate (Baulieu *et al.*, 1971). In any event, it appears that the sedimentation behavior can vary in relation to concentration and ionic conditions and, although probably artifactual, it is a useful and reliable method for identifying steroid-binding proteins (Stancel *et al.*, 1973). At present there is no way to determine the exact size or configuration of a receptor as it exists in the unbroken cell.

B. Progesterone

The chick oviduct, a specific target tissue for progesterone, exemplifies a similar type of interaction with progestational steroids. After an *in vivo* injection of [^{3}H]progesterone to an estrogen-pretreated chick, the major fraction of labeled steroid was detected in the cytoplasm and nucleus of the oviduct (O'Malley *et al.*, 1971). Within this tissue, the cytoplasmic radioactivity appeared to exist bound to a macromolecular complex that did not dissociate on passage of the cytosol (105,000 *g* supernatant extract) through a Sephadex G-200 column. As shown in Fig. 1 the cytosol–progesterone-binding macromolecule had a sedimentation coefficient of 3.8 during sucrose gradient centrifugation in the presence of 0.3 *M* KCl and aggregated to 5 S and 8 S under low salt (no KCl) conditions (Sherman *et al.*, 1970). The oviduct-binding component showed a striking affinity for progesterone ($K_d = 8 \times 10^{-10}$ at 4°C) and appeared to comprise only 0.02% of the cytosol protein. The cytosol–progesterone-binding macromolecule was unequivocally distinguished from plasma transcortin by agarose gel chromatography, discontinuous polyacrylamide electrophoresis, isoelectric gradient chromatography and protamine sulfate precipitation. Incubation of preparations of cytosol with various en-

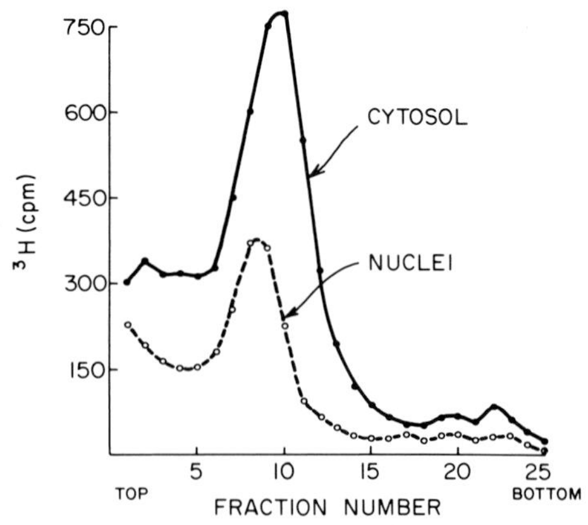

Fig. 1 Sucrose gradient centrifugation analysis of progesterone-binding components of chick oviduct cytoplasm (cytosol) and nuclei. Animals were sacrificed and oviduct preparations made at 8 min following *in vivo* injection of the steroid (90 μCi [^{3}H]progesterone). (From O'Malley *et al.*, 1970.)

zymes revealed the molecular binding to be destroyed by proteolytic enzymes but no RNase or DNase. This result suggested that at least the active molecular-binding site may be protein in nature. Further experiments showed the receptor to be a heat-sensitive, nondialyzable, and ammonium sulfate-precipitable protein with an isoelectric point of pH 4.0–4.5. Indirect physicochemical calculations suggested that the molecule existed in the shape of a prolate ellipsoid with a monomeric molecular weight of approximately 90,000 daltons. The calculated axial ratio revealed the protein to exist as a macromolecule with a length 14–18 times greater than its width (Sherman *et al.*, 1970). This explained an apparent discrepancy of molecular weight upon analysis by different techniques.

The cytosol-binding protein from oviduct showed very little affinity for estrogens (estradiol, estrone), mineralocorticoids (aldosterone), glucocorticoids (cortisol), or progesterone precursors and inactive metabolites. The tissue concentration of progesterone-binding protein was increased tenfold by prior estrogen treatment (O'Malley *et al.*, 1971). This estrogen-mediated increase in progesterone-binding protein correlated quite closely with the estrogen-induced quantitative enhancement of the oviduct progesterone response (avidin synthesis). The progesterone-binding protein was only found in target tissue (i.e., oviduct). For these reasons, it was felt that the combined evidence was compatible with the concept that this progesterone-binding molecule was in fact a physiological receptor for the hormone and that the formation of this steroid hormone–receptor complex is an obligatory initial step in steroid hormone action.

III. Transfer of the Hormone–Receptor Complex to the Nucleus

A. *Estrogen*

After exposure of uterine tissue to radioactive 17β-estradiol, two intracellular sites of hormone binding were noted: one in the cytoplasm and a second in the nucleus (Noteboom and Gorski, 1965). At later time points, nuclear binding appeared to predominate. The pioneering work of Jensen and associates (1968) and Gorski *et al.* (1968) led to the concept that, following an estrogen-induced conformational change in the uterine cytoplasmic–receptor protein, a translocation of this complex to the nucleus occurs. This translocation process appeared as an attractive hypothesis since it placed the hormone–receptor complex in a compartmental location adjacent to the site of hormone-induced changes in gene expression.

The demonstration of a temperature-dependent (37°C) intracellular transfer of protein-bound estradiol from the cytosol to the nucleus was initially accomplished in the rat uterus (Jensen *et al.*, 1968). Incubation of

[^{3}H]estradiol-17β with uterine tissue *in vitro* led to accumulation of a salt-extractable (0.3 *M* KCl) form of the estrogen–receptor complex from a preparation of nuclei. This nuclear hormone–receptor complex sedimented at 5 S during sucrose gradient centrifugation and was undetectable in nuclei of target tissue not previously exposed to estrogen. As nuclear 5 S receptor appeared during exposure to hormone, an apparent concomitant depletion in the total quantity of cytoplasmic 8 S receptor occurred. Cell-free exposure of preparations of nuclei to [^{3}H]estradiol and cytoplasmic receptor led to accumulation of extractable 5 S hormone–receptor complex from the nuclei, but no extractable 5 S complex was noted when [^{3}H]estradiol alone was incubated with uterine nuclei. These observations led to the "two-step" hypothesis of Jensen (Jensen *et al.*, 1968) who suggested that the 5 S estradiol–receptor complex extracted from nuclei may simply represent an altered form of the cytoplasmic receptor. Recent evidence has suggested that, upon interacting with estrogen, the binding protein undergoes a physical change permitting it to relocate to a nuclear state (Gorski *et al.*, 1973). A new equilibrium is then achieved in which up to 90% of the estrogen bound to receptor is in the nuclear state. Because of this equilibrium, correlations between tissue response and estrogen binding must take into account that both states of the estrogen–receptor complex will be equally correlated with response after equilibrium is reached.

Most of the conclusions discussed above have arisen from experiments with uterine tissue *in vitro* or under cell-free conditions. An important recent methodological advance has permitted a better understanding of estrogen–receptor translocation to the nuclear compartment under *in vivo* conditions and thus permitted correlations of this response with the biological responses of uterine tissue to various estrogenic compounds (Anderson *et al.*, 1972). This quantitative assay for determination of the concentration of nuclear estrogen–receptor sites relies on the exchange of [^{3}H]estradiol with unlabeled estradiol bound to target cell nuclei under various *in vivo* physiological states. Clark *et al.* (1972) have consequently demonstrated that the translocation of estrogen–receptor to its nucleus occurs under the influence of endogenous estrogen during the estrous cycle. The quantity of estrogen–receptor is a dose-dependent phenomenon and shows a striking positive correlation with early uterine physiological responses to estrogen administration (Clark *et al.*, 1973).

B. *Progesterone*

Because progesterone was thought to act in the nucleus to influence gene transcription, it became pertinent to establish whether this hormone was also bound to a macromolecular receptor in the nucleus of the target cell. Following stimulation with the hormone, purification of oviduct nuclei and extraction with high salt (0.3 *M* KCl) revealed the presence of such a nuclear receptor that appeared almost identical to that found in the cytoplasm of the same cells (Fig. 1) (O'Malley *et al.*, 1971). In experiments similar to those performed with estrogen in rat uterus, it was demonstrated that no appreciable quantities of nuclear receptor were found in purified oviduct nuclei from unstimulated animals. However, upon exposure to progesterone, an increase in extractable nuclear receptor protein occurred coordinately with a diminution in the amount of cytoplasmic receptor (O'Malley *et al.*, 1971). It appeared again that this steroid hormone was also capable of initiating a translocation process which resulted in an accumulation of receptor-bound intracellular [^{3}H] progesterone in the target cell nucleus.

A subsequent series of cell-free incubations proved to be of considerable interest. As with estrogen and uterus, it was noted that direct incubation of [^{3}H]progesterone with purified oviduct nuclei led to little bound steroid, and progesterone binding was shown to be dependent on the simultaneous presence of its receptor in the incubation. However, it was further shown (Fig. 2) that the progesterone–receptor complex could only bind well to nuclei of oviduct target tissue and that nuclei of nontarget tissue such as lung, spleen, heart, and intestine demonstrated little capac-

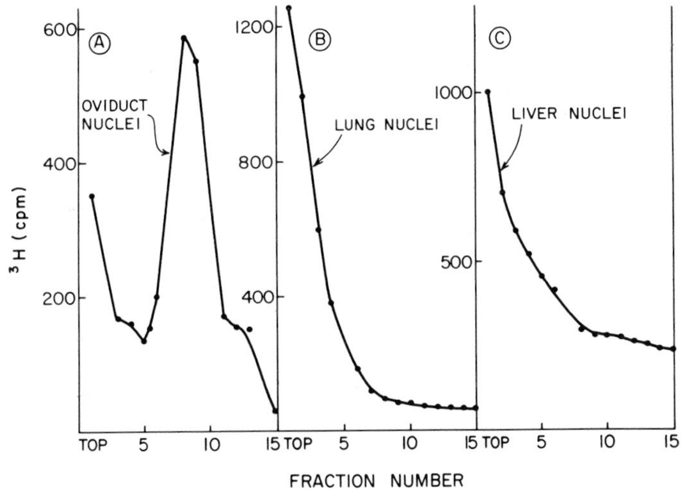

Fig. 2 Requirement for oviduct nuclei to accept [^{3}H]progesterone from oviduct cytosol *in vitro*. Gradient centrifugation of KCl extracts of oviduct (A), lung (B), and liver (C), nuclei incubated at 4°C with progesterone-labeled oviduct cytosol. (From O'Malley *et al.*, 1971.)

ity to "accept" and retain the hormone–receptor complex (O'Malley *et al.*, 1971).

This observation led to generation of the "nuclear acceptor hypothesis" which proposed the existence in target cell nuclei of acceptor sites having a specific affinity for the receptor molecules. Experiments performed concomitantly with radioactive androgens in rat prostate tissue confirmed the generality of this hypothesis for sex steroids and their target cell nuclei (Liao *et al.*, 1971).

It seemed that for the progesterone receptor to bind to the oviduct nuclei, it had to be complexed first with a progestational steroid and then allowed to undergo a time- and temperature-dependent transformation, presumably to a structural form capable of interacting with the nuclear acceptor sites. The intranuclear location of these acceptor sites was delineated when upon fractionation of nuclei exposed to [^{3}H]hormone–receptor it was found that the radioactive complex could be found attached to the chromatin (Spelsberg *et al.*, 1971; O'Malley *et al.*, 1971; 1972).

IV. Hormone–Receptor Binding to DNA

A. *Estrogen*

It has been reported that DNase treatment would release bound estradiol from uterine nuclei, thus implicating DNA in the nuclear binding of hormone–receptor complex (Harris, 1971). The binding of uterine estrogen receptors to DNA has also been demonstrated to occur under *in vitro* conditions (Toft, 1972; King and Gordon, 1972). The binding is of sufficient strength to withstand centrifugation of the DNA through sucrose gradients under ionic conditions up to 0.1 *M* KCl. These observations have been confirmed using DNA–cellulose chromatography (Toft, 1973; Yamamoto and Alberts, 1972). Both cytosol and nuclear receptor forms were shown to bind to DNA and partial purification of the receptor by ammonium sulfate precipitation appeared to enhance binding to DNA (Toft, 1973). This reaction has some of the properties of nuclear binding of estradiol *in vivo* and *in vitro* in that binding to DNA is disrupted by 0.3 *M* salt and there exists only a limited number of high-affinity binding sites on DNA for the receptor. In fact, DNA–cellulose chromatography has been used to effect receptor purification (Toft, 1973; Yamamoto, 1972). A recent report calculated that, assuming a 1:1 interaction between estradiol and receptor, binding sites are saturated at two binding sites per

10^7 nucleotides (DNA) or 500 receptor molecules per quantity of DNA found in a single nucleus (King and Gordon, 1972). Although DNA has many of the characteristics one would expect of the nuclear acceptor, a lack of binding specificity is evident. In fact, estrogen–receptor from rat uteri can interact with DNA from calf thymus, salmon sperm, *Escherichia coli* and *Bacillus subtilis.* It appears then that the acceptor site is more complex and that other components of chromatin may act to modify receptor binding to DNA.

B. *Progesterone*

Purified chick DNA has the capacity to bind and retain the [^{3}H]progesterone–receptor complex (Schrader *et al.*, 1972). At high concentrations of DNA, essentially all of a limited quantity of hormone receptor could be bound to DNA. This binding affinity did not appear to be uniquely sequence-specific as DNA from a heterologous eukaryotic species (calf) showed a similar capacity to bind the chick progesterone–receptor complex (O'Malley *et al.*, 1973).

Following elution of either crude or purified progesterone receptor from a DEAE column, the molecule can be resolved into two apparent subfractions or subunits, termed A and B (Schrader and O'Malley, 1972). Both subunits are specific, high-affinity, low-capacity binders for biologically active progestins. It was of considerable interest that only subunit A exemplified binding to purified DNA (Schrader *et al.*, 1972). The B unit had no such capacity.

It is possible, of course, that this DNA interaction was simply a nonspecific adsorption phenomenon. Such a possibility would be less plausible if the DNA binding was found to be a high-affinity effect with a limited number of binding sites. To study this, a more quantitative DNA-binding assay using gel filtration on an Agarose A–15M column was devised which separated DNA–receptor complexes from unbound receptor and also resolved free [^{3}H]progesterone from the receptor itself (O'Malley *et al.*, 1973). This technique was used to measure the equilibrium constant for the oviduct DNA-binding reaction by the Scatchard plot method. The equilibrium constant yielded $K_d = 3 \times 10^{-10}$ *M*. The binding-site molarity corresponded to approximately one receptor binding site for every 10^6 nucleotide pairs of double-stranded DNA. Thus, the receptor (A subunit) binding to oviduct DNA is both high affinity and also has a limited number of sites.

Studies were carried out to learn whether this binding was nucleotide sequence-specific. Calf thymus DNA appeared to bind the progesterone–

receptor complex in a fashion similar to that of the homologous chick DNA. However, when *B. subtilis* DNA was used, a dramatically reduced K_d for the interaction of progesterone–receptor complex to DNA was obtained (O'Malley *et al.*, 1973). Thus, the binding of receptor to DNA appears to involve a limited degree of species specificity.

V. Hormone–Receptor Interaction with Chromatin Nonhistone Proteins

A. *Estrogen*

Incubation of [^{3}H]estradiol directly with preparations of target tissue chromatin *in vitro* led to very little binding of [^{3}H]estradiol to the chromatin. However, incubation of the preformed [^{3}H]estradiol–receptor complex from uterus with uterine chromatin resulted in significant retention of the complex on chromatin (Steggles *et al.*, 1971). Removal of histones exposed even more receptor-binding sites (King and Gordon, 1972). Nontarget tissue chromatins appeared to bind quantitatively less estrogen–receptor complex. Similar results have been reported by three laboratories for androgen–receptor interactions with male target (e.g., prostate) and nontarget tissues (Liao *et al.*, 1973; Steggles *et al.*, 1971; Mainwaring and Peterken, 1971).

B. *Progesterone*

Cell-free recombination studies have been carried out which reveal very little binding of [^{3}H]progesterone to target cell (oviduct) chromatin in the absence of receptor protein (Spelsberg *et al.*, 1971; Steggles *et al.*, 1971; O'Malley *et al.*, 1972a). Substantial amounts of [^{3}H]progesterone were recovered bound to oviduct chromatin when the [^{3}H]progesterone was first combined with the oviduct cytosol receptor. Unmetabolized [^{3}H]progesterone–receptor complex could be reextracted intact by 0.3 *M* KCl treatment of chromatin following these *in vitro* incubations. Binding was specific for progesterone receptor alone as the binding of steroid could not be enhanced by extracellular-binding proteins such as transcortin or even by other cytosol preparations from nontarget tissues. Furthermore, these results demonstrated that progesterone–receptor complexes bind to oviduct chromatin to a much greater degree than to nontarget chromatin such as that from spleen, heart, lung, or hen erythrocytes (Spelsberg *et al.*, 1971).

Attempts were made to determine the important fraction of chromatin responsible for supporting binding of the hormone–receptor complex to chromosomes. Selective dissociation of histone from chromatin was carried out in the presence of 2.0 *M* NaCl, 5.0 *M* urea at pH 6.0. The chromatin could be reconstituted by sequential dialysis to remove first the NaCl and finally the urea. The reconstituted chromatins were shown to have the same chemical composition and histone species as their respective native chromatins. In this way, "hybrid" chromatins could also be prepared in which histones from other tissues or species were substituted during reconstitution. Binding of receptor to this reconstituted chromatin was similar to binding to the intact native chromatin of oviduct. Moreover, the capacity to bind the steroid–receptor complex was completely retained by hybrid chromatins containing histones from a nontarget tissue of a different species, e.g., calf thymus. Finally, it was noted that completely dehistonized oviduct chromatin still displayed a more extensive binding than dehistonized spleen chromatin. It was therefore felt that histones themselves were not primarily responsible for the specificity of receptor binding (Spelsberg *et al.*, 1971; O'Malley *et al.*, 1972).

Preliminary experiments in which the dissociated chromatin was treated with RNase prior to reconstitution suggested that chromosomal RNA was also not of major importance for the extensive binding. However, if during reconstitution the nonhistone (acidic) proteins of chromatin were removed, the chromatin lost most of its capacity to bind the progesterone–receptor complex (Spelsberg *et al.*, 1971).

More detailed experiments were subsequently performed to test the importance of the nonhistone proteins in regulating receptor binding. Acidic proteins (and histones) were dissociated from the chromatin of oviduct and erythrocytes by a single treatment using 2.0 *M* NaCl–5.0 *M* urea at pH 8.3. Approximately 90% of the acidic nonhistone proteins could be dissociated from the DNA by this procedure. These proteins were then separated from the DNA by ultracentrifugation. It was possible to reconstitute most of the acidic proteins back to the DNA by gradient dialysis in the presence of sodium bisulfite. In certain instances the acidic proteins and histones of the chromatin of one tissue were reconstituted with the DNA of another tissue to form hybrid chromatins. An immunochemical method employing specific antisera against chromatin nonhistone proteins of various tissues was used to monitor the nonhistone protein fraction during reconstitution experiments to substantiate formation of hybrid chromatins (Spelsberg *et al.*, 1972).

Oviduct cytosol, containing the [^{3}H]progesterone–receptor complex, was then incubated separately with the intact, reconstituted, or hybrid chromatins. Reconstituted oviduct chromatin binds progesterone–receptor complex in a manner quantitatively similar to that of intact native oviduct chromatin. However, when the nonhistone protein of erythrocyte is attached onto the oviduct DNA during reconstitution, this hybrid loses its enhanced ability to bind receptor and resembles native erythrocyte chromatin. Conversely, insertion of nonhistone protein from oviduct into erythrocyte chromatin bestows binding capacity to this hybrid chromatin resembling that of native oviduct (Spelsberg *et al.*, 1972; O'Malley *et al.*, 1972). These experiments demonstrate that the inherent acceptor capacity of target tissue chromatin for the hormone–receptor of that tissue can be transferred to a nontarget DNA through transfer of the nonhistone (acidic) protein fraction. Additional experiments have localized this "acceptor capacity" to a subfraction of the total nonhistone proteins of the target cell chromatin (Spelsberg *et al.*, 1972).

As with the binding to DNA discussed above, only one of the putative progesterone–receptor units, subunit B, has this capacity to interact with the nonhistone proteins of oviduct tissue (Schrader *et al.*, 1972). Thus, only if one considers the combined characteristics of both the A subunit (affinity for DNA) and the B subunit (affinity for nonhistone protein) of the progesterone receptor can the full potential of crude intact receptor be accounted for. It appears then that both DNA and a nonhistone protein fraction of target cell chromatin play a positive role in forming the acceptor sites for uterine receptor in uterine chromatin and progesterone receptor in oviduct chromatin (Fig. 3).

In summary, it can be speculated that, following entry of the steroid–receptor complex (S:Rn) into the nuclear compartment, the initial molecular interaction of the steroid–receptor complex with chromatin may occur in two parts, a high-affinity reaction between the receptor subunit A and chromatin DNA and a similar reaction between a specifier subunit B of the intact native receptor and the nonhistone acceptor proteins (O'Malley *et al.*, 1973). If this hormone–receptor complex is actually the inducer

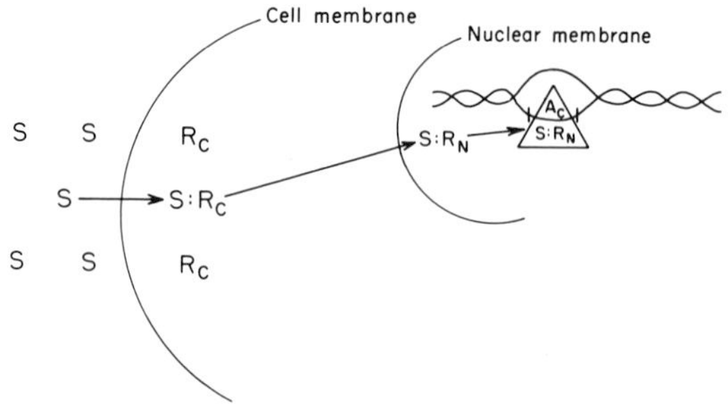

Fig. 3 Translocation of steroid hormone complex from cytoplasm to nucleus and attachment onto chromatin.

unit for steroid hormone modulation of nuclear RNA transcription, then this initial binding to the genome may prove to be of major importance to steroid hormone action.

VI. Steroid Hormone Effects on the Cell Cycle

A. *Estrogen*

A necessary prerequisite of the study of hormone effects on mitosis in a target tissue is a knowledge of the level and chronology of mitotic activity and cell cycle parameters in the absence of hormonal influence. In the immature oviduct, mitoses are infrequent (Kohler *et al.*, 1969). The mean mitotic index (MI) of the surface epithelium was 0.43 in 7-day-old chicks (Socher and O'Malley, 1973). This reflects the slow natural growth of the oviduct that continues until sexual maturation. The level of mitoses remained constant over a 24-hr period. Thus changes observed after hormone treatment should be due to the hormones and not to any natural mitotic rhythms within the tissue.

DNA synthetic inhibitors, 5-fluorodeoxyuridine (FUdR) and hydroxyurea (HU) have been used to estimate the duration of G_2 in the unstimulated oviduct. Since these antimetabolites block cells in S, the only cells that enter mitosis after treatment should be G_2 cells. FUdR and HU both produce a drop in MI. The time interval between injection of the inhibitor and the time at which the MI drops to 50% of the control value provides an estimate of the mean duration of G_2 + mitosis/2 (Socher and Davidson, 1971). In the chick oviduct, the duration of G_2 + mitosis/2 = 2.25 hrs. If it is assumed that the duration of mitosis is 1 hr, then G_2 = 1.75 hr. Another group of chicks received a single injection of [^{3}H]TdR. Labeled mitoses were found 1.5 to 2 hr after administration of the pulse label. Both inhibitor treatments and [^{3}H]TdR experiments indicate that G_2 is short in the immature unstimulated oviduct, i.e., 1.5 to 2 hours.

Unstimulated chicks were given a single injection of [^{3}H]-TdR to determine the number of cells in S, i.e. synthesizing DNA. Labeling indices were determined 0.5 to 2 hours after the pulse. Grain intensity of labeled nuclei was similar at all times examined. The mean labeling index of the surface epithelium was 3.16, indicating that a small fraction of the cells were synthesizing DNA. The duration of the S phase can be estimated if the following conditions apply to the population of cells making up the epithelium: (*a*) The dividing cells constitute a single population of cells; (*b*) the duration of cell cycle and of its subphases are constant; and (*c*) the proportion of dividing cells is constant. If these conditions apply to the epithelium, the following relationships hold:

$$\mathrm{MI} = d_m/C \tag{1}$$
$$\mathrm{LI} = d_s/C \tag{2}$$

and therefore

$$d_m/\mathrm{MI} = d_s\mathrm{LI} \tag{3}$$

where MI = mitotic index, LI = labeling index, d_m = duration of mitosis, d_s = duration of S and C = mean cycle duration. Solving Eq. (3) for d_s (assuming $d_m = 1$ hr) reveals that $d_s = 7.3$ hr in the unstimulated oviduct. (Socher and O'Malley, 1973).

Immature chicks (7 day) were treated with estrogen to examine the action of this hormone on cell proliferation and the effects of single and repeated injection of hormones on mitotic activity can be seen in Fig. 4 (Socher and O'Malley, 1973). A single injection of estrogen markedly stimulated mitosis in the oviduct. There was a rapid rise in MI between 9 and 12 hr. The MI reached a peak at 18 hours after treatment and the level of mitotic activity began to fall at 24 hr and continued to drop until 42 hr. The frequency of cells in mitosis 48 hr after estrogen was more than twice that observed in the unstimulated oviduct.

When chicks are given a second injection of estrogen 24 hr after the first, a second rise in MI is observed (Fig. 4). The patterns of changes in MI following single and double treatments could be explained in terms of a hormone-induced stimulation of division in a single population of cells. The cells appeared to progress through the cycle in a parasynchronous fashion. Estrogen treatment resulted in a similar stimulation of the rate of DNA synthesis in oviduct, leading to a dramatic increase in total DNA

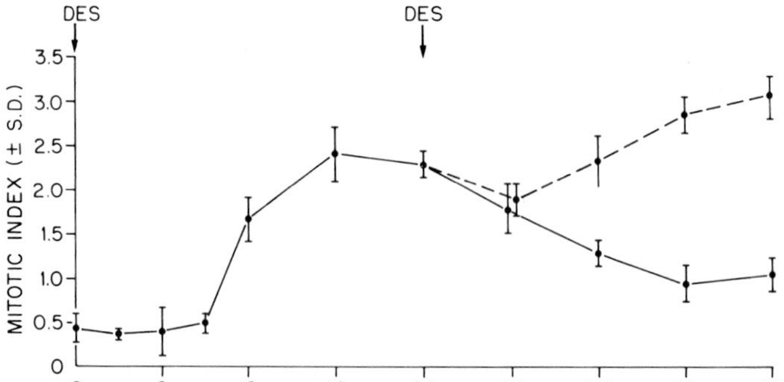

Fig. 4 Changes in mitotic index of oviduct following injection of estradiol-17β to immature chicks. Hormone (DES) was administered at 0 hours (●——●) or 0 and 24 hours (●----●). (From Socher and O'Malley 1973.)

consistent with the hormonally induced cell proliferation (Socher and O'Malley, 1973).

B. Progesterone

Progesterone also stimulated a small fraction of cells to undergo mitosis. A small, but significant rise in MI was observed 12 and 18 hr after treatment. The frequency of cells in mitosis dropped below control level at 24 hr and remained low until 48 hr, with or without a second injection of progesterone (Socher and O'Malley, 1973). These data indicate that although progesterone can act as a proliferative stimulus for a small population of cells in the surface epithelium of the immature oviduct, this steroid appeared to exert its predominant inhibiting action by the progress through the cycle of normally proliferating cells. The basis and relationship between these two actions of progesterone are not clear.

The initial response to a combination treatment of estrogen and progesterone was similar to that observed with estrogen alone. However, between 18 and 24 hr after treatment with estrogen and progesterone, a sharp fall in MI was observed. The timing of this drop in mitotic activity was similar to that observed with progesterone alone. With or without a second injection, only small fluctuations in MI are observed 24 to 48 hr after the beginning of treatment. During this time the level of mitotic activity was higher than the control level. It appears that initially estrogen and progesterone in combination act together to stimulate cells to enter mitosis. However, after the initial stimulation, a dramatic inhibition of mitosis was exerted by progesterone. This inhibition appears to be the primary action of progesterone.

Within a variety of tissues it has been shown that after an appropriate stimulus, nonproliferating cells can be stimulated to divide. In general, proliferative stimuli act to stimulate G_1 cells to enter S, G_2, and then mitosis, or G_2 cells to undergo mitosis. To ascertain at which stage the hormone-responsive cells exist in the immature oviduct, experiments were performed in which chicks received simultaneous injections of hormone and either FUdR or HU. If the hormones acted to stimulate G_1 cells to enter S, the inhibitors of DNA synthesis should have blocked the hormone-induced stimulation of mitosis. If these hormones stimulated a G_2 population to undergo mitosis, the inhibitors should not block the early hormone-induced rise in MI. Both FUdR and HU inhibit the stimulation of mitosis that normally occurred following treatment with estrogen and progesterone, alone or in combination. This demonstrates that the proliferative stimuli acted prior to the completion of DNA synthesis, that is, in G_1 or S. In the unstimulated oviduct $G_2 = 1.75$ hr and $S = 7.3$ hr. Furthermore a rise in MI was not observed until 12 hr after administration of the hormones. It appears, therefore, that estrogen and progesterone, alone or in combination, stimulate G_1 cells to enter S. These results are similar to the action of estrogen as a proliferative stimulus in the uterine and vaginal epithelia in the mouse (Perrotta, 1962).

VII. Steroid Hormone Effects on Chromatin Composition and Conformation

Estrogen

In chick oviduct, differentiation of epithelial cells and synthesis of specific proteins such as ovalbumin and lysozyme occur in response to estrogen administration (O'Malley *et al.*, 1969). Previous data have suggested transcriptional control of both the differentiation process and of cell-specific protein synthesis. Based upon the hypothesis that changes in gene transcription during differentiation reflect, in part, changes in the tissue-specific pattern of gene restriction, changes occurring in the chemical composition and physical properties of oviduct chromatin were investigated during estrogen-mediated tissue differentiation.

Quantitative analysis of the chromatin from various stages of oviduct development demonstrated that while the levels of histones varied randomly, the levels of the total acidic chromatin proteins increased during the first few days of differentiation followed by a gradual decrease until completion of development (15 days of estrogen treatment) (Spelsberg *et al.*, 1973). The levels of chromatin associated with RNA followed a similar pattern. Moreover, the capacity of the intact chromatins to serve as templates for *in vitro* RNA synthesis using bacterial polymerase also increased during the first few days of differentiation and then decreased during the final stages of development.

To clarify the structural relationship and the role of the acidic proteins in the tissue-specific restriction of DNA, the antigenic properties of nonhistone protein–DNA complexes isolated from different organs were determined. The antigenic properties of the nonhistone protein–DNA complexes were then compared with those of the corresponding native (intact) chromatins. Antibodies were produced in rabbits against a preparation of oviduct dehistonized acidic protein complexed to DNA (nucleoacidic protein), which was prepared from oviduct chromatin of chicks which had been injected with an estrogen for 15 days. The method of quantitative complement fixation was chosen for testing the antigenic properties (Chytil and Spelsberg, 1971).

To test for specificity, nucleoacidic protein was isolated from the chromatins of chick heart, liver, and spleen and tested for antigenicity using the same antiserum prepared against oviduct nucleoacidic protein described above. Nucleoacidic proteins from other organs showed a very limited affinity for this antibody. This limited affinity suggested that a large number of the antigenic sites present in oviduct preparations were tissue specific and indicated a considerable general dissimilarity in the antigenic sites of acidic protein–DNA complexes present in the chromatins of different organs. Furthermore, preparations of nucleoacidic proteins from undifferentiated oviduct, i.e., oviducts of unstimulated chicks, showed very little antigenicity using antiserum prepared against nucleoacidic protein from estrogen-stimulated, differentiated oviduct. However, fixation of complement by nucleoacidic proteins that were isolated from oviducts of chicks injected with estrogen for various periods of time showed a gradual appearance of antigenicity with length of estrogen administration (Chytil and O'Malley, 1973). Thus, development of antigenicity of the nucleoacidic proteins showed a developmental change which coincided with the estrogen-induced morphological development of the organ.

In conclusion: (*a*) The nucleoacidic proteins isolated from chromatin complexed to DNA are good immunogens which give complement-fixing antibody; (*b*) the antibodies do react strongly with the preparations from the homologous organ (chick oviduct), whereas the affinity for preparations of nucleoacidic proteins from heterologous sources (liver, heart, spleen) is very low; this is indicative that the arrangement of the antigenic sites inherent to the acidic proteins in chromatin is organ specific; (*c*) during development of chick oviduct the antigenic sites for acidic proteins undergo marked alterations which probably involve changes in the species of nonhistone proteins in addition to possible structural alterations of already existing proteins.

Structural analysis of oviduct chromosomal protein–DNA complexes have been carried out using CD* analysis under standardized conditions. Data were expressed as changes in the magnitude of mean residue ellipticity at 275 nm which primarily reflect changes in DNA conformation. The magnitude of the ellipticity at 275 nm for intact chromatin was much less than that for pure DNA at all stages of estrogen-mediated development. DNA complexed with proteins as in chromatin has previously been shown to have a reduced positive band at 275 nm (Simpson and Sober, 1970; Wagner and Spelsberg, 1971; Fasman *et al.*, 1970). This decreased ellipticity has been suggested to be caused by an altered geometry of the

* CD = circular dichroism.

DNA when complexed with proteins. However, as estrogen-stimulated development of the oviduct progresses, the chromatin DNA displayed a gradual increase in the magnitude of ellipticity. Other studies suggest that this increase in ellipticity may represent an "opening" of the DNA, i.e., removal of proteins from areas of the DNA (Wagner and Spelsberg, 1971).

Thus, CD analysis of chromatin during conformation of target cell supports the concept that an alteration of the composition and/or steric conformation of target cell chromatin occurs during steroid hormone-induced differentiation. Coupled with the data showing major quantitative and qualitative changes in the nucleoacidic proteins of chromatin, these studies provide additional evidence that differentiation represents progressive alterations in chromatin biochemistry which may result in changes in cell structure and function (Spelsberg *et al.*, 1973).

VIII. Steroid Hormone Effects on RNA Synthesis

Since it is now clear that the steroid hormone–receptor complex enters the nucleus and binds to target cell chromatin, it seems logical to assume that synthesis of RNA plays a major role in the primary mechanism of action. It has not yet been possible to demonstrate directly that the chromatin binding results in an increased rate of nuclear transcription. However, much evidence exists to support the hypothesis that steroid hormones do act upon the nucleus of target cells.

Estrogen

The first demonstration that estrogen stimulates the incorporation of precursor into total cell RNA came from Mueller *et al.* (1958). Initially the responses to estrogen were measured in terms of hours. These studies and those from the laboratories of Gorski and Hamilton led to an argument as to whether the stimulation of RNA synthesis was a cause or consequence of estrogen-mediated translation (Gorski *et al.*, 1965; Hamilton, 1963, 1964). Within 2 min following a single injection of estrogen a 40% increase in the synthesis of rapidly labeled nuclear RNA was observed (Means and Hamilton, 1966a). The estrogen stimulation of this activity was biphasic (Fig. 5), peaking initially at 20–30 min, falling abruptly until 2 hr, and then increasing once again and remaining elevated for at least 24 hr (Hamilton *et al.*, 1968). Moreover, the first phase of this response was prevented by actinomycin D administered 30 min prior to estrogen but was not blocked by cycloheximide (Means and

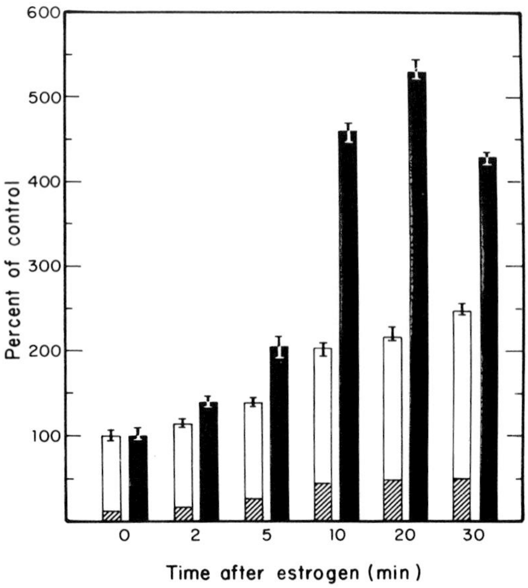

Fig. 5 Rapid enhancement of rapidly labeled uterine nuclear RNA synthesis by estradiol-17β. All animals were adult ovariectomized rats and were killed 10 min after an intraperitoneal injection of 100 μCi [^{3}H]uridine. Estradiol-17β (10 μg) was injected I P at the times shown in the Figure. Data are presented as percentage of the saline-injected control values, which were 6,920 cpm/mg DNA for uptake of [^{3}H]uridine (□); 11.0% for the fraction of tissue radioactivity that was acid-insoluble (▨); and 1,204 cpm/mg nuclear RNA (■). (From Means and Hamilton, 1966a.)

Hamilton, 1966b). These data strongly suggested that the initial response of the nuclear apparatus was not dependent upon the continued synthesis of protein.

One of the problems associated with the investigation of precursor incorporation into nucleic acid is to dissociate an effect on precursor uptake from a direct effect on RNA synthesis. In fact, in the experiments mentioned above increased incorporation of [^{3}H]uridine into RNA was accomplished by an increased uptake of the nucleotide (Means and Hamilton, 1966b). The difficulty comes in dissociating the estrogen effect on transcription from the well-known histamine response mediated by uterine β-receptors (Szego, 1965). Recently, several investigators have reported that estrogen does not actually increase RNA synthesis for a number of hours, and that the apparent early effects on RNA synthesis can be accounted for by changes in nucleotide synthesis and/or uptake (Billing *et al.*, 1969a, b; Szego, 1971). The most extensive and critical papers in this area, however, are by Oliver (Oliver and Kellie, 1970; Oliver, 1971). Such data, together with studies to be discussed below on estrogen-stimulated RNA polymerase activity and chromatin template *in vitro,* clearly show that the early transcriptional events result from a direct effect on synthesis (Hamilton *et al.*, 1971; Means and O'Malley, 1972). The uptake and vasodilatory responses only serve to augment the magnitude of the enhancement by allowing a continuous supply of the appropriate precursors.

The nature of the rapidly labeled nuclear RNA formed during the first few minutes of estrogen action seems still to be uncertain. Initially base-composition studies suggested that it was ribosomal-like (Hamilton *et al.*, 1968). However, more recent studies on the sedimentation characteristics of this RNA and kinetics of stimulation of nucleoplasmic RNA polymerase II yield evidence of a more DNA-like nature (Knowler and Smellie, 1971; Luck and Hamilton, 1972; Glasser *et al.*, 1972). Moreover, stimulation of synthesis of this rapidly labeled nuclear RNA seems to be mandatory for the appearance in the cytoplasm of a group of specific estrogen-induced proteins (De Angelo and Gorski, 1970). Within a few hours after injection, estrogen has resulted in stimulation of both ribosomal precursor RNA, 28 S, 18 S, and 5 S ribosomal RNA and 4 S transfer RNA (Hamilton *et al.*, 1968, 1971; Means and O'Malley, 1972; Knowler and Smellie, 1971). Thus the synthesis of all classes of uterine RNA are eventually enhanced by estrogen.

Another manner by which to investigate changes in RNA synthesis is to utilize chromatin template activity. This procedure gives an estimate of the percent of the total genome that is available for transcription using chromatin as the sole source of DNA but an excess of exogenous RNA polymerase (usually from *E. coli*). Teng and Hamilton (1968) described a 25% stimulation of template capacity of rat uterine chromatin within 30 min following injection of estrogen to adult ovariectomized animals. Peak enhancement was observed at 8 hr but some stimulation was still apparent for 2 days. Similar observations were made by Church and McCarthy (1970) using chromatin isolated from endometrium of castrate rabbits. In these experiments, however, effects were claimed as early as 10 min after estrogen administration and activity remained elevated for at least 2 hr. Enhancement of rat uterine chromatin template activity has also been observed 2 hr after estrogen administration by Barker and Warren (1966). More recently template capacity in the rat uterus has been reported by Glasser *et al.* (1973). In these experiments, care was taken to monitor the chemical composition of chromatin during the isolation procedure. Under their conditions of assay, the first demonstrable effects of estrogen were noted at 1 hr. It is interesting to note that the effect of estrogen on chromatin activity was paralleled by the stimulation of RNA polymerase I.

Chromatin template has also been shown to be stimulated by estrogen in chick oviduct (O'Malley and McGuire 1968, Spelsberg *et al.*, 1971). Template activity is markedly stimulated by 4 days and remains elevated during at least the first 20 days of estrogen treatment. These qualitative changes have also been shown to occur in concert with quantitative differences in the nature of RNA products. The technique of nearest-neighbor base analysis of the RNA was used for this purpose (O'Malley and McGuire 1968). Comparison of the base frequencies of RNA synthesized by chromatin from control and 6-day estrogen-treated chicks revealed considerable differences. In general the RNA tended to become more AU-rich during hormone treatment. Here again the data suggest that estrogen must promote the synthesis of new species of RNA which appear during differentiation of the oviduct.

All classes of RNA are eventually increased in the oviduct in response to estrogen. This is true both for rate of synthesis and for accumulation in oviduct tissue. Within 1 day following stimulation by estrogen of the undifferentiated oviduct, changes occur in the pattern of ribosomal precursor RNA as judged by polyacrylamide gel electrophoresis (Kapadia *et al.*, 1971). By 2–4 days ribosomal 28 S and 18 S RNA species are increased considerably (Means *et al.*, 1971). Marked effects on synthesis of 4 S and 5 S RNA's are also seen by this time (O'Malley *et al.*, 1968). These increased rates of synthesis of ribosome-associated RNA species are accompanied for the first 7 days after estrogen treatment by a continuous increase in the cytoplasmic accumulation of newly formed ribosomes (Means *et al.*, 1971; Means and O'Malley, 1971; Palmiter *et al.*, 1970).

One of the earliest events to be stimulated in the oviduct following estrogen administration to immature chicks is the synthesis of rapidly labeled nuclear RNA (O'Malley *et al.*, 1969). This is true whether studying estrogen-mediated differentiation of the oviduct or investigating the temporal sequence of events in the oviduct of chicks first differentiated with estrogen and then withdrawn from the hormone. Progesterone also affects the incorporation of uridine into rapidly labeled nuclear RNA in oviducts of estrogen-primed chicks (O'Malley *et al.*, 1969). Unlike estrogen, however, the first noticeable effect is a consistent decrease in the rate of synthesis seen between 2 and 6 hr following a single injection of progesterone. This initial decline is followed by a rise which peaks between 24 and 48 hr and precedes the major effect of progesterone, that is, the induction of avidin synthesis. Another difference between the responses of estrogen and progesterone is that whereas estrogen eventually results in a marked increase in RNA mass, no such change is noted following progesterone (O'Malley *et al.*, 1969). Changes in chromatin template activity parallel the alterations of rapidly labeled nuclear RNA in response to progesterone. An initial decline at 1 hr is followed by a considerable increase over control values by 3–4 hr.

Limited use has been made of the technique of DNA–RNA hybridization for studying effects of the female sex steroids on the uterine transcriptional apparatus. Church and McCarthy (1970) employed this method to examine the effect of estrogen on the appearance of new populations of nuclear RNA sequences. Unlabeled uterine nuclear RNA from control or 1-hr estrogen-treated rats was used to competitively anneal to DNA-binding sites with [^{3}H]RNA isolated from uteri of ovariectomized rabbit. The unlabeled RNA from untreated animals did not compete effectively with uterine RNA isolated from rabbits treated for 1 hr with estrogen. These hybridization competition experiments provide good evidence that estrogen very rapidly induces synthesis of different populations of RNA molecules. The limitations of this study, and indeed the reasons more work on hybridization is not available for hormone-sensitive tissues are that under the conditions normally utilized only changes in repetitive sequences of DNA can be analyzed. Thus the unique portion of the genome, responsible for synthesis of most mRNA molecules, is not assayed under the conditions utilized by Church and McCarthy (1970). Recent methodological advances have allowed the genome to be effectively subdivided into unique and repeating sequences of DNA (Britten and Kohne, 1968). As yet, however, no one has attempted to apply these techniques to investigate the effects of estrogen on the mammalian uterus. However, our laboratory has utilized hydroxylapatite chromatography to subdivide the chick oviduct genome into unique and repetitive DNA sequences (Liarakos *et al.*, 1973; Rosen *et al.*, 1973). No detectable differences were found in the renaturation profiles of oviduct DNA at various times of estrogen stimulation. The data suggest that estrogen is not acting through major gene duplication or deletion and offer evidence that the new proteins required for oviduct growth may arise from differential gene transcription. It has also been found that unique sequence DNA is transcribed in the oviduct with 25–30% of the resulting RNA being processed into the cytoplasm. Moreover, estrogen apparently causes an increase in the extent of unique DNA transcription. Thus, although the amount of RNA transcribed from unique sequence DNA that is processed into the cytoplasm does not appear to vary as a result of estrogen treatment, a qualitative difference seems to exist in the cytoplasmic mRNA populations at different stages of estrogen-induced differentiation.

IX. Steroid Hormone Effects on RNA Polymerase Activity

A. *Estrogen*

Gorski was the first to report an effect of estrogen on RNA polymerase activity assayed in a crude nuclear pellet obtained from uteri of immature rats (Gorski, 1964). Within 1 hr following a single injection of estradiol there was an increase in the activity of Mg^{2+}-dependent RNA polymerase. Subsequently these observations were confirmed using a highly purified preparation of nuclei which had been stripped of the outer membrane by treatment with detergent (Hamilton *et al.*, 1965, 1968). A second polymerase activity was also shown to be stimulated by estrogen but required 12 hr of hormone treatment (Hamilton *et al.*, 1965, 1968). This enzyme activity required high ionic strength and showed a preference of Mn^{2+}. It was subsequently demonstrated by autoradiography that the Mg^{2+}-dependent enzyme was restricted to the nucleolus. The base composition of the RNA product was ribosomelike. On the other hand, the high salt enzyme activity was located in the nucleoplasm and synthesized a product with a DNA-like base composition (Maul and Hamilton, 1967).

Much speculation followed suggesting that the high salt polymerase was nonphysiological and simply resulted from a nonspecific removal of histones from the chromatin. However, Roeder and Rutter (1970) were able to isolate and partially purify the two enzyme forms which were shown to be separate forms. The nucleolar form was termed polymerase I (A) and synthesizes ribosomal RNA, whereas the nucleoplasmic enzyme (polymerase B, II) catalyzes synthesis of high-molecular-weight heterodisperse nuclear RNA (see Chapter 7). Thus, it appeared that estrogen stimulated synthesis of ribosomal RNA before any apparent effect on DNA-like RNA formation.

Barry and Gorski (1971) have further investigated the effect of estrogen on uterine polymerase I activity. The methods employed in these experiments were capable of distinguishing between hormone-mediated increase in the number of growing nucleotide chains and the rate of chain elongation. The results obtained suggested that estrogen evokes an increase in the rate of chain elongation within 1 hr but does not affect the number of growing chains. The implication of such experiments is that estrogen stimulates the activity but not the number of polymerase molecules. Such data were interpreted to mean that the estrogen effect at this early time point is only minimally due to transcription of additional template.

Estrogen has also been shown to increase the activity of RNA polymerase in the chick oviduct during hormone-mediated differentiation (McGuire and O'Malley, 1968; O'Malley *et al.*, 1969). In this tissue there seems to occur a concomitant increase in polymerase I and II first demonstrable at about 6 hr after estrogen administration. By 48 hr stimulation has reached eight- to tenfold. It seems certain that the increase must occur both by enhancing the activity of existing polymerase molecules and by increasing the number of initiation sites on the DNA owing to an increased number of active polymerase molecules. This interpretation stems from the fact that during differentiation many additional cell-specific proteins appear and have been shown to be under transcriptional control. Therefore, although estrogen stimulates RNA polymerase activity in both uterus and oviduct, different mechanisms may be operative.

An enigma existed for some years in regard to the rapid but transient stimulation of uterine RNA synthesis by estrogen since neither chromatin template nor RNA polymerase seemed to be increased at the same time. This problem has recently been solved (Glasser *et al.*, 1972) by reevaluation of uterine RNA polymerase activity in response to estrogen utilizing stringent kinetic conditions of assay. Utilizing low temperature (15°C) to prevent RNase activity and a rate point for assay (10 min) it was possible to demonstrate a rapid but transient increase in polymerase II activity first demonstrable by 15 min after estrogen administration (Fig. 6). Moreover, the kinetics of this increase occur in concert with the previously documented stimulation of rapidly labeled nuclear RNA synthesis. The initial enhancement of polymerase II activity occurs prior to any detectable increase in template activity of polymerase I. In fact, stimulation of polymerase I first occurs at 1 hr, reaching maximal activity at 4 hr. These data confirm the results of several investigators regarding the time-course of activation of the nucleolar RNA polymerase I by estrogen (Gorski, 1964; Hamilton *et al.*, 1965).

The studies of Glasser *et al.* (1973) also offer evidence that may help to solve another problem of estrogen action in the uterus: whether some or all of the effects on transcription require continued protein synthesis. The activation of polymerase II is biphasic (Fig. 6). When actinomycin D is injected 30 min before the estrogen, both rises in activity are abolished. However, cycloheximide pretreatment does not prevent the initial rapid rise in polymerase II. In fact, there is an increase over the effect seen with estrogen alone. This is also the case for the effects of cycloheximide on rapidly labeled RNA (Means and Hamilton, 1966b). On the other hand, the secondary increase in polymerase II, which begins 2–3 hr after estrogen administration, is completely abolished by cycloheximide. These data could explain why in the earlier studies from Gorski's laboratory (Gorski et al., 1965), and in a more recent report from Nicolette and

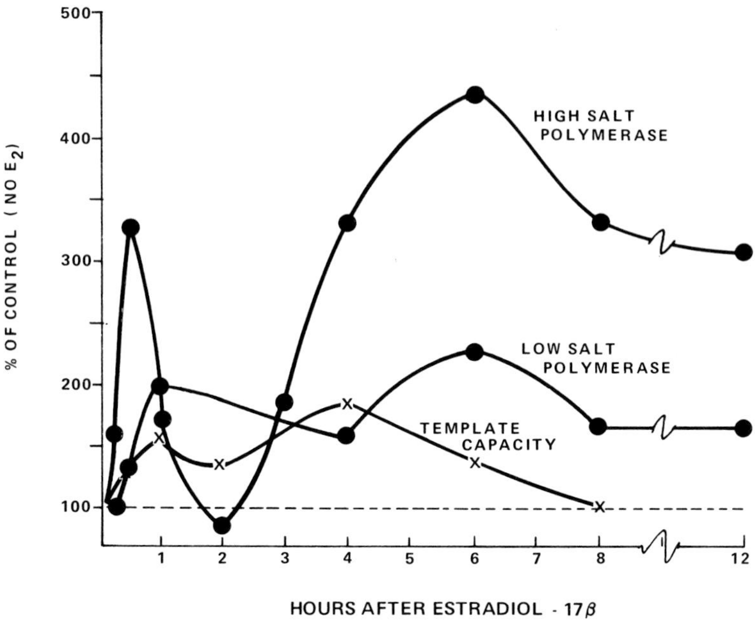

Fig. 6 Effects of estradiol-17β on uterine RNA polymerase and chromatin template activity. Endogenous RNA polymerase was assayed in low salt + α-amantin (●) which gives a measure of nucleolar polymerase I or high salt (●) under conditions optimal for nucleoplasmic polymerase II. Template capacity of uterine chromatin (X) was determined by methods previously reported. Nuclei and chromatin were isolated from uteri of ovariectomized adult rats at various times after a single saphenous vein injection of 1.0 μg estradiol-17β. (From Glasser *et al.*, 1972.)

Babler (1972), it was maintained that stopping uterine protein synthesis prevented an estrogen-induced enhancement of RNA synthesis. The usual time periods investigated were 2–6 hr after hormone administration, considerably after the rapid but transient first phase of the RNA stimulation curve. Recent results from DeAngelo and Gorski (1970) also revealed that a very early and specific event in estrogen action—induction of a specific protein (IP)—is blocked by pretreatment with actinomycin D but not by puromycin or cycloheximide. Thus it seems that an early increase in the activity of RNA polymerase II is necessary for the subsequent biochemical events in the action of estrogen on the uterus (Glasser *et al.*, 1973).

Few studies have succeeded in yielding data obtained *in vitro* which relate to the temporal sequence of metabolic events elicited by estrogen *in vitro*. Raynaud-Jammett and Baulieu (1969) were the first to demonstrate an *in vitro* effect of estrogen on uterine RNA polymerase activity. Estradiol-17β was preincubated with calf uterine cytosol, nuclei were then added, and RNA polymerase activity was measured. These experiments showed that indeed estrogen could stimulate polymerase activity in a cell-free system. Preincubation was absolutely necessary in order to show response. Presumably this treatment allowed the interaction of the steroid with its specific receptor. Similar observations have been made by Arnaud *et al.* (1971a) and by Mohla *et al.* (1972). The article by Arnaud *et al.* (1971a) suggested that only the 5 S form of the estrogen–receptor complex would stimulate uterine RNA synthesis *in vitro* and acted specifically on nucleolar RNA polymerase I. Moreover, these same workers (Arnaud *et al.*, 1971b) suggested that phosphorylation of the 5 S receptor resulted in enhanced ability of the complex to stimulate RNA synthesis.

Mohla *et al.* (1972) also reported a 40–60% stimulation of uterine polymerase activity *in vitro*. Again it was revealed that in order to demonstrate the stimulation, preincubation must be carried out under conditions which allow the "transformation" of the cytosol receptor to the 5 S form. Recently Jensen *et al.* (1973) have proceeded one step further and shown that the transformed receptor protein, uncomplexed with steroid, also stimulates RNA synthesis in uterine nuclei. These results were used to support their hypothesis that receptor transformation is an important step in estrogen action. Furthermore, it was suggested that one of the biochemical functions of estradiol may be to induce conversion of the receptor protein to an active form that can enter the nucleus, bind to acceptor molecules, and induce RNA synthesis.

B. *Progesterone*

Again the studies on the effect of progesterone on RNA polymerase activity are primarily those using the chick oviduct as a model system (McGuire and O'Malley, 1968; O'Malley *et al.*, 1969). Administration of progesterone to immature chicks not previously treated with estrogen produces little effect for the first 5 hr. Between 5 and 10 hr a considerable increase is seen in the activities of both polymerase I and polymerase II. Maximal effect occurs at about 24 hr. Thus, the increase in polymerase activity anticipates the induction of avidin synthesis.

The effect of progesterone on polymerase activity in oviducts from chicks previously treated with estrogen for 14 days is considerably different. A transient but significant fall is seen 2 hr after a single injection of progesterone (O'Malley *et al.*, 1969). This is followed by a rise which peaks at 24 hr and again the increased polymerase activity precedes the induction of avidin synthesis. This same type of response curve is seen for both rapidly labeled nuclear RNA and the polymerase which is presumably responsible for its synthesis (O'Malley *et al.*, 1969). The reason for the initial decrease in activity cannot be due to estrogen withdrawal. It is more likely that progesterone acts in two ways: first, to specifically induce the mRNA for avidin, and second, as an estrogen antagonist. At any rate, the decrease of estrogen-mediated events by progesterone is widespread among the biochemical events examined to date.

X. Steroid Hormone Effects on Production and Translation of Messenger RNA

A. *Estrogen*

One of the most elusive problems in studies concerned with the action of steroid hormones on the nucleus has been to obtain direct evidence that a rate-limiting step is the production of mRNA. Measurements of transcriptional changes have indicated that estrogen does, in fact, play a role at this level of cell regulation. In this regard most of the major changes induced in estrogen-sensitive target tissues by administration of the hormone are blocked by pretreatment with actinomycin D (Hamilton *et al.*, 1971; Means and O'Malley, 1972; O'Malley *et al.*, 1969). Moreover, as outlined previously, estrogen stimulates synthesis of rapidly labeled nuclear RNA, and RNA polymerase activity is altered, as is the template capacity of nuclear chromatin. Analysis of the RNA products by hybridization and base frequency methodology also reveals marked changes in response to estrogen (O'Malley *et al.*, 1969; Church and McCarthy, 1970, Liarakos *et al.*, 1973; Rosen *et al*,. 1973; Hahn *et al.*, 1968). None of these studies, however, constitute direct proof of alterations in the transcription of specific structural genes.

Some of the most convincing indirect data for effects of estrogen on mRNA come from studies with target tissue polyribosomes (Means *et al.*, 1971; Means and O'Malley, 1971; Palmiter *et al.*, 1970). In the oviduct or uterus of hormone-deficient animals, the majority of cytoplasmic ribosomes exist in monomeric or dimeric forms. Administration of estrogen results in a dramatic change in the polysomal profile. Within 4 days of estrogen treatment over 90% of chick oviduct ribosomes are present in aggregated forms. Moreover, at 10 days, once the tissue differentiation is complete, there is again a shift in polysome pattern. Further demonstration that formation of polysomes is estrogen dependent comes from the fact that following withdrawal of hormone for a few days, there is a shift back to the unaggregated condition. Finally, we have examined total nuclear mRNA activity during estrogen-induced differentiation.

Chick oviduct nuclear RNA was isolated at various times after estrogen injections and assayed for template activity in a cell-free system, as first described by Nirenberg (1963). This assay assesses the ability of RNA preparations to direct the synthesis of [^{14}C]phenylalanine into polyphenylalanine using a translation system derived from a mutant of *E. coli* which is defective in ribonuclease. A threefold stimulation of such activity was revealed within 24 hr following a single injection of estrogen to immature chicks. Maximal activity was noted at 3 days; by 6 days the messenger activity had begun to decline. Thus, the activity of nuclear RNA follows the changes noted in polysome profiles. These data offer further evidence that estrogen must promote the synthesis of new RNA species which precede the appearance of cell-specific proteins (Means *et al.*, 1971).

The oviduct offers a particular advantage to study estrogen-mediated mRNA production (Fig. 7), because this steroid hormone specifically induces the production of the egg-white protein ovalbumin and ovalbumin constitutes some 60% of the total protein synthesized by the oviduct. When oviduct minces from estrogen-treated chicks are incubated for short time periods with radioactive ribonucleotides, a heterogeneous pattern of labeled RNA is noted on the polyribosomes (Means *et al.*, 1972). A large proportion of the label was associated with polysomes containing 10–16 ribosomes. Ovalbumin, which has a molecular weight of 45,000, would be translated, on the average, by a ribosome to mRNA ratio of approximately 13:1. Thus it seemed possible that since a major proportion of protein being synthesized was ovalbumin, the radioactivity might

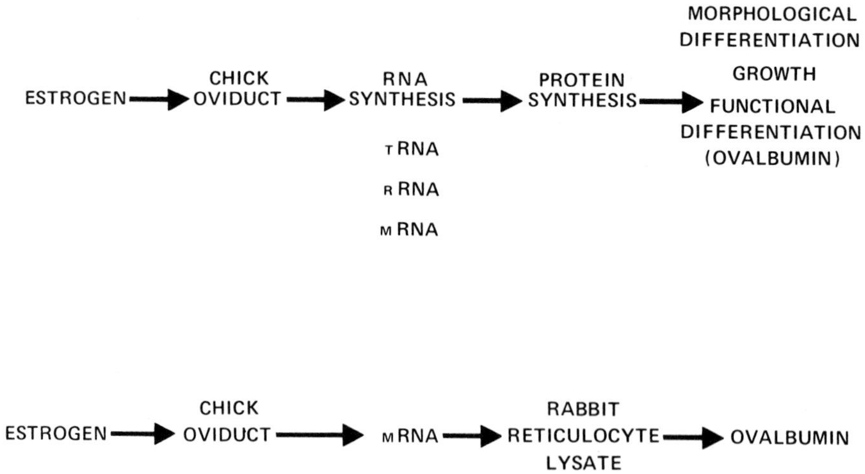

Fig. 7 Diagram of the nuclear events in chick oviduct which occur in response to estrogen and are presumably required for the subsequent physiological responses (upper line). The assay for mRNA activity is shown in the bottom line.

serve as a crude marker for the specific mRNA. Extraction of the polysomes with detergent, followed by sucrose gradient centrifugation of the RNA, yielded a single broad peak of radioactivity that peaked at 16–18 S. Again this closely corresponds to the expected sedimentation value for the ovalbumin messenger.

The only definitive way to prove the existence of the ovalbumin mRNA was to demonstrate that it would support the unambiguous translation of ovalbumin in a cell-free protein synthesis system. It would then be necessary to show the absolute dependence on estrogen administration. A modified rabbit reticulocyte lysate was used as the protein synthesis system. Indeed, an 8–17 S RNA fraction isolated from hen oviduct polysomes was shown to contain ovalbumin mRNA activity (Means *et al.*, 1972; Rosenfeld *et al.*, 1972a). Proof that the reaction product was authentic ovalbumin was gained by several procedures: (1) interaction with a specific antisera raised against purified ovalbumin; (2) solubilization of the immunoprecipitate and analysis on SDS-gels; (3) ion-exchange chromatography on carboxymethyl cellulose followed by reprecipitation with antiovalbumin; and (4) peptide maps (Means *et al.*, 1972; Rhoads *et al.*, 1971).

Table 1 shows that the ovalbumin mRNA activity was specific for RNA isolated from oviduct of estrogen-stimulated chicks and was restricted to the 8–17 S fraction of polysomal RNA. The amount of synthesis was increased by addition of protein synthesis initiation factors. Moreover, inhibitors of chain initiation, such as edeine or aurintricarboxylic acid, or of general protein synthesis, such as puromycin or cycloheximide, completely block ovalbumin synthesis directed by the oviduct mRNA fraction. Ribonuclease destroys the messenger activity whereas deoxyribonuclease does not. Neither steroids nor cyclic AMP have any effect on the translation of ovalbumin mRNA.

This system for translating the ovalbumin mRNA with fidelity allowed us to analyze the hormonal regulation of this specific messenger. Figure 8 shows that the laying hen oviduct in which ovalbumin is being synthesized at its maximal rate contains the greatest amount of ovalbumin mRNA activity. On the other hand, there is no detectable mRNA for ovalbumin in the unstimulated, immature oviduct of the 7-day-old chick. Stimulation of these animals with estrogen for 4, 10, or 16 days leads to increasing activity of the extractable messenger. However, when chicks treated with estrogen for 16 days are subsequently withdrawn from hormone for 16 days, the ovalbumin mRNA activity again returns to very low levels. Finally, readministration of estrogen to the withdrawn animal for 1, 2, or 4 days leads once more to a progressive increase in ovalbumin messenger. These data reveal that indeed the levels of extractable oval-

TABLE I
Ovalbumin Synthesis in Reticulocyte Lysate: Specificity of mRNA and Effects of Various Compounds[a]

Addition to lysate	Cpm ovalbumin synthesized
None	0
Total brain RNA	0
Total liver RNA	0
4 S oviduct RNA	0
18–28 S oviduct RNA	165
8–17 S oviduct RNA	5410
Total oviduct RNA	6952
Total oviduct RNA + edeine	327
Total oviduct RNA + aurintricarboxylic acid	388
Total oviduct RNA + puromycin	233
Total oviduct RNA + cycloheximide	361
Total oviduct RNA + RNase	390
Total oviduct RNA + DNase	6765
Total oviduct RNA + estradiol-17β	6481
Total oviduct RNA + estrogen receptor	6663
Total oviduct RNA + progesterone	6365
Total oviduct RNA + progesterone receptor	6870
Total oviduct RNA + cyclic AMP	7170
Total oviduct RNA − initiation factors	3386

[a] Methods for extraction of RNA, partial purification on Millipore filters, and the protein synthesis system have been previously described (Means *et al.*, 1972; Rosenfeld *et al.*, 1972b). The complete lysate system normally contained exogenous initiation factors prepared from rabbit reticulocytes (Means *et al.*, 1972). Filter-treated RNA (10 μg) was added to the system and samples were incubated for 60 min at 37°C. Ovalbumin synthesis was measured by a specific immunochemical assay (O'Malley *et al.*, 1969). Additions to the complete system were added to the following concentrations: edeine, 7×10^{-6} *M*; aurintricarboxylic acid, 10^{-4} *M*; RNase, 0.1 μg/ml; DNase, 0.1 μg/ml; estradiol-17β or progesterone, 10^{-9} *M*; estrogen receptor, 25 μl oviduct cytosol labeled with estradiol-17β; progesterone receptor, 25 μl oviduct cytosol labeled with progesterone; cyclic AMP, 10^{-7} *M*.

bumin mRNA from oviduct are directly dependent upon estrogen stimulation. Moreover, the changes in ovalbumin mRNA in response to estrogen paralleled or slightly anticipated the changes in oviductal accumulation of ovalbumin (Comstock *et al.*, 1972).

To better assess the changes in mRNA after estrogen treatment kinetic analysis of mRNA activity and rate of ovalbumin synthesis was used

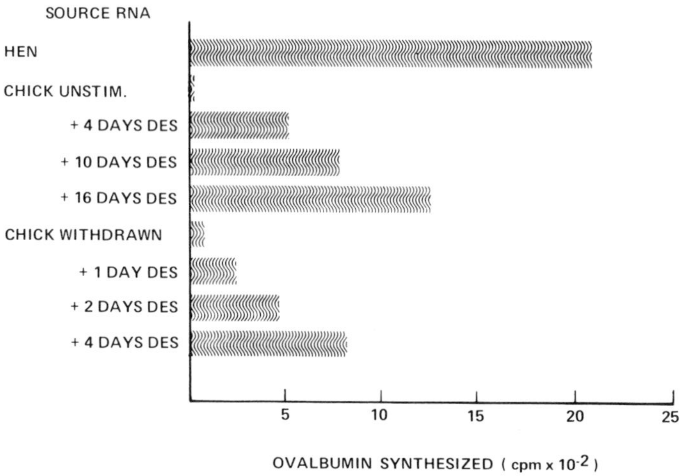

Fig. 8 Effect of estrogen on ovalbumin mRNA activity in chick oviduct. Total RNA was isolated from oviduct as previously described (Rosenfeld *et al.*, 1972a) and 100 μg of each preparation was incubated in the reticulocyte lysate system for protein synthesis (Means *et al.*, 1972). Incubation was for 1 hr at 37°C. Ovalbumin synthesis was assayed using a specific antibody procedure (O'Malley *et al.*, 1969).

(Chan *et al.*, 1973). At various times after injection of estrogen to previously withdrawn chicks, the rate of ovalbumin synthesis was assayed *in vitro* using minces of oviduct. The rate of synthesis of this specific protein was time-dependent, peaking at 18 hr after the injection of steroid. The approximate initial half-life time for the messenger was calculated from the descending limb of the induction curve to be 8–10 hr. Similar studies were then performed except that ovalbumin mRNA was extracted and the activity quantified in the translation system. A remarkable parallelism was shown to exist between the rate of ovalbumin synthesis and the available mRNA. Again mRNA was detectable prior to ovalbumin synthesis. Moreover, the half-life of the mRNA was again calculated to be 8–10 hr.

Thus is appears that estrogen does act in the nucleus to promote the synthesis of mRNA's which code for the cell-specific proteins. Since ovalbumin represents nearly 60% of the protein synthesized in oviduct gland cells under influence of estrogen, it was questioned whether the ovalbumin mRNA was transcribed from single copy DNA or whether this might represent an instance of gene amplification. The ovalbumin mRNA was purified by a variety of procedures until its activity, tested in the protein synthesis system, had been increased some thousandfold compared to a preparation of total oviduct RNA. The partially purified mRNA was incubated with RNA-directed DNA polymerase from avian myeloblastosis virus in order to produce a copy of radioactively labeled DNA which would be the complement of ovalbumin mRNA. The [^{3}H]-DNA produced was sized on alkaline sucrose gradients and shown to contain fragments up to 2000 nucleotides in length. When the [^{3}H]DNA was reacted with excess ovalbumin mRNA, 90% of the labeled DNA formed a stable hybrid with the mRNA, indicating that the [^{3}H]DNA was indeed a complementary copy. A fraction of the [^{3}H]DNA was then used in a DNA-excess hybridization experiment (Melli and Bishop, 1969). Whole chick DNA was sheared to 400 nucleotide lengths and incubated with the [^{3}H]DNA as an excess of 10^7:1. Complementary [^{3}H]DNA hybridized to chick DNA with a $Cot_{1/2}$ of 560. Under similar conditions of second-order kinetics single copy or unique sequence DNA hybridizes with a $Cot_{1/2}$ of 600 (Liarakos *et al.*, 1973; Rosen *et al.*, 1973). Thus the complementary [^{3}H]DNA has hybridization properties which suggest that the ovalbumin gene is only represented one time in the oviduct genome. These data suggest that estrogen may act at the level of transcription to stimulate production of numerous copies of a single gene. This type of hormonal regulation would lead to a high intracellular concentration of ovalbumin mRNA and subsequently of ovalbumin itself.

B. Progesterone

Available data strongly suggest that the estrogen acts in the nucleus to promote the synthesis of mRNA's which are necessary for the subsequent actions of this steroid on growth and differentiation. An important question that remained unanswered, however, was the general applicability of this mechanism in the actions of other hormones. In the chick oviduct progesterone has been shown to specifically control the synthesis of the egg-white protein, avidin (O'Malley *et al.*, 1969; O'Malley, 1967). Unlike estrogen, no marked changes are seen in total cell RNA synthesis and polysome profiles are seemingly unaltered (O'Malley *et al.*, 1969; Means and O'Malley, 1971). Avidin represents only 0.1% of the total egg-white protein. Consequently, it followed that the mRNA for this protein might also be present in small amounts. Extraction of total RNA from estrogen-stimulated hen oviducts proved to be less than satisfactory as a means of quantitation. When such RNA preparations were tested in the heterologous protein synthesis system it was not always possible to demonstrate avidin synthesis by a specific immunoprecipitation procedure. In order to assure reproducible results it was necessary to effect a partial purification of the messenger fraction. We were able to take advantage of the fact that many mRNA's, including the one for avidin, contain at the 3′-terminal end a rather extensive sequence of polyadenylate residues. The presence of a poly(A) sequence was shown by Brawerman *et al.* (1972) to allow the mRNA to be selectively adsorbed to nitrocellulose filters. Application of this procedure to oviduct RNA results in a one-step fifty-fold purification of avidin (and ovalbumin) mRNA (Rosenfeld *et al.*, 1972b). This simple procedure allowed us to routinely and consistently measure the avidin mRNA activity that appears in oviduct in response to progesterone (O'Malley *et al.*, 1972).

Avidin mRNA is only present in the 8–17 S fraction of oviduct polysomal RNA. Moreover, it has been shown by sucrose gradient analysis to have an average sedimentation of 9 S. This would be expected if the message was to code for a protein of approximately 18,000 daltons. This is, in fact, the molecular weight of a single subunit of avidin. Avidin mRNA activity is also abolished by RNase, and no avidin is synthesized when inhibitors of peptide chain initiation or elongation are present in the cell-free system. Direct addition of progesterone, estrogen, or cAMP has no effect on the translation process.

Avidin mRNA activity is highest in oviducts of mature laying hens where progesterone stimulation is maximal. No activity can be demonstrated in the unstimulated immature chicks or in oviducts from animals which have received multiple injections of estrogen. However, following a single injection of progesterone, avidin mRNA activity can be detected within 6 hr (Chan et al., 1973). Maximal concentrations are achieved by 18 hr, considerably before maximal tissue levels of avidin are achieved at 48 hr. From the decline of mRNA activity it can be calculated that the half-life of avidin mRNA is in the order of 12–15 hr. This makes the avidin mRNA somewhat longer lived than the messenger for ovalbumin. At any rate these data suggest that both estrogen and progesterone act on the nucleus of target cells to promote the synthesis of mRNA's. This response seems to be a rate-limiting step in the subsequent production of specific proteins.

XI. Summary

The experimental observations discussed in this chapter demonstrate the great variation in target-tissue response that can occur after administration of steroid hormones. The female sex steroids can exert regulatory effects on the synthesis, activity, and possibly even the degradation of tissue enzymes and structural proteins. Each response, nevertheless, appears to be dependent on the synthesis of nuclear RNA. In many instances, the steroid actually promotes a qualitative change in the base composition and sequence of the RNA synthesized by the target cell, implying a specific effect on gene transcription. Most importantly, we now have direct quantitative evidence that sex steroids cause a net increase in the intracellular levels of specific messenger RNA molecules in target tissues.

It thus appears that there is an evolving pattern of steroid hormone action which consists of (1) uptake of the hormone by the target cell and binding to a specific cytoplasmic receptor protein; (2) transport of the steroid–receptor complex to the nucleus; (3) binding of this "active" complex to specific "acceptor" sites on the genome (chromatin DNA and acidic protein); (4) activation of the transcriptional apparatus resulting in the appearance of new RNA species which includes specific mRNA's; (5) transport of the hormone-induced RNA to the cytoplasm resulting in synthesis of new proteins on cytoplasmic ribosomes; (6) the occurrence of the specific steroid-mediated "functional response" characteristic of that particular target tissue (Fig. 9).

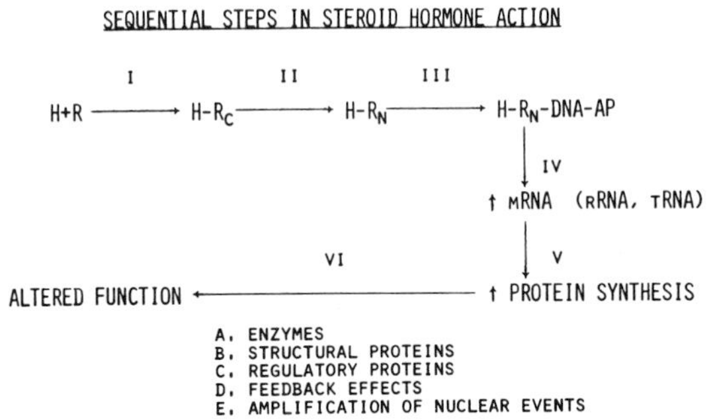

Fig. 9 Sequential events in the early action of steroid hormones on target cells.

A final objective in elucidation of the mechanism of steroid hormone action must involve a careful delineation of the biochemistry of the transfer of information held by the steroid hormone complex to the nuclear transcription apparatus. If the assumptions discussed in the beginning of this chapter are correct, we should ultimately learn how this hormone–receptor complex exerts a specific regulatory effect on nuclear RNA metabolism: (1) by direct effects on chromatin template leading to increased gene transcription and thus RNA synthesis; (2) by activation of the polymerase complex itself; (3) by inhibition of RNA breakdown; or (4) by regulation of intranuclear processing of large precursor molecules to biologically active sequences and transport of RNA from the nucleus to the cytoplasmic sites of cellular protein synthesis.

REFERENCES

Anderson, J., Clark, J. H., and Peck, E. J. (1972). *Biochem. J.* **126,** 561.

Arnaud, M., Beziat, Y., Guilleux, J. C., Hough, A., Hough, D., and Mousseron-Canet, M. (1971a). *Biochim. Biophys. Acta* **232,** 117.

Arnaud, M., Beziat, Y., Borgna, J. L., Guilleux, J. C., and Mousseron-Canet, M. (1971b). *Biochim. Biophys. Acta* **254,** 241.

Barker, K. L., and Warren, J. C. (1966). *Proc. Nat. Acad. Sci. U.S.* **56,** 1298.

Barry, J., and Gorski, J. (1971). *Biochemistry* **10,** 2384.

Baulieu, E. E. *et al.* (1971). *Rec. Progr. Horm. Res.* **27,** 351.

Billing, R. J., Barbiroli, B., and Smellie, R. M. S. (1969a). *Biochim. Biophys. Acta* **190,** 52–59.

Billing, R. J., Barbiroli, B., and Smellie, R. M. S. (1969b). *Biochim. Biophys. Acta* **190,** 60–65.

Brawerman, G., Mendecki, J., and Lee, S. Y. (1972). *Biochemistry* **11,** 637.

Britten, R. J., and Kohne, D. E. (1968). *Science* **161,** 529.

Chan, L., Means, A. R., and O'Malley, B. W. (1973). *Proc. Nat. Acad. Sci. U.S.* **70,** 1870.

Church, R. B., and McCarthy, B. J. (1970). *Biochim. Biophys. Acta* **199,** 103.

Chytil, F. C., and O'Malley, B. W. (1973). Submitted.

Chytil, F. C., and Spelsberg, T. C. (1971). *Nature (London) New Biol.* **233,** 215.

Clark, J. H., Anderson, J., and Peck, E. J. (1972). *Science* **176,** 528.

Clark, J. H., Anderson, J. N., and Peck, E. J. (1973). *In* "Receptors for Reproductive Hormones" (B. W. O'Malley, and A. R. Means, eds.), pp. 15–59. Plenum Press, New York.

Comstock, J. P., Rosenfeld, G. C., O'Malley, B. W., and Means, A. R. (1972). *Proc. Nat. Acad. Sci. U.S.* **69,** 2377–2380.

De Angelo, A. B., and Gorski, J. (1970). *Proc. Nat. Acad. Sci. U.S.* **66,** 693.

Edelman, I. S., and Fanestil, D. D. (1970). *In* "Biochemical Actions of Hormones" (J. Litwack, ed.), Vol. I, pp. 324–331, Academic Press, New York.

Fasman, G. D., Schaffenhausen, B., Goldsmith, L., and Adler, A. (1970). *Biochemistry* **9,** 2814.

Glasser, S. R., Chytil, F. C., and Spelsberg, T. C. (1972). *Biochem. J.* **130,** 947.

Gorski, J. (1964). *J. Biol. Chem.* **239,** 889–892.

Gorski, J., Noteboom, W. D., and Nicolette, J. A. (1965). *J. Cell. Comp. Physiol.* **66,** 91–109.

Gorski, J., Toft, D., Shyamala, G., Smith, D., and Notides, A. (1968). *Rec. Progr. Horm. Res.* **24,** 45.

Gorski, J., Williams, D., Giannopoulos, G., and Stancel, G. (1973). *In* "Receptors for Reproductive Hormones" (B. W. O'Malley and A. R. Means, eds.), pp. 1–14. Plenum Press, New York.

Greenman, D. L., Wicks, W. D., and Kenney, F. T. (1965). *J. Biol. Chem.* **240,** 4420.

Hahn, W. E., Church, R. H., Gorbman, A., and Wilmot, L. (1968). *Gen. Comp. Endocrinol.* **10,** 438.

Hamilton, T. H. (1963). *Proc. Nat. Acad. Sci. U.S.* **49,** 373–379.

Hamilton, T. H. (1964). *Proc. Nat. Acad. Sci. U.S.* **51,** 83–89.

Hamilton, T. H. (1968). *Science* **161,** 649.

Hamilton, T. H., Widnell, C. C., and Tata, J. R. (1965). *Biochim. Biophys. Acta* **108,** 168–172.

Hamilton, T. H., Teng, C.-S., and Means, A. R. (1968). *Proc. Nat. Acad. Sci. U.S.* **59,** 1265–1272.

Hamilton, T. H., Widnell, C. C., and Tata, J. R. (1968). *J. Biol. Chem.* **243,** 408–417.
Hamilton, T. H., Teng, C.-S., Means, A. R., and Luck, D. N. (1971). *In* "The Sex Steroids" (K. W. McKerns, ed.), pp. 197–240. Appleton, New York.
Harris, G. S. (1971). *Nature New Biol.* **231,** 246.
Jensen, E. V., and DeSombre, E. R. (1972). *Ann. Rev. Biochem.* **41,** 203.
Jensen, E. V., and Jacobson, H. I. (1962). *Rec. Progr. Horm. Res.* **18,** 387.
Jensen, E. V., DeSombre, E. R., Jungblut, P. W., Stumpf, W. E., and Roth, L. J. (1969). *In* "Autoradiography of Diffusible Substances" (L. J. Roth and W. E. Stumpf, eds.), p. 81. Academic Press, New York.
Jensen, E. V. *et al.* (1968). *Proc. Nat. Acad. Sci. U.S.* **59,** 632.
Jensen, E. V., Mohla, E. V., Brecher, P. I., and DeSombre, E. R. (1973). *In* "Receptors for Reproductive Hormones" (B. W. O'Malley, and A. R. Means, eds.), pp. 60–79. Plenum Press, New York.
Kapadia, G., Means, A. R., and O'Malley, B. W. (1971). *Cytobios* **3,** 33–42.
King, R. J., and Gordon, J. (1972). *Nature (London)* **240,** 185.
Knowler, J. T., and Smellie, R. M. S. (1971). *Biochem. J.* **125,** 605–614.
Kohler, P. O., Grimley, P. M., and O'Malley, B. W. (1969). *J. Cell Biol.* **40,** 8.
Korenman, S. G., and Rao, B. R. (1968). *Proc. Nat. Acad. Sci. U.S.* **61,** 1028.
Liao, S., and Fang, S. (1969). *Vitam. Horm.* **27,** 17.
Liao, S., Tymoczko, J. L., Liang, T., Anderson, K. M., and Fang, S. (1971). *Advan. Biosci.* **7,** 213–234.
Liao, S., Liang, T., Shao, T. C., and Tymoczko, J. L. (1973). *In* "Receptors for Reproductive Hormones" (B. W. O'Malley, and A. R. Means, eds.), Plenum Press, New York (in press).
Liarakos, C. D., Rosen, J. M., and O'Malley, B. W. (1973). *Biochemistry* **12,** 2809.
Luck, D. N., and Hamilton, T. H. (1972). *Proc. Nat. Acad. Sci. U.S.* **69,** 157.
Mainwaring, W. I. P., and Peterken, B. M. (1971). *Biochem. J.* **125,** 285.
Maul, G. G., and Hamilton, T. H. (1967). *Proc. Nat. Acad. Sci. U.S.* **57,** 1371.
McGuire, W. L., and O'Malley, B. W. (1968). *Biochim. Biophys. Acta* **157,** 187.
Means, A. R., and Hamilton, T. H. (1966a). *Proc. Nat. Acad. Sci. U.S.* **56,** 1594–1598.
Means, A. R., and Hamilton, T. H. (1966b). *Proc. Nat. Acad. Sci. U.S.* **56,** 686–693.
Means, A. R., and O'Malley, B. W. (1971). *Biochemistry* **10,** 1570.
Means, A. R., and O'Malley, B. W. (1971). *Acta Endocrinol. Suppl.* **153,** 318.
Means, A. R., and O'Malley, B. W. (1972). *Metabolism* **21,** 357–370.
Means, A. R., Abrass, I. B., and O'Malley, B. W. (1971). *Biochemistry* **10,** 1561.
Means, A. R., Comstock, J. P., Rosenfeld, G. C., and O'Malley, B. W. (1972). *Proc. Nat. Acad. Sci. U.S.* **69,** 1146.
Melli, M., and Bishop, J. O. (1969). *J. Mol. Biol.* **40,** 117.
Mohla, S., DeSombre, E. R., and Jensen, E. V. (1972). *Biochim. Biophys. Res. Commun.* **46,** 661.
Mueller, G. C. (1965). *In* "Mechanisms of Hormone Action" (P. Carlson, ed.), p. 228. Academic Press, New York.
Mueller, G. C., Herranen, A. M., and Jervell, K. J. (1958). *Rec. Progr. Horm. Res.* **14,** 95–139.
Nicolette, J. A., and Babler, M. (1972). *Arch. Biochem. Biophys.* **149,** 183–188.
Nirenberg, M. W. (1963). *Methods Enzymol.* **6,** 17.
Noteboom, W. D., and Gorski, J. (1965). *Arch. Biochem. Biophys.* **111,** 559.
Oliver, J. M. (1971). *Biochem. J.* **121,** 83–88.
Oliver, J. M., and Kellie, A. E. (1970). *Biochem. J.* **119,** 187–191.
O'Malley, B. W. (1967). *Biochemistry* **6,** 2546.
O'Malley, B. W., and McGuire, W. L. (1968). *Proc. Nat. Acad. Sci. U.S.* **60,** 1527.

O'Malley, B. W., Aronow, A., Peacock, A. C., and Dingman, C. W. (1968). *Science* **162,** 567.
O'Malley, B. W., McGuire, W. L., Kohler, P. O., and Korenman, S. G. (1969). *Rec. Progr. Horm. Res.* **25,** 105.
O'Malley, B. W., Sherman, M. R., and Toft, D. O. (1970). *Proc. Nat. Acad. Sci. U.S.* **67,** 501.
O'Malley, B. W., Sherman, M. R., Toft, D. O., Spelsberg, T. C., Shrader, W. T., and Steggles, A. W. (1971). *Advan. Biosci.* **7,** 213–234.
O'Malley, B. W., Toft, D. O., Sherman, M. R. (1971). *J. Biol. Chem.* **246,** 1117.
O'Malley, B. W., Rosenfeld, G. C., Comstock, J. P., and Means, A. R. (1972a). *Nature (London) New Biol.* **240,** 45–48.
O'Malley, B. W., Spelsberg, T. C., Schrader, W. T., Chytil, F., and Steggles, A. W. (1972b). *Nature (London)* **235,** 141.
O'Malley, B. W., Comstock, J. C., Rosen, J. R., Liarakos, C., Chan, L., Rosenfeld, G. C., Harris, S., and Means, A. R. (1973). *Cold Spring Harbor Symp. Quant. Biol.* (In press).
Palmiter, R. D., Christensen, A. K., and Schimke, R. T. (1970). *J. Biol. Chem.* **245,** 833.
Perotta, C. A. (1962). *Amer. J. Anat.* **111,** 195.
Raynaud-Jammet, M., and Baulieu, E. E. (1969). *C. R. Acad. Sci. Paris* **268,** 3211–3214.
Rhoads, R. E., McKnight, B. S., and Schimke, R. T. (1971). *J. Biol. Chem.* **246,** 7407.
Rochefort, H., and Baulieu, E. E. (1968). *C. R. Acad. Sci. Paris* **D267,** 662.
Roeder, R. G., and Rutter, W. J. (1970). *Biochemistry* **9,** 2543.
Rosen, J. M., Liarakos, C. D., and O'Malley, B. W. (1973). *Biochemistry* **12,** 2803.
Rosenfeld, G. C., Comstock, J. P., Means, A. R., and O'Malley, B. W. (1972a). *Biochem. Biophys. Res. Commun.* **46,** 1695.
Rosenfeld, G. C., Comstock, J. P., Means, A. R., and O'Malley, B. W. (1972b). *Biochem. Biophys. Res. Commun.* **47,** 387–392.
Schrader, W. T., and O'Malley, B. W. (1972). *J. Biol. Chem.* **247,** 51.
Schrader, W. T., Toft, D. O., and O'Malley, B. W. (1972). *J. Biol. Chem.* **247,** 2401.
Segal, S. J., and Scher, W. (1967). *In* "Cellular Biology of the Uterus" (R. M. Wynn, ed.), p. 114. Appleton, New York.
Sherman, M. R., Corvol, P. L., O'Malley, B. W. (1970). *J. Biol. Chem.* **245,** 6085.
Simpson, R. T., and Sober, H. A. (1970). *Biochemistry.* **9,** 3103.
Socher, S. H., and Davidson, D. (1971). *J. Cell Biol.* **48,** 248.
Socher, S. H., and O'Malley, B. W. (1973). *Develop. Biol.* **30,** 2.
Spelsberg, T. C., Steggles, A. W., and O'Malley, B. W. (1971). *Biochim. Biophys. Acta* **240,** 888.
Spelsberg, T. C., Steggles, A. W., O'Malley, B. W. (1971). *J. Biol. Chem.* **246,** 4188.
Spelsberg, T. C., Steggles, A. W., Chytil, F., and O'Malley, B. W. (1972). *J. Biol. Chem.* **247,** 1368.
Spelsberg, T. C., Mitchell, W., Chytil, F. C., and O'Malley, B. W. (1973). *Biochim. Biophys. Acta* (In press).
Stancel, G., Leung, K. M. T., and Gorski, J. (1973). *Biochemistry* **12,** 2130.
Steggles, A. W., Spelsberg, T. C., Glasser, S. R., and O'Malley, B. W. (1971). *Proc. Nat. Acad. Sci. U.S.* **68,** 1479.
Szego, C. M. (1971). *In* "The Sex Steroids" (B. W. McKerns, ed.), p. 1–52. Appleton, New York.
Szego, C. M. (1965). *Fed. Proc.* **24,** 1343–1352.
Teng, C.-S., and Hamilton, T. H. (1968). *Proc. Nat. Acad. Sci. U.S.* **60,** 1140.

Toft, D. O. (1972). *J. Steroid Biochem.* **3,** 515–522.

Toft, D. O. (1973). *In* "Receptors for Reproductive Hormones" (B. W. O'Malley, and A. R. Means, eds.), pp. 85–96. Plenum Press, New York.

Toft, D., and Gorski, J. (1966). *Proc. Nat. Acad. Sci. U.S.* **55,** 1574.

Tomkins, G. M., Thompson, E. B., Hayrashi, S., Gelehrter, T., Granner, D., and Peterkofsky, B. (1966). *Cold Spring Harbor Symp. Quant. Biol.* **31,** 349.

Tomkins, G. M., Gelehrter, T. D., Granner, D., Martin, D., Samuels, H. H., and Thompson, E. B. (1969). *Science* **166,** 1474.

Wagner, T., and Spelsberg, T. C. (1971). *Biochemistry* **10,** 2599.

Williams-Ashman, H. G., and Reddi, A. H. (1971). *Ann. Rev. Physiol.* **33,** 31.

Yamamoto, K. R., and Alberts, B. M. (1972). *Proc. Nat. Acad. Sci. U.S.* **69,** 2105.

11

The Nucleus During Avian Erythroid Differentiation

N. R. Ringertz and L. Bolund

I. Introduction

In recent years avian erythropoiesis has been used extensively as a model for cell differentiation. Much of this popularity is based on the fact that erythropoiesis is a well-defined differentiation process characterized by the synthesis of specific macromolecules and by marked changes in nuclear function. During erythropoiesis the rate of RNA synthesis gradually decreases at the same time as the pattern of transcription and protein synthesis changes. Cells representing different stages of erythropoiesis can be isolated by gradient centrifugation from embryonic and adult blood as well as from normal and anemic bone marrow. This makes it possible to analyze the gradual inactivation of the cell nucleus with biochemical methods. At the final stage of erythroid differentiation the nucleus is virtually inactive with respect to RNA and DNA synthesis and protein synthesis is greatly reduced. Contrary to the situation in mammals, the nucleus is retained in an inactive state in the mature erythrocytes of birds, amphibians, and reptiles. Nuclei from these cells can be obtained in very large quantities making it possible to analyze in great detail the biochemical composition of a diploid eukaryotic genome in a repressed state.

Chick erythropoiesis has been analyzed both *in vivo* (Dantschakoff, 1909; Dawson, 1936; Lemez, 1964; Wilt, 1967; Bank *et al.*, 1970; Campbell *et al.*, 1971; Weintraub *et al.*, 1971) and in organ cultures of chick blastoderm (Spratt and Hass, 1960a, b; Hell, 1964; Levere and Granick, 1967; Wilt, 1967; Miura and Wilt, 1969, 1970, 1971; Hagopian and Ingram, 1971; Reynolds and Ingram, 1971). In this chapter, the discussion will be focused on the nucleus during avian erythropoiesis. For more detailed information on the cell kinetics, biochemistry, and morphology of erythroid differentiation in birds, reviews by Romanoff (1960), Lucas and Jamroz (1961) and Wilt (1967) should be consulted. However, since the function of the avian erythrocyte nucleus is intimately reflected in its transcription pattern and in the synthesis of specific proteins, we need to summarize briefly the main features of erythroid differentiation during avian embryogenesis. One important point in this context is that there are at least two different erythroid cell lines, the primitive erythroid cells found in very early embryos and the definitive erythroid cells found in later embryonic stages and in adult animals. Both erythroid cell lines mature in a similar way and can both be considered as models for genome repression. Our discussion will be focused on definitive erythroid cells because these cells offer the greatest possibilities for biochemical work and for use in somatic cell hybridization (for review see Ringertz and Bolund, 1974).

II. Primitive Erythropoiesis

The first erythroid precursor cells (Sabin, 1920) arise in the area opaca vasculosa of the early avian blastoderm. These cells form clusters, the blood islands, in which erythroid differentiation begins. In chick embryos, hemoglobin-containing cells can be detected after about 32 hr of incubation (Wilt, 1967). These early erythroid cells are referred to as primitive (or primary) erythroid cells and differ in morphology and types of hemoglobin from the definitive erythroid cells found in older embryos and after hatching. In the following we shall define only briefly some stages of primitive erythropoiesis (for a detailed review of avian erythropoiesis, see Romanoff, 1960).

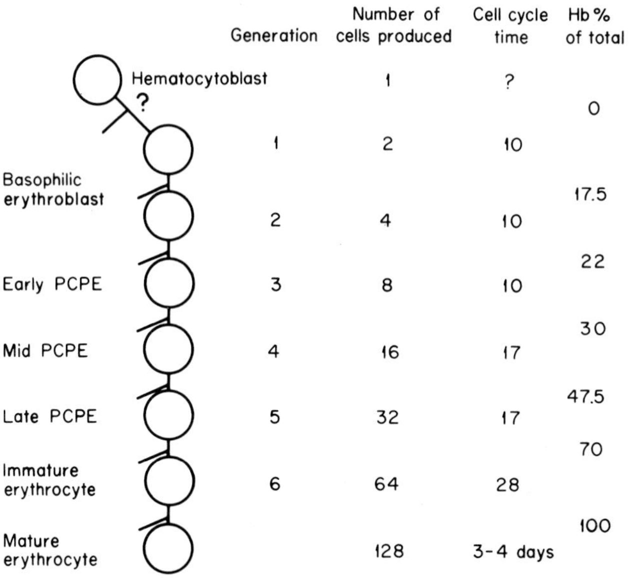

Fig. 1 Schematic diagram of primitive erythroid development expressed in terms of morphological stages of development, the number of generations, and the cell cycle time for each generation. Hb value for each mitosis expressed as a percentage of the Hb content of the mature primitive erythrocyte. These data summarize the findings of Weintraub *et al.* (1971). PCPE, polychromatic primitive erythrocyte. (From Campbell *et al.*, 1971.)

The earliest cell type recognized in the primitive erythroid cell line is the erythroblast. Measurements of the hemoglobin content of mitotic primitive erythroid cells (Fig. 1) suggest that there are six consecutive cell cycles (Weintraub *et al.*, 1971) during which the hemoglobin content increases. The earliest erythroblast generation, basophilic erythroblasts, are believed to arise from stem cells (hematocytoblasts). Through a series of divisions the basophilic erythroblasts become early, mid-, and late polychromatic erythrocytes,* which finally give rise to immature and mature erythrocytes (Fig.1). The maturation of the primitive chick erythrocytes is well synchronized so that at any given time point the primitive erythroid cell population consists mainly of one cell type. By day 5 the primitive cells have stopped dividing and by day 13 they have probably disappeared from the circulation being replaced by the definitive type of cells.

The primitive chick erythroid cells produce two major types of hemoglobin (hemoglobin P and E) which are different from those of definitive cells (which make hemoglobins A and D) (D'Amelio, 1966; Wilt, 1967;

* These cells are also referred to as polychromatic erythroblasts by some authors (see, e.g., Small and Davies, 1972).

Ingram, 1972; Fraser *et al.*, 1972). The synthesis of hemoglobin is most rapid in the dividing cells (primitive erythroblasts and polychromatic erythrocytes), but hemoglobin also accumulates at a reduced rate for 3–4 days after cell division has stopped (Campbell *et al.*, 1971). Hemoglobin is found both in the cytoplasm and the nucleus, the nuclear concentration being somewhat lower than the cytoplasmic (Small and Davies, 1972). The messenger RNA for the primitive hemoglobins is made at an early stage of the differentiation process since after the head fold stage (19 hr of incubation) blocking of RNA synthesis by actinomycin D fails to stop hemoglobin synthesis (Wilt, 1965). The rate of RNA synthesis diminishes gradually during primitive erythropoiesis and between days 5 and 6 there is a dramatic decrease in RNA synthesis (Weintraub *et al.*, 1971). Parallel to this the RNA content (Thorell and Raunich, 1966) and the number of ribosomes (Small and Davies, 1972) in the cytoplasm decrease markedly. The mature primitive erythrocytes have lost almost all ribosomes and contain very little RNA.

Nuclear size and the nuclear–cytoplasmic ratio decrease during primitive erythropoiesis as the nucleus undergoes a marked condensation. The mature primitive erythrocyte has the form of a round, biconvex lens and contains a small, round nucleus where the chromatin is organized into tightly packed clumps oriented along the nuclear membrane. Nucleoli are present in mature erythrocytes (Dawson, 1936) but are much smaller than in earlier maturation stages. The nucleolus in mature erythrocytes has been estimated to be approximately 0.5 μm in diameter and consists mainly of fibrillar material (Small and Davies, 1972). The number of nuclear pores decreases by a factor of about 2 during the maturation of the primitive erythroid cells (Small and Davies, 1972).

III. Definitive Erythropoiesis

A. *Maturation Stages*

From day 7 the primitive erythrocytes are gradually replaced by definitive red cell lines (Fig. 2). The stem cells for these erythrocytes are first found in the yolk sac, later in liver and spleen, and eventually in the bone marrow as the principal definitive site. The different stages during the maturation of definitive erythroid cells in the chick have been studied by Dantschakoff (1909), Dawson (1936), and others. In the following, cell stages will be defined according to Lucas and Jamroz (1961).

The earliest cell type that can be recognized on the basis of morphology is the erythroblast. These cells are the largest in the erythroid matu-

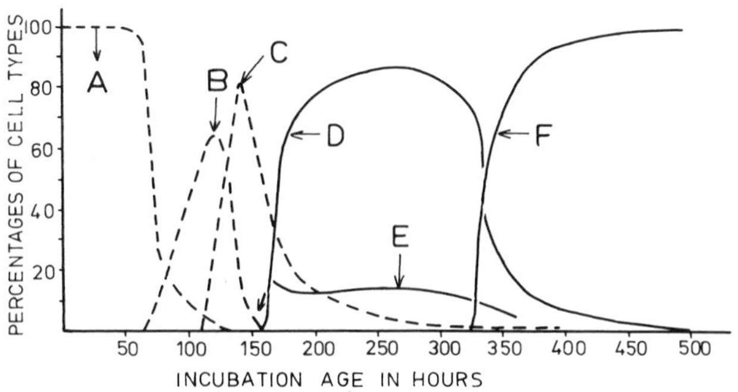

Fig. 2 Curves showing the percentage rise and fall of three stages of maturity of the primary and later embryo erythrocytes in the blood. (A) Primary erythroblasts and early primary polychromatic erythrocytes; (B) late primary polychromatic erythrocytes; (C) mature primary erythrocytes; (D) late polychromatic erythrocytes of later definitive generations; (E) mid-polychromatic erythrocytes of later definitive generations; (F) mature definitive erythrocytes. (From Lucas and Jamroz, 1961.)

ration series. Some authors (Small and Davies, 1972), in place of the erythroblast stage, distinguish a proerythroblast and a basophilic erythroblast. The erythroblast nuclei are rounded and contain a large nucleolus. The chromatin is finely distributed except for a few condensed clumps. Erythroblasts synthesize DNA, RNA, and protein, some of which is hemoglobin. Even in the later generations of erythroblasts (probably the smallest erythroblasts) the quantity of hemoglobin is, however, very small. Hemoglobin-containing cells are seen in mitosis but the exact number of erythroblast generations during definitive avian erythropoiesis is not known. The erythroblasts give rise to early polychromatic erythrocytes. Contrary to the situation during primitive erythropoiesis the polychromatic definitive erythrocytes are considered to be nondividing cells (Williams, 1972a) but they still show an active synthesis of both RNA and protein. Although the absence of DNA replication is a convenient criterion for distinguishing definitive polychromatic erythrocytes from erythroblasts in biochemical experiments, it is not quite certain that this agrees with the cytological definitions for these two cell types in the hematological literature.

The "polychromatic" appearance of these cells in many of the standard hematological stains derives from the fact that they stain for both RNA (basic dyes) and hemoglobin (acidic dyes). According to the cytological definitions (Lucas and Jamroz, 1961) the nucleus of the early polychromatic erythrocyte contains a small nucleolus and its chromatin is condensed into clumps. The next maturation stage, the mid-polychromatic erythrocyte, is rounded or oval and smaller than the preceding stage. The nucleus is small, rounded, lacks a nucleolus, and has a clumped chromatin. The rate of RNA synthesis has decreased (Cameron and Prescott, 1963) and considerable quantities of hemoglobin have accumulated in the cytoplasm. The late polychromatic erythrocyte is also round or slightly oval with an eosinophilic cytoplasm rich in hemoglobin. The clumping of the chromatin is marked and the shape of the nucleus is beginning to shift from rounded to oval. The reticulocytes have an oval nucleus but can be identified only after certain staining reactions on the basis of basophilic cytoplasmic inclusions. The mature erythrocyte is oval in shape and has an oval or rod-shaped nucleus with a tightly condensed chromatin. The cytoplasm is very rich in hemoglobin although practically all the RNA has been degraded. The cytoplasm appears homogeneous and strongly eosinophilic. The mature erythrocyte represents the terminal stage in a cell differentiation process during which DNA replication and mitotic activity stop completely and where RNA and protein synthesis are also arrested.

B. Isolation of Erythroid Cells and Their Nuclei

1. SEPARATION OF DIFFERENT CELL TYPES

The cell population found in normal, adult, hen blood consists almost exclusively (99.3%) of mature erythrocytes (Table I). In hens made anemic by repeated bleedings or by phenylhydrazine-induced hemolysis, large numbers of polychromatic erythrocytes and a small number of erythroblasts can be found in addition to remaining nonlysed, mature erythrocytes. Erythroblasts constitute a considerable proportion of the cells found in bone marrow preparations from normal or anemic hens, but bone marrow preparations also contain large numbers of nonerythroid

TABLE I
Cell Types in Unfractionated Erythroid Cell Populations[a]

	Percentage of total cells			
Cell type	Normal blood	Anemic blood	Normal bone marrow	Anemic bone marrow
Erythroblasts	0.0	0.1	4.0	39.0
Early and mid-polychromatic erythrocytes	0.0	40.0	5.9	37.5
Late polychromatic erythrocytes	0.0	55.0	14.7	7.5
Mature erythrocytes	99.3	3.0	53.0	0.0
Nonerythroid cells	0.7	2.0	22.4	16.0

[a] From Williams (1972a).

cells. In order to obtain reasonably pure suspensions of polychromatic erythrocytes and erythroblasts from anemic blood and bone marrow, respectively, the different cell types have to be separated. This can be done by equilibrium density gradient centrifugation on bovine serum albumin gradients (Kabat and Attardi, 1967; Williams, 1972a). With this method cells are separated primarily on the basis of differences in their buoyant densities. The purity of the stages obtained is illustrated in Table II. Early maturation stages characterized by their large size, low hemoglobin, and high RNA content are obtained in low density fractions while the hemoglobin-rich but small, mature erythrocytes are found in the high density fractions (Figs. 3 and 4).

TABLE II
Purity of Cell Fractions Obtained by Buoyant Density Centrifugation of Avian Erythroid Cells[a]

Cell fraction	Buoyant density range (gm/cm^3)	Erythroblasts (%)	Polychromatic erythrocytes (%)	Erythrocytes (%)	Nonerythroid cells (%)	Cells in S phase (%)
Erythroblasts[b]	1.061–1.064	86	4.5	<1	9.5	63
Mid- and late polychromatic erythrocytes[c]	1.071–1.083	<1	97	3	<1	<1
Erythrocytes[d]	>1.091	<1	<1	100	<1	<0.5

[a] Modified from Appels *et al.* (1972).
[b] From anemic bone marrow.
[c] From anemic blood.
[d] From normal blood.

2. ISOLATION OF GHOSTS AND NUCLEI

Nuclei have been isolated from erythroblasts as well as from polychromatic and mature erythrocytes. Erythroblasts do, however, present special difficulties owing to a high content of proteolytic enzymes and a marked instability of the nuclear material (Appels, 1972).

The first step in the isolation of nuclei from mature erythrocytes normally has been to lyse the cells. Lysis can be induced by osmotic shocks, treatment with detergents, hemolyzing viruses, etc., and frequently results in a more or less complete loss of cytoplasmic material. Mature avian erythrocytes lack an endoplasmic reticulum and the only cytoplasmic structures which can be identified after lysis are a few mitochondria and

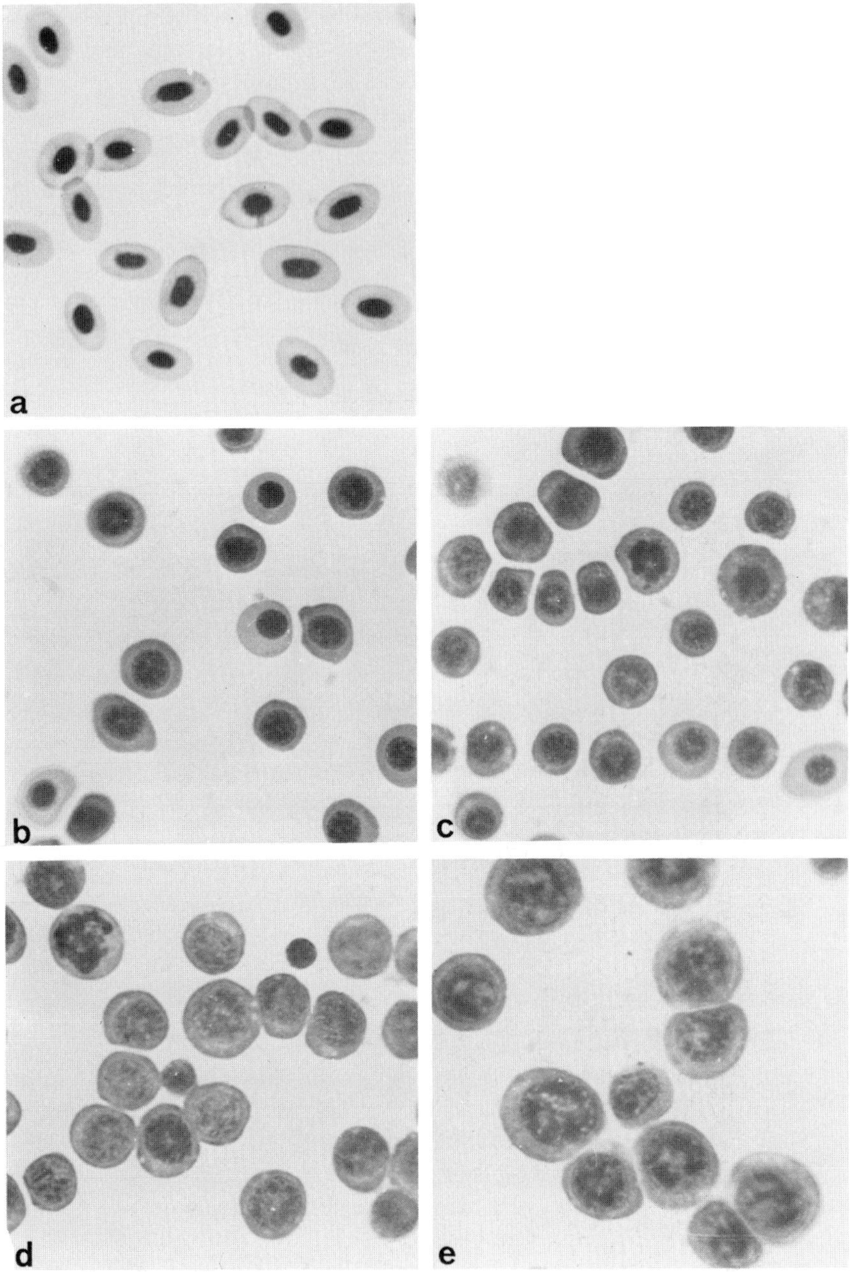
a
b
c
d
e

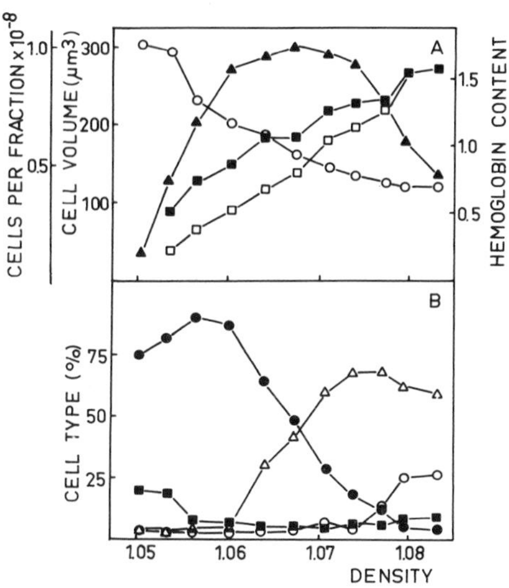

Fig. 4 Centrifugation of anemic bone marrow cells on a BSA density gradient. The following parameters were determined: (A) ▲, Cells per fraction; ■, hemoglobin (mg × 10^8) per erythroid cell; □, hemoglobin concentration (mg × 10^8) per 120 μm^3 in erythroid cell; ○, median cell volume. (B) Differential cell counts from Leishman's stained smears: ●, erythroblasts; △, early and mid-polychromatic erythrocytes; ○, late polychromatic erythrocytes; ■ nonerythroid cells. (From Williams, 1972a.)

some microfilamentous material (Zentgraf *et al.*, 1969; Harris and Brown, 1971). The structure left after lysis is referred to as a ghost and consists of little more than a nucleus surrounded by a plasma membrane.

The isolation of avian erythrocyte nuclei, free of plasma membrane contamination, has been found to be surprisingly difficult. Contamination of nuclear preparations with membrane fragments is easily detected with electron microscopy and can also be observed in a well-adjusted phase-contrast microscope. Membrane contamination is, however, difficult to observe by regular light microscopy and there is reason to believe that many analytical data reported in the biochemical literature are in fact based on ghosts and not as claimed, on nuclei. To obtain pure nuclei very high concentrations of detergent and/or rotating knife homogenization have to be employed. Among the different methods described so far the

Fig. 3 Smears of mature erythrocytes from adult hen blood (a) and of anemic marrow cells of different fractionated densities (b–e) stained with Leishman's stain. (b) Early polychromatic erythrocytes with some mid-polychromatic erythrocytes, density 1.081–1.085 gm cm^{-3}; (c) mainly early polychromatic erythrocytes, density 1.072–1.076 gm cm^{-3}; (d) mainly small erythroblasts, density 1.061–1.065 gm cm^{-3}; (e) large erythroblasts, density 1.052–1.055 gm cm^{-3} (approx. × 800). (From Williams, 1972a.)

one described by Zentgraf *et al.* (1969) probably results in the best nuclear preparations.

C. *Nuclear Function and Erythroid Differentiation*

As mentioned above, erythroblasts replicate their DNA and divide. Erythroblasts give rise to early polychromatic erythrocytes which do not divide. The rest of the erythroid maturation takes place without replicative DNA synthesis.

With increasing degree of maturation the rate of RNA synthesis declines to virtually zero in both duck and chick erythrocytes (Hammarsten *et al.*, 1953; Cameron and Prescott, 1963; Scherrer *et al.*, 1966). The decreased rate of RNA synthesis is paralleled by a decreased RNA content and RNA concentration in the cytoplasm (Thorell and Raunich, 1966). This is similar to what has been observed (Thorell, 1947; Yataganas *et al.*, 1970) in species where the nucleus is eliminated during the terminal stage of erythroid maturation.

The type of RNA produced by immature avian erythroid cells varies depending on the stage of maturation. Both ribosomal RNA (rRNA), high-molecular-weight, heterogeneous nuclear RNA (HnRNA), and other RNA species are produced in cells from anemic blood (mainly polychromatic erythrocytes and some erythroblasts) (Bruns *et al.*, 1965; Attardi *et al.*, 1966; 1970; Scherrer *et al.* 1966; Scherrer and Marcaud, 1968). The labeling of rRNA fractions is slow and relatively insignificant in comparison with the labeling of HnRNA.

The HnRNA produced by immature duck erythrocytes gives sedimentation coefficients of 30–80 S, is highly unstable, and has a DNA-like base composition. Two main functions have been suggested for this RNA. It may contain sequences participating in the regulation of transcription (Britten and Davidson, 1969) and/or sequences which are precursors of cytoplasmic messenger RNA, thus specifying the synthesis of protein. The latter notion is supported mainly by nucleic acid hybridization studies indicating the presence of sequences in HnRNA very similar to those in cytoplasmic messenger RNA (Scherrer *et al.*, 1970; Melli and Pemberton, 1972). Nucleic acid hybridization experiments (Attardi *et al.*, 1970) have also shown that the HnRNA produced by duck erythroid cells is homologous to a smaller fraction of the genome than is the population of RNA species synthesized by active nonerythroid cells. Since it is likely that this type of RNA contains messenger RNA sequences this may be an indication that gene expression during erythropoiesis is at least in part regulated by mechanisms specifying the pattern of transcription. The nucleic acid hybridization techniques used in the latter study, however, have a rather low sensitivity and will only detect genes which are repeated many times. Furthermore, the HnRNA may be a transcript of many genes and/or contain nonsense or regulatory sequences which are trimmed away before translation occurs.

Recent results by Fantoni *et al.* (1972) show that the transcription of ribosomal genes is 20 times faster in less differentiated erythroblasts of 10-day-old chick embryos than in the more mature erythroid cells of day 13. Interestingly, the processing of preribosomal RNA seems completely conservative at day 13, whereas in day 10 erythroblasts, the processing of 32 S to 28 S RNA to a great extent is inhibited and 18 S RNA in excess of 28 S RNA is destroyed before reaching the state of mature ribosomes. Existing evidence from adult hens suggests (Scherrer *et al.*, 1966; Attardi *et al.*, 1966, 1970) that the synthesis of rRNA is switched off before HnRNA synthesis stops.

A general characteristic of RNA synthesis in immature red cells is that the transport of labeled RNA from the nucleus to the cytoplasm is very slow (Bruns *et al.*, 1965; Scherrer *et al.*, 1966) and may be more or less nonexistent in mid- and late polychromatic erythrocytes, i.e., in those cells which still synthesize some RNA but have lost their normal nucleolus.

It has been suggested that the nucleolus may play a role in the stabilization and transport of nuclear RNA to the cytoplasm (Harris *et al.*, 1969; Sidebottom and Harris, 1969; Deák *et al.*, 1972). This is thus another factor which must be elucidated in order to understand how gene expression is regulated during erythroid differentiation. Since RNA transport from the nucleus to the cytoplasm more or less stops in mid- and late polychromatic erythrocytes, the RNA which provides the messenger sequences, used in cytoplasmic hemoglobin synthesis must be synthesized, processed, and transported out into the cytoplasm some considerable time before translation. This also implies that most of the HnRNA synthesized at late stages of erythropoiesis undergoes intranuclear degradation without ever being translated. Avian erythropoiesis therefore offers an extreme example of the "poor RNA economy" of eukaryotic cell nuclei discovered by Harris (1963) in other types of cells.

Other factors which may affect the phenotype of differentiating erythroid cells are (*a*) selective breakdown of different mRNA species, and (*b*) selective translation of certain mRNA molecules. There is some evidence for both these forms of regulation. Actinomycin D inhibition experiments suggest that the globin messenger has a longer half-life than other messengers (Spohr and Scherrer, 1972). Specific translation factors have been found necessary for efficient globin synthesis in cell-free systems by some workers (Heywood, 1969, 1970; Cohen, 1971; Pritchard *et al.*, 1971; Conconi *et al.*, 1972) but not by others (Lane *et al.*, 1971; Sampson *et al.*, 1972). Although selective degradation of specific mRNA molecules and regulation at the translation level by messenger-specific initiation factors have been implicated as factors controlling gene expression during erythroid differentiation this falls outside the scope of the present review and the reader is referred to the papers cited above.

D. Changes in Nuclear Composition during Erythropoiesis

1. DECREASE IN NUCLEAR PROTEIN CONTENT

The drastic decrease in nuclear size during erythroid maturation is paralleled by a decrease in nuclear dry mass (Kernell *et al.*, 1971). The difference in nuclear dry mass between erythroblasts and later maturation stages is in part due to the fact that many erythroblasts are in S or G_2 phase. These cells therefore have a higher mean DNA content than the late polychromatic and mature erythrocytes which are arrested in the G_1 phase. But the dry mass/DNA ratio falls progressively during erythroid maturation (Table III) and continues to fall after DNA synthesis has stopped and therefore can only be explained in terms of a decreased protein content. A decrease in nuclear protein content has also been observed during erythropoiesis in the rabbit (Grasso *et al.*, 1963). Biochemical analysis of chromatin and nuclei from cells representing different stages of maturation has shown that the histone/DNA ratio remains more or less constant during erythroid differentiation (Appels, 1972). The decrease in nuclear dry mass as polychromatic erythrocytes differentiate into mature cells therefore appears to be due to a marked decrease in nonhistone proteins. The protein chemistry of nuclei isolated from mature erythrocytes will be discussed further in Section III,B.

2. PROTEIN CHANGES

a. Changes in the Histone Pattern. Avian erythropoiesis is of particular interest as a model for genome repression in a eukaryotic cell system because the DNP* complex contains a characteristic histone component, the F2C histone, not found in nonerythroid tissues (Neelin and Butler, 1961; Neelin *et al.*, 1964). It has been suggested that this specific histone could be responsible for the complete inactivation of the avian erythrocyte genome. Recent studies by Stevely (1971), Sotirov and Johns (1972), Appels (1972), Appels *et al.* (1972), and Appels and Wells (1972) show, however, that the F2C histone is already present in erythroblasts, i.e., at a stage when the erythroid cells still have a dispersed chromatin and are

* DNP = deoxyribonucleoprotein.

TABLE III
Cytochemical Properties of the Nucleus at Different Stages of Erythroid Maturation in the Chick[a]

	Erythroblasts (Mean ± S.E.M.)	Polychromatic erythrocytes (Mean ± S.E.M.)			Mature erythrocytes (Mean ± S.E.M.)
	(n)	Early (n)	Mid (n)	Late (n)	(n)
Acridine orange binding	3.17 ± 0.08	2.30 ± 0.05	1.82 ± 0.04	1.31 ± 0.02	1.00 ± 0.01
(relative to mature cells)	(84)	(102)	(113)	(113)	(123)
Feulgen–DNA	1.31 ± 0.07	1.21 ± 0.09	1.03 ± 0.07	1.01 ± 0.05	1.00 ± 0.04
(relative to mature cells)	(40)	(20)	(20)	(21)	(20)
Ratio acridine orange/Feulgen	2.4	1.9	1.8	1.3	1
Nuclear dry mass	263 ± 21	228 ± 12			146 ± 5
(arbitrary units)	(20)	(18)			(20)

[a] From Kernell *et al.* (1971).

synthesizing large quantities of RNA. However, the relative amount of F2C histone seems to increase gradually during maturation, whereas the amount of the normal lysine-rich histone F1 diminishes (Dick and Johns, 1969; Sotirov and Johns, 1972; Appels *et al.*, 1972).

Although it has not been possible to link the very appearance of the F2C histone to the cessation of RNA synthesis or to chromatin condensation, the F2C histone shows some unusual metabolic properties which may be related to the inactivation phenomenon. In polychromatic erythrocytes, the F2C histone appears to be rapidly synthesized and to be in a state of flux, combining and detaching from the chromatin at the same time as the other histones appear to be integrated in the DNP complex in a more stable way (Appels, 1972). Another phenomenon, the role of which remains to be explained, is enzymatic modification of histones during erythropoiesis. In the immature erythroid cells the arginine-rich histones F2A1 and F3 are acetylated at a high rate. As maturation proceeds the rate of acetylation of the ϵ-amino group of lysine residues in these histones drops dramatically (Allfrey, 1970). Similarly the rate of phosphorylation of serine in the more lysine-rich histones (mainly F2C and F2A2) decreases with maturation (Seligy and Neelin, 1971) in goose erythroid cells. It is not clear whether the decrease of the enzymatic modification occurs when cell proliferation ceases or later during maturation.

Evidence for histone changes during erythropoiesis has also been obtained by cytochemical methods. Meetz and MacRae (1968) studied chick erythroid cells with the aid of the ammoniacal silver reaction for histones. The chemical basis for this method is not known in detail, but ϵ-amino groups of lysine are thought to produce a yellow staining reaction while a brown-black staining is produced by guanidino groups of arginine residues. The "arginine" type of staining reaction was found to increase significantly during erythroid differention and was strongest in mature erythrocytes. Similar observations were also made using an electron microscopic modification of this technique (MacRae and Meetz, 1970); it was interpreted as reflecting either the accumulation of arginine-rich histones or changes in the availability of reactive sites in preformed histones.

b. Changes in Nonhistone Protein and Enzyme Patterns. The decrease in nuclear nonhistone proteins during avian erythropoiesis may be paralleled by a decrease in the absolute number of these protein species. These changes have, however, not yet been analyzed in detail (see also Section III, E,2,d). Like the histones the nonhistone proteins also undergo enzymatic modification, primarily phosphorylation. The rate of phosphorylation of nuclear nonhistone proteins has been found to decrease during avian erythroid maturation (Gershey and Kleinsmith, 1969).

RNA polymerase activity decreases as erythroblasts differentiate into mature erythrocytes but some activity can be demonstrated in the terminally differentiated cells (Appels and Williams, 1970; Moore *et al.*, unpublished observations). DNA polymerase (Williams, 1972b) and thymidine kinase activity (Williams, 1972c) is present in dividing erythroblasts but decreases so as to become undetectable in mature erythrocyte nuclei.

3. DEOXYRIBONUCLEOPROTEIN CHANGES

The decrease in nuclear size during avian erythropoiesis is paralleled not only by an increased condensation of the interphase chromatin but also by altered properties of the deoxyribonucleoprotein (DNP). Using quantitative cytochemical methods, Kernell *et al.* (1971) examined the properties of DNP in ethanol–acetone-fixed erythroid cells from anemic chick bone marrow. With increasing degree of erythroid maturation the chromatin bound progressively less basic fluorochromes (acridine orange, Table III) and showed an increased stability toward thermal denaturation (Fig. 5) and to acid hydrolysis in the Feulgen reaction (Fig. 6). These

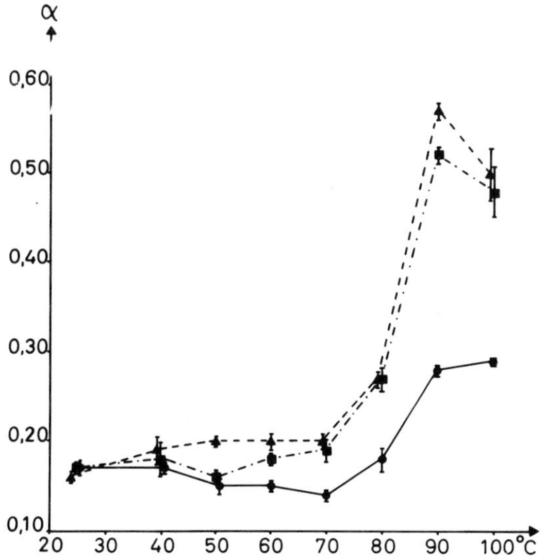

Fig. 5 Heat denaturation curves of erythroblasts (▲), early polychromatic erythrocytes (■), and mature erythrocytes (●). The curves illustrate the heat denaturation process as studied by measuring the F_{590}/F_{530} ratio of acridine orange stained preparations. In comparison with erythroblasts and polychromatic erythrocytes, the rate of increase as well as the maximum value of this ratio are lower for the mature erythrocytes. Abscissa: temperature, °C; ordinate: α (F_{590}/F_{530}). (From Kernell *et al.*, 1971.)

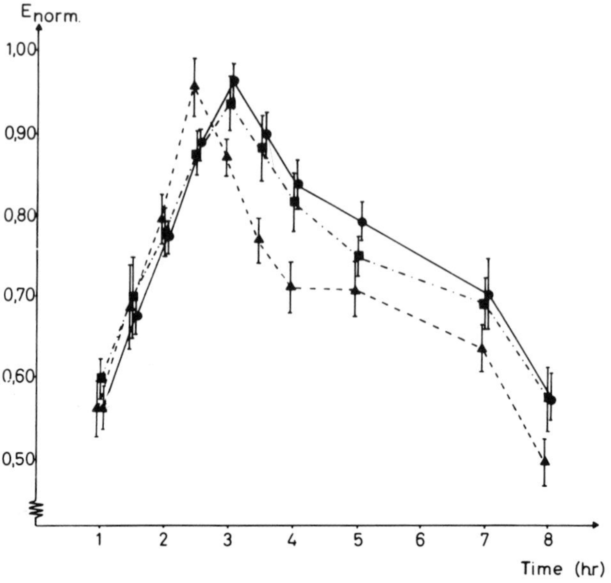

Fig. 6 Feulgen hydrolysis curves for erythroblasts (▲), early polychromatic erythrocytes (■), and mature erythrocytes (●), obtained by varying the hydrolysis time in 1 *N* TCA at 37°C. Each curve is derived from five separate experiments which have been normalized. Erythroblast nuclei are maximally stained after shorter hydrolysis than are nuclei of mature erythrocytes. The decrease of dye binding after the maximum is also faster for the erythroblasts. Abscissa: time (hr); ordinate: E_{tot} 546 nm (relative to maximum for each cell type in each experiment). (From Kernell *et al.*, 1971.)

DNP alterations indicate that the interaction between negatively charged DNA phosphate groups and basic proteins is modified during erythropoiesis, negative groups becoming masked by protein (and/or other cationic molecules) so as to be inaccessible to acridine orange binding. The increased stability of DNP to thermal denaturation and acid hydrolysis supports this hypothesis since biophysical studies have demonstrated that partly deproteinized DNP preparations are more susceptible to denaturation than the native DNP complex.

The altered physical properties of the DNP complex during avian erythropoiesis may be the molecular basis for the marked chromatin condensation and for the inactivation of the nucleus as a whole. Changes in the ionic milieu, hemoglobin accumulation, and decreased water content in the nuclear compartment could be other and probably rather unspecific factors involved in the inactivation of the erythrocyte genome. It is of interest, therefore, that Appels *et al.* (1973) found that chromatin isolated from polychromatic chick erythrocytes was a better template for

RNA synthesis with added bacterial RNA polymerase than was chromatin from mature erythrocytes. When compared with chromatin from other tissues, avian erythrocyte chromatin is a very poor template for RNA synthesis with exogenous RNA polymerase (Hoare and Johns, 1970). This indicates that the cessation of RNA synthesis *in vivo* may be due to changes in the molecular architecture of the chromatin itself and that some of these changes persist also when the chromatins have been isolated and are compared under identical ionic conditions and in the absence of hemoglobin. Exchange of the lysine-rich histones of erythrocyte chromatin (including hen erythrocyte specific fraction F2C) for lysine-rich histones from calf thymus and a similar exchange of nonhistone chromosomal proteins extractable with 0.35 *M* NaCl do not seem to change the low template activity of the hen erythrocyte chromatin. Neither does the ability of calf thymus chromatin to serve as a template for exogeneous RNA polymerase decrease significantly when its F1 histone is exchanged for erythrocyte F2C (Bolund and Johns, 1973). This may indicate that the low template activity *in vitro* of erythrocyte chromatin cannot simply be ascribed to any one of the specific protein fractions tested.

E. The Nucleus of the Mature Erythrocyte

1. MORPHOLOGY

Following staining by conventional hematological methods the nucleus of the fully mature erythrocyte is found to contain a series of tightly condensed chromatin clumps. The nucleus is centrally positioned and has an oval or rodshaped form. The length of the rodshaped nucleus found in adult chick erythrocytes of peripheral blood is approximately 3.3 μm, the diameter 2.5 μm, and the volume is approximately 13 μm^3. Although UV microscopy offers some possibilities of studying the structure of avian erythrocyte nuclei *in situ* the detailed information about the morphology of this nucleus is based on electron microscopy of sectioned material. The ultrastructure of avian erythrocytes has been studied by Davies (1961, 1968), Schneeberger and Harris (1966), Harris and Brown (1971), Zentgraf *et al.* (1969), and Small and Davies (1972). Nucleated erythrocytes from amphibian species have been examined by Davies and Tooze (1964, 1966) and Davies (1968). After heavy metal staining the heterochromatic clumps seen in the light microscope show up as very electron-dense bodies lining the nuclear membrane (Fig. 7). The DNA concentration in such clumps may be as high as 0.3–0.4 gm/ml (Olins and Olins, 1972). Nucleoli can be detected in mature definitive erythrocytes but are

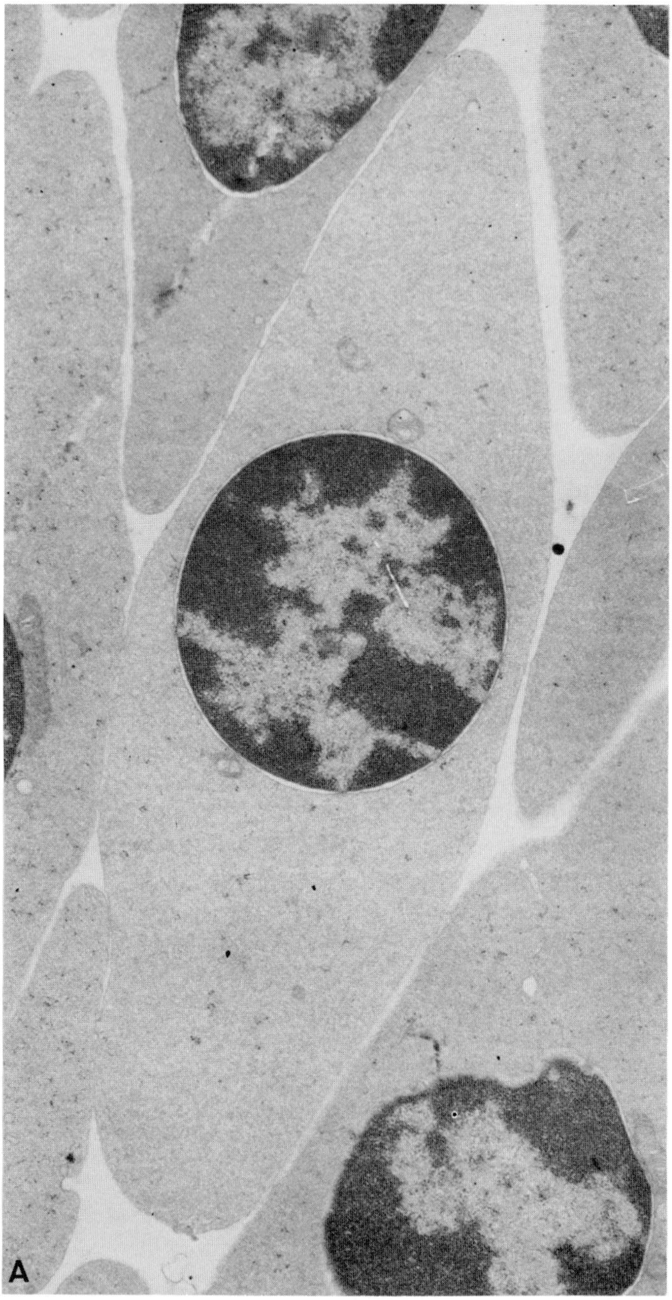

Fig. 7 (A) Definitive chick erythrocyte from 9-day chick embryo (× 14,000). (B) Enlarged nucleus of a definitive erythrocyte from a 17-day chick embryo (× 39,000). (From Small and Davies, 1972.)

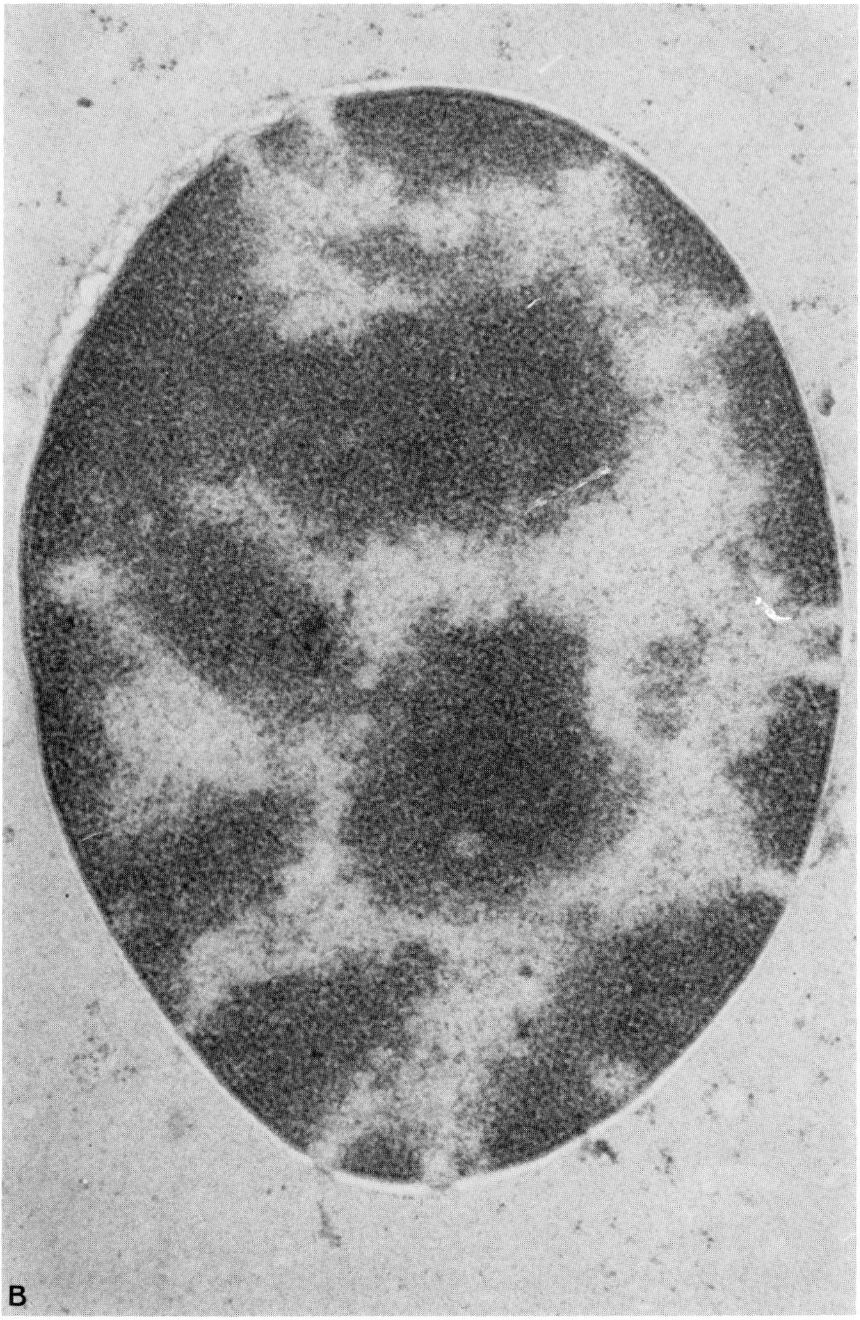
B

very small and consist only of fibrillar components (Small and Davies, 1972). Sometimes two fibrillar zones with different electron density are observed. The granular components characteristic of nucleoli in active cells are missing, however. The heterochromatin appears to consist of fibrils with a diameter of approximately 170 Å. At the periphery of the nucleus these fibrils are oriented in parallel arrays (Fig. 8) along the inner surface of the nuclear membrane (Everid *et al.*, 1970) while in the interior parts of the nucleus these fibrils are arranged in a more random pattern. The central nucleoplasmic regions between the heterochromatic

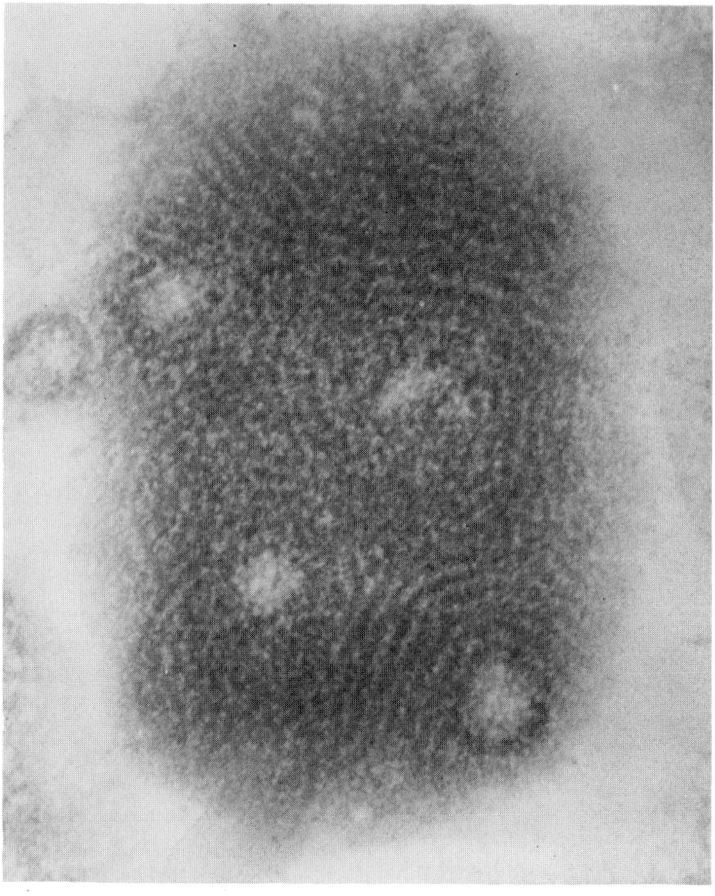

Fig. 8 Electron micrograph showing regular arrangement of 170 Å chromatin fibers in a tangential section of an adult chick erythrocyte nucleus. The chromatin fibers are arranged around the nuclear pores in arcs or circles which have the same center as the nuclear pores (× 80,000). (From Everid *et al.*, 1970.)

clumps appear homogeneous and less electron-dense. Microspectrophotometric measurements on sectioned nuclei indicate that hemoglobin is present in the nucleoplasm of both avian (Small and Davies, 1972) and amphibian erythrocyte nuclei (Tooze and Davies, 1963).

The ultrastructure of the nuclear membrane is difficult to resolve in intact erythrocytes but can be studied after lysis of the cells. Ghosts tend to maintain the shape of the intact erythrocyte and the nucleus appears to be anchored in a central position by microfilaments running from the outer layer of the nuclear membrane to the plasma membrane (Harris and Brown, 1971). The nuclear membrane consists of an outer and an inner layer similar to the situation in other cells. Nuclear pores are present but fewer than in normal mammalian cells (Zentgraf *et al.*, 1971). More detailed studies of the ultrastructure of the chromatin fibers have been performed by spreading erythrocyte chromatin (avian and amphibian) on water surfaces and then transferring the chromatin filaments to electron microscope grids. Such chromatin fibers vary from 37 to 313 Å in diameter depending on the technique used (Solari, 1971). The presence of hemoglobin tends to give thicker fibers, supporting the suggestion by Tooze and Davies (1963) that the accumulation of hemoglobin may be a factor influencing the structure of the nucleoprotein and possibly causing the marked chromatin condensation typical of mature avian and amphibian erythrocytes.

The diameter of the thinnest fibril observed by Solari (1971) agrees fairly well with the dimensions expected from a double helical DNA molecule coated with some protein. The thicker chromatin fibrils, which can be seen both in sectioned material and in chromatin stretched on water, may form by supercoiling and/or irregular folding of a 30–40-Å deoxyribonucleoprotein filament. The low angle X-ray diffraction pattern observed for both erythrocyte nuclei and isolated chromatin fibers (Wilkins *et al.*, 1959; Olins and Olins, 1972; Bradbury *et al.*, 1972) indicates the presence of some form of superhelical structure and is similar to that observed by Pardon *et al.* (1967) and Pardon and Wilkins (1972) in calf thymus nucleohistone. For a more detailed discussion of the molecular aspects of the chromatin fiber structure, reviews by DuPraw (1968) or Pardon and Richards (1972) should be consulted.

2. CHEMISTRY

a. Overall Composition. Zentgraf *et al.* (1969, 1971) studied the gross chemical composition of pure chick erythrocyte nuclei. DNA represented approximately 33% and protein 63% of the total dry mass (Table IV). Lipid components and RNA together only accounted for a few percent of

TABLE IV
Contents of Hen Erythrocyte Whole Cells and Isolated Nuclei[a,b]

	DNA	RNA	Protein	Phospholipids	Cholesterol	Total weight
Cell	1.72	0.19	50.50	0.75	0.275	53.44
Isolated nucleus	1.70	0.13	3.02	0.08	0.016	4.95
Recovery in nuclear fraction (%)	98.8	68.4	6.0	10.7	5.8	9.3

[a] From Zentgraf *et al.* (1971).
[b] Results given as picograms per cell and nucleus, respectively; mean values from three determinations.

the dry mass which is much less than what is found in more active cell nuclei from birds and other animals. The absolute amount of DNA in chick erythrocyte nuclei has previously been found to be 2.3 pg (Vendrely, 1955) and more recently 1.7 pg (Zentgraf *et al.*, 1971). This discrepancy seems to require some further investigations. The total dry weight of chick erythrocyte nuclei is approximately 5 pg, which represents approximately 9–10% of the total dry mass of the intact cell. The very high dry mass per unit volume ($\approx$ 0.4 gm/cm^3) implies a high density and a low water content of the erythrocyte nucleus.

Using various fractionation techniques avian erythrocyte nuclei have been separated into chromatin and nuclear membrane fractions. So far no reports mention nuclear sap fractions or nucleoplasmic fractions similar to those which can be isolated from more active cell nuclei. This probably has to do with the condensed state of the avian erythrocyte nuclei, the nucleus being reduced to its deoxyribonucleoprotein components, having lost most of the nonchromosomal proteins. The chromatin fractions reported by Seligy and Neelin (1970), Dingman and Sporn (1964), Elgin and Bonner (1970), and others give protein/DNA ratios varying between 1.1 and 1.6, suggesting that the chromatin components make up some 80% of the nuclear dry mass.

b. DNA. On the basis of our present knowledge there is no reason to believe that the DNA molecules in erythrocyte nuclei differ from the DNA of other tissues within the same organism. It should, however, be pointed out that gene modification (recombination, amplification, looping out, and excision, etc.) has been implied not only to explain changes in immunoglobulin production but also as a mechanism involved in erythroid cell differentiation. Thus Kabat (1972) has suggested that the switch from the embryonic type to the adult type of hemoglobin synthesis is caused by a process where genes coding for embryonic types of hemoglobin are looped out and excised.

c. Histones. Histones are directly linked to the DNA molecules mainly by ionic binding and have been implicated as unspecific repressors of transcription (for reviews see Huang, 1971; Phillips, 1971). In view of the highly differentiated nature of the avian erythrocyte and the condensed inactive state of its nucleus, it is of great interest to know if the histone pattern of avian erythrocytes differs from that of other avian tissues.

The histone/DNA ratio of chicken erythrocyte chromatin has in general been reported to be among the lowest in the range of values (0.9–1.2) commonly found for chromatin of eukaryotic cells (e.g., Vendrely, 1955; Dingman and Sporn, 1964; Elgin and Bonner, 1970; Seligy and Neelin, 1970; Fredericq, 1971). The individual histones can be separated by acrylamide gel electrophoresis (Fig. 9) into five main fractions which are designated as follows: F1 is very lysine-rich, F2B is relatively lysine-rich, F3 and F2A1 are arginine-rich histones, and F2A2 is an intermediate type having a molar ratio of lysine to arginine of about 1 (Johns, 1971). As illustrated in Fig. 9 the histone pattern of chick erythrocyte chromatin differs from that of calf thymus chromatin in that the F1 histone is to some extent replaced by a sixth component, the so-called F2C histone. This fraction appears to be unique for avian erythrocytes in much the same way as protamines are for sperm. The possibility that the F2C histone is in some way involved in the repression of the erythrocyte genome and/or the marked condensation of the nuclear chromatin has stimulated detailed biochemical studies of erythrocyte histones.

The arginine-rich histone F3 of chick erythrocytes has been sequenced and found practically identical to calf thymus F3 (Brandt and von Holt, 1972). One of the lysines is completely methylated and many of the others

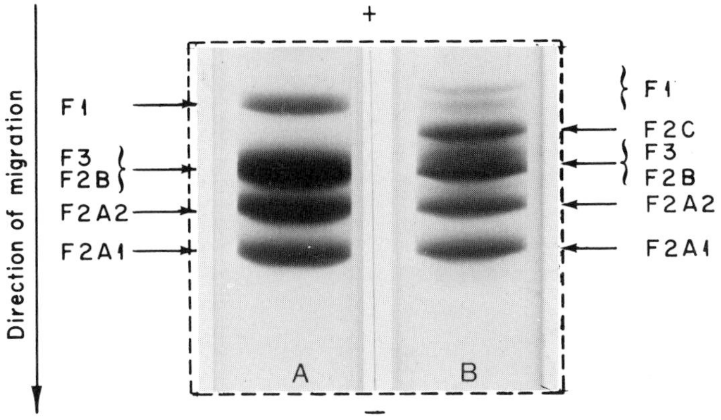

Fig. 9 Acrylamide gel separation of histones from (A) calf thymus and (B) chick erythrocyte chromatin. A histone fraction specific for avian erythrocytes, the F2C, migrates between the F1 and F3 histone fractions. (From Bolund and Johns, 1973.)

are partially acetylated which leads to a microheterogeneity in the fraction (Brandt and von Holt, 1972). The other arginine-rich histone F2A1, which has been shown to be virtually identical (except for two conservative replacements) in species as wide apart as calf and pea (deLange *et al.*, 1969), looks electrophoretically the same in chicken erythrocytes as well. Judging from the electrophoretical behavior, histones F2A2 and F2B are also similar to the corresponding fractions in mammalian chromatin. The very lysine-rich histones, however, are very different in the avain erythrocyte. A unique lysine, arginine, and serine-rich histone is found in the erythroid chromatin (Neelin and Butler, 1961) but not in other bird tissues (Neelin, 1964). This avian histone F2C (Hnilica, 1964) constitutes more than a fourth of the total chicken erythrocyte histone and seems to have replaced mainly the other lysine-rich histone F1 (Sotirov and Johns, 1972). In some avian erythrocytes there is hardly any F1 histone left (Dick and Johns, 1969), but in hen it still amounts to some 10% of the total erythrocyte histone (Sotirov and Johns, 1972).

The chicken erythrocyte F1 exhibits a marked heterogeneity. In polyacrylamide gel electrophoresis two or three fractions are resolved (Fig. 9) and chromatography on Amberlite IRC-50 reveal five or six components (Kinkade, 1969). The F1 pattern is qualitatively different from F1 of other species but probably only quantitatively different from F1 of other hen tissues (Kinkade, 1969).

The chicken erythrocyte F2C, which has been partially sequenced (Greenaway and Murray, 1971; Greenaway, 1971), is found to be polymorphic (Greenaway and Murray, 1971). On Amberlite CG-50 two forms can be separated, which differ in one amino acid residue and seem to be coded for by allelic genes (Greenaway and Murray 1971).

O-phosphoserine is found in chick histone F2C in significant and variable amounts (Murray and Milstein, 1967; Greenaway and Murray, 1971). Phosphoserine is also found in the other lysine-rich histones of chicken erythrocytes but in much smaller, although still significant, amounts (Murray and Milstein, 1967).

On the basis of the data cited it can be concluded that the histone pattern of avian erythrocytes differs from that of other tissues mainly by the presence of the F2C histone. Since in general, histones appear to be the same in nuclei of widely different tissues and animal species, this is a noteworthy finding. As has already been pointed out in the discussion of histone changes during avian erythropoiesis, there is, however, no definite proof that the unusual histone pattern alone is responsible for the repressed and/or condensed state of avian erythrocyte chromatin.

d. Nonhistone Proteins. The nonhistone proteins of erythrocyte nuclei can be divided into two main groups, those which are linked to chromatin and those which are not ("nonchromosomal proteins"). Unfortunately this distinction is somewhat vague and has not been made identically in all laboratories.

The nonhistone proteins of avian erythrocyte chromatin seem fewer in number (Loeb and Creuzet, 1969, 1970; Allfrey, 1970; Kostraba and Wang, 1970; Shelton and Neelin, 1971) and include relatively more high-molecular-weight proteins (Elgin and Bonner, 1970; Shelton and Neelin, 1971) than do equivalent fractions from other tissues. Judging from electrophoretic separations, some nonhistone proteins appear to be identical to those found in nonerythroid tissues of the hen. Differences, however, exist between the quantity of these proteins in erythrocyte chromatin as compared to other chromatin preparations (Elgin and Bonner, 1970; Shelton and Neelin, 1971). A considerable number of the nonhistone proteins may be specific for avian erythrocyte chromatin (Loeb and Creuzet, 1969; Elgin and Bonner, 1970; Shelton and Neelin, 1971) since they are not found in other tissues.

The nonchromosomal proteins of the erythrocyte nucleus probably account for less than 20% of the nuclear mass (see above). In the case of more active nuclei this value generally exceeds 50%. Cytochemical data (Tooze and Davies, 1963; Hammel and Bessman, 1964; Small and Davies, 1972) suggest that hemoglobin is present in the nuclei of intact erythrocytes and could make up a considerable part of the nonchromosomal proteins. This is of interest from the point of view of gene regulation. Hemoglobin is a basic protein and easily combines with DNA and DNP and causes conformational changes in the DNP molecules (Solari, 1971). It should, however, be pointed out that biochemical work (e.g., Zentgraf *et al.*, 1971) has not verified the presence of hemoglobin in isolated nuclei. The explanation for the discrepancy between the results obtained by microspectrophotometry on sectioned erythrocytes (Small and Davies, 1972) and isolated nuclei may be that the isolation procedures elute hemoglobin from the nuclei.

e. Nuclear Enzymes. Low levels of RNA polymerase have been found (Thompson and McCarthy, 1968; Appels and Williams, 1970; Moore *et al.*, unpublished observations) in avian erythrocytes. DNA polymerase and thymidine kinase activities (Williams, 1972b, c) appear to be lost as the erythroblasts stop DNA synthesis and mitosis and cannot be detected in fully mature chick erythrocytes. Histone acetyltransferase (Allfrey, 1970) and protein kinase activity can, however, be detected (see also Section III,D,2).

f. Nuclear Membrane Components. The nuclear membrane of the hen erythrocyte has a characteristically low cholesterol/phospholipid ratio, a relatively high content of protein, and enzyme activities characteristic of the endoplasmic reticulum of other cells (Zentgraf et al., 1971). The chemical difference between the nuclear and plasma membranes is, however, still as obvious in the hen erythrocyte as in any other cell (Zentgraf *et al.*, 1971).

IV. Concluding Remarks

The avian erythrocyte nucleus offers a useful model for studies of gene regulation in eukaryotic cells. During erythropoiesis the nucleus undergoes a gradual inactivation at the same time as the cell as a whole undergoes marked morphological and biochemical changes typical of this form of cell differentiation.

The regulatory mechanisms determining the erythroid phenotype seem to operate at several different levels. Marked changes in the properties of the template during erythropoiesis in combination with indications of a relative paucity of RNA sequences in cells of the erythroid series suggest control at the level of transcription. The processing and transport of the main product of transcription, HnRNA (DNA-like heterogeneous nuclear RNA), may be other steps in the process of gene expression controlled in the nucleus. In the cytoplasm further control of RNA processing and/or degradation as well as specific regulation of the translation of mRNA into protein have been reported to occur, but this falls outside the scope of the present article.

At the level of transcription several different control mechanisms have been postulated. The DNA–protein complex undergoes marked changes in its physicochemical or cytochemical properties during erythropoiesis; parallel to this the activity of endogenous RNA and DNA polymerase decreases to a very low level. A histone unique to avian erythroid cells, F2C, is present in all erythroid cells but increases in relative abundance as maturation proceeds. The great evolutionary stability of the genes specifying the structure of histones in general suggests that the histones play an important functional role. Mutations leading to changes in histone structure must have been rapidly eliminated during evolution. The only situation other than avian (and amphibian) erythropoiesis where nature has permitted "experimentation" with the structure of histones is during spermiogenesis where protamines of widely differing compositions replace the somatic histones. However, no close relationship between the histone changes and the inactivation of the nucleus has been found in either system. A correlation may still exist between the histone changes in erythrocytes and sperm and the marked condensation of their chromatin which in turn may affect the transcription. It is likely, however, that other factors, e.g., the accumulation of hemoglobin, changes in the ionic environment, the increased concentration of macromolecules, and lowered water content, etc., are also involved in this process.

A decrease in the amounts of nonhistone chromosomal or nuclear proteins occurs during erythropoiesis; the mature erythrocyte seems to contain fewer species of these proteins than cells from other tissues. It has therefore been postulated that these proteins are necessary for the maintenance of the dispersed state of the chromatin and/or for the transcription of the genome.

Enzymatic modification (such as acetylation and phosphorylation) of histones as well as nonhistones decreases during erythroid maturation. This may also have consequences for the structure and function of the template.

The majority of the RNA synthesized in the erythroid cells is HnRNA. Most of this HnRNA is broken down in the nucleus but still it contains sequences very similar to, and probably precursors of, cytoplasmic messenger RNA. The transport of this RNA from nucleus to cytoplasm is relatively slow even in the immature erythroid cells and not detectable in the more mature cells which still synthesize some HnRNA but have lost their normal nucleoli. This loss of functional nucleoli may thus also be of regulatory significance for gene expression.

The differentiating avian erythroid cells have been very useful in the study of gene regulation of the eukaryotic cell and will probably continue to be so in the future. That the dormant nucleus of the mature erythrocyte can be reactivated (see review by Ringertz and Bolund, 1974) adds to its usefulness. None of the postulated regulatory mechanisms mentioned here has been proved to exist beyond doubt, but it is our hope that the steps between suggestive evidence and definite proof will soon be taken.

REFERENCES

Allfrey, V. (1970). *Fed. Proc.* **29,** 1447.

Appels, R. (1972). Thesis, Univ. of Adelaide.

Appels, R., and Wells, J. R. E. (1972). *J. Mol. Biol.* **70,** 425.

Appels, R., and Williams, A. F. (1970). *Biochem. Biophys. Acta* **217,** 531.

Appels, R., Wells, J. R. E., and Williams, A. F. (1972). *J. Cell Sci.* **10,** 47.

Appels, R., Harlow, R., Tolstoshev, P., and Wells, J. R. E. (1973). *In* "Biochemistry of Gene Expression in Higher Organisms" (J. K. Pollack and J. Wilson Lee, eds.). Australian and New Zealand Book Co.

Attardi, G., Parnas, H., Hwang, M. J. H., and Attardi, B. (1966). *J. Mol. Biol.* **20,** 145.

Attardi, G., Parnas, H., and Attardi, B. (1970). *Exp. Cell Res.* **62,** 11.

Bank, A., Rifkind, R. A., and Marks, P. A. (1970). *In* "Regulation of Hematopoiesis" (A. S. Gordon, ed.), Vol. 1, p. 701. Appleton, New York.

Bolund, L., and Johns, E. W. (1973). *Eur. J. Biochem.* **40,** 591.

Bradbury, E. M., Molgaard, H. V., Stephens, R. M., Bolund, L., and Johns, E. W. (1972). *Eur. J. Biochem.* **31,** 474.
Brandt, W. F., and von Holt, C. (1972). *FEBS Lett.* **23,** 357.
Britten, R. J., and Davidson, E. M. (1969). *Science* **165,** 349.
Bruns, G. P., Fischer, S., and Lowy, B. A. (1965). *Biochem. Biophys. Acta* **95,** 280.
Cameron, T. L., and Prescott, D. M. (1963). *Exp. Cell Res.* **30,** 609.
Campbell, G., Le, M., Weintraub, H., Mayall, B. H., and Holtzer, H. (1971). *J. Cell Biol.* **50,** 669.
Cohen, B. B. (1971). *Biochem. Biophys. Acta* **247,** 133.
Conconi, F., Rowley, P. T., del Senno, L., Pontremoli, S., and Volpato, S. (1972). *Nature (London) New Biol.* **238,** 83.
D'Amelio, V. (1966). *Biochem. Biophys. Acta* **127,** 59.
Dantschakoff, V. (1909). *Arch. Mikrosk. Anat.* **73,** 117.
Davies, H. G. (1961). *J. Biophys. Biochem. Cytol.* **2,** 671.
Davies, H. G. (1968). *J. Cell Sci.* **3,** 129.
Davies, H. G., and Tooze, J. (1964). *Nature (London)* **203,** 990.
Davies, H. G., and Tooze, J. (1966). *J. Cell Sci.* **1,** 331.
Dawson, A. B. (1936). *Z. Zellforsch. Mikrosk. Anat.* **24,** 256.
Deák, I., Sidebottom, E., and Harris, H., (1972). *J. Cell Sci.* **11,** 379.
deLange, R. J., Fambrough, D. M., Smith, E. L., and Bonner, J. (1969). *J. Biol. Chem.* **244,** 5669.
Dick, C., and Johns, E. W. (1969). *Biochem. Biophys. Acta* **175,** 414.
Dingman, C. W., and Sporn, M. B. (1964). *J. Biol. Chem.* **239,** 3484.
DuPraw, E. J. (1968). "Cell and Molecular Biology." Academic Press, New York.
Elgin, S. C. R., and Bonner, J. (1970). *Biochemistry* **9,** 4440.
Everid, A. C., Small, J. V., Davies, H. G. (1970). *J. Cell Sci.* **7,** 35.
Fantoni, A., Bordin, S., and Lunadei, M. (1972). *Cell Differentiation* **1,** 219.
Fraser, R., Horton, B., Dupaurque, D., and Chernoff, A. (1972). *J. Cell Phys.* **80,** 79.
Fredericq, E. (1971). *In* "Histones and Nucleohistones" (D. M. Phillips, ed.), p. 135. Plenum Press, New York.
Gershey, E., and Kleinsmith, L. J. (1969). *Biochem. Biophys. Acta* **194,** 519.
Grasso, J. A., Woodard, J. W., and Swift, H. (1963). *Proc. Nat. Acad. Sci.* **50,** 134.
Greenaway, P. J. (1971). *Biochem. J.* **124,** 319.
Greenaway, P. J., and Murray, K. (1971). *Nature (London) New Biol.* **229,** 233.
Hagopian, H. K., and Ingram, V. M. (1971). *J. Cell. Biol.* **51,** 440.
Hammarsten, E., Thorell, B., Åquist, S., and Eliasson, N. (1953). *Exp. Cell. Res.* **5,** 404.
Hammel, C. L., and Bessman, S. P. (1964). *J. Biol. Chem.* **239,** 2228.
Harris, H. (1963). *Proc. Roy. Soc. Ser. B* **158,** 179.
Harris, H., Sidebottom, E., Grace, D. M., and Bramwell, M. E. (1969). *J. Cell Sci.* **4,** 499.
Harris, J. R., and Brown, J. N. (1971). *J. Ultrastruct. Res.* **36,** 8.
Hell, A. (1964). *J. Embryol. Exp. Morphol.* **12,** 600.
Heywood, S. M. (1969). *Cold Spring Harbor Symp. Quant. Biol.* **34,** 799.
Heywood, S. M. (1970). *Proc. Nat. Acad. Sci.* **67,** 1782.
Hnilica, L. S. (1964). *Experientia* **20,** 13.
Hoare, T. A., and Johns, E. W. (1970). *Biochem. J.* **119,** 931.
Huang, P. C. (1971). *Progr. Biophys. Mol. Biol.* **23,** 103.
Ingram, V. (1972). *Nature (London)* **235,** 338.
Johns, E. W. (1971) *In* "Histones and Nucleohistones" (D. M. P. Phillips, ed.), p. 1. Plenum Press, New York.

Kabat, D. (1972). *Science* **175,** 134.
Kabat, D., and Attardi, G. (1967). *Biochem. Biophys. Acta* **138,** 382.
Kernell, A-M., Bolund, L., and Ringertz, N. R. (1971). *Exp. Cell Res.* **65,** 1.
Kinkade, J. M. (1969). *J. Biol. Chem.* **244,** 3375.
Kostraba, N. C., and Wang, T. Y. (1970). *Int. J. Biochem.* **1,** 327.
Lane, C. D., Marbaix, G., and Gurdon, J. B. (1971). *J. Mol. Biol.* **61,** 73.
Lemez, L. (1964). *Advan. Morphogen.* **3,** 197.
Levere, R. D., and Granick, S. (1967). *J. Biol. Chem.* **242,** 1903.
Loeb, J. E., and Creuzet, C. (1969). *Fed. Eur. Biochem. Soc. Lett.* **5,** 37.
Loeb, J. E., and Creuzet, C. (1970). *Bull. Soc. Chim. Biol.* **52,** 1007.
Lucas, A. M., and Jamroz, C. (1961). U.S. Dept. of Agr. Monogr., Washington, D.C.
MacRae, E. K., and Meetz, G. D. (1970). *J. Cell Biol.* **45,** 235.
Meetz, G. D., and MacRae, E. K. (1968). *J. Histochem. Cytochem.* **16,** 148.
Melli, M., and Pemberton, R. E. (1972). *Nature (London) New Biol.* **236,** 172.
Miura, Y., and Wilt, F. H. (1969). *Develop. Biol.* **19,** 201.
Miura, Y., and Wilt, F. H. (1970). *Exp. Cell Res.* **59,** 217.
Miura, Y., and Wilt, F. H. (1971). *J. Cell Biol.* **48,** 523.
Murray, K., and Milstein, C. (1967). *Biochem. J.* **105,** 491.
Neelin, J. M. (1964). *In* "The Nucleohistones" (J. Bonner and P. Ts'o, eds.), pp. 66–71. Holden-Day, San Francisco.
Neelin, J. M., and Butler, G. L. (1961). *Can. J. Biochem. Physiol.* **39,** 485.
Neelin, J. M., Callahan, P. Y., Lamb, D. C., and Murray, K. (1964). *Can. J. Biochem.* **42,** 1743.
Olins, D. E., and Olins, A. L. (1972). *J. Cell Biol.* **53,** 715.
Pardon, J. F., and Richards, B. M. (1972). *In* "Chromosomes Today" (C. D. Darlington and K. R. Lewis, eds.). Vol. III, pp. 38–46. Hafner, New York.
Pardon, J. F., and Wilkins, M. H. F. (1972). *J. Mol. Biol.* **68,** 115.
Pardon, J. F., Wilkins, M. H. F., and Richards, B. M. (1967). *Nature,* **215,** 508.
Phillips, D. M. P. Ed. (1971). "Histones and Nucleohistones." Plenum Press, New York.
Pritchard, P. M., Picciano, D. J., Laycock, D. G., and Andersson, W. F. (1971). *Proc. Nat. Acad. Sci. U.S.* **68,** 2752.
Reynolds, L. W., and Ingram, V. M. (1971). *J. Cell Biol.* **51,** 433.
Ringertz, N. R., and Bolund, L. (1974). *Int. Rev. Exp. Pathol.* **13,** 83.
Romanoff, A. L. (1960). "The Avian Embryo." Macmillan, New York.
Sabin, F. R. (1920). *Carnegie Inst. of Wash. Publ. No. 272 Contribut. Embryol.* **9,** 213.
Sampson, J., Mathews, M. B., Osborn, M., and Borghetti, A. F. (1972). *Biochemistry* **11,** 3636.
Scherrer, K., and Marcaud, L. (1968). *J. Cell Physiol.* **72,** 181.
Scherrer, K., Marcaud, L., Zajdela, F., London, I. M., and Gros, F. (1966). *Proc. Nat. Acad. Sci. U.S.* **56,** 1571.
Scherrer, K., Spohr, G., Granboulan, N., Morel, C., Grosclaude, J., and Ghezzi, C. (1970). *Cold Spring Harbor Symp. Quant. Biol.* **35,** 539.
Schneeberger, E. E., and Harris, H. (1966). *J. Cell Sci.* **1,** 401.
Seligy, V. L., and Neelin, J. M. (1970). *Biochem. Biophys. Acta* **213,** 380.
Seligy, V. L., and Neelin, J. M. (1971). *Can. J. Biochem.* **49,** 1062.
Shelton, K. R., and Neelin, J. M. (1971). *Biochemistry* **10,** 2342.
Sidebottom, E., and Harris, H. (1969). *J. Cell Sci.* **5,** 351.
Small, J. V., and Davies, H. G. (1972). *Tissue and Cell* **4,** 341.
Solari, A. J. (1971). *Exp. Cell Res.* **67,** 161.

Sotirov, N., and Johns, E. W. (1972). *Exp. Cell Res.* **73**, 13.
Spohr, G., and Scherrer, K. (1972). *Cell Differentiation* **1**, 53.
Spratt, N. T., and Hass, H. (1960a). *J. Exp. Zool.* **144**, 139.
Spratt, N. T., and Hass, H. (1960b). *J. Exp. Zool.* **145**, 97.
Stevely, W. S. (1971). *Biochem. J.* **124**, 48.
Thompson, L. R., and McCarthy, B. J. (1968). *Biochem. Biophys. Res. Commun.* **30**, 166.
Thorell, B. (1947). "Studies on Blood Cell Formation." Kimpton, London.
Thorell, B., and Raunich, L. (1966). *Ann. Med. Exp. Fenn.* **44**, 131.
Tooze, J., and Davies, H. G. (1963). *J. Cell Biol.* **16**, 501.
Vendrely, R. (1955). *In* "The Nucleic Acids, Chemistry and Biology." (E. Chargaff and J. N. Davidson, eds.), Vol. II, p. 155. Academic Press, New York.
Weintraub, H., Campbell, G. L., and Holtzer, H. (1971). *J. Cell Biol.* **50**, 652.
Wilkins, M. H. F., Zubay, G., and Wilson, H. R. (1959). *J. Mol. Biol.* **1**, 179.
Williams, A. F. (1972a). *J. Cell Sci.* **10**, 27.
Williams, A. F. (1972b). *J. Cell Sci.* **11**, 785.
Williams, A. F. (1972c). *J. Cell Sci.* **11**, 777.
Wilt, F. H. (1965). *J. Mol. Biol.* **12**, 331.
Wilt, F. H. (1967). *Advan. Morphogen.* **6**, 89.
Yataganas, X., Gahrton, G., and Thorell, B. (1970). *Exp. Cell Res.* **62**, 254.
Zentgraf, H., Deumling, B., and Franke, W. W. (1969). *Exp. Cell Res.* **56**, 333.
Zentgraf, H., Deumling, B., Jarasch, E. D., and Franke, W. W. (1971). *J. Biol. Chem.* **246**, 2956.

12

Inhibitors as Tools in Elucidating the Structure and Function of the Nucleus

René Simard, Yves Langelier, Rosemonde Mandeville,
Nicole Maestracci, and André Royal

I. Introduction

The major biosynthetic events that characterize cell type and cell function are under the immediate control of the DNA of the interphase chromatin. Active loci of the chromosomal material are the site of genetic expression and the immediate gene products—messenger, ribosomal, and transfer RNA—then direct the synthesis of proteins which are characteristics of the cell type, cell function, and cell specialization. Since only a small percentage of DNA in eukaryotic cell is expressed, the problems of regulation at the chromosomal level are further complicated by selection mechanisms that decide which genetic loci will be active and which will not. This problem is not only of fundamental interest in the molecular biology of the nucleus but also is probably highly relevant for the study of control mechanisms in cancer cells.

However, when one turns to consider nuclear transcription, even the complexity of bacterial transcriptional systems dwindles considerably. First, the mass of nuclear DNA is extraordinarily large; second, the genetic sequences themselves are complex; third, many sequences are repeated extensively and may never be used as functional unit of transcription. Moreover, the RNA transcripts are heterogeneous and their presence and formation is an intrinsic feature of the nuclear transcription system. There is also good evidence that the nucleus and the nucleolus have their own genetic information and their own transcriptional machinery. Since genetic studies and valuable biochemical techniques used for prokaryotes are also of limited value in eukaroytic systems, much knowledge has been obtained through the use of inhibitors that block specific steps of transcription or interfere with nucleic acid synthesis in various ways. This review will deal mainly with the ultrastructural modifications of these inhibitors and correlation with the biochemical events will be emphasized. A similar review was published earlier (Simard, 1970); and this review contains more recent advances on this subject.

II. Agents That Primarily Affect the Nucleolus

A large number of drugs have been used as cytological probes to integrate ultrastructural and biochemical studies of the nucleolus. The study of the molecular events involved with the action of those drugs has been facilitated in some cases by the possibility to purify isolated nucleoli. In Fig. 1, nucleoli were isolated from rat liver under normal conditions (Fig. 1a), after the action of thioacetamide induced a striking hypertrophy (Fig. 1b), following the action of aflatoxin which cause segregation of nucleolar components (Fig. 1c) and after exposure to supranormal temperature for nucleolar degranulation (Fig. 1d). A fourth lesion characteristic of nucleoli does not permit successful isolation since it results in fragmentation of nucleoli; it is observed in cells treated with ethionine and other related drugs.

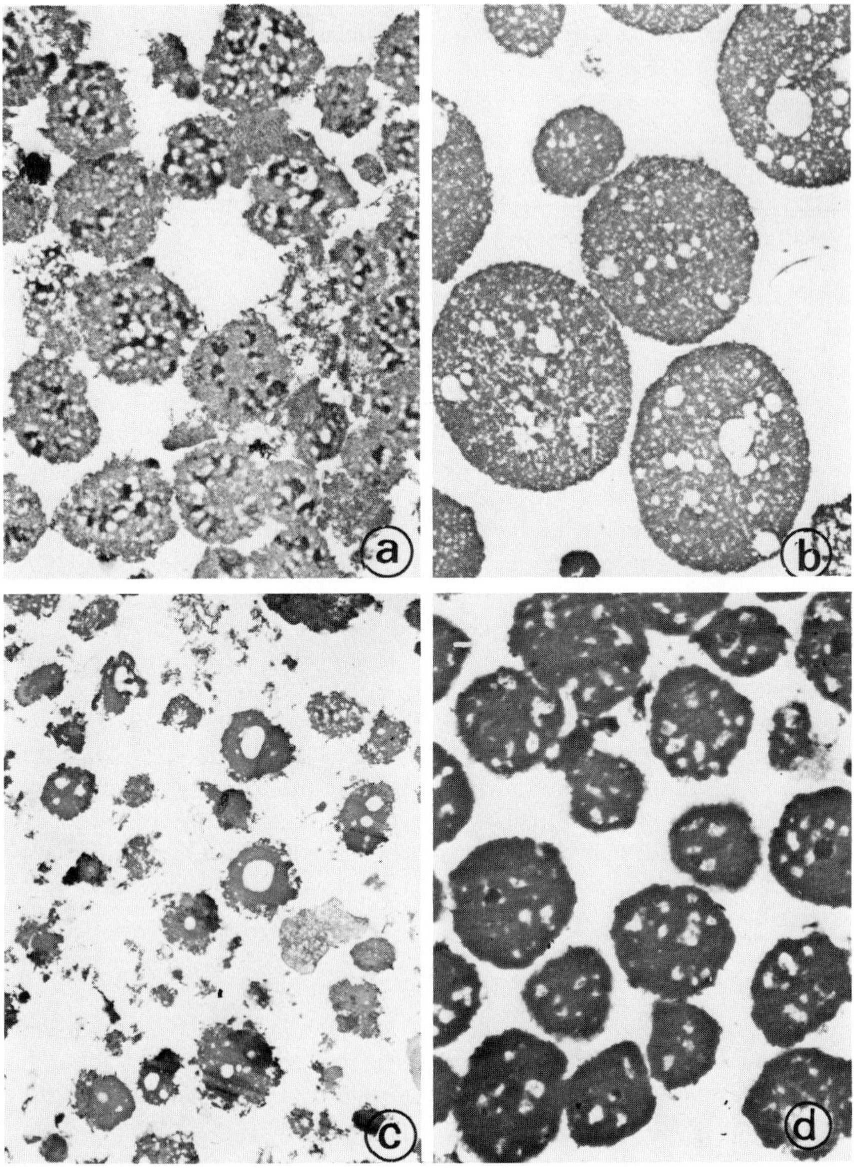

Fig. 1 Nucleolar subfractions obtained from rat liver cells. The nucleoli were isolater in 2.2 *M*, sucrose 3.3 m*M* Mg^{2+} by sonication of the nuclei. Normal nucleoli (a) can be compared to other drug-induced pathological conditions. Hypertrophy of the nucleolus (b) was induced after intoxication with thioacetamide for 11 days. Segregation of nucleolar components (c) is observed 2 days after a single injection of aflatoxin (500 μg/kg). Degranulation of nucleoli (d) is caused by a 30-min exposure to a supranormal temperature of 43°C. Osmium teroxide; Epon (× 4250).

A. *Nucleolar Segregation and Actinomycin D*

1. THE EFFECT OF ACTINOMYCIN D ON THE NUCLEOLUS

The first cytological observations relating the action of actinomycin to the nucleolus were performed with time-lapse cinematography (Robineaux *et al.*, 1958) and phase-contrast microscopy (Bierling, 1960). The development of actinomycin-resistant and -sensitive HeLa cells enabled Goldstein *et al.* (1960) and Journey and Goldstein (1961) to describe nucleolar disruption and what is now known as nucleolar segregation affecting only the sensitive strain. Reynolds *et al.* (1963, 1964) observed the same phenomenon: they described it as "nucleolar cap" formation induced by both the carcinogen 4-nitroquinoline-*N*-oxide and actinomycin D and suggested that the lesions could represent the morphological expression of a specific biochemical action. Cytochemical studies made by Schoefl (1964) demonstrated that actinomycin D causes coalescence of three nucleolar components: (*a*) RNP granules embedded in a protein matrix; (*b*) the nucleolonema; and (*c*) an amorphous matrix. The effect of actinomycin D has since been shown to be more or less identical on various types of cells: rat liver (Stenram, 1965; Smuckler and Benditt, 1965; Oda and Shiga, 1965; Smetana *et al.*, 1966; Shankar Narayan *et al.*, 1966) and pancreas (Jézéquel and Bernhard, 1964; Rodriguez, 1967; Yamaguchi *et al.*, 1971); salivary gland of rat (Takahama and Barka, 1967), *Chironomus* (Stevens, 1964), and *Smittia* (Jacob and Sirlin, 1964); amphibian cell (Eakin, 1964; Jones and Elsdale, 1964; Simard and Duprat, 1969) and leukemic myoblasts (Heine *et al.*, 1966). One can generalize therefore that the first cytological target effect of actinomycin D is separation and redistribution of nucleolar components; the term "nucleolar segregation" is now accepted to describe this type of lesion (Bernhard *et al.*, 1965) (Fig. 2).

2. SEQUENTIAL DESCRIPTION OF THE LESION

In most instances, whether induced by actinomycin D or other substances, nucleolar segregation follows the same sequential steps if careful attention is given to both dosage and duration of treatment. (*a*) At very low doses the nucleolus first takes the form of a compact sphere with condensation of the fibrillar portion and migration toward the periphery; the granular zone is of greater importance proportionally; (*b*) segregation of the nucleolar components then occurs, resulting in distinct granular, fibrillar, and amorphous portions; (*c*) longer exposure to the inhibitor causes dispersion of the granular zone, which is seen to migrate from the nucleolus toward the nucleus; (*d*) subsequently all that is left is a mass of closely packed fibrils with an occasional amorphous zone (Fig. 3).

This sequence has been observed repeatedly regardless of the agent

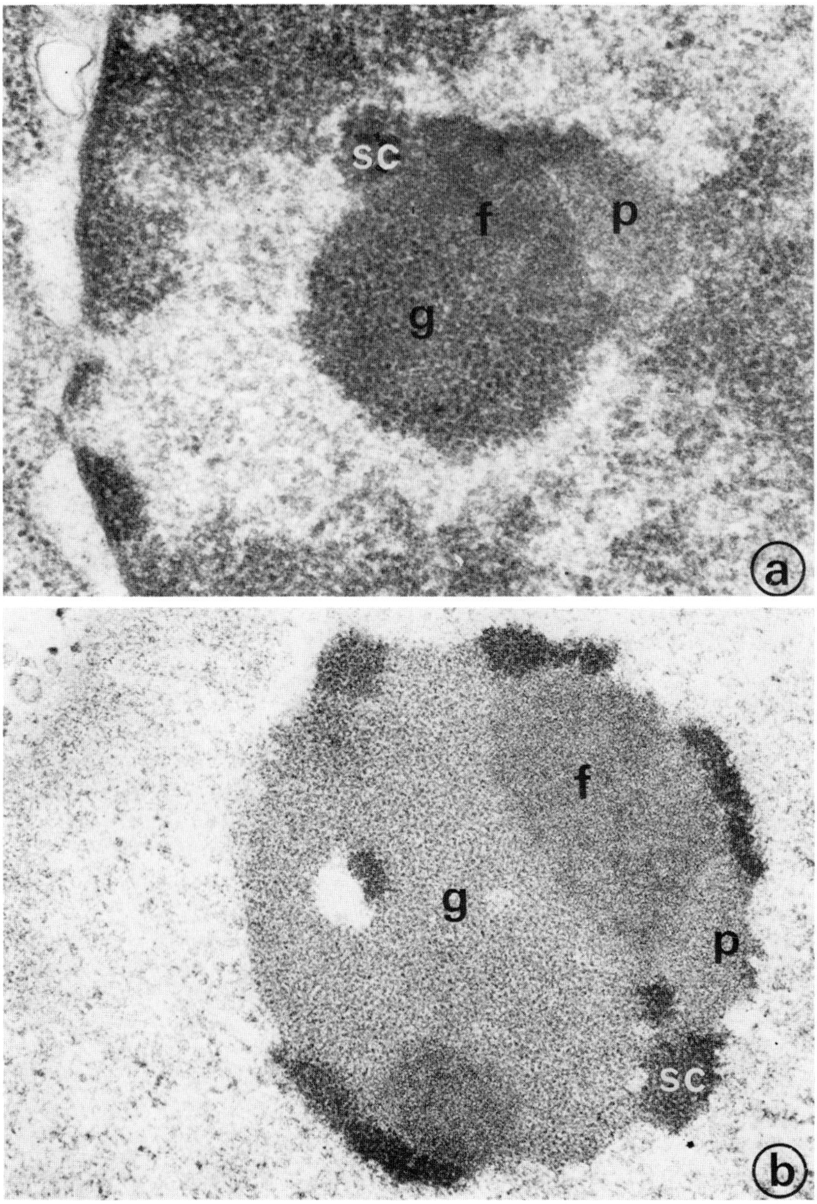

Fig. 2 Segregation of nucleolar components of two different types. In (a), the lesion is observed in rat pancreatic acinar cell 2 hr after injection of actinomycin D (1 mg/kg). The granular (g), fibrillar (f), amorphous (p), and contrasted zone (sc) can be easily identified as segregated regions. Glutaraldehyde; osmium tetroxide; Epon (× 60,000). In (b), the segregation of nucleolar components are observed after treatment of cultured rat embryo cells with chromomycin A_3 (0.5 μg/ml) for 1 hr. Same legend as in (a). Osmium tetroxide; Epon (× 36,000).

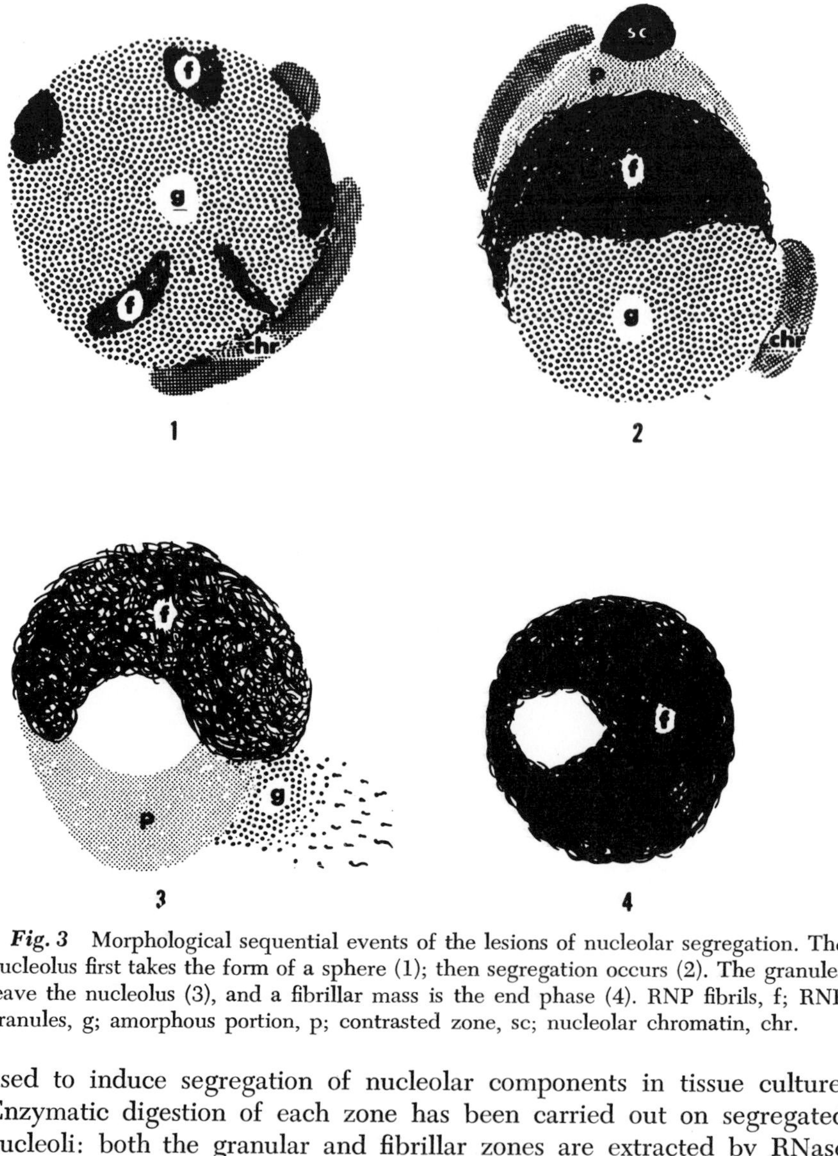

Fig. 3 Morphological sequential events of the lesions of nucleolar segregation. The nucleolus first takes the form of a sphere (1); then segregation occurs (2). The granules leave the nucleolus (3), and a fibrillar mass is the end phase (4). RNP fibrils, f; RNP granules, g; amorphous portion, p; contrasted zone, sc; nucleolar chromatin, chr.

used to induce segregation of nucleolar components in tissue culture. Enzymatic digestion of each zone has been carried out on segregated nucleoli: both the granular and fibrillar zones are extracted by RNase while the amorphous zone is pepsin sensitive (Jézéquel and Bernhard, 1964).

Occasionally a fourth zone is seen in the periphery of segregated nucleoli. This zone is electron dense and granular and has been successively referred to as "blebs" or "satellites," (Schoefl, 1964) "dense plaques or caps," (Reynolds *et al.*, 1964) "new peripheral dense substance," (Stevens,

1964) "contrasted fourth zone of unknown etiology," (Bernhard *et al.*, 1965; Simard and Bernhard, 1966) and "microspherules" (Unuma and Busch, 1967). Agreement has been reached as to the content of the zone which has been shown to react strongly to pepsin and RNase (Schoefl, 1964; Stevens, 1964; Unuma and Busch, 1967). However, Monneron (1968) recently presented convincing evidence that these dense masses actually represent a clustering of perichromatin granules (see Section III,B).

More recent reviews on the subject (Goldblatt and Sullivan, 1970; Recher *et al.*, 1971a; b) are in essential agreement with this sequence of events, except that in tissue culture the end stage is more easily obtained than in hepatocytes. The end stage obtained by Goldblatt and Sullivan (1970) in hepatic cells still contains granules with a large proportion of "dense plaques." For these authors the "plaques" probably consist of twisting and retraction of nucleohistone fibers which could be responsible for the segregation of nucleolar components. Prior inhibition of protein synthesis with cycloheximide does not interfere with the induction of nucleolar segregation by actinomycin in rat liver or intestinal crypt cells (Goldblatt *et al.*, 1970).

3. SPECIFICITY OF NUCLEOLAR SEGREGATION

In view of the variety of chemically unrelated substances capable of causing nucleolar segregation, a systematic study was undertaken to determine the level of the biochemical block responsible for the morphological lesions. Cultured cells of rat embryos were treated with various antimetabolites, analogs, and antibiotics. The compounds were chosen because of their wide use, their known biochemical action, and their site of attack of nucleic acid or protein synthesis. Substances were employed at the lowest concentration capable of inducing a characteristic cytological lesion. In most instances, higher doses resulted in nonspecific cytotoxic effects (Fig. 4).

Two conclusions could be drawn from this study (Simard and Bernhard, 1966). First, nucleolar segregation is a specific lesion. Neither antimetabolites acting at the level of nucleotide precursors nor those interfering with polynucleotide incorporation or protein synthesis were associated with the characteristic nucleolar lesion. One exception to this, although more apparent than real, was azaserine. This antibiotic is a known glutamic acid antagonist (Hartman *et al.*, 1955) but possesses an unsaturated diazo group capable of nucleophilic substitution. Furthermore, it behaves like an alkylating agent at appropriate doses, assuming mutagenic (Iyer and Szybalski, 1958), antimitotic (Maxwell and Nickel, 1954), and radiomimetic properties (Terawaki and Greenberg, 1965). All

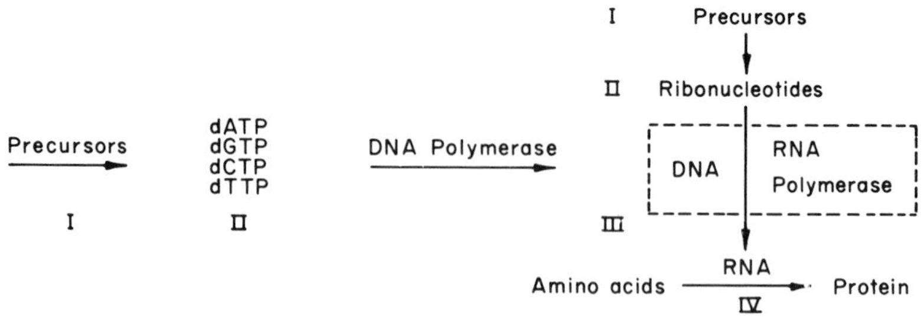

	I	II	III	IV
A	Aminopterin Azaserine (weak dose) 6-Mercaptopurine Alkaloids of vinca rosea	5-Iododeoxyuridide 5-Bromodeoxyuride Cytosine arabinoside 5-Fluorouracil Thymidine (excess)	Alkylating agents Antibiotics Amino acridines Ethidium bromides Histones (excess) Polylysine	Azaguanine Chloramphenicol Puromycin Alkaloids of vinca rosea
SN			Proflavin Chromomycin A_3 Echinomycin Nogalamycin Ethidium bromides Azaserine Daunomycin	

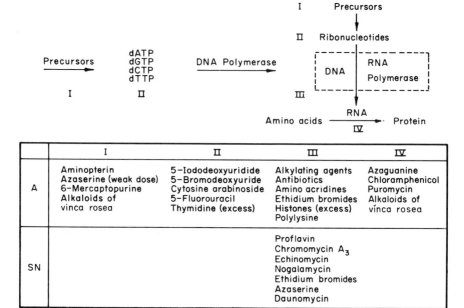

Fig. 4 The antimetabolites (A) used were classified into four groups according to their action at the level of precursors of DNA and RNA (I), ribo- and deoxyribonucleotides (II), DNA–RNA template system (III), and protein synthesis (IV). The substances causing nucleolar segregation (SN) were all part of group III.

other substances causing nucleolar segregation were part of a group of compounds binding directly to the DNA molecule and interfering with its template activity.

Second, nucleolar segregation corresponds to a blocking of nucleolar RNA synthesis. Among the 24 substances used, the compounds causing nucleolar segregation block the enzymatic synthesis of RNA by RNA polymerase by the formation of complexes with DNA like actinomycin D (Reich, 1964) or proflavin (Hurwitz *et al.*, 1962). These compounds are echinomycin (Ward *et al.*, 1965), nogalamycin (Bhuyan and Smith, 1965), chromomycin A (Ward *et al.*, 1965), daunomycin (Di Marco *et al.*, 1963; Calendi *et al.*, 1964), proflavin, and ethidium bromide (Lerman, 1961; Luzzati *et al.*, 1961; Waring, 1966).

Support for this hypothesis has recently been obtained. Using varying doses of actinomycin D, Goldblatt *et al.* (1969a) observed that nucleolar segregation is a reflection or response of some alteration of DNA rather than the consequence of inhibition of RNA synthesis. Interference with RNA synthesis alone is insufficient to explain the nucleolar damage (Goldblatt *et al.*, 1969b). Other authors have reviewed the effects of selected hepatocarcinogens and stressed the coincidence of nucleolar segregation with the blocking of RNA synthesis in rat liver by lasiocarpine, 3-methyl-4-dimethylaminoazobenzene, dimethylnitrosamine, tannic acid (Reddy and Svoboda, 1968), and urethan (Lombardi, 1972).

4. NUCLEOLAR SEGREGATION INDUCED BY OTHER AGENTS

An increasing number of substances are known to affect the nucleolus in a specific manner. The potent carcinogen, aflatoxin, causes rapid segregation of nucleolar components in rat liver (Bernhard *et al.*, 1965; Svoboda *et al.*, 1966; Terao *et al.*, 1971). *In vitro* binding of the toxin to DNA has been suggested (Sporn *et al.*, 1966; Clifford and Rees, 1967) to explain its action on RNA synthesis (Lafarge *et al.*, 1965; Clifford and Rees, 1967) particularly at the nucleolar level (Lafarge *et al.*, 1966). The reversibility of nucleolar segregation with aflatoxin has been well studied; 24 hr after injection in rats, the liver recovers its ability to transcribe RNA (first nuclear RNA, then nucleolar RNA), but is unable to replicate DNA although the DNA polymerase system is still active. This was interpreted as support for the hypothesis that aflatoxin acts on the chromatin (Lafarge and Frayssinet 1970). This is accompanied by restoration of normal nucleolar structure (Bernhard *et al.*, 1965; Pong and Wogan 1970).

Another carcinogen, 4-nitroquinoline-*N*-oxide (Endo, 1958), similarly affects the nucleolus (Reynolds *et al.*, 1963); recent work suggests that it reacts with nascent DNA (Malkin and Zahalsky, 1966) and blocks the action of RNA polymerase (Paul *et al.*, 1967). The alkylating antibiotic mitomycin C (Iyer and Szybalski, 1963) induces the same nucleolar lesions (Lapis and Bernhard, 1965) but only after long exposure, probably because its action on RNA synthesis is secondary (Kuboda and Furuyama, 1963). Ribonuclease (Robineaux *et al.*, 1967) and ultraviolet flying spot irradiation (Montgomery *et al.*, 1966) also belong to the same group of agents primarily affecting the nucleolus and RNA synthesis. The antibiotic mythramycin (Kume *et al.*, 1967) and camptothecin (Recher *et al.*, 1972) can also be added to this list. Hydroxyurea, a substance that reacts with DNA *in vivo* (Eisenberg *et al.*, 1965), induces the same nucleolar lesions in amphibian embryos (Geuskens, 1968).

Other stimuli for nucleolar segregation cannot be related to a direct action on RNA synthesis because of their as yet undefined mode of action; the antibiotic anthramycin (Harris *et al.*, 1968a), cycloheximide (Harris *et al.*, 1968b), cordycepin (Stockert *et al.*, 1970), puromycin aminonucleoside (Lewin and Moscarello, 1968), and amino sugars (Molnar and Bekesi, 1972a, b) have recently been related to this phenomenon. Micrographs of typical nucleolar segregation have been obtained in cultured cells infected by herpesvirus (Sirtori and Bosisio-Bestetti, 1967) and mycoplasma

(Jézéquel *et al.*, 1967). Natural segregation can also be observed in newts (Reddy and Svoboda, 1972) but in this case it appears to be more related to a successive appearance of fibrils and granules in the course of development than to anything else.

5. RNA SYNTHESIS IN PRESENCE OF ACTINOMYCIN D

Light microscopic autoradiography has revealed severely decreased incorporation of RNA precursors following actinomycin D treatment in tissue culture (Schoefl, 1964) and rat liver (Stenram, 1965). High resolution autoradiography further demonstrated that when pulses of [^{3}H]uridine precede actinomycin D treatment, accumulation of the radioactivity is found over the granular portion of the segregated nucleolus, although a persistent labeling of the fibrillar zone subsists (Geuskens and Bernhard, 1966).

Microbeam experiments (Perry *et al.*, 1961) and low doses of actinomycin D (Perry, 1962, 1963) provided biochemical evidence that the 45 S RNA synthesized in the nucleolus is a precursor of ribosomal RNA (Perry, 1963), a conclusion reached earlier by Scherrer and Darnell (1962). Although synthesis of the 45 S is blocked by the antibiotic, transformation of the previously labeled 45 S RNA can still proceed, resulting in the accumulation of 28 S and 6 S in the nucleolus (Muramatsu *et al.*, 1966; Unuma *et al.*, 1972) since the 18 S fraction is not affected (Steele and Busch, 1966b). It seems that the action of actinomycin D is mediated through a preferential binding to the guanine residues of DNA (Reich, 1964). Recently, however, Sobell and Jain (1972) proposed a model of actinomycin binding in which the chromophore group intercalates between base pairs and the amino rings project into the minor groove of helicoidal DNA.

Figure 5 shows the action of increasing doses of actinomycin on the profiles of nucleolar RNA for the same duration of treatment in ascites tumor cells. The amount of 45 S RNA decreases considerably at high concentration of the drug owing to the blocking of its synthesis. However, the nucleolar products of the 45 S, 28 S, and 32 S RNA accumulate at high levels indicating that actinomycin D interferes not only with the synthesis but also with the processing of ribosomal RNA.

Langelier (1973) has measured the kinetics of incorporation of [^{3}H] actinomycin D into the chromatin fraction of hamster cultured cells. At saturation, there is 1 molecule bound per 220 base pairs of DNA whereas it requires only 1 molecule per 6000 base pairs to block nucleolar RNA synthesis by 70%, according to the data of Perry and Kelley (1970). The saturation plateau is obtained after 12 hrs; it is interesting to note that

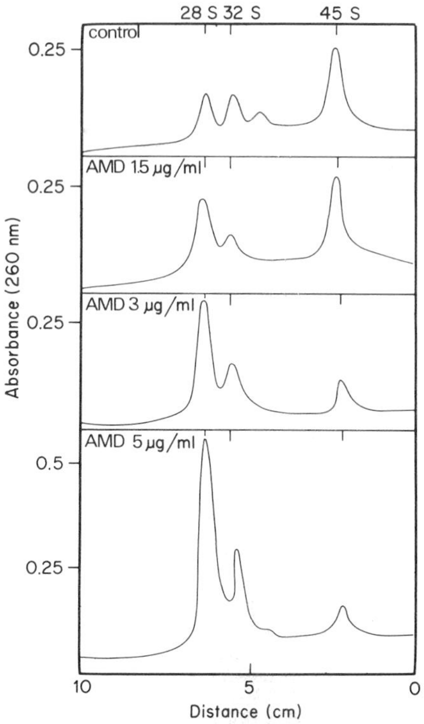

Fig. 5 Gel electropherogram of nucleolar RNA run 6 hr on a 2.7 acrylamide gel at 5 mA. Nucleoli from ascites tumor cells were isolated 60 min after treatment with increasing doses of actinomycin D. The increase in 28 S RNA is grossly proportional to the decrease in 45 S RNA at higher concentrations of the antibiotic.

the penetration of the drug in sensitive cell lines is twice as fast as the binding to DNA. Therefore, in sensitive cell lines, there should exist twice as many free actinomycin molecules than bound ones. The total amount present in the cell is proportional to the concentration in the culture medium.

Nucleolar DNA has been shown to contain a high proportion of guanine and cytosine (McConkey and Hopkins, 1964). The expectation that actinomycin D would bind preferentially to nucleolar DNA has been partly confirmed by high resolution autoradiography. Incorporation of [^{3}H]actinomycin D in cultured BHK cells resulted in accumulation of the radioactivity in the condensed portion of the chromatin whether associated to the nucleolus or not, but with occasional ring formation around the nucleolar body (Fig. 6). The amount of labeling was found to be time dependent but concentration seemed to have little effect (Simard, 1967; Simard and Cassingena, 1969).

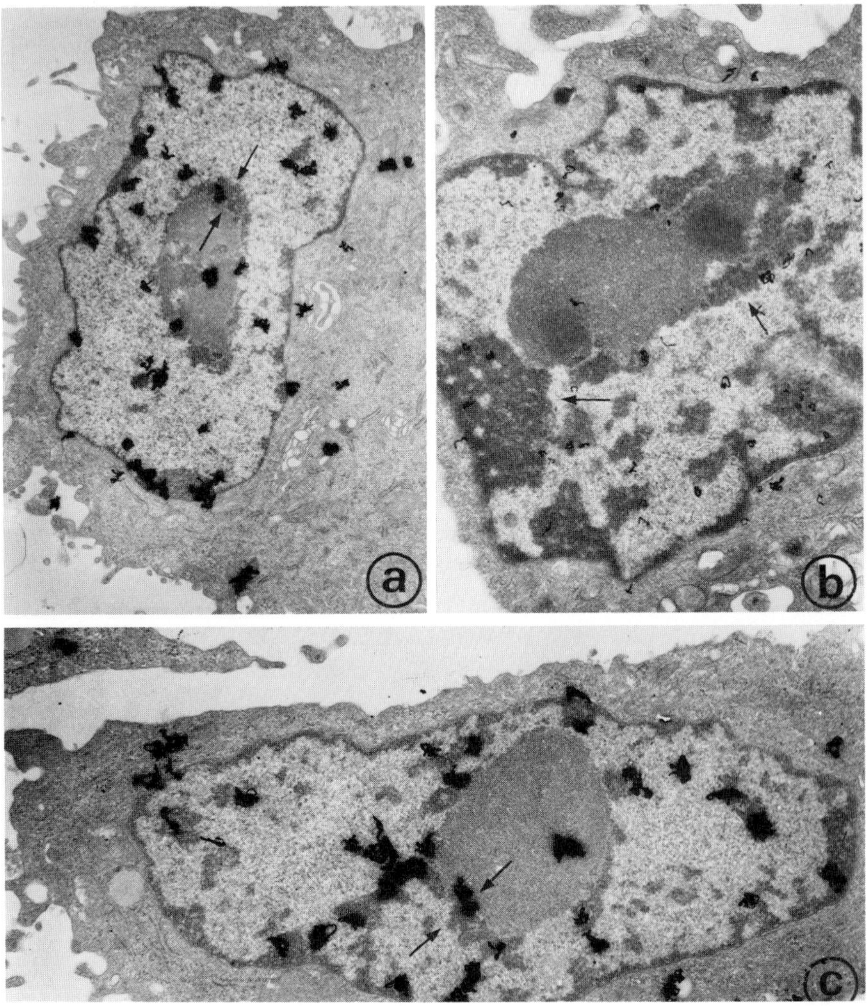

Fig. 6 Hamster fibroblasts treated with [^{3}H]actinomycin for 20 (a and b) and 30 min (c). The silver grains are mostly localized on the condensed portion of the nuclear chromatin whether located around the nucleolus or not (arrows) [(a) × 4500; (b) × 9000; (c) × 12,000].

However, Bernier *et al.*, (1972) studied the "staining" of hepatocytes by [^{3}H]actinomycin D by floating ultrathin, frozen sections on the antibiotic. These authors were able to show by autoradiography the specificity of the binding to the nuclear regions and the increasing amount of radioactivity as a function of time. Kupffer cells are labeled more heavily than hepatocytes, but no precise nucleolar localization was noticed.

B. *Nucleolar Degranulation and Supranormal Temperature*

1. CYTOLOGICAL EFFECT OF SUPRANORMAL TEMPERATURE

Supranormal temperature has been used for a long time to synchronize cell cultures (Juul and Kemp, 1933) following the work of Bucciante (1928) relating the effect of temperature to the cell cycle, particularly to mitosis (Rao and Engleberg, 1965; Sisken *et al.*, 1965). Fusion and condensation of nucleoli in *Transdescentia* after temperature exposure have been reported by phase-contrast microscopy (Snoab, 1955).

Systematic electron microscopic and cytochemical studies were carried out on cultured hamster fibroblasts (BHK strain) after exposure to supranormal temperatures (Simard and Bernhard, 1967). At temperatures of 38°, 39°, and 40°C, no noticeable lesion occurs even after an incubation of 120 min. The cells continue to grow normally when transferred to a new medium and cultured at 37°C for 24–48 hr. At 41°C early changes appear in some nucleoli after 1 hr of incubation; there is a fading out of the nucleolar reticular aspect and a decrease in the granular RNP particles which appear as fuzzy and cloudy spots.

A critical point is reached at 42°C with the appearance of striking nucleolar lesions. As early as 15 min after treatment, but more markedly after 1 hr, there is complete loss of the granular RNP component and disappearance of the nucleolar reticulum, associated with a complete retraction of the intranucleolar chromatin. The remaining material in the morphologically homogeneous nucleolus is a large amount of RNase-sensitive, closely packed fibrillar RNP. The lesions remain identical as temperature increases to 45°C. These alterations prove to be reversible when the cells are returned to 37°C with the reappearance of an exaggerated amount of intranucleolar chromatin and granular RNP resulting in nucleoli of considerable size. These nucleolar lesions occur in otherwise well-preserved cells.

Identical lesions were observed in normal diploid rat embryonic cells and liver cells (Fig. 7) subjected to the same treatment. Ascites tumor cells were found to react similarly, but the critical temperature was 44.5°C. Degranulation of nucleoli was observed in embryonic differentiating cells of the amphibian *Pleurodeles waltlii* incubated at 37°C for 5 hr (Duprat, 1969). In a systematic study, Heine *et al.* (1971) reported similar results in HeLa cells except that they observed a remarkable increase of perichromatin granules with occasional threadlike structures. Love *et al.* (1970) also observed the formation of cytoplasmic inclusions during hyperthermia in HeLa cells and confirmed the reversibility of the lesions; there seems to be evidence that the inclusions originated from the nucleolus in their system.

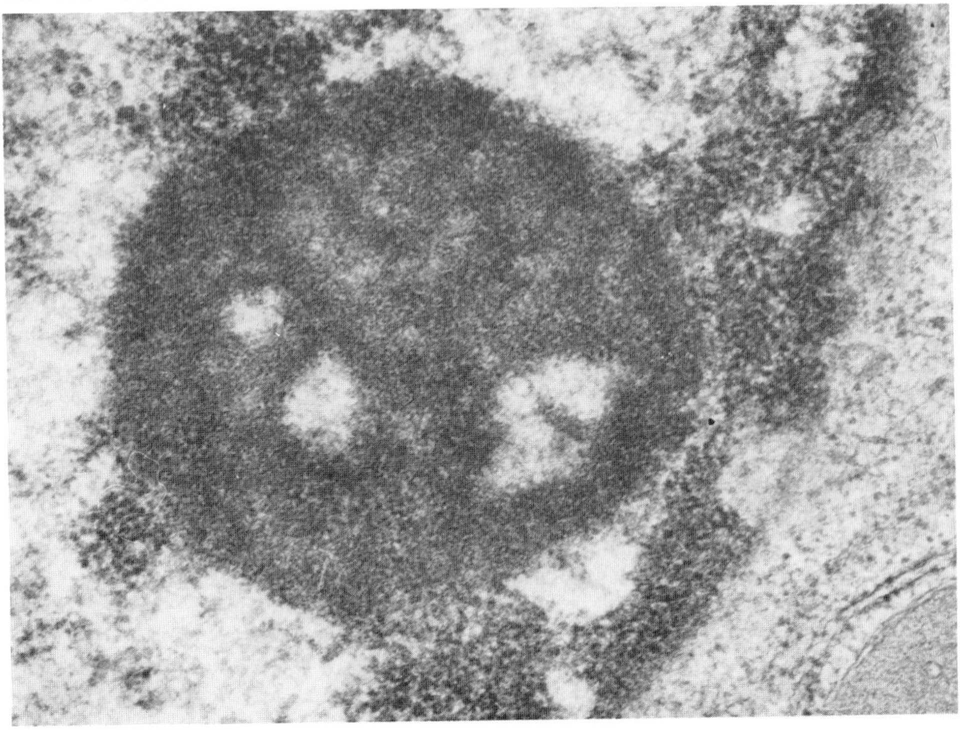

Fig. 7 Nucleolar degranulation induced by supranormal temperature (42°C for 30 min) in rat liver cells. The nucleolus has rounded up and lost its reticular aspect as well as its granular component. It is now homogeneous and consists of closely packed electron-dense fibrils (× 50,000).

2. SEQUENTIAL APPEARANCE OF THE LESIONS

In ascites tumor cells, complete degranulation is obtained after only 30 min of treatment at 45°C. Systematic cytochemical and morphological studies were carried out on the same batches of cells incubated at 44.5°C for periods of time varying from 5 to 30 min. After 10 min of heat shock, intranucleolar chromatin is completely absent from the nucleolus, while the granular RNP are still present; the granules disappear after 30 min.

A similar sequence of events was found after 30 min of incubation at increasing temperatures of 39°–45°C. Retraction of intranucleolar chromatin was completed at 43°C, while the granular RNP disappeared as previously at 44.5°C only.

It seems, therefore, that whenever cells are exposed to supranormal temperatures, retraction of intranucleolar chromatin precedes the degranulation of the nucleolus, an observation that has proved to be of significant importance for functional studies (Simard *et al.*, 1969).

3. SPECIFICITY OF NUCLEOLAR DEGRANULATION

Whether or not supranormal temperatures affect the nucleolus in a specific manner cannot yet be ascertained. For most biological reactions there exists an optimum temperature responsible for a given equilibrium (Lwoff and Lwoff, 1961; Lwoff, 1962). However, the ultrastructural lesions induced by thermic shock are striking and concern only the nucleolus; apart from a slight clumping of chromatin, the nucleus is not altered and no lesions have ever been observed in the cytoplasm.

The lesions observed in the nuclei of degenerating and dead cells have been described by Trump *et al.* (1965). The reticular aspect of the nucleolus becomes blurred after 4 hr of autolysis and the nucleolar RNP granules disappear after 8–12 hr. But these changes take place concomitantly with severe nuclear and cytoplasmic damage and are not reversible. Moreover, they involve the whole cell, not a specific organelle.

4. NUCLEOLAR DEGRANULATION INDUCED WITH OTHER AGENTS

Depletion of the granular component has been reported to occur in all nucleoli that undergo segregation of their components. In most instances an end phase is reached with entirely fibrillar but small nucleoli resembling those exposed to supranormal temperatures. Recently, Ganotte and Rosenthal (1968) have shown that methylazoxymethanol, a hepatotoxin derived from cycasin, causes an aborted nucleolar segregation rapidly followed by degranulation of nucleoli which then assume a clumped reticulated pattern with wide open meshes. The picture closely resembles the lesions obtained with supranormal temperatures. The lesions observed after treatment of ascites tumor cells with amino sugars (Molnar and Bekesi, 1972a, b) also appear to consist mostly of degranulation of nucleoli.

Entirely fibrillar nucleoli are also observed during amphibian embryogenesis. The formation of nucleolus is observed during gastrulation with the appearance of dense fibrous bodies within the chromatin material (Karasaki, 1965, 1968). These fibrous bodies resemble primary nucleoli of early gastrula in the anucleolate mutant of *Xenopus,* which lacks the nucleolar organizer (Jones, 1965; Hay and Gurdon, 1967).

5. RNA SYNTHESIS AT SUPRANORMAL TEMPERATURE

Incorporation of [^{3}H]uridine was studied by high-resolution autoradiography in hamster fibroblasts. Following a thermic shock of 43°C for

1 hr, the uptake of [^{3}H]uridine by the altered nucleolus is almost absent after 5-min or 30-min pulses while the incorporation over the nuclear dispersed chromatin is much less affected. An approximate grain count revealed a 90% reduction in the nucleolar incorporation of the precursor of heat-treated cells as compared with control cells, whereas the extranucleolar nuclear uptake is lowered by only 20%. When the nucleoli were given a 30-min pulse of [^{3}H]uridine just prior to heat treatment to label both the fibrillar and the granular RNP, subsequent chases for 1 hr with cold uridine at both 37° and 43°C resulted in heavy nucleolar labeling in both experiments without any appreciable difference in grain count. Similar results were obtained in ascites tumor cells maintained at 44.5°C for 30 min. No incorporation of [^{3}H]uridine was found after such treatment, and retention of radioactivity in the nucleolus was observed when incorporation of the precursor preceded the thermic shock (Simard and Bernhard, 1967).

Density gradiant analysis of nucleolar RNA's was performed systematically on ascites tumor cells. In correlation with the sequential events in nucleolar degranulation induced by thermic shock, the specific activity of nucleolar 45 S RNA was recorded after a thermic shock of 44.5°C for increasing periods of time and after a thermic shock of 30 min at increasing temperatures. In both cases, a rapid decrease was observed with an inflection point corresponding to the disappearance of intranucleolar chromatin.

Qualitative analysis of nucleolar RNA's performed after thermic shock failed to show any differences with the control cells even when the granular RNP component had disappeared ultrastructurally. The same curves showed that while the synthesis of rapidly labeled 45 S is severely affected by thermic shock, a labeling of the 8–10 S region persists throughout the experiment. Ribosomal RNA was not labeled after a thermic shock except once in the 8–10 S region.

When cells were pulse-labeled 30 min before a heat shock of 30 min at 44.5°C, accumulation of radioactivity was found in the nucleolus and no transport was observed in the cytoplasmic ribosomes. However, when the same pulse-labeling was followed by a heat shock of 30 min at 43°C (the granular RNP being still present, while nucleolar RNA synthesis is blocked to the extent of 80%) the transport appeared unimpaired in the ribosomes (Amalric *et al.*, 1969; Simard *et al.*, 1969) (Fig. 8).

Warocquier and Scherrer (1969) studied RNA metabolism in HeLa cell nuclei heated at 42°C. These authors were able to show that the 45 S RNA precursor is still formed at this temperature but that its subsequent transformation to functional ribosomal RNA is impaired since no new RNA appears in cytoplasmic ribosomes. However, the synthesis of nascent messengerlike RNA appears less affected. They suggest that the shift in temperature selectively destroyed G + C-rich RNA while the polydisperse messenger is not affected. Their conclusions provide support for the existence of a heat-sensitive cellular function located in the nucleolus where the G + C-rich RNA is synthesized. Blocking of the synthesis of the 45 S RNA and of the processing was confirmed by Heine *et al.* (1971). There also appears to be an inhibition of protein synthesis at elevated temperature (McCormick and Penman, 1969; Pouchelet *et al.*, 1971).

6. CONCLUSIONS

Several conclusions can be drawn from this model of nucleolar degranulation at supranormal temperature.

A. The synthesis of nucleolar RNA is heat sensitive, and is a reversible matter, provided the system permits recovery harvestings of cells (in tissue culture, for example). Other authors have observed the thermosensitivity of RNA synthesis in different biological systems (Moner, 1967; Gharpure, 1965; Stevens, 1966, 1967). The precise level at which temperature affects the nucleolar RNA synthesis has still to be determined and probably results from simultaneous alterations of several factors. In any case, the critical points of variation of the specific activity of 45 S nucleolar RNA (10 min at 44.5°C and 30 min at 43°C) are concomitant with the retraction of intranuclear chromatin, the presence and integrity of which appears to be necessary for the synthesis of 45 S nucleolar RNA.

B. The granular RNP are probably transitory configurational forms, unraveling after thermic shock. One of their main functions appears to be the transport of nucleolar RNA to the cytoplasm. This function is blocked at supranormal temperatures, even if the nucleolus has been previously heavily labeled.

C. The fibrillar RNP would then consist of a stable pool of RNA's where all ribosomal and possibly nonribosomal precursors could be stored. No modifications are indeed observed in qualitative analysis of nucleolar RNA after thermic shock when nucleoli are entirely fibrillar. Similar conclusions have been reached by other authors. Geuskens and Bernhard (1966) explained the persistent labeling of the fibrillar zone after actinomycin treatment as being consistent with the hypothesis that fibrils contain a stable pool linked with ribosomal and nonribosomal nucleolar functions. A similar conclusion was reached by Jones (1965) and by Hay and Gurdon (1967) to explain the presence of a fibrillar pseudonucleolus in the homozygote anucleolate mutants of *Xenopus* which do not synthesize ribosomal RNA owing to deletion of ribosomal cistrons.

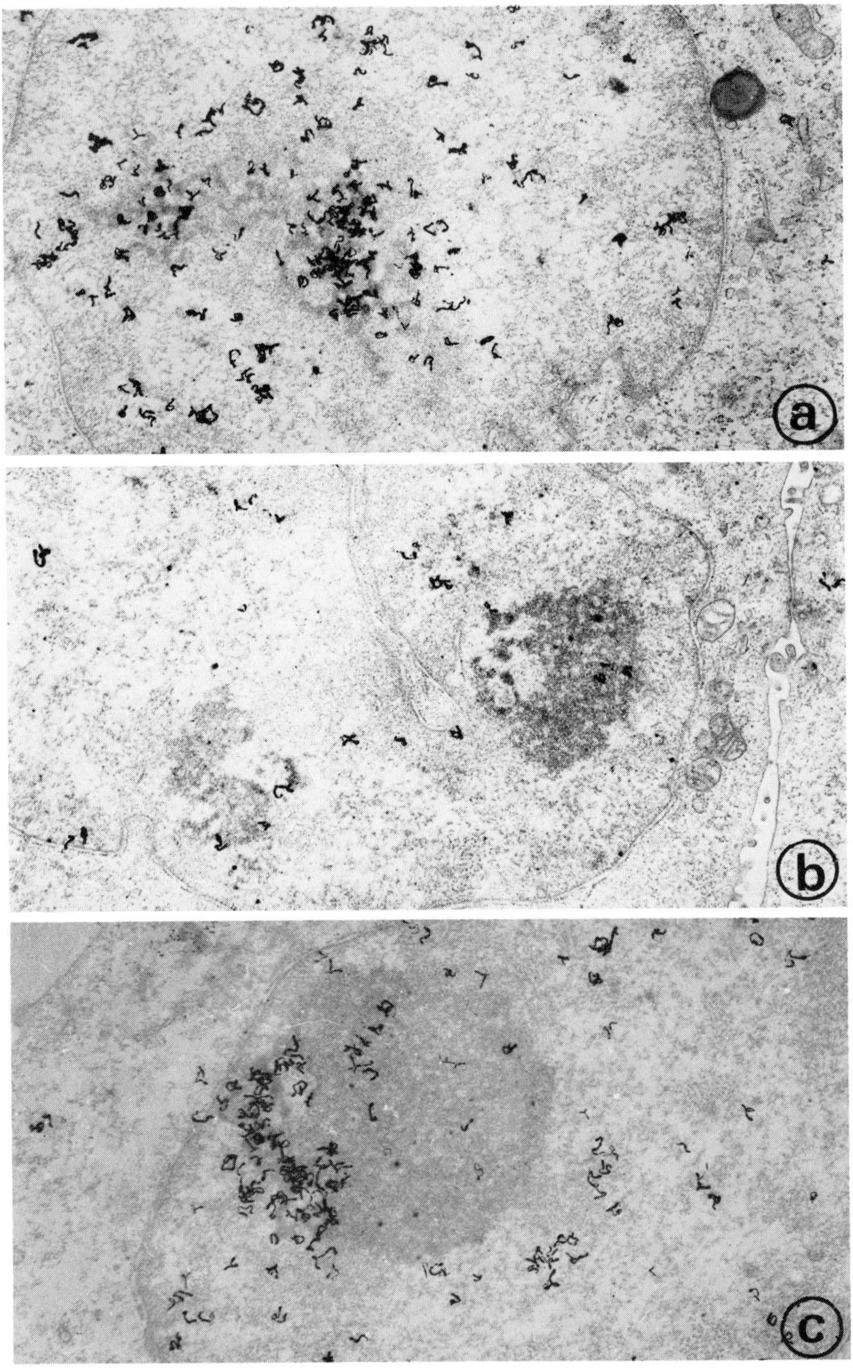

C. *Nucleolar Hyperthrophy and Thioacetamide*

1. CYTOLOGICAL ACTION OF THIOACETAMIDE (TA)

The effect of TA on rat liver is characterized by hepatic cell damage that progresses to centrolobular necrosis and cirrhosis with sufficient dosage (Kleinfeld, 1957; Ruttner and Rondez, 1960; Gupta, 1956). The hepatocarcinogenic properties of TA were first described by Fitzhugh and Nelson in 1948 and later by various authors (Gupta, 1955; Jackson and Dessau, 1961). The first detectable cytological action of TA in hepatocytes is a remarkable nucleolar hypertrophy (Rather, 1951; Kleinfeld, 1957) that is related to an increase in nuclear RNA synthesis (Rather, 1951; Laird, 1953) although the RNA itself was not demonstrably chemically different from normal RNA (Kleinfeld and Von Haam, 1959).

Electron microscopic studies following TA administration have emphasized that sublethally injured rat liver undergoes rapid nuclear enlargement with frequent cytoplasmic inclusions and striking hypertrophy of the nucleolus (Rouiller and Simon, 1962; Salomon, 1962; Salomon *et al.*, 1962; Thoenes and Bannasch, 1965) (Fig. 9). Cytoplasmic lesions include an abnormal increase of agranular endoplasmic reticulum, hypertrophy of the Golgi complex, and mitochondrial lesions. The giant nucleoli were found to be particularly rich in RNP granules by planimetry (Shankar Narayan *et al.*, 1966), but it was later demonstrated that the proportion between granular and fibrillar RNP remained approximately the same when segregation of nucleolar components was induced in giant nucleoli by actinomycin D (Suter and Salomon, 1966).

The mechanism by which TA causes hypertrophy of the nucleolus and hepatocellular injury is not understood. Furthermore, it constitutes the sole example of such nucleolar enlargement. Therefore, the question of the drug specificity of TA-induced lesions cannot be discussed here. Previous studies with isotopically labeled [^{35}S]TA have shown that, instead of binding with protein or alkylating nucleic acids, the compound is rapidly broken down since no increase in radioactivity is found in rat liver fed [^{35}S]TA (Nygaard *et al.*, 1954; Maloof and Soodak, 1961). Rees *et al.*

Fig. 8 Incorporation of [^{3}H]uridine by ascites tumor cells. In control cells (a) exposed to [^{3}H]uridine for 30 min at 37°C, the radioactivity is heavy over the nucleolus but extranucleolar regions are labeled as well. In (b), the cells were heated at 44.5°C for 30 min followed by a 30-min incorporation of [^{3}H] uridine at 37°C: almost no labeling is seen over the nucleolus but the extranucleolar regions are less affected. In (c), the incorporation of [^{3}H]uridine was done simultaneously with a heat shock of 44.5°C for 30 min. Incorporation occurred during the first minutes but the radioactivity is localized over only one portion of the nucleolus (isotopic segregation) (× 15,000).

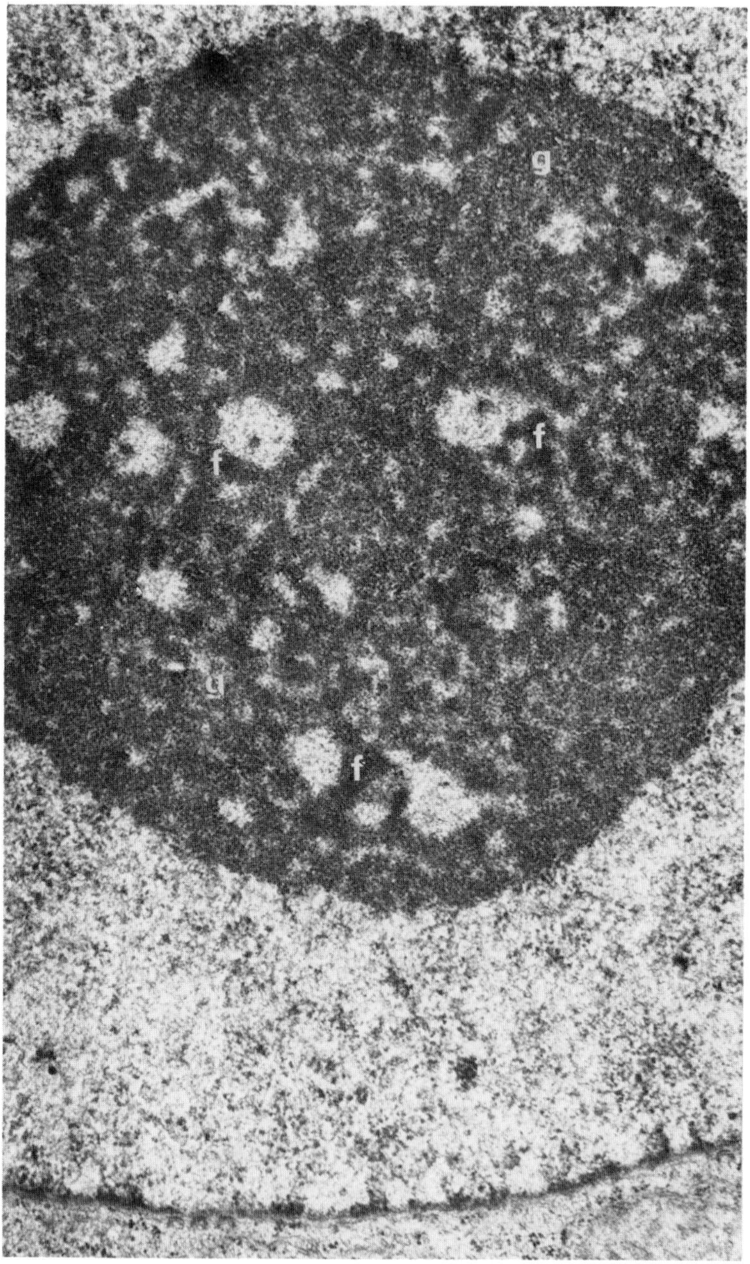

Fig. 9 Nucleolar hypertrophy in rat liver cell induced by administration of thioacetamide for 11 days. The granular (g) and fibrillar component (f) are present in this nucleolus that has more than doubled its size (× 30,000).

(1966) studied the metabolism of TA labeled with the ^3H methyl group instead of ^{35}S and found that the carcinogen is metabolized within 24 hr and converted to acetate and hydrogen sulfate. Many carcinogens have been found to undergo metabolic changes to a form in which interaction with cellular components can take place. In the case of TA, there still remains the problem of identifying an active form of its metabolite capable of interaction in order to explain the hepatocellular lesions among which nucleolar hypertrophy is the most striking.

2. NUCLEOLAR HYPERTROPHY IN OTHER CONDITIONS

Besides TA, there are other causes of nucleolar hypertrophy. MacCarty (1928, 1936) was the first to recognize nucleolar hypertrophy as a pathognomonic sign of cancer cells. The ultrastructure of malignant cell nucleoli exhibits wide variations in the proportion of their components (Bernhard and Granboulan, 1963; Busch *et al.,* 1963) with inclusions of all sorts (Haguenau, 1960; Thoenes, 1964) and occasionally granular material of unknown origin (Shankar Narayan and Busch, 1965). Hyperplastic liver nodules induced by carcinogens exhibit enlarged nucleoli (Yasuzumi *et al.,* 1970). But none of these lesions can be considered as typical of cancer cells. Hypertrophy of the nucleolus is a common feature in rapidly growing tissues, such as embryonic cells or regenerating liver cells, following partial hepatectomy (Higgins and Anderson, 1931; Bucher, 1963; Stenger and Confer, 1966).

Enlargement of nucleoli has also been described following chronic ethionine intoxication in liver cells (Miyai and Steiner, 1965; Svoboda *et al.,* 1967), isoproterenol administration in salivary glands (Takahama and Barka, 1967), and protein deficiency (Svoboda *et al.,* 1966). Starved animals fed a low protein diet (Stenram, 1958; 1963) or a threonine-devoid diet (Shinozuka *et al.,* 1968b) show enlargement of liver nucleoli. In most of these cases the changes appear to be related to an increase in RNA and protein synthesis (Barka, 1966; Stenram, 1958; Sidransky and Farber, 1958; Sidransky *et al.,* 1964; Kleinfeld, 1966).

3. RNA SYNTHESIS IN NUCLEOLAR HYPERTROPHY INDUCED WITH TA

The effect of TA on RNA synthesis is characterized by a rapid increase in nucleolar RNA synthesis; at the same time a decrease in cytoplasmic ribosomal RNA is observed (Laird, 1953; Kleinfeld and Von Hamm, 1959; Koulish and Kleinfeld, 1964). The increased rate of RNA synthesis in the nucleolus corresponds to high-molecular-weight ribosomal precursors (Steele *et al.,* 1965) as the sedimentation profile of RNA of nucleolar fractions shows a six- to eight-fold increase in the relative amount of 45 S, 35 S, and 28 S RNA with the appearance of a new peak of 55 S RNA

(Steele and Busch, 1966a). However, most of this RNA does not reach the cytoplasmic ribosomes (Kleinfeld, 1966). It is now known that TA considerably alters the nuclear enzymes. The increased nucleolar RNA synthesis is associated with an increased activity of the nucleolar RNA polymerase system *in vitro* (Villalobos *et al.*, 1964a) and latent ribonuclease (Villalobos *et al.*, 1964b). Because of its property of increasing the number of nucleolar elements, TA is frequently used as a model to provide a richer source of ribonucleoprotein particles for isolation. It is interesting to note that these particles have the same characteristics as those obtained from normal nucleoli (Koshiba *et al.*, 1971; Bachellerie *et al.*, 1971).

A clear interpretation of these facts leading to giant nucleoli and increased nucleolar RNA content has yet to be proposed to explain the action of TA. Kleinfeld (1966) considers the possibility that TA does not alter the biosynthetic process per se but activates the initial phase of ribosomal RNA transcription simply by increasing the number of sites open for transcription. The subsequent events would then only be part of regulatory control mechanisms resulting in a piling up of this RNA in nucleoli and its subsequent breakdown as only a normal amount is transported to the cytoplasm. The use of this model would, in any case, be rewarding if it yielded more information regarding the active form or metabolite of TA capable of interacting with nucleic acids (for the effects of TA on RNase activity, see Busch and Smetana, 1970).

D. Nucleolar Fragmentation and Ethionine

1. CYTOLOGICAL LESIONS INDUCED BY ETHIONINE

The ultrastructural changes induced by ethionine, a methionine analog, have been known for some time (Herman and Fitzgerald, 1962; Herman *et al.*, 1962) but only recently has it been documented systematically in liver cells (Miyai and Steiner, 1965; Shinozuka *et al.*, 1968a). Within the first hours after administration of ethionine there is a progressive decrease in the nucleolar size with appearance of electron-opaque masses punctuating the nucleolonema. Within 6–8 hr after ethionine injection, fragmentation and disorganization of the nucleolar architecture occur with preservation of both the fibrillar and granular components. Nucleolar remnants can take the form of round or rug-shaped, electron-opaque fibrillar masses with peripheral granular aggregates; in other cases, they stimulate aborted segregation of nucleolar components. These changes are observed in most hepatocytes and end up in complete dispersion of nucleolar components bearing little resemblance to what is accepted as a normal nucleolus (Fig. 10). Other lesions in the nucleus include clumping of interchromatin gran-

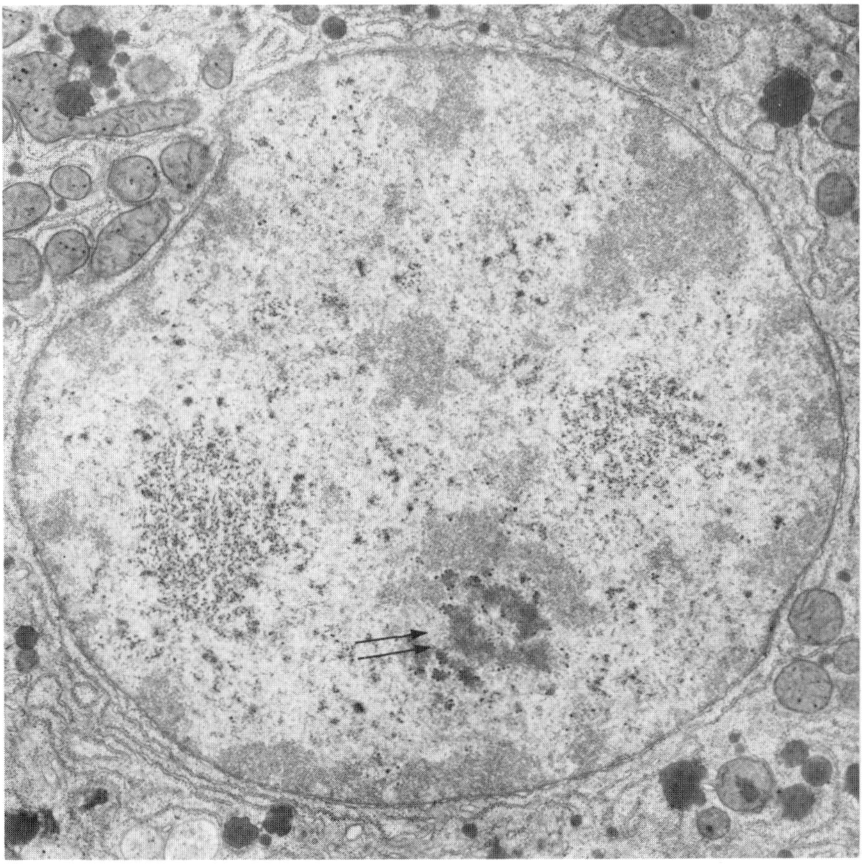

Fig. 10 Nucleolar fragmentation in rat liver cell 12 hr after ethionine injection. Nucleolar remnants are indicated by arrows (× 20,000). (From Shinozuka *et al.*, 1968a.)

ules and condensation of chromatin. Both methionine and adenine prevent fragmentation; administration of adenine 8 hr after ethionine completely reverses the nucleolar lesions within 4 hr (Shinozuka *et al.*, 1968a). Protein synthesis does not appear to be essential for the restoration of nucleolar structure after fragmentation, but RNA synthesis proved to be important (Shinozuka and Farber, 1969).

2. NUCLEOLAR FRAGMENTATION INDUCED BY OTHER SUBSTANCES

There are few instances of similar changes being induced by other substances. The term fragmentation has been used to describe light microscopic observations of nucleolar changes induced by ribonuclease (Chèvremont *et al.*, 1956) and 5-fluorodeoxyuridine (Love *et al.*, 1965). Treatment of monkey kidney cultured cells with adenosine resulted in a dissociation of nucleolar structure somewhat similar to ethionine-induced nucleolar fragmentation (Stenram, 1966a).

The action of 5-fluorouracil on nucleolar ultrastructure is somewhat similar to that of ethionine. The first lesions to appear are an increase in nucleolar size, with formation of dense, granular aggregates (Stenram, 1966b; Lapis and Benedeczky, 1966). Longer exposure to the antimetabolite in tissue culture leads to degranulation of the nucleolus and fragmentation of the fibrillar reticular network (Simard, 1968).

Of considerable interest is the action of two drugs which in recent years have been used extensively as valuable tools for the study of transcription. Toyocamycin, an analog of adenosine induces a spotted appearance of the nucleoli which become uncoiled and fragmented within 30 min. High-resolution autoradiography reveals a definitive decrease in RNA synthesis in the nucleoli, the only labeled parts of which are the fibrillar rods and the spots. Similar lesions were observed with adenosine (Monneron *et al.*, 1970). These observations are of considerable interest since these two compounds, like ethionine, interact with the nucleotide pool. Similar lesions were described in avian leukemia myoblasts treated with toyocamycin (Bonar *et al.*, 1970) and in a human cell line treated with adenosine (Recher, 1970).

α-Amanitin, a toxin obtained from *Amanita phalloides,* also induces fragmentation of nucleoli in mouse kidney and rat liver cells. The lesions are observed after only 30 min of treatment and were described as a breakup of nucleoli with subsequent separation and redistribution of nucleolar components in different areas. Chromatin condensates at the border of the nucleus and there is an increase in the number of perichromatin granules forming perichromatin bodies (Fiume *et al.*, 1969; Marinozzi and Fiume 1971). The lesions shown are similar to fragmentation of nucleoli induced with ethionine, toyocamycin, and adenosine.

3. THE EFFECT OF ETHIONINE ON RNA SYNTHESIS

The mode of action of ethionine is complex since it involves several levels in the synthesis of RNA and proteins. Being a substitute for methionine, it can be incorporated in its place or competitively inhibit metabolic reactions requiring methionine (Farber, 1963).

Villa-Trevino *et al.* (1963, 1966) have shown that ethionine severely inhibits RNA synthesis in rat liver and that this inhibition follows a decrease in adenosine triphosphate concentration but precedes the inhibition of protein synthesis. The administration of adenine and methionine prevented this inhibition or reversed it when administered after the injection of ethionine. According to several authors it is through an excessive trapping of the adenosine moiety of ATP as S-adenosylethionine that ethionine affects RNA polymerase requirements of ATP for the conduct of RNA synthesis (Farber, 1963; Smith and Salmon, 1965; Raina *et al.*, 1964).

The parallelism between the biochemical and utrastructural studies is striking and enabled Shinozuka *et al.* (1968a) to suggest that the disorganization of nucleoli was either a reflection of the disturbance of cell metabolism due to ATP deficiency or a consequence of it resulting in a reduction of RNA synthesis.

Toyocamycin, on the other hand, does prevent, at low concentration, the transcription of the various RNA's on their DNA template, but the cleavage of 45 S nucleolar RNA is completely inhibited. At higher concentration, the overall RNA synthesis is drastically reduced. Toyocamycin is known to be incorporated into the 45 S RNA as an analog of adenosine (Tavitian *et al.*, 1968, 1969). There is certainly a strong analogy between the molecular action of ethionine, toyocamycin, and adenosine and it is clear that the end result is similar at the ultrastructural level.

The mechanism by which 5-fluorouracil affects RNA synthesis is well known. This uracil analog is incorporated as a fraudulent base into RNA. It also affects incorporation of normal precursors into nucleic acids and inhibits thymidilate synthetase activity (Heidleberger, 1963). Severe inhibition of RNA synthesis is observed after 5-fluorouracil treatment (Heidelberger and Ansfield, 1963) but low doses give rise to high incorporation into heavy RNA molecules (Stenram, 1966b). The possibility exists that fraudulent RNA first accumulates in the nucleolus, causing an enlargement of this organelle, followed by a fragmentation of the reticulated fibrillar component owing to subsequent blocking of nucleolar function.

The mode of action of α-amanitin is far from being understood. The toxin blocks extranucleolar RNA synthesis (Fiume and Stirpe, 1966) through specific inhibition of RNA polymerase B (Kedinger *et al.*, 1970). In rat liver, impairment of the synthesis of nucleolar 45 S RNA is also observed to a lesser extent although α-amanitin has no effect when applied directly to isolated nucleoli (Jacob *et al.*, 1970; Tata *et al.*, 1972). In cultured CHO cells, the drug inhibits heterogenous extranucleolar RNA, but the synthesis and processing of nucleolar RNA is unaltered as demonstrated by radioautography and polyacrylamide gel analysis (Kedinger and Simard, 1974). Similar results have been published with chick embryo fibroblasts (Hastie and Mahy, 1973). The fact that in cultured cells, nucleolar RNA synthesis and processing can occur normally within completely fragmented nucleoli remains to be explained.

Therefore, there is a possibility that fragmentation of nucleoli represents a more specific lesion than originally believed. Only a more systematic study of that lesion could establish the precise relationship with a known biochemical action.

III. Agents That Primarily Affect the Nucleus

Compared with the nucleolus, which is ultrastructurally highly organized, the interphase nucleus appears somewhat chaotic. Granules dispersed in chromatin fibrils in various stages of extension or coiling leave little room for cytological indications that could relate morphology to activity, chormosomal structures to duplication, and transcription of genetic informations. Recently, a number of drugs have been shown to induce characteristic lesions on the nuclear fine structure; some of these compounds affect the nucleolus as well and have therefore already been mentioned in the preceding sections. Their action on the nuclear structures will be stressed in the following one.

A. *Margination of Chromatin and Proflavin*

1. CYTOLOGICAL LESIONS INDUCED BY PROFLAVIN

At low doses, proflavin completely modifies the ultrastructural aspect of the nucleus. When cultured cells are treated for 6 hr at 10 μg/ml, the bulk of the chromatin mass forms electron-dense osmiophilic aggregates standing out in a nucleus that otherwise keeps its size and shape. There is an unusual unsticking and margination of the chromatin clumps from the nuclear membrane. After 24 hr of proflavin treatment, the margination and clumping increase as the nucleoplasm has lost an appreciable quantity of material and electron density. The perichromatin granules disappear during the treatment while the interchromatin granules are grouped in clusters (Fig. 11). At this moment nucleolar segregation has taken place.

As the treatment progresses in time the sparseness of the chromatin markedly increases, leaving sticky-looking aggregates in a poorly defined fibrillar network. These striking nuclear lesions are observed together with nonspecific cytoplasmic alterations such as disorganization of endoplasmic reticulum and presence of large cytoplasmic inclusions with myelin figures and osmiophilic material. No mitosis has ever been observed in treated cells (Simard, 1966).

Other authors reported nucleolar segregation induced by proflavin in cultured cells (Reynolds and Montgomery, 1967) and liver cells (Stenram

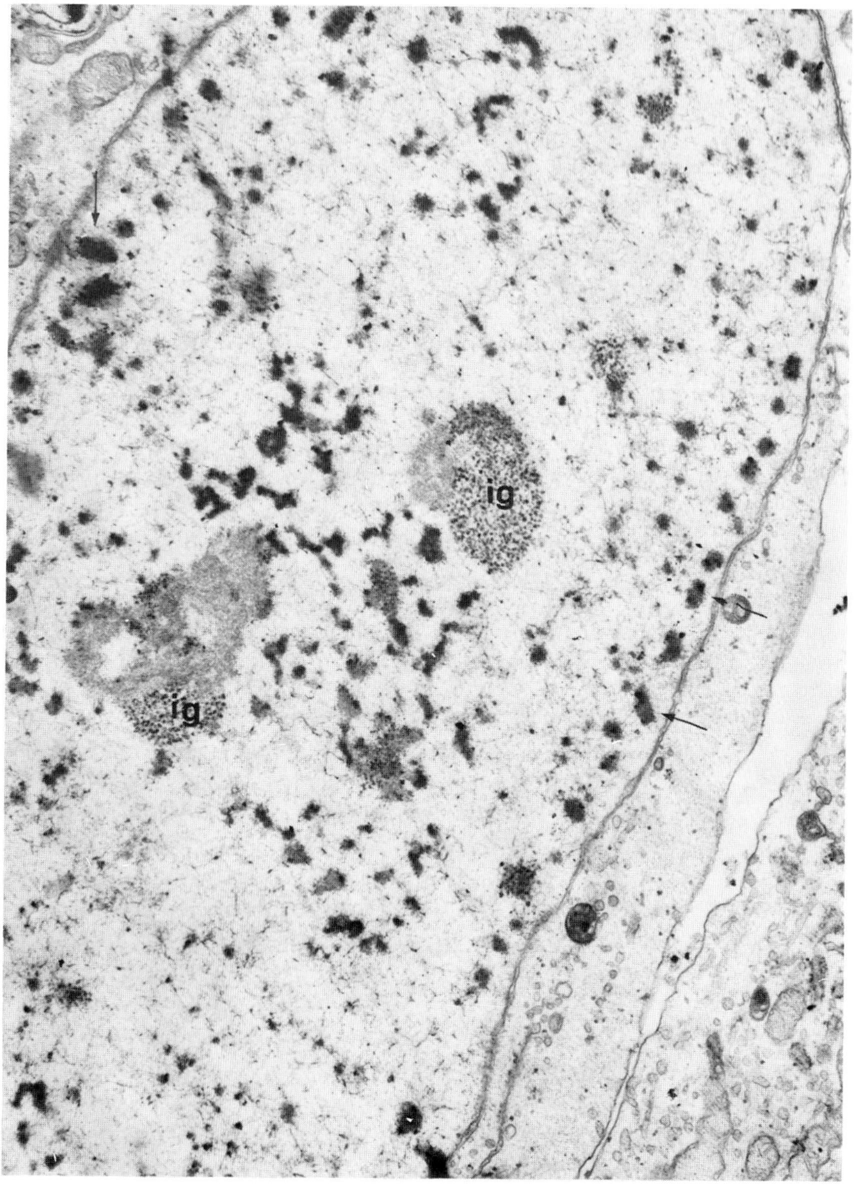

Fig. 11 Margination of chromatin induced by proflavin (10 μg/ml for 24 hrs in cultured cells. The whole appearance of the nucleus is changed. The chromatin forms electron-dense clumps with margination from the nuclear membrane (arrows). The nucleoplasm has lost its electron density and clustering of interchromatin granules (ig) is seen. Osmium tetroxide; Epon (× 12,000).

and Willen, 1968) but did not emphasize the nuclear lesions. Recher *et al.*, (1971b) compared the effects of proflavin and actinomycin D on mammalian cell nucleoli: They observed fibrillar nodules closely associated with interchromatin granules. The formation of these nodules is prevented by treatment with actinomycin D.

2. MARGINATION OF CHROMATIN INDUCED BY OTHER AGENTS

Similar lesions were produced by the antibiotic daunomycin and the trypanocidal drug ethidium bromide with a different sequence. In cells treated with these two compounds, nucleolar segregation was the first lesion to appear, followed by margination and clumping of chromatin after a longer period of treatment (Simard, 1966). Ethionine has also been reported to cause margination of chromatin (Herman *et al.*, 1962; Shinozuka *et al.*, 1968a). Recently, we studied new antibiotics that induce the same lesions after 1 hr in cultured cells. In this instance, the nuclear alterations appear so rapidly that no cytoplasmic modifications can be observed. The antibiotic U-12241 (Bhuyan, 1967), for instance, causes margination of chromatin and clumping of interchromatin granules after 1 hr at 10 μg/ml. The changes are even more striking than with proflavin as only interchromatin granules within a fibrillar network can be seen in the nucleus along with chromatin masses (Fig. 12a). Since this compound provided a model for the possible isolation of interchromatin granules, it was studied in detail. The preferential stain for ribonucleoproteins was found to be strongly positive for the clumps of interchromatin granules (Fig. 12b) and pronase followed by RNase caused a disappearance of the granules (Fig. 12c). But all attempts at the isolation of the granules failed to come through, and the only successful results were fractions enriched in 18 S RNA obtained from isolated granules (Simard and Brailovsky, 1969).

3. MODE OF ACTION OF PROFLAVIN AND TENTATIVE CORRELATIONS

Proflavin is a mitotic inhibitor (Balis *et al.*, 1963) and potent mutagen (Freese, 1959; Lerman, 1964; Orgel and Brenner, 1961) that interferes with nucleic acids *in vivo* (Bubel and Wolf, 1965; Franklin, 1958; Schaffer, 1962; Scholtissek and Rott, 1964) and *in vitro* (Hurwitz *et al.*, 1962). Proflavin binds to DNA by intercalation between adjacent base pairs (Lerman, 1961; Luzzati *et al.*, 1961) and blocks, without specificity, the enzymatic reaction leading to RNA and DNA synthesis (Hurwitz *et al.*, 1962). The antibiotic drug daunomycin possesses cytotoxic and antimitotic activity (Di Marco *et al.*, 1963) and is also believed to bind to DNA by intercalation between base pairs (Calendi *et al.*, 1964); daunomycin inhibits RNA synthesis regardless of the base composition of the DNA template (Ward *et al.*, 1965). Ethidium bromide is also known to form reversible complexes with both DNA and RNA and inhibits nucleic acid synthesis and nucleic acid polymerases (Elliott, 1963); this compound binds with DNA without preferences for any base composition (Ward *et al.*, 1965) probably by intercalation between base pairs (Waring, 1966).

The mode of action of antibiotic U-12241, a new antimicrobial agent isolated from cultures of *Streptomyces bellus* var. *cirolerosus*, is postulated to result from the formation of stable complexes with DNA with secondary inhibition of nucleic acid polymerase activity (Bhuyan, 1967).

The relationship between the morphological effect of these compounds and their biological and biochemical action can only be speculated upon. It seems, however, that a certain class of DNA-binding agents alters the structural organization of the nucleus in a characteristic and reproducible manner. Alteration of the physicochemical properties of chromatin, including its affinity for the binding sites on the nuclear membrane, stainability, and distribution, has been proposed as a tentative explanation (Simard, 1966). Other possibilities such as competitive displacement of histone-rich basic proteins from the chromosomal DNA by those agents that react strongly with nucleic acids, also appear to be attractive hypotheses.

B. Perichromatin Granules: Aflatoxin and Lasiocarpine

1. CYTOLOGICAL ACTION OF AFLATOXIN AND LASIOCARPINE

The ultrastructural lesions induced by aflatoxin in rat liver have been described by Bernhard *et al.* (1965) and others (Svoboda *et al.*, 1966). When injected in low doses in hepatectomized animals during the regeneration phase, this potent carcinogen induced nucleolar segregation within 30 min–1 hr. In association with the nucleolus a "contrasted zone" of unknown nature was noted and interpreted as a disruption of the nucleolar granular component (Svoboda *et al.*, 1966). Another potent carcinogen, lasiocarpine, was found to produce the same lesions. This pyrrolizidine alkaloid causes segregation of nucleolar components within 30 min in rat liver (Svoboda and Soga, 1966) with, once again, the appearance of contrasted granular masses that were interpreted as disrupted nucleolar components (Reddy *et al.*, 1968; Reddy and Svoboda, 1968). Both compounds are associated with the early appearance of helical polysomes in the cytoplasm of liver cells (Monneron, 1968, 1969).

A recent ultrastructural and cytochemical study showed that both aflatoxin and lasiocarpine induce an increase of perichromatin granules

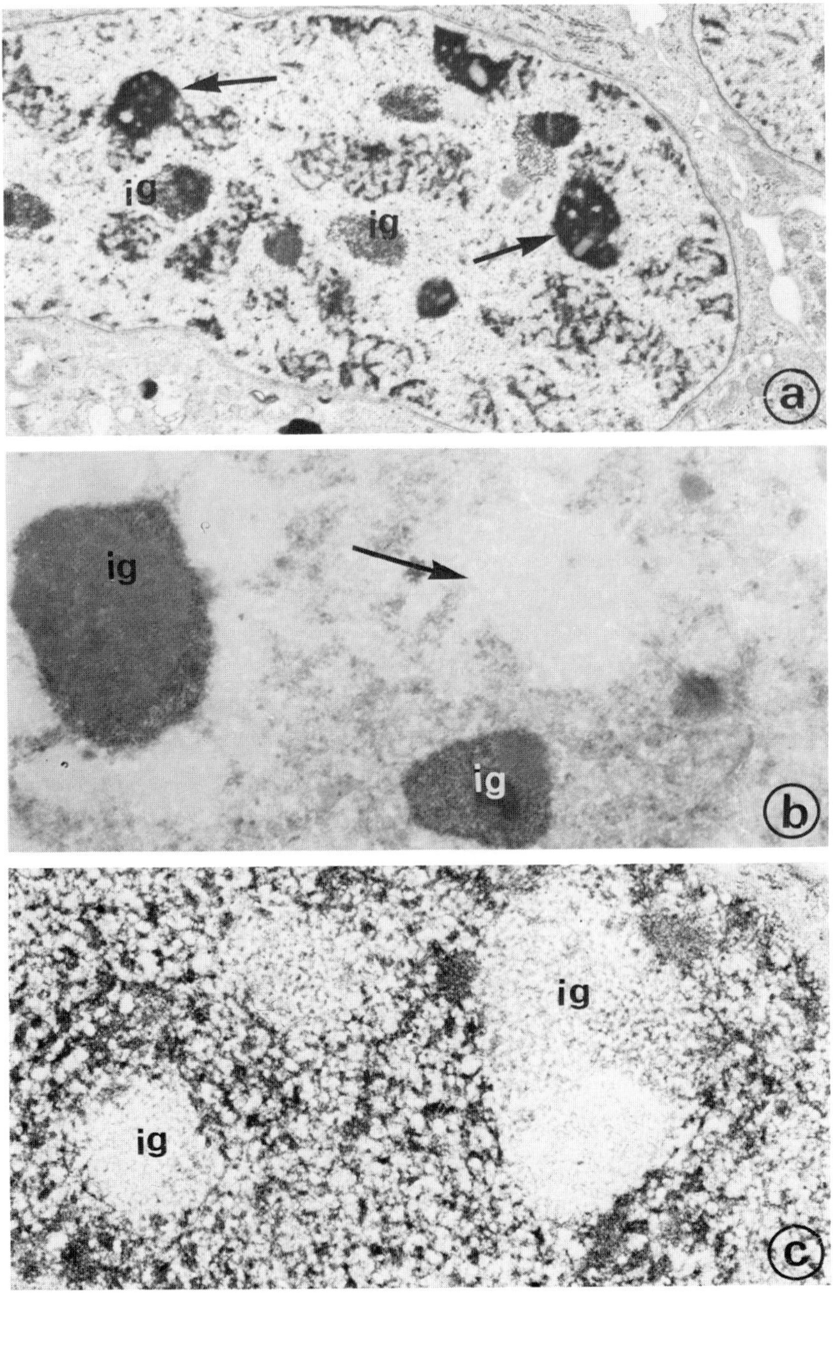
ig
ig
a
ig
ig
b
ig
ig
c

in rat liver, with early formation of dense masses measuring 0.2 to 1 μm in diameter (Monneron *et al.*, 1968). These masses are pepsin- and pronase-sensitive in their center but their granular cortex is formed by ribonucleoprotein particles similar in their structural and cytochemical properties to perichromatin granules. According to Monneron *et al.* (1968), the formation of these dense masses is independent of the nucleolar lesions, but is closely related to the increase of perichromatin granules (Fig. 13).

2. MODE OF ACTION OF AFLATOXIN AND LASIOCARPINE

These two carcinogens almost completely inhibit nucleolar RNA synthesis (Clifford and Rees, 1967; Lafarge *et al.*, 1966; Reddy *et al.*, 1968; Moulé and Frayssinet, 1968) which is accompanied as seen previously by a segregation of nucleolar components. Little is known of the action of these drugs on non-nucleolar nuclear RNA, the synthesis of which persists significantly after injection of the drugs (Lafarge *et al.*, 1965; 1966). Monneron *et al.* (1968) proposed that the increase of perichromatin granules and formation of dense masses are linked to the persistent synthesis of certain types of ribonucleoproteins and the blocking of their transport to the cytoplasm. These authors conclude that, since no RNP particles are observed in the nuclear pores, the perichromatin granules are either stocked *in situ* or aggregate in dense masses as degradation proceeds owing to a long sequestration in the nucleus.

C. *Interchromatin Granules*

Clumping of interchromatin granules has been reported in such a large variety of pathological conditions that it now appears to represent part of a nonspecific reaction related to cytotoxicity resulting from various injurious agents. In proflavin-treated cultured cells, there seems to be an increase of interchromatin granules with cluster formation (Fig. 11). This could eventually permit their isolation and characterization although the granules are somewhat larger than normal (Simard, 1966). In recent studies, increased granules have been observed in rat liver treated with aflatoxin, lasiocarpine, tannic acid, and TA in acute stages of intoxication and with dimethylnitrosamine and ethionine in chronic stages (Svoboda and Higginson, 1968; Miyai and Steiner, 1965; Shinozuka *et al.*, 1968b).

Fig. 12 Lesions induced by U12–241 (10 μg/ml for 2 hr). In (a), clumps of chromatin (arrows) and clustering of interchromatin granules are evident. In (b), the granules (ig) are strongly positive to a preferential stain for ribonucleoproteins but the chromatin clumps are bleached (arrows). In (c), the granules (ig) have been digested by the combined action of pronase on ribonuclease [(a) × 9000; (b) × 14,400; (c) × 22,500].

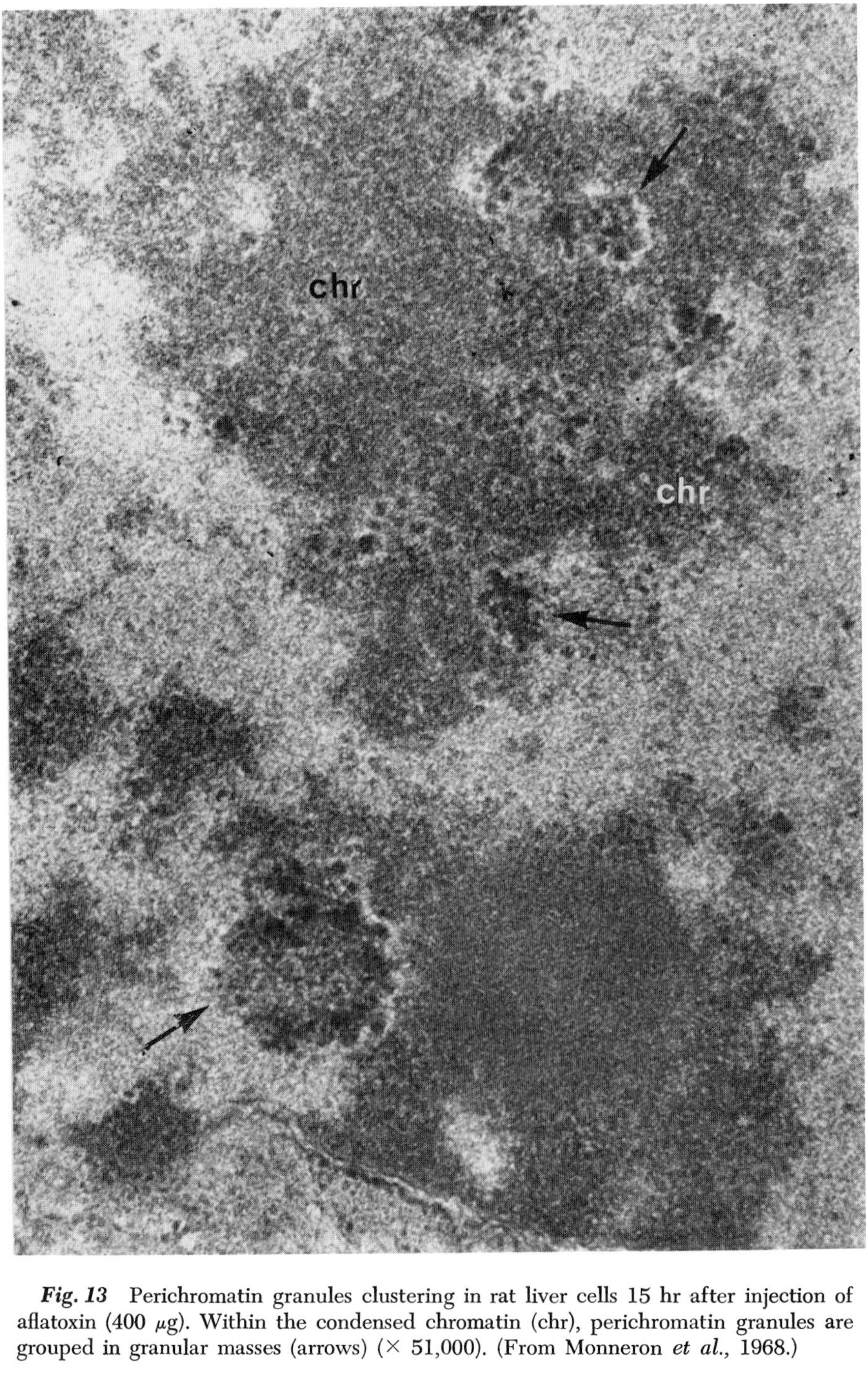

Fig. 13 Perichromatin granules clustering in rat liver cells 15 hr after injection of aflatoxin (400 μg). Within the condensed chromatin (chr), perichromatin granules are grouped in granular masses (arrows) ($\times$ 51,000). (From Monneron *et al.*, 1968.)

Clustering was reported in several normal and cancer cells (Bernhard and Granboulan, 1963; Granboulan and Bernhard, 1961; Swift, 1959) in degenerating and dead cells (Trump *et al.*, 1965) or in cells exposed to irradiation (Andres, 1963) and supranormal temperature (Simard and Bernhard, 1966).

It has been suggested that the interchromatin granules represent extrachromosomal RNA (Swift, 1963; Bernhard and Granboulan, 1963) or nucleolar RNA migrating to the cytoplasm (Smetana *et al.*, 1963). The possibility exists that alterations of the interchromatin granules reflect activation or impairment of RNA synthesis at extranucleolar sites (Reddy and Svoboda, 1968). Since their biochemical nature, RNA and protein content, and function are not known, such conclusions at the moment are of limited significance.

D. Nuclear Inclusions

The term "nuclear inclusions" is descriptive and is used mostly to characterize the appearance of new structures in the interphase nucleus that are not known to result from a redistribution of normal nuclear structures or from a trapping of cytoplasmic material. The dense masses appearing after aflatoxin or lasiocarpine treatment and the clustering of nuclear granules after proflavin treatment are nuclear inclusions; but since they are clearly related to the peri- and interchromatin granules, respectively, they have been discussed in preceding sections. Nuclear inclusions are not to be confused with nuclear bodies (Weber and Froomes, 1963; Weber *et al.*, 1964) which were first described by de Thé *et al.* (1960). These bodies are found in normal and pathological conditions under different morphological forms; they are probably normal cellular organelles related to cellular hyperactivity (Bouteille *et al.*, 1967). On the basis of cytochemical and histochemical studies, the nuclear bodies do not contain DNA or RNA but may have proteins in their structure (Krishan *et al.*, 1967).

The presence of nuclear inclusions has been consistently reported in cultured cells treated with the carcinogen 4-nitroquinoline-*N*-oxide (Endo *et al.*, 1959, 1961; Levy, 1963; Reynolds *et al.*, 1963, 1964; Lazarus *et al.*, 1966). These inclusions appear as distinct spherical areas of low electron density after permanganate fixation for electron microscopy. Histochemical studies have stressed the presence of RNA in these inclusions (Endo *et al.*, 1961), a conclusion also reached from studies with acridine orange fluorescence. Levy (1963) suggested that nuclear inclusions do not represent RNA in transit to the cytoplasm but rather RNA trapped in the nucleoplasm by an inhibiting effect of these drugs on the transport of RNA. The inclusions would then originate from the redistribution of an already synthesized material (Lazarus *et al.*, 1966).

Nuclear inclusions were observed in embryonic cells of amphibians cultured in presence of actinomycin D (Duprat *et al.*, 1965). These inclusions are numerous (40–50 per nucleus after 8 hr of actinomycin treatment) but they seem to become confluent as treatment progresses in time; cytoplasmic differentiation is not affected by the presence of the antibiotic (Duprat *et al.*, 1966). A study at the ultrastructural levels has shown that the inclusions are formed by agglomeration of coarse fibrillar elements 400–600 Å long located in the interchromatin space (Fig. 14). Cytochemical studies suggest the presence of RNA and proteins in these inclusions but no significant incorporation of [^{3}H]uridine could be obtained by high-resolution autoradiography with various pulse-chase experiments. The RNP content is, therefore, not newly synthesized, but is probably the product of a stable "pool" or the result of accumulation of an RNA originating from excessive degradation. The appearance of the inclusions coincides with the development of nucleolar segregation and proceeds independently of it. They are believed to originate from a rearrangement in the ribonucleoprotein network of the nucleoplasm following actinomycin treatment (Simard and Duprat, 1969).

Inclusions of another type were described by Jones and Elsdale (1964) and Jones (1967) on embryonic cells of *Rana pipiens* following actino-

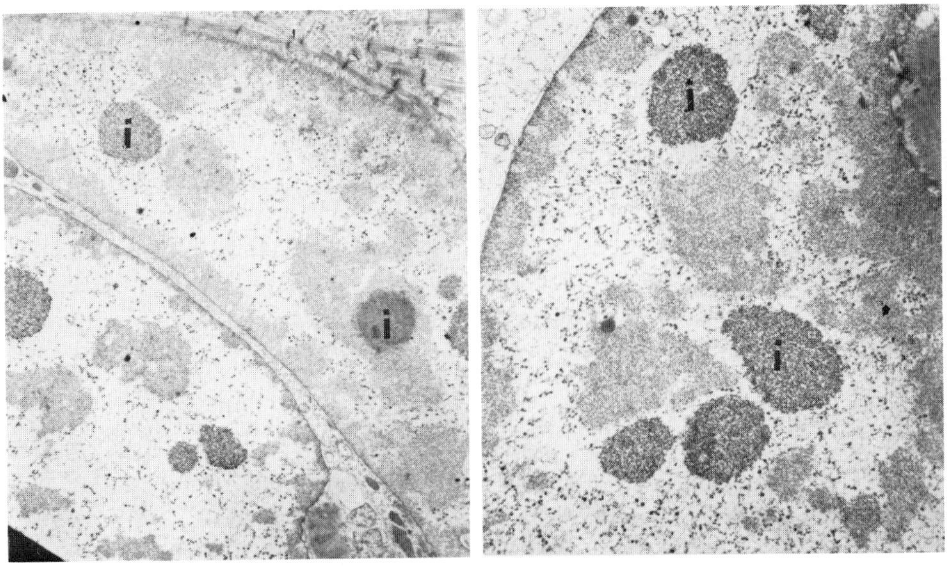

Fig. 14 Nuclear inclusions induced by actinomycin D (1 μg/ml for 12 hr) in embryonic cells of amphibians. The inclusions (i) are numerous and located in the interchromatin space. Osmium tetroxide; Epon (× 4000).

mycin treatment. The antibiotic gives rise to the formation of bundles of relatively coarse threads measuring 200–500 Å in thickness. The threads are arranged in crystalline arrays and several aggregates can be seen in a single section in the interchromatin region of the nuclei. The same lesions were induced by nogalamycin, chromomycin A3, and olivomycin, but not by ethidium bromide, daunomycin, and proflavin. Therefore, the thread formation appears to be connected with the binding site on the DNA molecule. It also seems to be related to the special nuclear structural organization of *Rana* species since actinomycin did not produce the same lesions in other amphibious species.

Intranuclear fibers resembling tonofibrils have been described in guinea pig epidermis following treatment with 4-hydroxyanisole, a compound that reacts preferentially with melanocytes and causes rapid depigmentation effect (Riley and Seal, 1969).

IV. Summary and Conclusions

Inhibitors of various types have been used to study a variety of ultrastructural lesions, only a few of which could be discussed here. Examples of more general interest were selected and each lesion was associated with the mechanisms of action of one typical compound. We would like to draw the following conclusions.

1. The nucleolus reacts in a distinctive, sometimes specific, manner. This is certainly true for nucleolar segregation which can be considered as a morphological marker of inhibition of nucleolar RNA synthesis induced by substances forming complexes with DNA. Fragmentation of nucleoli appears to be associated with interference at the nucleolar processing level whereas hypertrophy of the nucleolus is undoubtedly related to an increase in ribosomal RNA precursors. Degranulation of nucleoli will need more correlative studies to be associated with a precise molecular function, but already it is used as a valuable tool to study the nature of ribonucleoprotein particles in the nucleolus.

2. No situation of that sort is known in the extranucleolar region of the nucleus. This lack of data is probably the reflection of our poor knowledge of the chromosomal arrangement of the genetic material and its transcriptional machinery. It is evident that much more is to be learned in this area and that the use of genetic markers will be of great importance in localizing the precise site of functional activity in the nucleus.

In summary, the inhibitors described here have been and will certainly continue to be useful for elucidating the arrangement and the functions of macromolecules in the nucleus.

ACKNOWLEDGMENTS

The authors gratefully acknowledge the priceless secretarial assistance of Miss Francine Dion and the invaluable help of Miss Claudette Laurendeau.

REFERENCES

Amalric, F., Zalta, J. P., and Simard, R. (1969). *Exp. Cell Res.* **55**, 370.
Andres, K. H. (1963). *Zellforsch. Microsc. Anat.* **60**, 560.
Bachellerie, J. P., Martin-Prevel, C., and Zalta, J. P. (1971). *Biochimie* **53**, 383.
Balis, M. E., Pecora, P., and Salser, J. P. (1963). *Exp. Cell Res. Suppl.* **9**, 472.
Barka, T. (1966). *Exp. Cell Res.* **41**, 573.
Bernhard, W., and Granboulan, N. (1963). *Exp. Cell Res. Suppl.* **9**, 19.
Bernhard, W., Frayssinet, C., Lafarge, C., and Le Breton, E. (1965). *C.R. Acad. Sci. Paris* **261**, 1785.
Bernier, R., Iglesias, R., and Simard, R. (1972). *J. Cell Biol.* **53**, 798.
Bhuyan, B. K. (1967). *Arch. Biochem. Biophys.* **120**, 285.
Bhuyan, B. K., and Smith, C. G. (1965). *Proc. Nat. Acad. Sci. Paris* **54**, 566.
Bierling, R. (1960). *Z. Krebsforsch,* **63**, 519.
Bonar, R. A., Chabot, J. F., Langlois, A. J., Sverak, L., Veprek, L., and Beard, J. W. (1970). *Cancer Res.* **30**, 753.
Bouteille, M., Kalifat, S. R., and Delarue, J. (1967). *J. Ultrastruct. Res.* **19**, 474.
Bubel, H. C., and Wolf, D. A. (1965). *J. Bacteriol.* **89**, 977.
Bucciante, L. (1928). *Arch. Exp. Zellforsch.* **5**, 1.
Bucher, N. L. R. (1963). *Int. Rev. Cytol.* **15**, 245.
Busch, H., and Smetana, K. (1970). "The Nucleolus," pp. 495–497. Academic Press, New York.
Busch, H., Byvoet, P., and Smetana, K. (1963). *Cancer Res.* **23**, 313.
Calendi, E., Di Marco, A., Reggiani, B., Scarpinato, B., and Valentini, L. (1964). *Biochim. Biophys. Acta* **103**, 25.
Chevremont, M., Chevremont-Combaire, S., and Firket, H. (1956). *Arch. Biol. (Liège)* **67**, 635.
Clifford, J. I., and Rees, K. R. (1967). *Biochem. J.* **102**, 65.
de Thé, G., Rivière, M., and Bernhard, W. (1960). *Bull. Cancer* **47**, 569.
Di Marco, A., Gaetani, M., Dorigotti, L., Soldatti, M., and Bellini, O. (1963). *Tumori* **49**, 203.
Duprat, A. M. (1969). *Exp. Cell Res.* **57**, 37.
Duprat, A. M., Beetschen, J. C., Zalta, J. P., and Duprat, P. (1965). *C.R. Acad. Sci. Paris* **261**, 5203.
Duprat, A. M., Zalta, J. P., and Beetschen, J. P. (1966). *Exp. Cell Res.* **43**, 358.
Eakin, R. M. (1964). *Z. Zellforsch. Mikrosk. Anat.* **63**, 81.
Elliott, W. H. (1963). *Biochem. J.* **86**, 562.
Eisenberg, H. W., van Praag, D., Rosenkranz, H. S., and Shemin, D. (1965). *Biol. Bull.* **129**, 403.
Endo, H. (1958). *Gann* **49**, 151.
Endo, H., Aoki, M., and Aoyama, Y. (1959). *Gann* **50**, 209.
Endo, H., Takayama, S., Kasuga, T., and Oyashi, M. (1961). *Gann* **52**, 173.
Farber, E. (1963). *Advan. Cancer Res.* **7**, 383.
Fitzhugh, D. G., and Nelson, A. D. (1948). *Science* **108**, 626.

Fiume, L., and Stirpe, F. (1966). *Biochim. Biophys. Acta* **123,** 643.
Fiume, L., Marinozzi, V., and Nardi, F. (1969). *Brit. J. Exp. Pathol.* **50,** 270.
Franklin, R. M. (1958). *Virology* **6,** 525.
Freese, E. (1959). *Proc. Nat. Acad. Sci. U.S.* **45,** 622.
Ganote, C. E., and Rosenthal, A. S. (1968). *Lab. Invest.* **19,** 382.
Geuskens, M. (1968). *Exp. Cell Res.* **52,** 621.
Geuskens, M., and Bernhard, W. (1966). *Exp. Cell Res.* **44,** 579.
Gharpure, M. (1965). *Virology* **27,** 308.
Goldblatt, P. J., and Sullivan, R. J. (1970). *Cancer Res.* **30,** 1349.
Goldblatt, P. J., Sullivan, R. J., and Farber, E. (1969a). *Cancer Res.* **29,** 124.
Goldblatt, P. J., Sullivan, R. J., and Farber, E. (1969b). *Lab. Invest.* **20,** 283.
Goldblatt, P. J., Verlin, R. S., and Sullivan, R. J. (1970). *Exp. Cell Res.* **63,** 117.
Goldstein, M. N., Slotnick, I. J., and Journey, L. J. (1960). *Ann. N.Y. Acad. Sci.* **89,** 474.
Granboulan, N., and Bernhard, W. (1961). *C.R. Soc. Biol.* **155,** 1767.
Gupta, D. N. (1955). *Nature (London)* **175,** 257.
Gupta, D. N. (1956). *J. Pathol. Bacteriol.* **72,** 183.
Haguenau, F. (1960). *Nat. Cancer Inst. Monogr.* **4,** 211.
Harris, C., Grady, H., and Svoboda, D. (1968a). *Cancer Res.* **28,** 81.
Harris, C., Grady, H., and Svoboda, D. (1968b). *J. Ultrastr.* **22,** 240.
Hartman, S. C., Levenberg, B., and Buchanan, J. M. (1955). *J. Amer. Chem. Soc.* **77,** 501.
Hastie, N. D., and Mahy, B. W. (1973). *FEBS Letters* **32,** 95.
Hay, E. D., and Gurdon, J. B. (1967). *J. Cell Sci.* **2,** 151.
Heidelberger, C. (1963). *Exp. Cell Res. Suppl.* **9,** 462.
Heidelberger, C., and Ansfield, F. J. (1963). *Cancer Res.* **23,** 3006.
Heine, U., Langlois, A. J., and Beard, J. W. (1966). *Cancer Res.* **26,** pt. 1, 1847.
Heine, U., Sverak, L., Kondratick, J., and Bonar, R. A. (1971). *J. Ultrastruct. Res.* **34,** 375.
Herman, L., and Fitzgerald, P. (1962). *J. Cell Biol.* **12,** 277.
Herman, L., Eber, L., and Fitzgerald, P. (1962). *In* "Electron Microscopy" (S. S. Breese, Jr. ed.), Vol. 2, pp. VV6. Academic Press, New York.
Higgins, G. M., and Anderson, R. M. (1931). *Arch. Pathol.* **12,** 186.
Hurwitz, J., Furth, J., Malamy, M., and Alexander, M. (1962). *Proc. Nat. Acad. Sci. U.S.* **48,** 1222.
Iyer, U. N., and Szybalski, W. (1958). *Proc. Nat. Acad. Sci. U.S.* **44,** 446.
Iyer, U. N., and Szybalski, W. (1963). *Proc. Nat. Acad. Sci. U.S.* **50,** 355.
Jacob, J., and Sirlin, J. L. (1964). *J. Ultrastruct. Res.* **11,** 315.
Jacob, S. T., Sajdel, E. M., Muecke, W., and Munro, H. N. (1970). *Cold Spring Harbor Symp. Quant. Biol.* **35,** 681.
Jackson, B., and Dessau, F. I. (1961). *Lab. Invest.* **10,** 909.
Jezequel, A. M., and Bernhard, W. (1964). *J. Microsc.* **3,** 279.
Jezequel, A. M., Shreeve, M. M., and Steiner, J. W. (1967). *Lab. Invest.* **16,** 287.
Jones, J. W. (1965). *J. Ultrastruct. Res.* **13,** 257.
Jones, J. W. (1967). *J. Ultrastruct. Res.* **18,** 71.
Jones, J. W., and Elsdale, T. R. (1964). *J. Cell Biol.* **21,** 245.
Journey, L. J., and Goldstein, M. N. (1961). *Cancer Res.* **1,** 929.
Juul, J., and Kemp, T. (1933). *Strahlentherapie* **48,** 457.
Karasaki, S. (1965). *J. Cell Biol.* **26,** 937.
Karasaki, S. (1968). *Exp. Cell Res.* **52,** 13.

Kedinger, R., and Simard, R. (1974). In preparation.
Kedinger, R., Gniazdowski, M., Mandel, J. L., Gissinger, F., and Chambon, P. (1970). *Biochem. Biophys. Res. Comm.* **38**, 165.
Kleinfeld, R. G. (1957). *Cancer Res.* **17**, 954.
Kleinfeld, R. G. (1966). *Nat. Cancer Inst. Monogr.* **23**, 369.
Kleinfeld, R. G., and Von Haam, E. (1959). *Cancer Res.* **19**, 769.
Koshiba, K., Thirumalachary, C., Daskal, I., and Busch, H. (1971). *Exp. Cell Res.* **68**, 235.
Koulish, S., and Kleinfeld, R. G. (1964). *J. Cell Biol.* **23**, 39.
Krishan, A., Uzman, B. G., and Hedly-Whyte, E. T. (1967). *J. Ultrastruct. Res.* **19**, 563.
Kuboda, Y., and Furuyama, J. (1963). *Cancer Res.* **23**, 682.
Kume, F., Maruyama, S., D'Agostino, A. N., and Shiga, M. (1967). *Exp. Mol. Pathol.* **6**, 254.
Lafarge, C., and Frayssinet, C. (1970). *Int. J. Cancer* **6**, 74.
Lafarge, C., Frayssinet, C., and de Recondo, A. M. (1965). *Bull. Soc. Chim. Biol.* **47**, 1724.
Lafarge, C., Frayssinet, C., and Simard, R. (1966). *C.R. Acad. Sci. Paris* **23**, 1011.
Laird, A. K. (1953). *Arch. Biochem. Biophys.* **46**, 119.
Langelier, Y. (1974). In preparation.
Lapis, K., and Benedeczky, I. (1966). *Acta Biol. Hung.* **17**, 199.
Lapis, K., and Bernhard, W. (1965). *Cancer Res.* **25**, 628.
Lazarus, S. S., Vethamany, V. G., Shapiro, S. H., and Amsterdam, D. (1966). *Cancer Res.* **26**, 2229.
Lerman, L. S. (1961). *J. Mol. Biol.* **3**, 18.
Lerman, L. S. (1964). *J. Cell. Comp. Physiol. Suppl. 1* **64**, 1.
Levy, H. B. (1963). *Proc. Soc. Exp. Biol. Med.* **113**, 886.
Lewin, P. K., and Moscarello, M. A. (1968). *Lab. Invest.* **19**, 265.
Lombardi, L. (1972). *Cancer Res.* **32**, 675.
Love, R., Studzinski, G. P., and Ellem, K. O. A. (1965). *Fed. Proc.* **24**, 1206.
Love, R., Soriano, R. Z., and Walsh, R. J. (1970). *Cancer Res.* **30**, 1525.
Luzzati, V., Masson, F., and Lerman, L. S. (1961). *J. Mol. Biol.* **3**, 634.
Lwoff, A. (1962). *Cold Spring Harbor Symp. Quant. Biol.* **27**, 159.
Lwoff, A., and Lwoff, M. (1961). *Ann. Inst. Pasteur* **101**, 490.
MacCarty, W. C. (1928). *J. Lab. Clin. Med.* **8**, 354.
MacCarty, W. C. (1936). *Amer. J. Cancer* **26**, 529.
Malkin, M. F., and Zahalsky, A. C. (1966). *Science* **154**, 1665.
Maloof, F., and Soodak, M. (1961). *Cancer Res.* **14**, 625.
Marinozzi, V., and Fiume, L. (1971). *Exp. Cell Res.* **67**, 311.
Maxwell, R. E., and Nickel, V. S. (1954). *Science* **120**, 270.
McConkey, E. H., and Hopkins, J. W. (1964). *Proc. Nat. Acad. Sci. U.S.* **51**, 1197.
McCormick, W., and Penman, S. (1969). *J. Mol. Biol.* **39**, 315.
Miyai, K., and Steiner, J. W. (1965). *Exp. Mol. Pathol.* **4**, 525.
Molnar, Z., and Bekesi, J. G. (1972a). *Cancer Res.* **32**, 380.
Molnar, Z., and Bekesi, J. G. (1972b). *Cancer Res.* **32**, 756.
Moner, J. G. (1967). *Exp. Cell Res.* **45**, 618.
Monneron, A. (1968). *Excerpta Med.* **166**, 66.
Monneron, A. (1969). *Lab. Invest.* **20**, 178.
Monneron, A., Lafarge, C., and Frayssinet, C. (1968). *C.R. Acad. Sci. Paris* **267**, 2053.
Monneron, A., Burglen, J., and Bernhard, W. (1970). *J. Ultrastruct. Res.* **32**, 370.

Montgomery, P. O'B., Reynolds, R. C., and McLendom, D. E. (1966). *Amer. J. Pathol.* **43**, 555.

Moule, Y., and Frayssinet, C. (1968). *Nature (London)* **218**, 93.

Muramatsu, M., Hodnett, J. L., Steele, W. J., and Busch, H. (1966). *Biochim. Biophys. Acta* **123**, 116.

Nygaard, O., Eldjarn, L., and Nakken, K. F. (1954). *Cancer Res.* **14**, 625.

Oda, A., and Shiga, M. (1965). *Lab. Invest.* **14**, 1419.

Orgel, H., and Brenner, S. (1961). *J. Mol. Biol.* **3**, 762.

Paul, J. S., Reynolds, R. C., and Montgomery, P. O'B. (1967). *Nature (London)* **215**, 749.

Perry, R. P. (1962). *Proc. Nat. Acad. Sci. U.S.* **48**, 2179.

Perry, R. P. (1963). *Exp. Cell Res.* **29**, 400.

Perry, R. P., and Kelley, D. E. (1970). *J. Cell Physiol.* **76**, 127.

Perry, R. P., Hell, A., Errera, M., and Durwald, H. (1961). *Biochim. Biophys. Acta* **49**, 47.

Pong, R. S., and Wogan, G. M. (1970). *Cancer Res.* **30**, 294.

Pouchelet, M., Beaure D'Augeres, C., and Robineaux, R. (1971). *C.R. Acad. Sci. Paris* **272**, 3333.

Raina, A., Janne, J., and Slimes, M. (1964). *Acta Chem. Scand.* **18**, 1804.

Rao, P. N., and Engleberg, J. (1965). *Science* **148**, 1092.

Rather, L. J. (1951). *Bull. Johns Hopkins Hosp.* **88**, 38.

Recher, L. (1970). *J. Ultrastruct. Res.* **32**, 212.

Recher, L., Briggs, L. G., and Parry, N. T. (1971a). *Cancer Res.* **31**, 140.

Recher, L., Parry, N. T., Briggs, L. G., and Whites Carver, J. (1971b). *Cancer Res.* **31**, 1915.

Recher, L., Chan, H., Briggs, L., and Parry, N. T. (1972). *Cancer Res.* **32**, 2495.

Reddy, J., and Svoboda, D. J. (1968). *Lab. Invest.* **19**, 1320.

Reddy, J., and Svoboda, D. J. (1972). *J. Ultrastruct. Res.* **38**, 608.

Reddy, J., Harris, C., and Svoboda, D. (1968). *Nature (London)* **217**, 659.

Rees, K. R., Rowland, G. F., and Varcoe, J. S. (1966). *Int. J. Cancer* **1**, 197.

Reich, E. (1964). *Science* **143**, 684.

Reynolds, R. C., and Montgomery, P. O'B. (1967). *Amer. J. Pathol.* **51**, 323.

Reynolds, R. C., Montgomery, P. O'B., and Karney, D. H. (1963). *Cancer Res.* **23**, 535.

Reynolds, R. C., Montgomery, P. O'B., and Hugues, B. (1964). *Cancer Res.* **24**, 1269.

Riley, P. A., and Seal, P. (1969). *Exp. Mol. Pathol.* **10**, 63.

Robineaux, R., Buffe, D., and Rimbaut, C. (1958). *In* "La Chimiothérapie des Cancers et des Leucémies," Colloq. Int. Paris, 1957. C.N.R.S., Paris.

Robineaux, R., Rosselli, L., and Moncel, C. (1967). *J. Microsc.* **6**, 80A.

Rodriguez, T. G. (1967). *J. Ultrastruct. Res.* **19**, 116.

Rouiller, C., and Simon, G. (1962). *Rev. Int. Hepatol.* **12**, 167.

Ruttner, J. R., and Rondez, R. (1960). *Pathol. Microbiol.* **23**, 113.

Salomon, J. C. (1962). *J. Ultrastruct. Res.* **7**, 293.

Salomon, J. C., Salomon, M., and Bernhard, W. (1962). *Bull. Ass. Fr. Cancer* **49**, 139.

Schaffer, F. L. (1962). *Virology* **18**, 412.

Scherrer, K., and Darnell, J. E. (1962). *Biochim. Biophys. Res. Commun.* **7**, 486.

Schoefl, G. (1964). *J. Ultrastruct. Res.* **10**, 224.

Scholtissek, C., and Rott, R. (1964). *Nature (London)* **204**, 39.

Shankar Narayan, K., and Busch, H. (1965). *Exp. Cell Res.* **38**, 439.

Shankar Narayan, K., Steele, W., and Busch, H. (1966). *Exp. Cell Res.* **43**, 483.

Shinozuka, H., and Farber, E. (1969). *J. Cell Biol.* **41**, 280.

Shinozuka, H. P., Goldblatt, P. J., and Farber, E. (1968a). *J. Cell Biol.* **36**, 313.
Shinozuka, H. P., Verney, E., and Sidransky, H. (1968b). *Lab. Invest.* **18**, 72.
Sidransky, H., and Farber, E. (1958). *A.M.A. Arch. Pathol.* **66**, 135.
Sidransky, H., Stachelin, T., and Verney, E. (1964). *Science* **146**, 766.
Simard, R. (1966). *Cancer Res.* **26**, 2316.
Simard, R. (1967). *J. Cell Biol.* **35**, 716.
Simard, R. (1968). Thesis, Univ. Paris, C.N.R.S. N° 2472.
Simard, R. (1970). *Int. Rev. Cytol.* **28**, 169.
Simard, R., and Bernhard, W. (1966). *Int. J. Cancer* **I**, 463.
Simard, R., and Bernhard, W. (1967). *J. Cell Biol.* **34**, 61.
Simard, R., and Brailovsky, C. (1969). *J. Cell Biol.* **43**, 133a.
Simard, R., and Cassingena, R. (1969). *Cancer Res.* **29**, 1590.
Simard, R., and Duprat, A. M. (1969). *J. Ultrastruct. Res.* **29**, 60.
Simard, R., Amalric, F., and Zalta, J. P. (1969). *Exp. Cell Res.* **55**, 359.
Sirtori, C., and Bosisio-Bestetti, M. (1967). *Cancer Res.* **21**, 367.
Sisken, J. E., Morasca, C., and Killy, S. (1965). *Exp. Cell Res.* **39**, 103.
Smetana, K., Steele, W. J., and Busch, H. (1963). *Exp. Cell Res.* **31**, 198.
Smetana, K., Narayan, K. S., and Busch, H. (1966). *Cancer Res.* **26**, 786.
Smith, R. C., and Salmon, W. D. (1965). *Arch. Biochem. Biophys.* **111**, 191.
Smuckler, E. A., and Benditt, E. P. (1965). *Lab. Invest.* **14**, 1699.
Snoab, B. (1955). *Exp. Cell Res.* **8**, 554.
Sobell, H. M., and Jain, S. C. (1972). *J. Mol. Biol.* **68**, 21.
Sporn, M. D., Dingman, C. W., Phelps, H. L., and Wogan, G. M. (1966). *Science* **151**, 1539.
Steele, W. J., and Busch, H. (1966a). *Biochim. Biophys. Acta* **119**, 501.
Steele, W. J., and Busch, H. (1966b). *Biochim. Biophys. Acta* **129**, 54.
Steele, W. J., Okamura, N., and Busch, H. (1965). *J. Biol. Chem.* **240**, 1742.
Stenger, R. J., and Confer, D. B. (1966). *Exp. Mol. Pathol.* **5**, 455.
Stenram, U. (1958). *Exp. Cell Res.* **15**, 174.
Stenram, U. (1963). *Exp. Cell Res. Suppl.* **9**, 176.
Stenram, U. (1965). *Z. Zellforsch. Mikrosk. Anat.* **65**, 211.
Stenram, U. (1966a). *Nat. Cancer Inst. Monogr.* **23**, 379.
Stenram, U. (1966b). *Z. Zellforsch. Mikrosk. Anat.* **71**, 207.
Stenram, U., and Willen, R. (1968). *Exp. Cell Res.* **50**, 505.
Stevens, B. J. (1964). *J. Ultrastruct. Res.* **11**, 329.
Stevens, J. G. (1966). *Virology* **29**, 570.
Stevens, J. G. (1967). *Virology* **32**, 654.
Stockert, J. C., Fernandez, M. E., Sogo, J. M., and Lopez, J. F. (1970). *Exp. Cell Res.* **59**, 85.
Suter, E., and Salomon, J. C. (1966). *Exp. Cell Res.* **43**, 248.
Svoboda, D., and Higginson, J. (1968). *Cancer Res.* **28**, 1703.
Svoboda, D., and Soga, J. (1966). *Amer. J. Pathol.* **48**, 347.
Svoboda, D., Grady, H., and Higginson, J. (1966). *Amer. J. Pathol.* **49**, 1023.
Svoboda, D., Racela, A., and Higginson, J. (1967). *Biochem. Pharmacol.* **16**, 651.
Swift, H. (1959). *Brookhaven Symp. Biol.* **12**, 134.
Swift, H. (1963). *Exp. Cell Res. Suppl.* **9**, 54.
Takahama, M., and Barka, T. (1967). *J. Ultrastruct. Res.* **17**, 452.
Tata, J. R., Hamilton, M. J., and Shields, D. (1972). *Nature New Biol.* **238**, 161.
Tavitian, A., Uretsky, S. C., and Acs, G. (1968). *Biochim. Biophys. Acta* **157**, 33.
Tavitian, A., Uretsky, S. C., and Acs, G. (1969). *Biochim. Biophys. Acta* **179**, 50.

Terao, K., Sokakibara, Y., Yamazaki, M., and Miyaki, K. (1971). *Exp. Cell Res.* **66,** 81.
Terawaki, A., and Greenberg, J. (1965). *Biochim. Biophys. Acta* **95,** 170.
Thoenes, W. (1964). *J. Ultastruct. Res.* **10,** 194.
Thoenes, W., and Bannasch, P. (1962). *Virchows Arch. Pathol. Anat. Physiol.* **335,** 556.
Trump, B. F., Goldblatt, P. J., and Stowell, R. E. (1965). *Lab. Invest.* **14,** 1969.
Unuma, T., and Busch, H. (1967). *Cancer Res.* **27,** 1232.
Unuma, T., Senda, R., and Muramatsu, M. (1972). *J. Electron. Microsc.* **21,** 61.
Villalobos, J. G., Steele, W. J., and Busch, H. (1964a). *Biochim. Biophys. Acta* **91,** 233.
Villalobos, J. G., Steele, W. J., and Busch, H. (1964b). *Biochim. Biophys. Res. Commun.* **17,** 723.
Villa-Trevino, S., Shull, K. H., and Farber, E. (1963). *J. Biol. Chem.* **238,** 1757.
Villa-Trevino, S., Shull, K. H., and Farber, E. (1966). *J. Biol. Chem.* **241,** 4670.
Ward, D. C., Reich, E., and Goldberg, I. H. (1965). *Science* **149,** 1259.
Waring, M. J. (1966). *Biochim. Biophys. Acta* **114,** 234.
Warocquier, R., and Scherrer, K. (1969). *Eur. J. Biochem.* **10,** 362.
Weber, A. F., and Frommes, S. P. (1963). *Science* **141,** 912.
Weber, A., Whipp, S., Usenik, E., and Frommes, S. (1964). *J. Ultrastruct. Res.* **11,** 564.
Yamaguchi, K., Kobayachi, Y., Sato, T., Herman, L., Marsh, W., Rosenstock, L., and Fitzgerald, P. (1971). *Amer. J. Pathol.* **64,** 337.
Yasuzumi, G., Sugihara, R., Ito, N., Konishi, Y., and Hiasa, Y. (1970). *Exp. Cell Res.* **63,** 83.

13

Intranuclear Viruses

Ursula I. Heine

I. Introduction

The term "intranuclear viruses" is to be used here for those virions the major components of which are formed or assembled inside the nucleus, giving rise to the accumulation of either incomplete or complete virus particles in this organelle. Virions of this kind have been found in different cells of various plants, and have been shown to be present in a multitude of cells of many species of the animal kingdom, from the oyster to man. Intranuclear viruses do not belong to a single class of virus, but are represented in both major classes, the ribonucleic acid (RNA)-containing and the deoxyribonucleic acid (DNA)-containing viruses. The majority of the intranuclear viruses are members of the DNA virus group, only a few members of the RNA virus group being morphologically identifiable inside the nucleus. The intranuclear viruses are of particular interest not only because they, like other viruses, are the causative agents for a variety of diseases, ranging from the common herpetic fever blister to malignant growths, but also because the study of their structure and of their interaction with the host cell has broadened our knowledge of biological structures and cellular metabolism.

It is beyond the scope of this chapter to discuss in detail all aspects of the intranuclear viruses. Since much of the information concerning their configuration and their relationship to the host cell has been obtained through the use of the electron microscope, special emphasis will be given to this approach. For further information on differing aspects the reader is advised to consult the excellent survey articles and books published in recent years (Beard, 1968; Fenner, 1968; Gross, 1970; Horne and Wildy, 1963; Maramorosch and Kurstak, 1971; Mayor and Jamison, 1966; Rivers and Horsfall, 1959; Smith *et al.*, 1968; Wildy, 1971, Wilner, 1964).

II. General Properties

A. *Classification*

In Table I are listed those virus groups in which intranuclear viruses have been observed. In addition, the kind of nucleic acid present in the virus, the symmetry of the particle, and some representative members of each group and their most common hosts are mentioned.

B. *Morphology of the Intranuclear Viruses*

1. SIZE OF THE VIRIONS

It is well known that the size and, in some respects also, the shape, of the mature virions may be dependent on the method of preparation. For example, in the electron microscope a number of virions appear to be larger in negatively stained or in metal-shadowed preparations than in ultrathin sections. This is especially true for those virions that are surrounded by a flexible envelope which may easily be subject to deformation. In some instances, the difference in size may amount to about 30% when mature viruses prepared according to one method are compared with those subjected to another method. The size of virions without envelopes, however, is more constant under different conditions. During the past years great efforts have been made to determine accurately the size of different virions (Crawford and Crawford, 1963; Crawford and Follett, 1969; Finch and Klug, 1965; Horne *et al.*, 1959; Horne and Wildy,

TABLE I
Virus Groups Containing Intranuclear Viruses[a]

I. Animal and human viruses
- A. Picodnaviruses (DNA, cubic)
  1. Autonomous replication: hamster osteolytic virus RV, X14, Kirk virus, minute virus of mice and dogs, feline panleukemia virus, mink enteritic virus, bovine and porcine picodna virus
  2. Replication in presence of "helper" viruses; adenoassociated satellite viruses
- B. Papovaviruses (DNA, cubic)
  1. Papillomaviruses: rabbit, human, canine, and equine papilloma virus
  2. Polyomaviruses: Polyoma, simian, and rabbit vacuolating virus, K virus, progressive multifocal leukoencephalopathy virus (mouse, rabbit, monkey, man)
- C. Adenoviruses (DNA, cubic): human, simian, bovine, porcine, canine, murine, and avian adenoviruses
- D. Herpesviruses (DNA, complex)
  1. Herpesvirus group A: herpesvirus simplex, pseudorabies virus
  2. Herpesvirus group B: cytomegalovirus, varicella-zoster virus (mouse, rat, man)
  3. Oncogenic herpesviruses: *Herpesvirus saimiri*, *H. sylvilagus*, *H. ateles*, Marek's disease virus, Lucké frog tumor virus, Epstein-Barr virus (?) (frog, chicken, rabbit, monkey, man)
- E. Myxoviruses (RNA, complex): measles virus, parainfluenza type viruses (swine, horse, man)

II. Insect viruses: polyhedrosis virus (DNA, helical)

III. Plant viruses
- RNA viruses
  1. Bacilliform viruses (potato yellow dwarf virus, lettuce necrotic yellow virus)
  2. Viruses with cubic symmetry (Tomato bushy stunt virus group)

[a] Including kind of viral nucleic acid, particle symmetry, representative members and their hosts.

1963; Karasaki, 1966; Lee *et al.*, 1969; Mayor and Melnick, 1966; Wildy *et al.*, 1960b).

Even with these uncertainties in mind, it is known to date that the size of the mature virion varies considerably among different groups. As seen in Table II, the smallest spherical particles are naked DNA viruses of about 18–25 nm diameter belonging to the picodnavirus group; the largest are complex particles found among the RNA-containing viruses, i.e., measles, exhibiting a diameter of about 300–700 nm. The diameter of the bacilliform viruses is 75–100 nm, and their length varies between 300 and 400 nm.

TABLE II
Size of Intranuclear Viruses

Virus	Particle structure				No. of capsomeres
	Cubic	Complex	Bacilliform	Rod-shaped	
I. DNA viruses					
Picodnaviruses	18–25 nm				32
Papovaviruses					72
Papillomavirus	50–55 nm				
Polyomavirus	40–45 nm				
Adenoviruses	70–90 nm				252
Herpesviruses		150–200 nm			162
Polyhedrosis virus				20 × 400 nm	
II. RNA viruses					
Myxoviruses		300–700 nm			
Plant viruses	24–30 nm		75 × 400 nm		

2. CONFIGURATION AND SYMMETRY

The intranuclear viruses exhibit a complex and often a most intricate configuration. Two large groups can be distinguished: the naked and the enveloped virions. The naked virions are composed of a nucleocapsid consisting of a nucleoprotein core representing the viral genome, surrounded by a capsid of proteinaceous subunits, the capsomeres, which are arranged in very regular patterns. The enveloped virions possess one or more membranes in addition to the core and capsid.

Metal shadowing of virions and the negative-staining technique are the most advantageous methods for the study of viral architecture. Using these techniques it was found that the capsomeres are arranged in either of two types of symmetry on the surface of the viral core: helical (rod-shaped and bacilliform viruses) or cubic (nearly spherical virions). The symmetrical packing of capsomeres into such regular structures as the nearly spherical virions represent can only be achieved with a few arrangements. Symmetrical bodies of such a kind are either tetrahedrons, dodecahedrons, or icosahedrons. As numerous investigations have shown, the capsomeres of virions with cubic symmetry are distributed in icosahedral fashion.

As shown in Table II, the number of capsomeres is constant within certain groups of DNA viruses, but varies considerably from one group to the other.

3. FINE STRUCTURE

a. Fine Structure of Negatively Stained Virions. The architecture of negatively stained naked virions is illustrated in Figs. 1–5.

i. Picodnaviruses. The fine structure of these small viruses is extremely difficult to resolve, especially that of the capsomeres (Fig. 1). At higher magnifications a hexagonal outline of the particles is apparent, and tiny knobs representing capsomeres are barely visible on their surface (Fig. 2) (Mayor *et al.*, 1963; Mayor and Jordan, 1966; Mayor *et al.*, 1967; Mayor, 1973; Siegel *et al.*, 1971). The minuteness of these structures still causes some confusion or doubt as to their exact number and arrangement. In addition to the icosahedron, the dodecahedral form has been proposed for these viruses with cubic symmetry (Mayor, 1973).

ii. Papovaviruses. Notwithstanding various difficulties encountered during the examination of the fine structure of these viruses, considerable knowledge of their composition is available at the present time. There is no doubt of the capsid having icosahedral symmetry (Bernhard *et al.*, 1962; Finch and Klug, 1965; Klug, 1965; Klug and Finch, 1965). Although these virions are, like picodnaviruses, very small (40–55 nm in diameter), their cores and capsomeres are readily recognizable in negatively stained preparations (Fig. 3). The latter are described as hollow rods about 6–8 nm in length, with an outer diameter of 6.5 nm; the width of their hollow center is quoted as 2.5–3.0 nm (Bernhard *et al.*, 1962; Wildy *et al.*, 1960b). Elongated, rodlike virions have been seen repeatedly in different virus preparations (Bernhard *et al.*, 1959; Howatson and Almeida, 1960a; ZuRhein, 1969; ZuRhein and Chou, 1965). These forms are thought to be assembled either by a faulty mechanism of the infected cells or they may represent precursors to the round forms.

iii. Adenoviruses. These round virions have an extremely high number of capsomeres: 252. There exist two kinds of capsomeres. Pentons are capsomeres located at the twelve icosahedral vertices and are thus surrounded by five other capsomeres. Each penton carries a projection called a fiber to the tip of which a small, round particle is attached. Hexons are capsomeres located on the faces of the icosahedron, with six other capsomeres in the most closely surrounding rank; there are 240 hexons present on the surface of the adenoviruses. Pentons, hexons, and fibers have each been isolated in purified form. They represent three distinct antigens: antigen A (hexon), B (penton with fiber), and C (fiber) (Horne, 1973; Norrby, 1969; Valentine and Perreira, 1965).

The architecture of the enveloped virions as seen in negatively stained preparations is illustrated in Figs. 6–8.

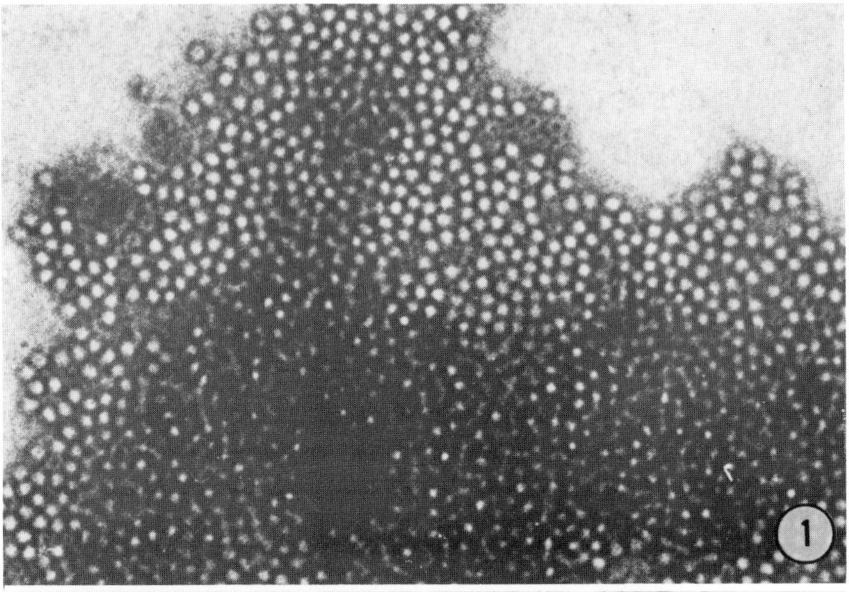
1

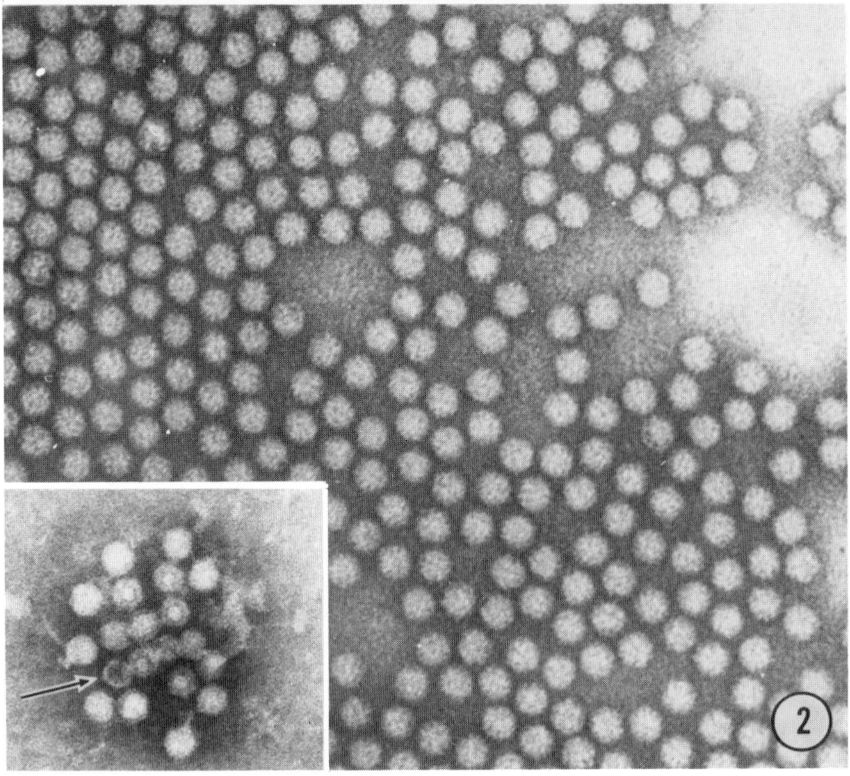
2

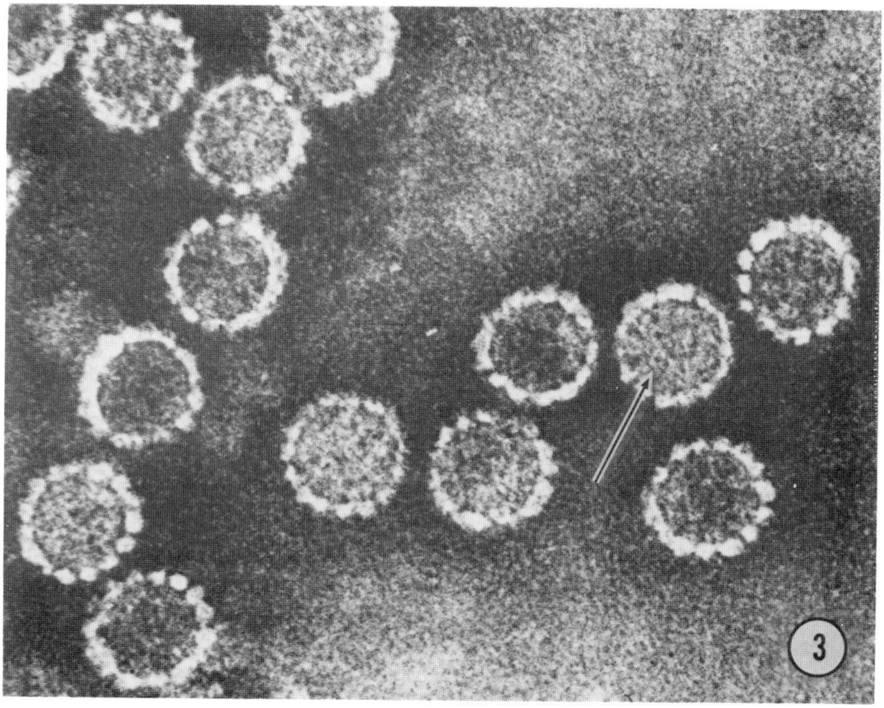

Fig. 3 SV40, a virus of the papova group, grown in monkey kidney cells and purified by centrifugation. Phosphotungstic acid stain. Capsomeres are readily recognizable on the surface of the virions. A few capsomeres exhibit a hollow center (arrow) (× 400,000). (From Bernhard *et al.*, 1962.)

Fig. 1 X14 virus, a virion of the picodna group, purified by banding in CsCl. Phosphotungstic acid stain (× 85,000). (Courtesy of Mayor and Jordan, 1966.)

Fig. 2 Adenoassociated satellite virus of type 4, a virion of the picodna group, negatively stained with phosphotungstic acid. The hexagonal outline of the particles and minute knobs (capsomeres) on their surface are apparent (× 210,000). Inset: The arrow points to an empty virion devoid of its nucleic acid (× 168,000). (From Torikai *et al.*, 1970.)

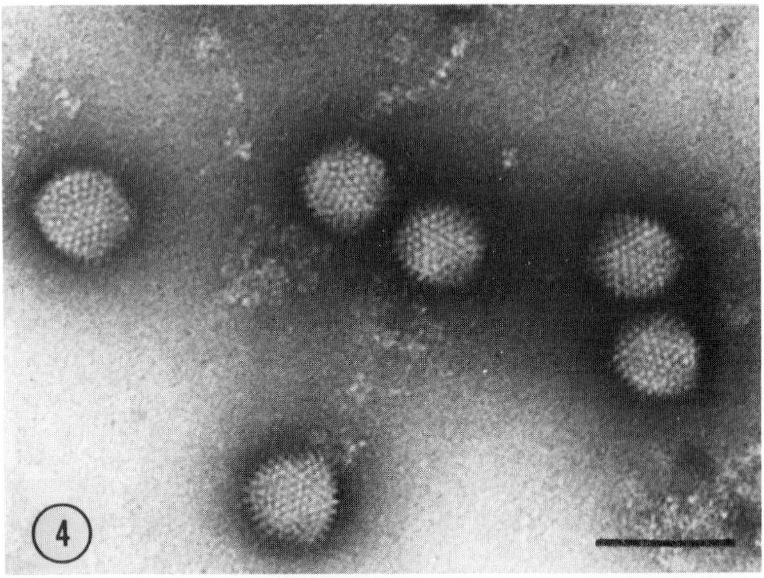

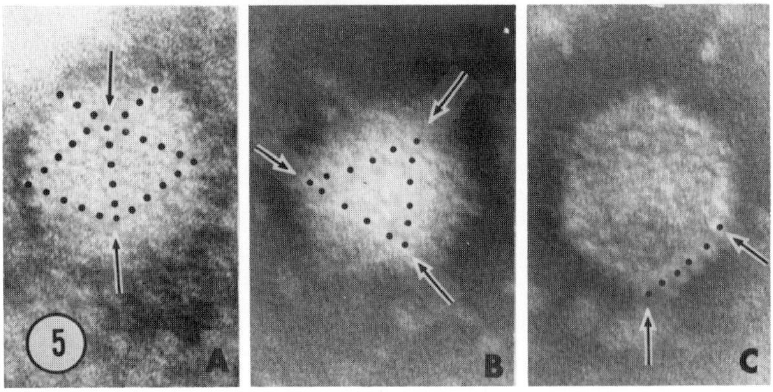

Fig. 4 Human adenovirus, SV15, negatively stained with phosphotungstic acid. The capsomeres are outstanding (× 185,500). (Courtesy of Dr. H. Mayor, unpublished observation.)

Fig. 5 Human adenoviruses with two- (A), three- (B), and fivefold (C) axes. Vertex pentamers are designated by arrows. Phosphotungstic acid stain (× 300,000). (From G. Schidlovsky, unpublished observation.)

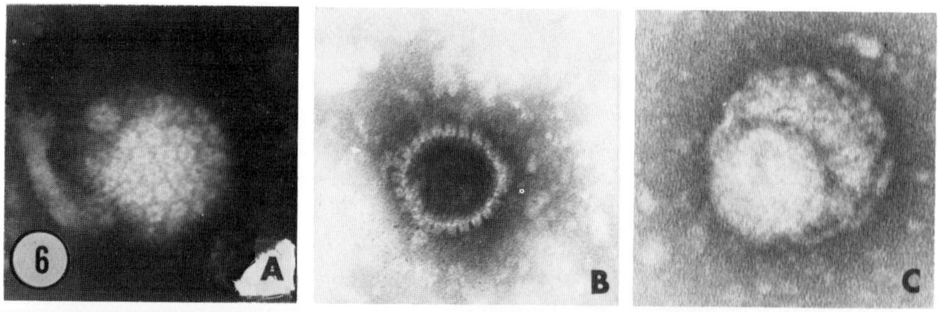

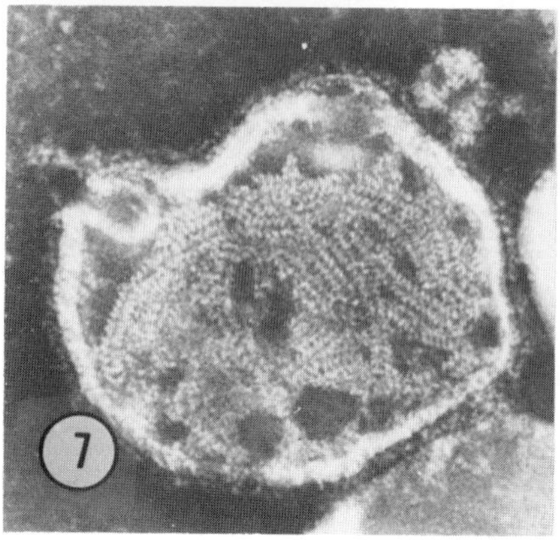

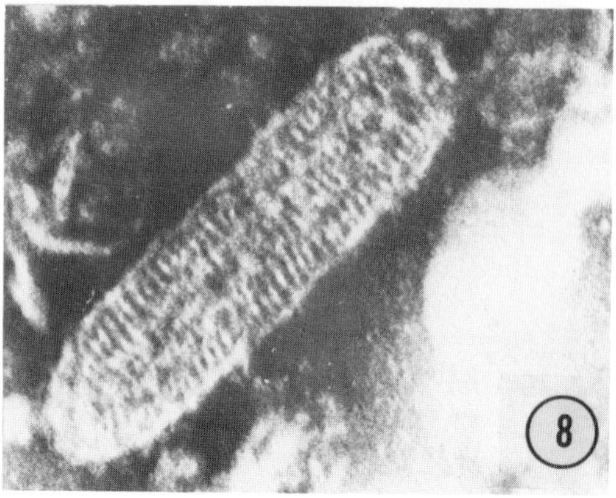

Fig. 6 *Herpesvirus saimiri* isolated from monkey kidney cells grown *in vitro* and negatively stained with phosphotungstic acid. (A) Complete core. The hollow centers of the capsomeres are clearly visible (× 160,000). (B) Empty core. (C) Complete, enveloped virion (B and C, × 120,000).

Fig. 7 Measles virus negatively stained with phosphotungstic acid. The particle is of pleomorphic configuration. The enclosing membrane is covered with projections and contains helical strands of the nucleoprotein (× 150,000). (From Nakai *et al.*, 1969.)

Fig. 8 Potato yellow dwarf virus. Phosphotungstic acid stain after fixation with glutaraldehyde and osmic acid. The particle is bullet shaped, and the transverse striations of its interior are clearly recognizable (× 200,000). (From McLeod, 1968.)

iv. Herpesviruses. Herpesviruses are large viruses which can be isolated in three different forms from infected cells (Fig. 6). These forms are (*a*) naked particles with electron-dense centers; their diameter is about 100 nm; they consist of a core which contains the viral DNA and a capsid; (*b*) naked particles with electron-lucent centers, so-called "empty particles," of 100 nm diameter; they consist of a capsid devoid of the core and thus lack the viral nucleoprotein; (*c*) complete particles consisting of a core, a capsid, and a pleomorphic envelope; their diameter varies between 150 and 200 nm. The capsomeres of the herpesviruses are elongated prisms with mean dimensions of 9 × 12 nm with a hollow center of approximately 4 nm width (Wildy *et al.*, 1960a). Thus, these capsomeres are somewhat larger than those of the papovaviruses. In herpes virions, evidence exists for the presence of a further membranous layer enveloping the capsid, a middle capsid situated underneath the outer capsid and an inner capsid (Roizman and Spear, 1973).

v. Myxoviruses. The measles virus will be described since it is a well-studied intranuclear paramyxovirus. Its fine structure is illustrated in Fig. 7. The virions are extremely pleomorphic, but spherical forms are the most common. They are membrane-bound structures of varying size. The majority of the particles have a diameter of 300–700 nm, but particles as large as 1 μm in diameter have been observed (Howatson, 1971; Nakai *et al.*, 1968, 1969). The enclosing membrane is covered with many projections about 10 nm long. The membranous bag is filled with numerous helical strands, the nucleoprotein, thus suggesting that each virion contains several copies of the viral genome. These strands have a diameter of about 15 nm; the periodicity of their serrations is approximately 5 nm.

vi. Bacilliform viruses. Two morphologically quite distinct groups of viruses can be found inside the nuclei of infected plants. These are either large bacilliform types or small isometric viruses with the icosahedral arrangement. The bacilliform viruses will be described here. These virions are very labile, subject to distortion and disruption, and special care has to be taken to preserve their fine structure during purification procedures (McLeod, 1968). As illustrated in Fig. 8, the surface of the virions is rather smooth in fixed preparations. However, the presence of surface projections has been described in unfixed, negatively stained material. Distinct transverse striations appear in those virions in which the negative stain has penetrated into the interior of the particle (Fig. 8). The presence of helices inside the particles has also been observed.

b. Fine Structure of Virions in Ultrathin Sections. The fine structure of mature intranuclear virions as it appears in fixed, plastic-embedded, and thin-sectioned preparations will be mentioned only briefly, since it will be illustrated repeatedly in the forthcoming passages.

i. Picodnaviruses. In thin sections, the naked virions of this group are represented by very small, electron-dense granules similar in size to cytoplasmic ribosomes (Atchison *et al.*, 1966; Bernhard *et al.*, 1963; Chandra and Toolan, 1961; Mayor *et al.*, 1967). In addition to the dense forms, particles with hollow centers (empty capsids) have been observed (Fig. 9) (Henry *et al.*, 1972).

ii. Papovaviruses. As illustrated in Fig. 10, these virions are characterized by small, electron dense, spherical or subspherical particles (Bierwolf, *et al.*, 1961; Dalton *et al.*, 1963; Gaylord and Hsiung, 1961; Graffi *et al.*, 1968, 1972; Granboulan *et al.*, 1963; Howatson and Almeida, 1960a, b; Howatson *et al.*, 1965; Melnick, 1962; Stone *et al.*, 1959; Tournier *et al.*, 1961; ZuRhein, 1969; ZuRhein and Chou, 1965). Hollow forms have been observed, although infrequently (Fig. 10, arrow). Arrays of tubular virions are shown in Fig. 11. They are represented by long, rather straight rods, and their structural details can be seen in cross-section images. They are apparently composed of three elements: an electron-lucent center, a dense inner sheath surrounding the center, and a less electron-dense outer layer in close proximity to the inner layer.

iii. Adenoviruses. The intricate and complex composition of these virions pictured so clearly in negative-stained preparations is not evident in ultrathin sections. They appear as extremely dark, round to ovoid particles in ultrathin sections (Fig. 12). In late stages of infection the virions are arranged in a crystalline pattern. Particles in such arrays do not contact each other, but are separated by electron-lucent areas of definite and similar widths, indicating the possible presence of an unstained component at their surface. It has been proposed that the fibers at the vertices of the virions are the cause for the lack of particle contact (Schlesinger, 1969).

iv. Herpesviruses. The different components of the herpesviruses so easily recognizable in negatively stained preparations can also readily be distinguished in ultrathin sections (Roizman and Spear, 1973). Figure 13, A–D, illustrates representative examples of virion species, which are the causative agents of different diseases. They, as other herpesviruses, have the following features in common: an electron-dense core containing the viral DNA, a faintly stained capsid, and an envelope with the characteristic triple layer of a unit membrane (Fig. 13C). Virions of different origin may exhibit variable amounts of dense material between envelope and capsid (Fig. 13D). This material is usually acquired by the virions during maturation in the cytoplasm. As recently shown (Chai, 1971; Furlong *et al.*, 1972), the herpesvirus DNA in the core may be arranged in a toroidal structure 50 nm high, with an outside diameter of 70 nm and an inside diameter of 18 nm penetrated by a less dense cylindrical mass (Fig. 14).

It is noteworthy that negatively stained preparations of purified capsids suggest that the toroid consists of DNA-containing threads spooled around the central, cylindrical mass. Thus far, cores of this kind have been observed in herpesvirus simplex (Chai, 1971; Furlong *et al.*, 1972), and in cytomegalovirus (Heine, unpublished observation).

III. Virus–Cell Interactions

A. *The Uptake of Virions*

Uptake of the virions into the cytoplasm and their transfer into the nucleus is usually a rather fast process, and the presence of virions inside the nucleus has been reported as early as 30 min after infection (Barbanti-Brodano *et al.*, 1970). Many aspects of the early virus–cell interactions are quite well understood, but some still require further investigation (Dales, 1965; Nii *et al.*, 1968; Philipson, 1963, 1967; Sussenbach, 1967). Since their morphological manifestations differ in certain respects among the different virus groups, the interaction of four viruses with their host cells will be discussed here as representative models of viral uptake. These viruses are the papovavirus SV40, the adenoviruses of type 5 and 7, and herpes simplex virus. The early events leading to infection can be divided into three steps: (1) attachment of the virion to the cellular membrane, (2) transfer into the cytoplasm, and (3) penetration into the nucleus.

1. ATTACHMENT TO THE CELL MEMBRANE

The attachment of the virion to the host is completed in a very short time and results in an unusually strong binding of the virion to the cellular membrane. For example, SV40 virions are irreversibly bound to plasma membranes as early as 10 min after infection (Hummeler *et al.*, 1970). Cellular as well as viral factors are responsible for the binding of the virions to the host. An interesting observation was the detection of genetically controlled differences in adsorption and penetration (Chardonnet and Dales, 1970b). Adenovirus of type 5, which exhibits extremely long fibers at the pentons, is adsorbed firmly to the host cells and penetrates rapidly to the nucleus. On the other hand, adenovirus of type 7 with short fibers is less firmly attached and preferentially transported into lysosomal bodies. Figures 15 and 19 are examples of virus attachment to the cellular surface; Fig. 15 shows the SV40 virus attachment, Fig. 19 the attachment of adenovirus 7.

2. TRANSFER INTO THE CYTOPLASM

At least two phenomena have been recognized that lead to the transfer of virions from the cell surface into the cytoplasm. One process frequently employed by naked virions consists of the tight envelopment of the particles by the cellular membrane, which is then pinched off, resulting in the formation of tiny pinocytic vesicles containing one or a few virus particles (Chardonnet and Dales, 1970a; Hummeler *et al.*, 1970). This process is called viropexis (Fazekas de St. Groth, 1948); it is illustrated in Figs. 15 and 19. Subsequently, the virions are transported towards the nucleus either inside these vesicles (SV40, Fig. 16) or as naked particles through the cytoplasm after shedding the cellular envelope (adenovirus, Figs. 20–22) (Chardonnet and Dales, 1970a; Dales, 1962; Morgan *et al.*, 1969).

On the other hand, enveloped particles, which can also be ingested by pinocytosis or phagocytosis, are frequently found fused to the plasma membrane (Fig. 28, A and B). In an elegant study, Miyamoto and Morgan (1971) described the entry and uncoating of herpesvirus simplex in HeLa cells. They presented evidence for the presence of two morphologically distinct types of virus cores in different virions. After entry, one type of nucleocapsid, containing a "dense" core, is disrupted inside the cytoplasm near the cellular surface, thus releasing its viral DNA early during infection. The other, "less dense" cores, appear to migrate toward the nucleus (Figs. 28C and 29). Apparently these particles shed their nucleoprotein during migration since they can be observed exhibiting different electron densities throughout the cytoplasm. At later stages of infection, empty cores are encountered near the nucleus. It was suggested that the two types of particles may play different roles during the viral uptake. The DNA released from the dense nucleocapsids may be transcribed to messenger RNA which, in turn, could code for an enzyme capable of altering the permeability of the second light core. Near the nucleus, this core would then release its DNA without visible disruption of the capsid (Fig. 29); and only this viral nucleoprotein would enter the nucleus giving rise to viral progeny.

3. PENETRATION INTO THE NUCLEUS

The events leading to the transfer of viral nucleoprotein from the cytoplasm into the nucleus may vary depending on the virus–cell system studied. To produce viral progeny, it appears to be sufficient to transmit only the genetic information, i.e., the viral core material (Chardonnet and Dales, 1970a; Miyamoto and Morgan, 1971; Morgan *et al.*, 1969). Such a phenomenon was observed in herpesvirus- and in adenovirus-infected

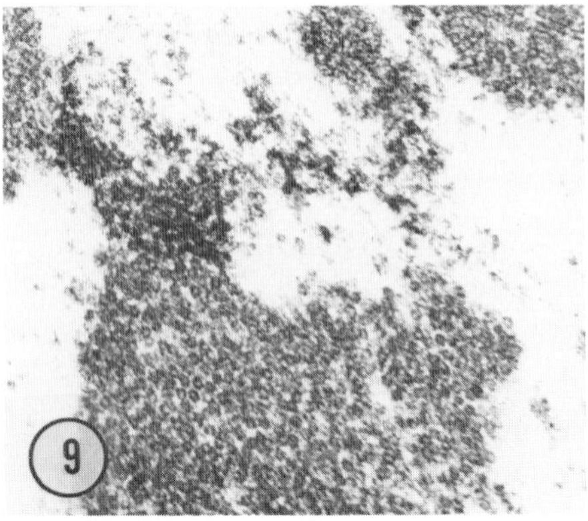

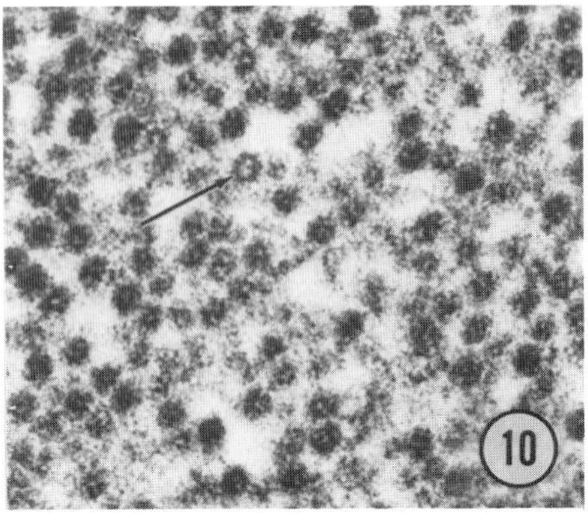

Fig. 9 Incomplete adenoassociated virus of type 1 in close association with the nucleolus, which is fragmented. The virion capsids exhibit hollow centers (× 40,000). (From Henry *et al.*, 1972.)

Fig. 10 K virus, a papovavirus, as seen in the nucleus of an endothelial cell of a pulmonary capillary of a mouse. For the most part the virions are electron-dense, but a few exhibit electron-lucent centers (arrow) (× 80,000). (From Dalton *et al.*, 1963.)

Fig. 11 Mouse cell infected with polyomavirus. Arrays of tubular forms, some in cross sections (arrow), are illustrated (× 65,000). (Courtesy of Dr. A. F. Howatson, unpublished observation.)

Fig. 12 KB cells infected with adenovirus of type 5. The virions are regular, round to ovoid structures of high electron density (× 50,000). (From Marusyk *et al.*, 1972.)

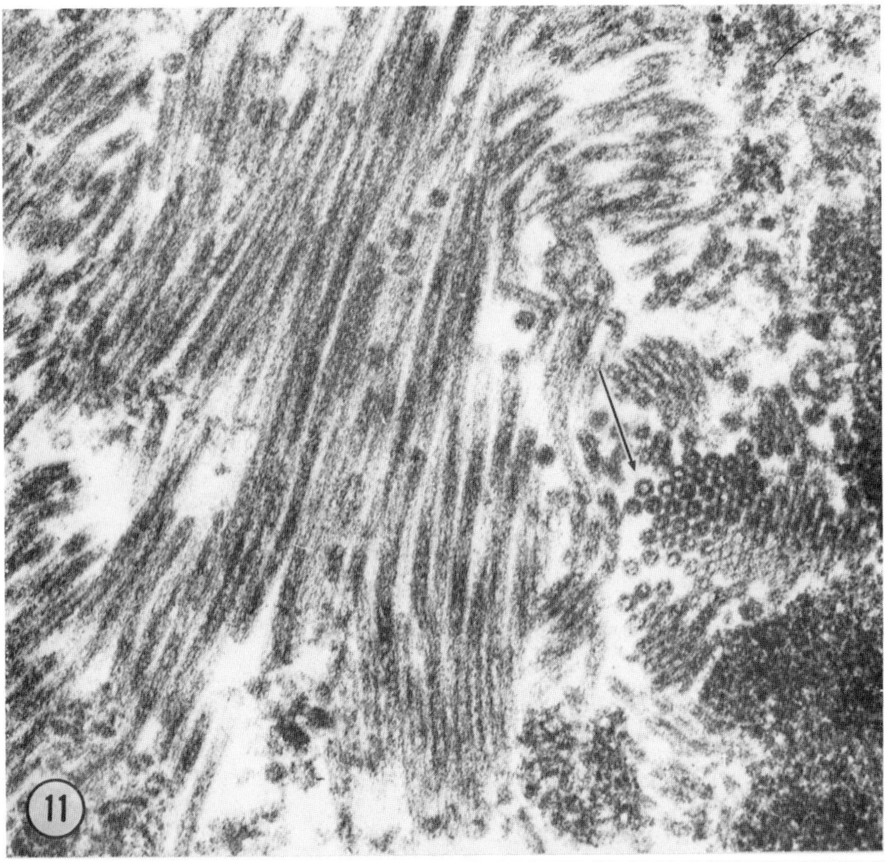
11

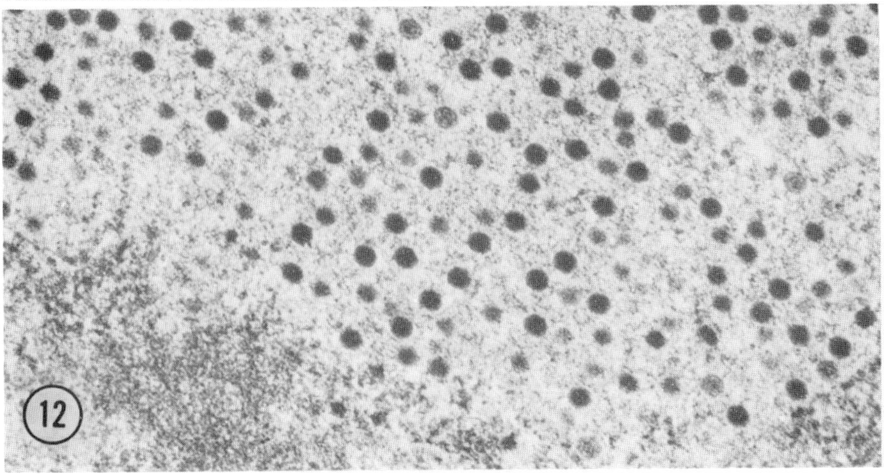
12

cells, as illustrated in Figs. 23–27 and indicated in Fig. 29. Such transfer is especially indicated in those instances in which complete virions are too large to enter the nucleus via the nuclear pores. Recent investigations suggest that an energy-requiring process located in the nuclear pores may be responsible for the migration of viral core material through the nuclear membranes (Chardonnet and Dales, 1972).

In addition, it has been shown that small virions, i.e., SV40, can migrate *in toto* into the nucleus (Hummeler *et al.*, 1970) (Figs. 16–18). These morphological observations are supported by radiochemical studies verifying the presence of intact SV40 virions inside nuclei shortly after infection. The uncoating of the virions takes place after penetration into the nuclei, resulting in a complete dissociation of the viral genome from the protein coat (Barbanti-Brodano *et al.*, 1970).

B. Nuclear Changes after Infection

1. GENERAL CONSIDERATIONS

After infection the morphological changes are frequently of an impressive magnitude and may, under certain circumstances, involve all components of the affected cells. They may be coded for by the viral or the host cell genome. The first visible alterations may occur shortly after

Fig. 13 Thin sections of virions of the herpes type, grown *in vitro* after isolation from different diseases. The three components of the particles are recognizable: the electron-dense core, the electron-lucent capsid (arrows), and the envelope. (A) Herpesvirus simplex (× 60,000). (B) Epstein-Barr virus (× 70,000). (C) Cytomegalovirus (× 70,000). (D) *Herpesvirus sylvilagus* (× 70,000). (From Heine and Dalton, 1972.)

Fig. 14 (A, B, and C) Thin sections through the cores of herpes virions cut at different angels, illustrating the toroid structures (× 200,000). (D, E, and F) Models of the viral cores cut in the same place as shown in A–C. (G) Thin section of a herpesvirus core treated with EDTA to remove the uranyl acetate binding to DNA (× 200,000). (From Furlong *et al.*, 1972.)

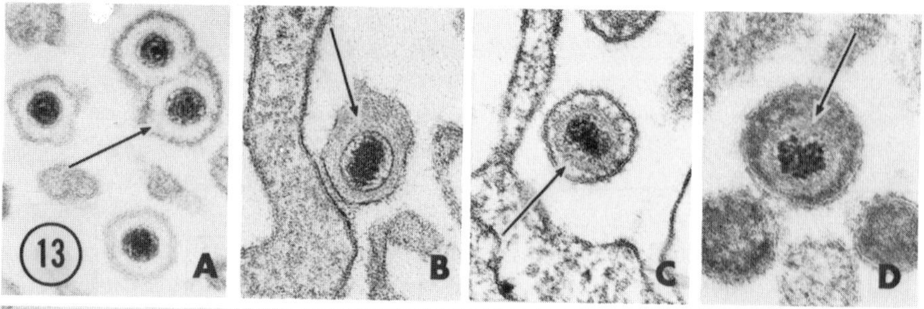
13
A
B
C
D

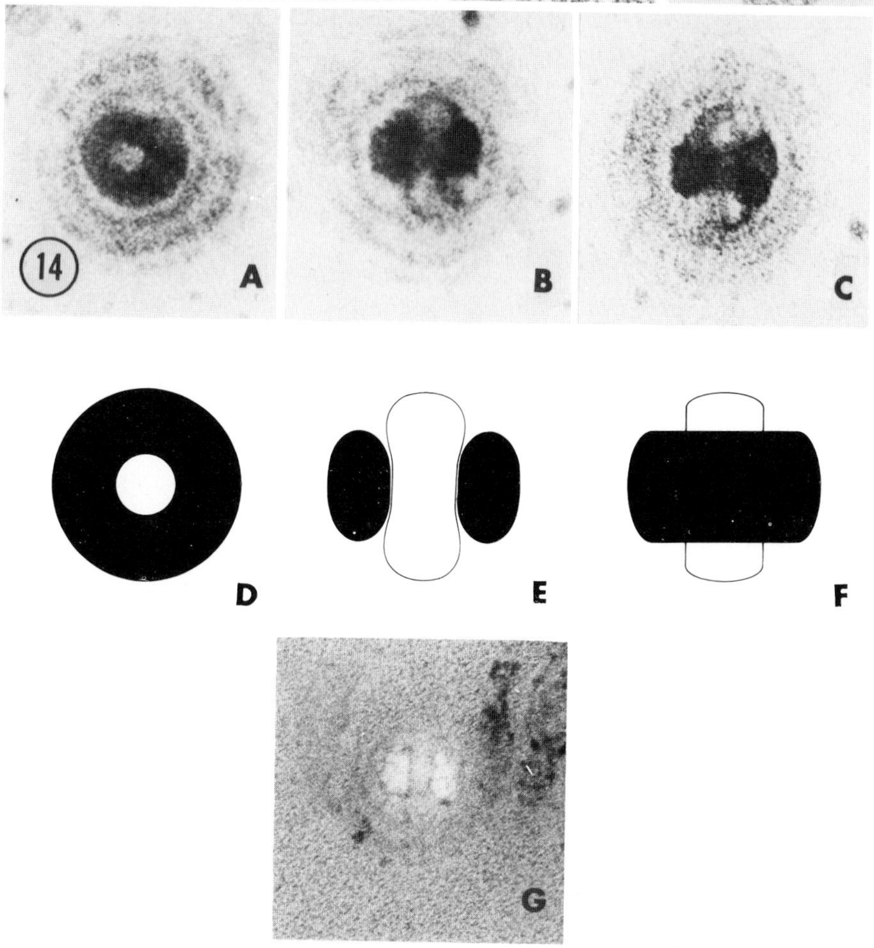
14
A
B
C
D
E
F
G

infection, i.e, in papova-, adeno-, or some herpesvirus-infected cells, or may be delayed for many days as is the case in cytomegalovirus-infected cells. Existent cellular components may become superfluous, and undergo redistribution or decomposition. The fragmentation and redistribution of the chromatin, often paralleled by chromatolysis [after papova- (Gaylord and Hsiung, 1961), adeno- (Mayor *et al.*, 1967), or herpesvirus (Heine and Dalton, 1972) infection], exemplifies this phenomenon. In addition, the accumulation of viral precursors, mature virions, and/or different, often virus-coded, metabolic products, may alter completely the appearance of the cells. In this respect, the intranuclear inclusion body of Cowdry type-A is a well-known example indicating viral infection at the level of the light microscope. These bodies are Feulgen positive early during infection, suggesting the presence of large amounts of virus. Later, after release of virus, the bodies are Feulgen negative, now representing the classic Cowdry type-A inclusion body (Scott, 1959). Other results of viral infections are the frequent formation of giant cells and polynucleated cells.

The extreme variations in appearance of virus-infected nuclei are illustrated in a series of electron micrographs taken at low magnifications (Figs. 30–34). In Fig. 30, representing tissue of *Nicotiana rustica* L. in-

Figures 15–18 illustrate the uptake of SV40 virions. (From Hummeler *et al.*, 1970.) ***Fig. 15*** African green monkey kidney cell 10 min after infection with SV40 virus. Numerous virions are attached to the cell surface. Two virions are ingested by pinocytosis (arrows). The plasma membrane envelops these particles tightly (× 73,500). ***Fig. 16*** Two hours after infection, three SV40 virions are in close proximity to the nuclear membrane. One particle (arrow) appears to fuse with the outer nuclear membrane, shedding its cytoplasmic envelope at this time (× 73,500). ***Figs. 17 and 18*** Two hours after infection, SV40 virions are found inside the nucleus. At the point of penetration the nuclear membrane is altered (Fig. 17, arrow). The virions are not enclosed by the cytoplasmic membrane. Some particles are less electron-dense, possibly indicating the start of uncoating (× 73,500).

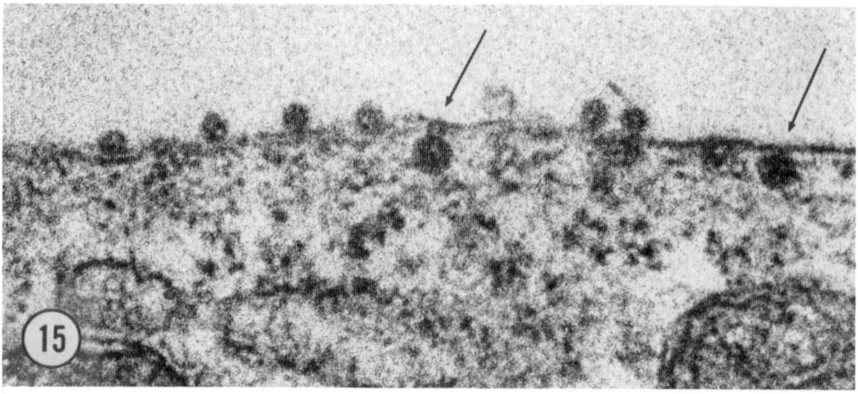
15

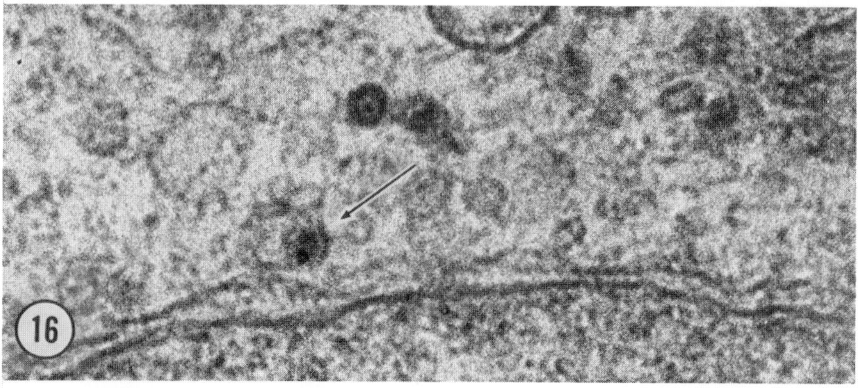
16

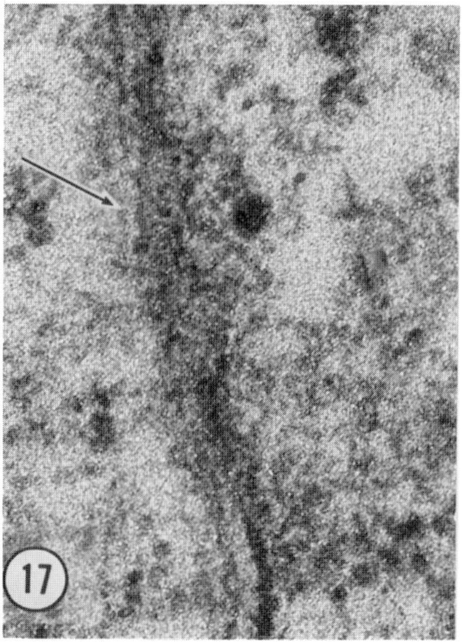
17

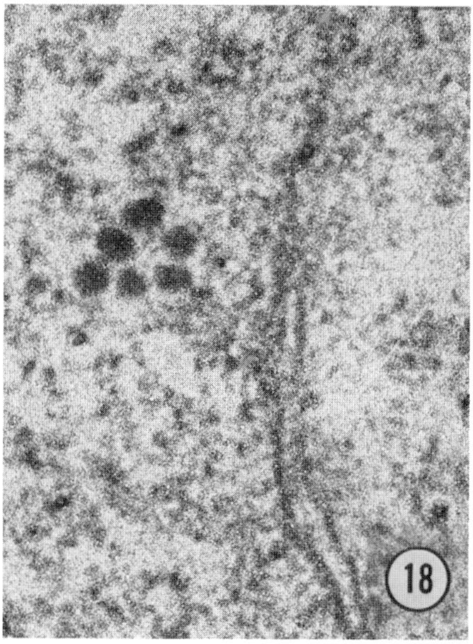
18

Figures 19–27 illustrate the uptake of adenovirus. ***Fig. 19*** HeLa cell infected with adenovirus, type 7. Forty minutes after infection, one virion can be seen adsorbed to the cell surface, another is inside a pinocytotic vacuole, and a third is seen free in the cytoplasm (× 150,000). (From Morgan *et al.*, 1969.) ***Fig. 20*** HeLa cell infected with adenovirus, type 5. One hour after infection, virions are found near the nuclear membrane lodged between fibrils and microtubules (× 60,000). (From Chardonnet and Dales, 1970a). ***Figs. 21 and 22*** HeLa cells infected with adenovirus, type 5. One hour after infection, virions are already adjacent to the nuclear pores. In Fig. 22, the viral capsid appears to be disrupted (arrow) (× 100,000). (From Chardonnet and Dales, 1970a). ***Figs. 23–26*** HeLa cells infected with adenovirus, type 7. After the first hour of infection the virions are situated near the nuclear membrane and are in the process of releasing their nucleoproteins into the nucleus (Fig. 25, arrow). The inner nuclear membrane is altered and forms a pocket, in which a small part of the cytoplasm containing the virion protrudes (Figs. 24–26). During the release of its nucleoprotein the virion loses its electron density. At the end of this process an empty virion remains in the nuclear pocket (Fig. 26, arrow) (× 150,000). (From Morgan *et al.*, 1969.) ***Fig. 27*** HeLa cell infected with adenovirus, type 7. Two hours after infection, several empty particles (arrows) are seen adjacent to the nuclear membrane (× 71,000). (From Chardonnet and Dales, 1970a.)

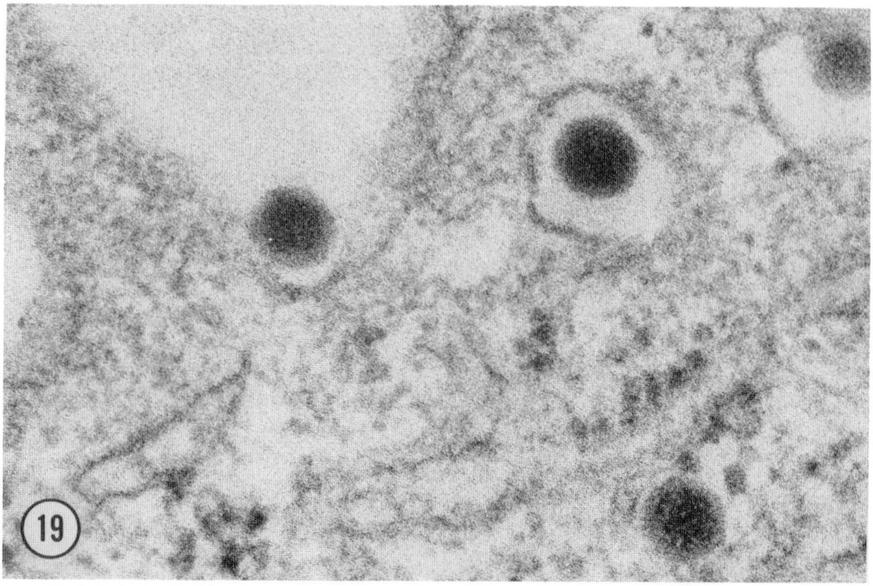
19

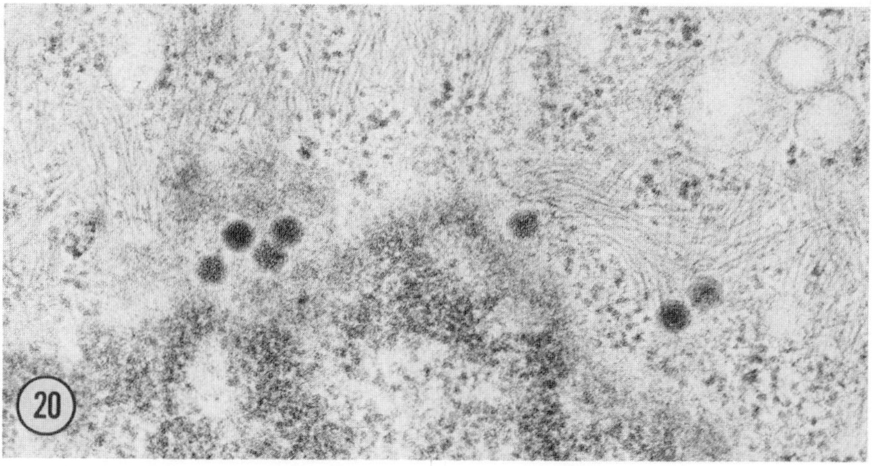
20

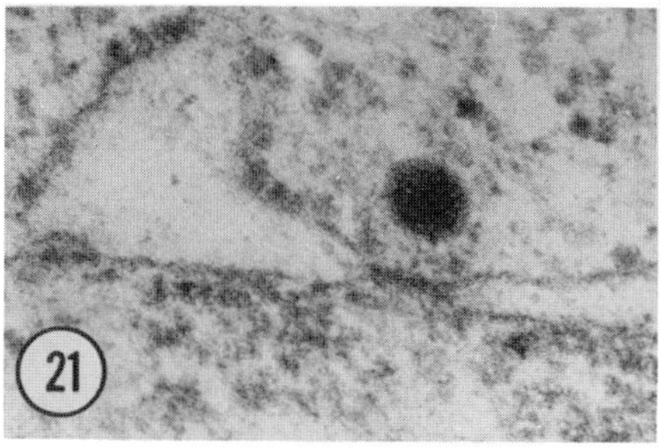
21

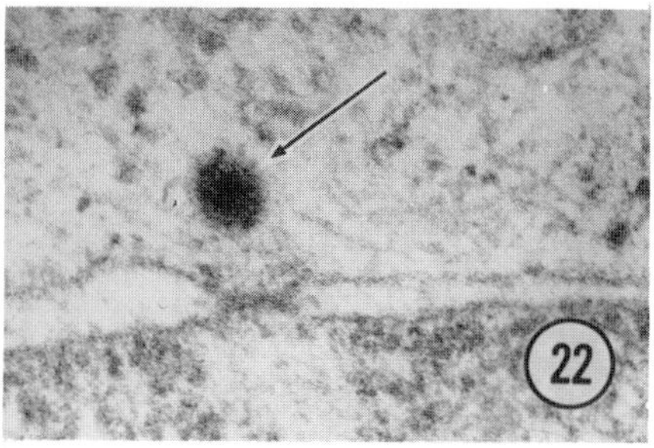
22

fected with potato yellow dwarf virus, the nucleoplasm is completely replaced by viroplasm. It consists of very homogeneous material, and virions are only present outside the nuclear envelope (McLeod *et al.*, 1966). In contrast, cells infected with polyoma virus exhibit an abundance of mature virions throughout the nuclei (Fig. 31). The changes in adenovirus-infected nuclei are manifold (Fig. 32). The nuclei are not only conspicuous owing to the large number of mature virions arranged in a crystallike pattern, but in addition because they contain a number of well-defined nuclear inclusions. These are fibers, most probably representing filamentous antigens, spherical granules containing viral structural antigens, proteinaceous inclusions associated with nuclear or viral DNA, and accumulations of viral nucleic acid (Norrby, 1971; Martinez-Palomo, 1968; Martinez-Palomo and Granboulan, 1967). Still other variations of nuclear fine structure are provoked by infection with viruses of the herpes type. As illustrated in Fig. 33, the infection of cells with herpesvirus simplex, belonging to herpesvirus group A, results in the fragmentation of the nucleus, the condensation and accumulation of chromatin adjacent to the nuclear membrane (leaving the interior of the nucleus free for virus production), and the growth of large sheaths of nuclear membranes throughout the cytoplasm. In contrast, the nuclei of cells infected with a cytomegalovirus, which belongs to herpesvirus group B, are characterized by an unusual, electron-dense inclusion body, differing in composition from the inclusion body of Cowdry type-A, the presence of a well-defined nucleolus, the lack of chromatolysis, and an abundance of viral progeny (Fig. 34).

2. VIRAL PRECURSORS AND MATURE VIRIONS

The length of the eclipse phase, which represents the time between infection and the appearance of the first viral progeny, may vary considerably as it is dependent on factors either coded for by the viral genome or influenced by the host cell. It may also be dependent on the degree of viral adaption to a specific host. Early morphological changes in infected nuclei are often quite subtle and may consist of the appearance of small patches of viroplasm distinguishable from the remaining nucleoplasm by its higher grade of homogeneity. The first virions appear in or adjacent to these areas (Fig. 35). In the case of naked viruses, the particles are usually immediately present in their mature form. During the course of infection, their number may increase significantly, coming to occupy large areas of the nucleus (Fig. 32).

An interesting phenomenon was observed in green monkey kidney cells infected with measles virus (Howatson, 1971; Nakai *et al.*, 1968, 1969).

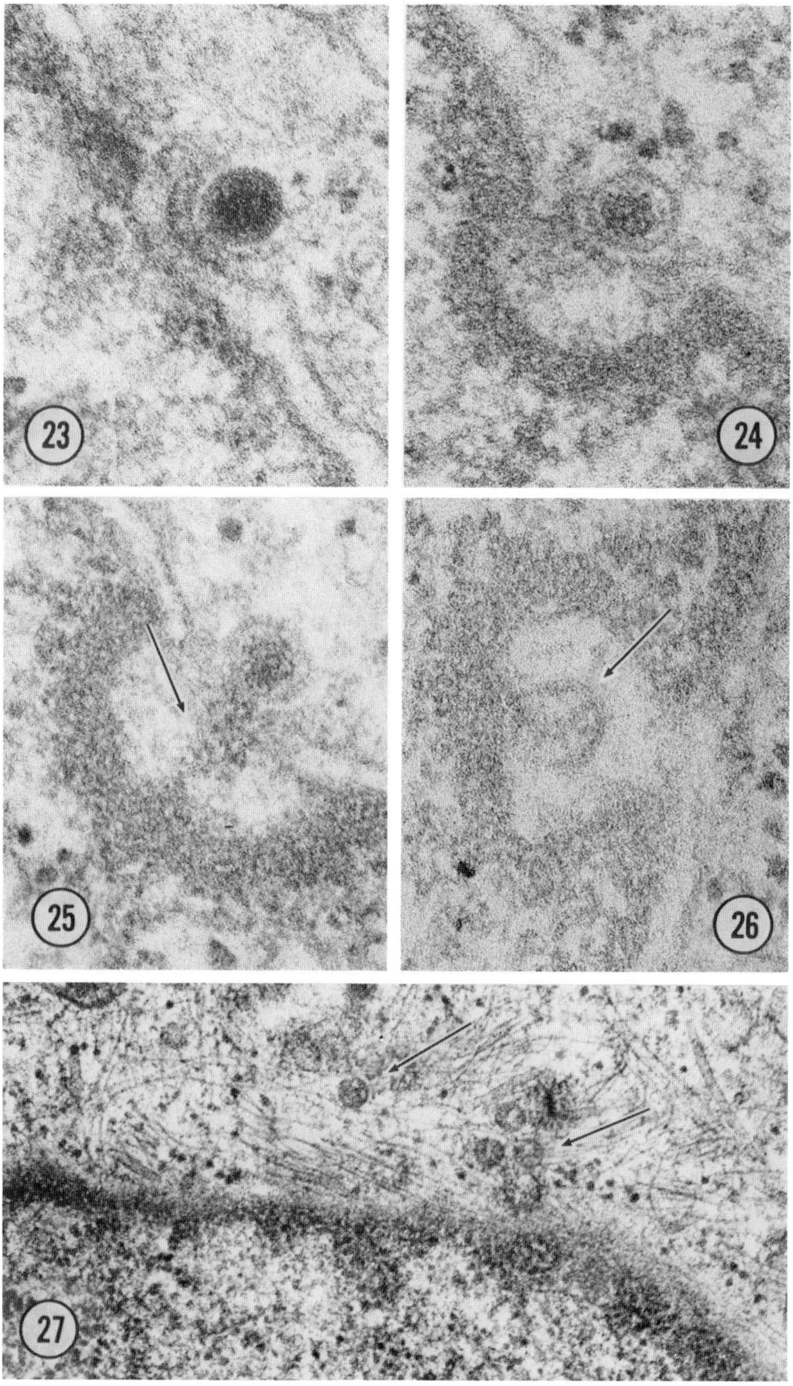

Figs. 23–27 See legend, p. 508.

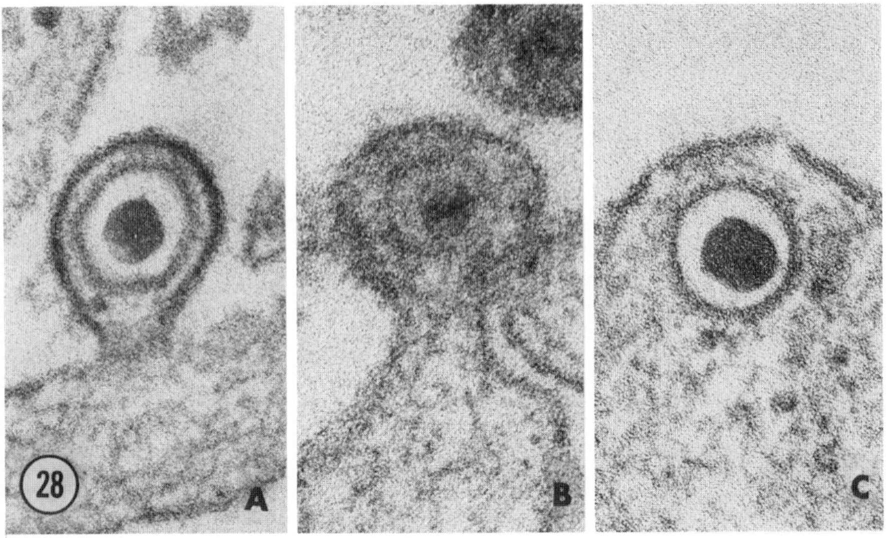

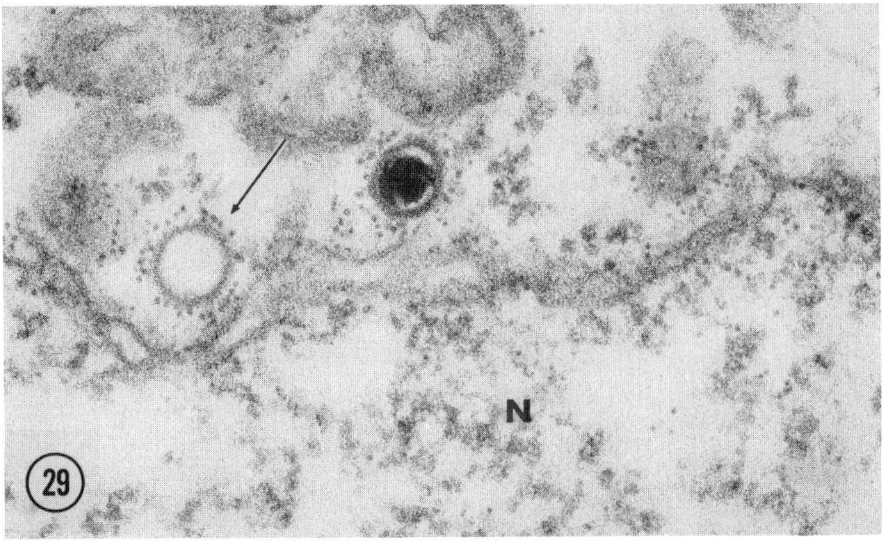

Figures 28 and 29 illustrate the uptake of herpesvirus simplex by HeLa cells. (From Miyamoto and Morgan, 1971.) ***Fig. 28*** (A and B) The fusion of the enveloped virion with the cellular membrane, 2 and 5 min after infection, respectively. (C) A naked capsid near the cell membrane after entry 5 min after infection. The core is very distinct (× 180,000). ***Fig. 29*** An empty capsid (arrow) near the nuclear membrane adjacent to a nuclear pore. Another, a complete nucleocapsid, is situated in the center of the photograph. Both particles are tagged with ferritin-conjugated 7 S antibody to verify their viral nature. N, nucleus (× 100,000).

This virus, containing an RNA genome, usually produces tubules of viral nucleoprotein in the cytoplasm. Later during infection the tubules are extruded by a budding process, giving rise to mature measles virions. In some preparations, large amounts of nucleoprotein strands were present almost solely in the nuclei of the majority of the cells (Fig. 36). It was found that infectivity in those preparations containing large numbers of cells with cytoplasmic and budding virus was high, but was low in preparations in which intranuclear formation of virus was predominant and budding was infrequent.

Enveloped particles often exhibit different, morphologically distinct steps in their maturation. For example, the emergence of circular or globular structures, representing parts of the viral nucleocapsids, is frequently the first sign of newly formed viral progeny after infection with viruses of the herpes type. These forms, illustrated in Fig. 37A, consist of incomplete capsids, sometimes present as early as 5 hr after infection (Nii *et al.*, 1968), complete but empty capsids as well as capsids containing a dense core. Numerous hexagonal shells of about 35 nm diameter may be present in large numbers throughout the interior of the nuclei (Stackpole, 1969) (Fig. 37B), and may become surrounded by viral cores (Nazarian *et al.*, 1971; Zeigel and Clark, 1972) (Fig. 37C). Other aspects of incomplete, herpes-type virions are tadpolelike forms with filamentous projections of nucleoprotein penetrating only halfway into the viral cores (Fawcett, 1956), cores in which the nucleoprotein appears to be spooled onto a central rod (Chai, 1971; Furlong *et al.*, 1972) and nucleocapsids with extremely large cores (Heine and Hinze, 1972; Heine *et al.*, 1971) (Figs. 37–39). To date, it is not yet clear if the observed differences in the fine structure of the nucleocapsids reflect true morphological differences or if they are caused by differences in preparative methods.

3. THE NUCLEOLUS

The nucleolus is one of the most important cellular organelles involved in the synthesis of major portions of cellular RNA (see Volume I, Chapter 2). Generally, its morphological appearance reflects variations in cellular metabolism. Thus, it is not surprising to learn that viral infections which frequently alter drastically the cellular metabolism can also cause profound changes in nucleolar fine structure. These changes may range from nucleolar hypertrophy (Bernhard and Granboulan, 1968; Granboulan and Tournier, 1965; Mayor *et al.*, 1962) to complete destruction (Almeida *et al.*, 1962; Shikata and Maramorosch, 1966). Autoradiographic, enzymatic, and high-resolution examinations of the nucleolus revealed its active role early during infection with some viruses, i.e., picodna- (Henry *et al.*, 1972), papova- (Granboulan and Tournier, 1965; Granboulan *et al.*, 1963; Tournier *et al.*, 1961), adeno- (Martinez-Palomo, 1968; Martinez-Palomo and Granboulan, 1967; Martinez-Palomo *et al.*, 1967; Matsui and Bernhard, 1967; Philips and Raskas, 1972), cytomegalovirus, and plant viruses (Esau and Hoefert, 1972a, b). The resulting morphological changes may be expressed by nucleolar hypertrophy (Fig. 34), the segregation of nucleolar components (Philips and Raskas, 1972; Sirtori and Bosisio-Bestetti, 1967), the increase in nucleolar chromatin (Granboulan and Tournier, 1965; Granboulan *et al.*, 1963; Martinez-Palomo and Granboulan, 1967), and the appearance of spotted nucleoli (Bernhard and Granboulan, 1968; Bierwolf *et al.*, 1968; Granboulan and Tournier, 1965; Martinez-Palomo, 1968; Martinez-Palomo *et al.*, 1967) (Fig. 32). The accumulation of dense condensations of granular and fibrillar material throughout or on the surface of the nucleolus is especially common in SV40- (Tournier *et al.*, 1961) (Figs. 40–42) and adenovirus-infected cells (Martinez-Palomo and Granboulan, 1967; Martinez-Palomo *et al.*, 1967), and it has been shown that some of these areas may be related to an enhanced synthesis of DNA and RNA (Granboulan and Tournier, 1965; Martinez-Palomo, 1968). One nucleolar alteration is worthy of special consideration. As illustrated above, large masses of empty capsids of two of the picodnaviruses are not only closely associated with but may actually replace the fragmented nucleoli (Fig. 44), thus suggesting the assembly and possibly the synthesis of viral capsids by this organelle (Al-Lami *et al.*, 1969; Henry *et al.*, 1972).

During later stages of viral infection, when cellular metabolism is deteriorating, the nucleoli degenerate completely. At that time, fragmentation (Fig. 43) and condensation of their remaining components are prevalent.

4. NUCLEAR INCLUSIONS OF VARIOUS ORIGINS

After some virus infections, various inclusions have been repeatedly observed in the nuclei but only infrequently in the cytoplasm of the infected cells. Their presence was described as early as 1957 (Morgan *et al.*,

---

Fig. 30 Part of a cell of *Nicotiana rustica* infected with potato yellow dwarf virus. Viroplasm replaces large areas of the nucleus (N), mature virions are only present in the cytoplasm (V) (approx. × 36,000). (Courtesy of Dr. McLeod, unpublished observation.)

Fig. 31 Nucleus of a cell infected with polyomavirus. The virions (V), round and tubular forms, are abundant in the nucleoplasm (× 50,000). (Courtesy of Dr. A. F. Howatson, unpublished observation.)

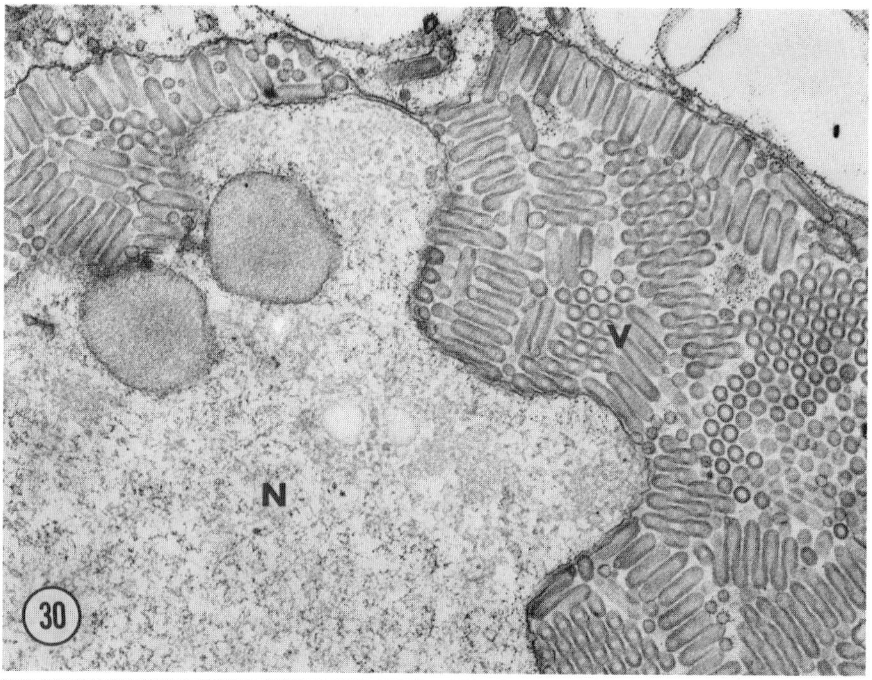
V
N
30

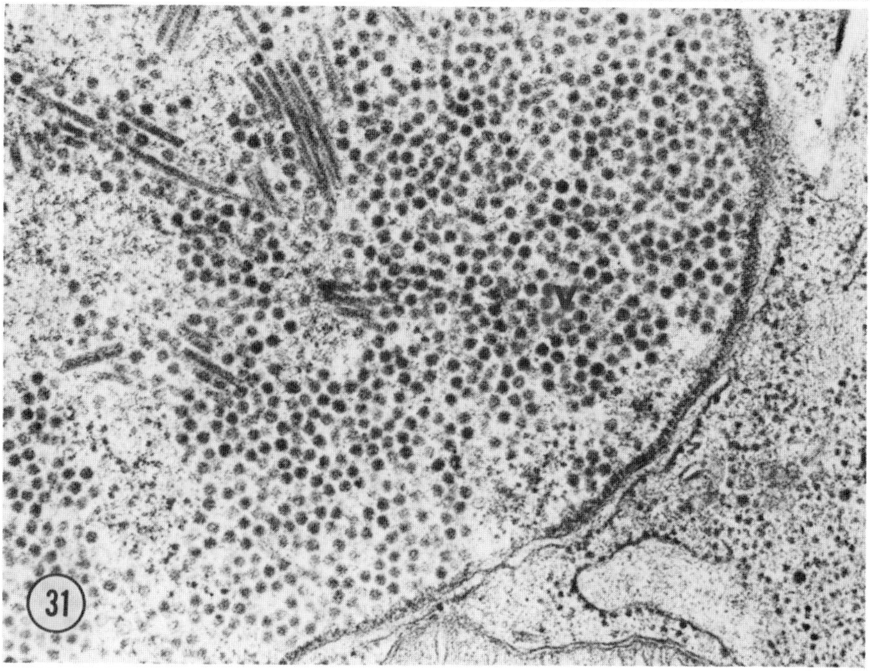
V
31

1957), and their most common forms are crystals, fibers, and/or tubules. All are composed mainly of protein (Boulanger *et al.*, 1970; Ginsberg and Dingle, 1965; Henry *et al.*, 1971; Martinez-Palomo, 1968; Marusyk *et al.*, 1972; Rouse and Schlesinger, 1972; Torpier and Petitprez, 1968). Taking into account the fact that large amounts of viral parts, i.e., about 90% of the capsid components in adenovirus-infected nuclei (Ginsberg and Dingle, 1965), are often not used in the assembly of virions, their accumulation into huge, sometimes complex structures is not difficult to understand.

Paracrystalline inclusions are especially frequent in adenovirus-infected cells. Their fine structure varies little from strain to strain (Figs. 45 and 46). High-resolution electron microscopy (Boulanger *et al.*, 1970) and immunofluorescent staining (Henry *et al.*, 1971) (Fig. 47) suggested their composition of viral capsid antigen being possibly a combination of hexons, pentons, and fibers. In contrast, recent investigations, illustrating the arginine dependence of the virus particles (Rouse and Schlesinger, 1972), the arginine-rich nature of the crystals, and their immunological relationship to core protein (Marusyk *et al.*, 1972) make it probable that the latter may constitute an integral part of the paracrystalline formations.

Fibers (as illustrated in Figs. 39 and 48) have been observed after infection with different viruses: Picodnavirus Hl (Bernhard *et al.*, 1963), SV40 virus (Granboulan *et al.*, 1963), adenovirus, type 12 (Martinez-Palomo *et al.*, 1967), canine hepatitis virus (Matsui and Bernhard, 1967), and *Herpesvirus saimiri* (Heine *et al.*, 1971). Digestion with pronase has given evidence of their proteinaceous nature (Martinez-Palomo *et al.*, 1967) and immunofluorescence studies suggest that they may represent viral structural antigens (Pope and Rowe, 1964; A. Rabson, personal communication).

Long tubules, repeatedly observed in herpesvirus-infected cells (Farley *et al.*, 1972; Heine and Hinze, 1972; Nii *et al.*, 1968; Stackpole and Mizell, 1968), may be caused by the faulty assembly of viral capsid material, since, in cross sections, they sometimes show a remarkable resemblance to incomplete virions lacking the DNA core (Fig. 49).

Other inclusion bodies of as yet unknown nature are cylindrical bodies formed by concentric arrays of dense lamellae, known to be of proteinaceous nature (Bierwolf *et al.*, 1968; Martinez-Palomo, 1968) (Fig. 50), bundles of fibrils with regular periodicity (Bierwolf *et al.*, 1968; Martinez-Palomo *et al.*, 1967) (Fig. 51), and intricate crystalloid structures of complex composition (Figs. 52 and 53). The latter are only rarely present in the nuclei of cytomegalovirus-infected cells (A. Howatson, personal communication), but have been seen more frequently in the cytoplasm of cells infected with canine hepatitis virus (Matsui and Bernhard, 1967).

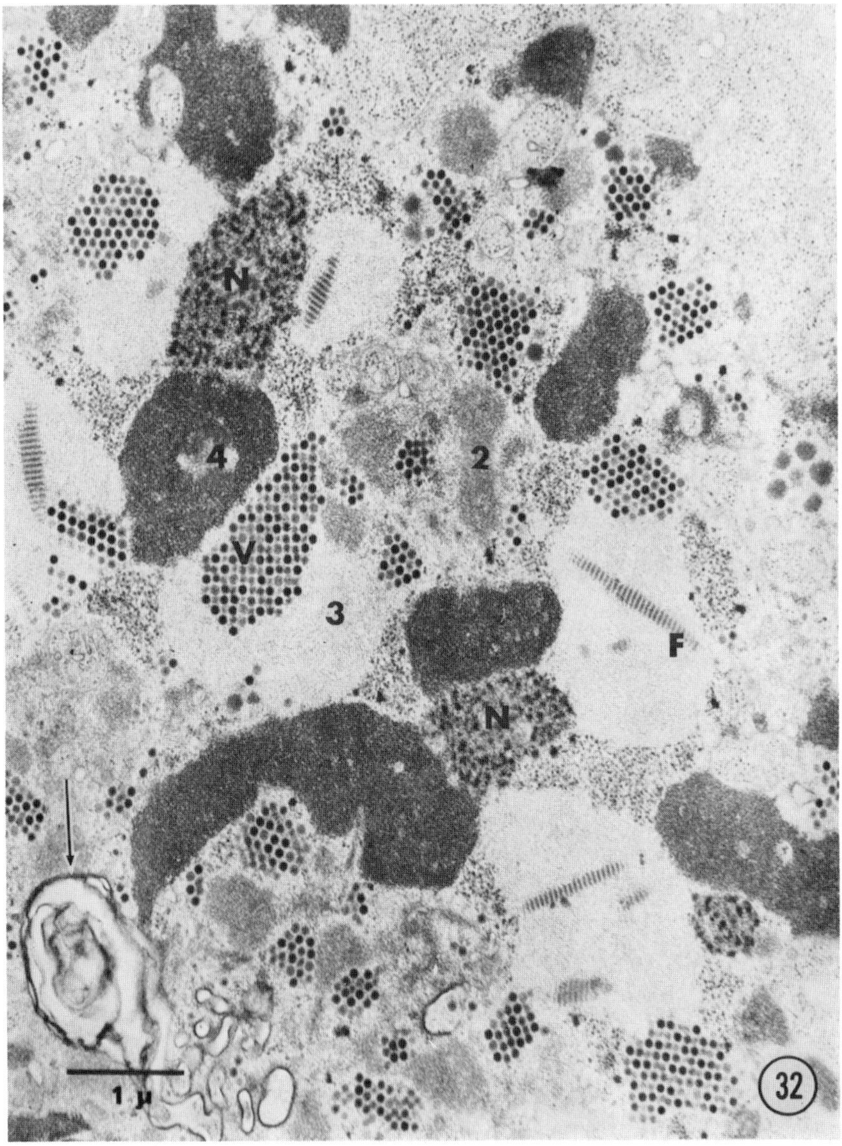

Fig. 32 Part of a nucleus of a KB cell 3 days after infection with adenovirus, type 2. Viral crystals (V) are numerous. The small nucleoli (N) have a spotted appearance. A myelin structure is indicated by an arrow. Inclusion bodies of three types are evident (2, 3, 4). One type (3) contains a number of striated protein fibers (F) (approx. × 15,000). (From Martinez-Palomo, 1968.)

C. The Egress

The mode of virus release differs among different virus strains.

1. NAKED VIRIONS

Small virions often "spill" through minor breaks of the nucleus into the cytoplasm (Mayor *et al.*, 1962, 1967; ZuRhein, 1969) and reach the extracellular space subsequently by being either transported directly through the cytoplasm to the cell surface (Fig. 54) or transferred to cytoplasmic vacuoles, which may later open to the extracellular space (Mayor *et al.*, 1962). In addition, cell lysis renders possible the release of large numbers of virions at one time (Granboulan *et al.*, 1963) (Fig. 55).

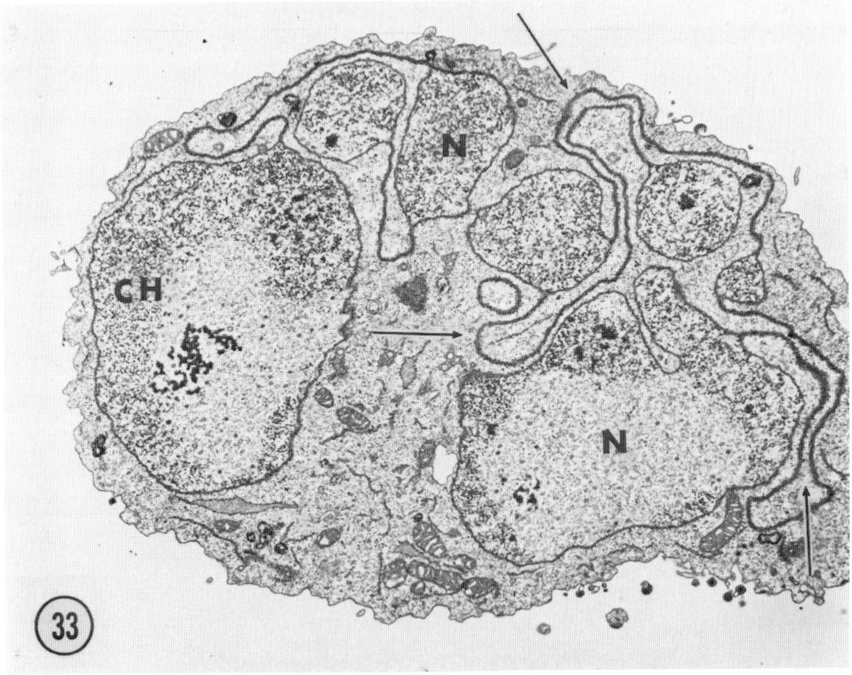

Fig. 33 W1–38 cell infected with herpesvirus simplex. The nucleus (N) is fragmented into numerous bodies of different size. The chromatin (CH) is adjacent to the nuclear membrane leaving the interior of the nucleus electron lucent and homogeneous in appearance. Nuclear membranes are present as large sheaths in different areas of the cytoplasm (arrows) (× 8000). (From Heine and Dalton, 1972.)

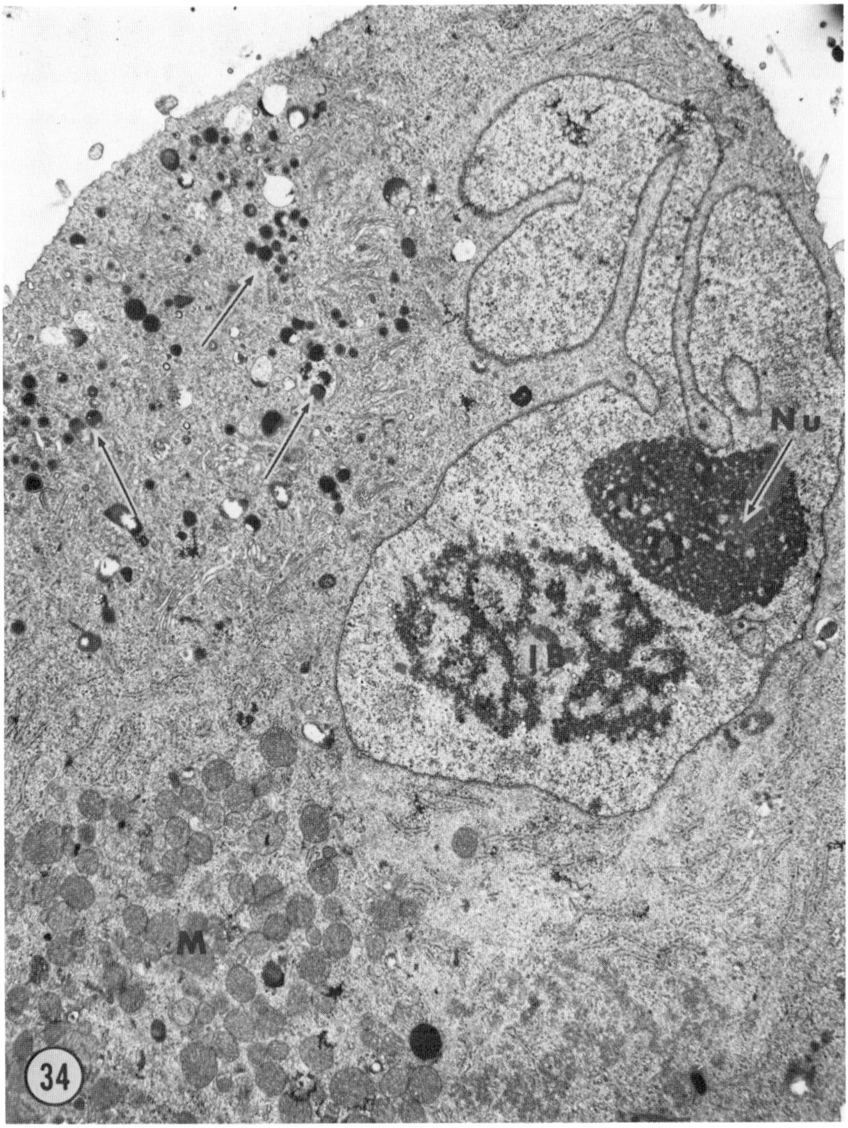

Fig. 34 Cell grown *in vitro* and infected with a cytomegalovirus. The distribution of the chromatin throughout the nucleus is of normal appearance. The nucleolus (Nu) is very large. The nucleus contains an electron-dense inclusion body of skeinlike formation (IB). Small, electron-dense inclusion bodies are numerous in the cytoplasm (arrows). Mitochondria (M) are of normal composition. Virions are barely visible in close association with the inclusion body (approx. × 6300). (Courtesy of Dr. F. Haguenau, unpublished observation.)

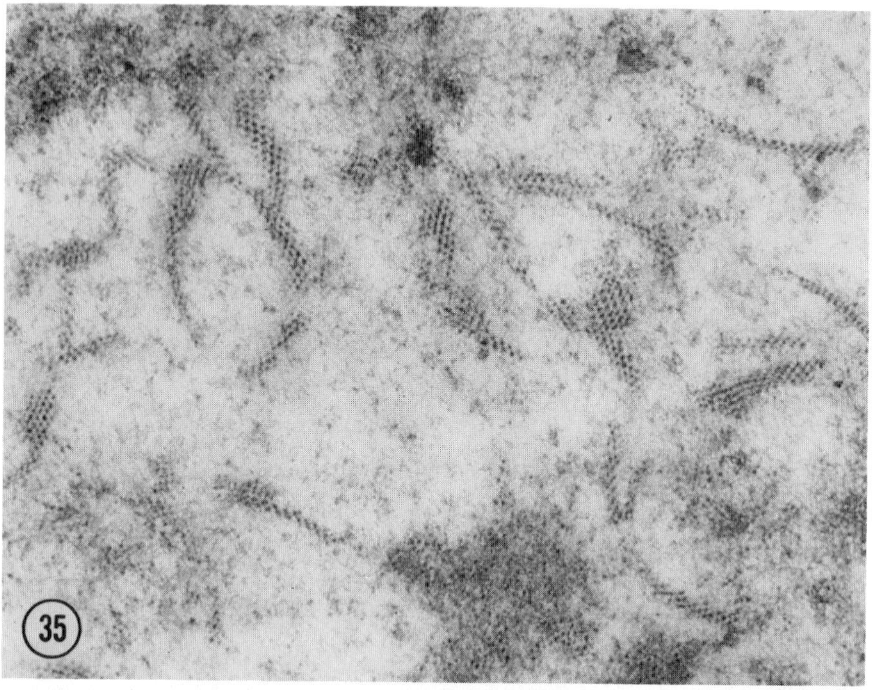
35

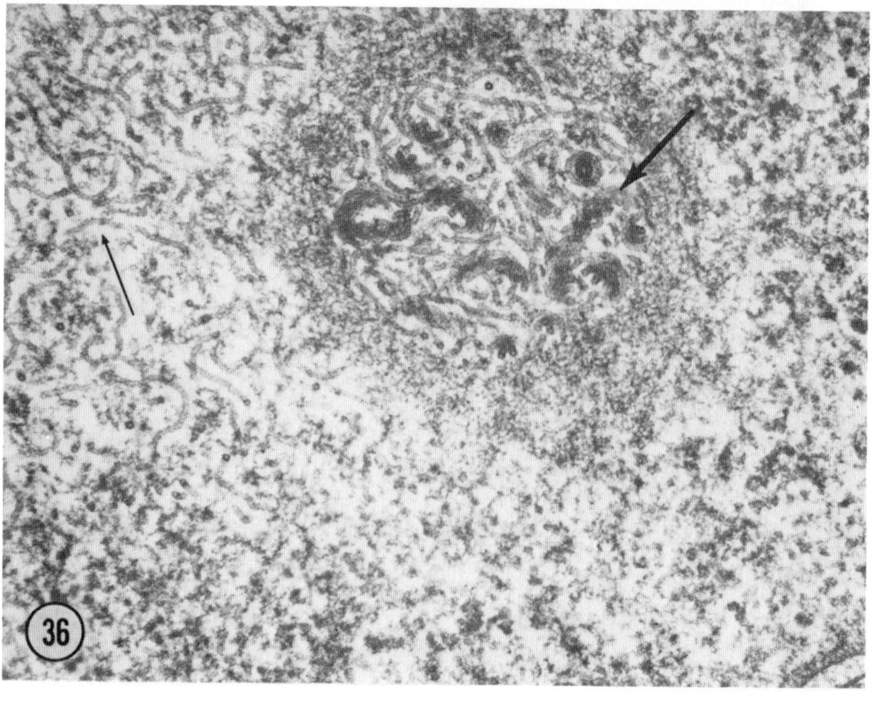
36

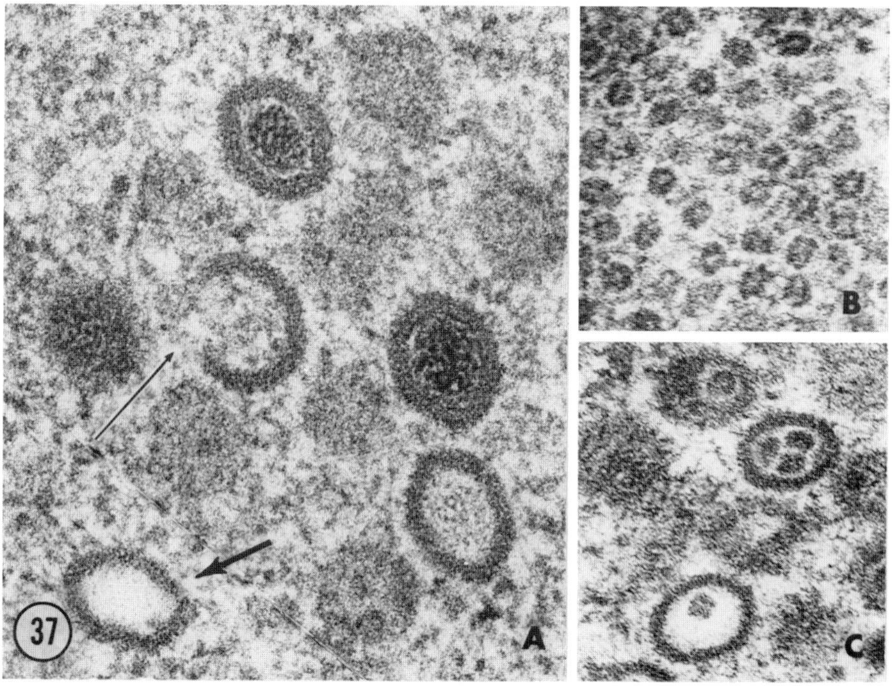

Fig. 37 Reptilian cells grown *in vitro* and infected with a herpes-type virus isolated from *Iguana iguana.* (A) Nucleocapsids at different stages of completion. An incomplete capsid (long arrow), an empty capsid (short arrow), and complete capsids with electron-dense cores are evident (× 132,000). (B) Cluster of 35 nm particles in the matrix of the infected nucleus. (C) Viral cores containing different numbers of 35 nm particles (× 140,000). (From Zeigel and Clark, 1972.)

Fig. 35 Green monkey kidney cell infected with satellite virus and then superinfected with adenovirus. Twenty-four hours after superinfection, featherlike arrays of mature satellite virus are present throughout the nucleus. The virions have a diameter of approximately 20 nm (× 52,800). (From Mayor *et al.*, 1967.)

Fig. 36 BSC-1 cell infected with measles virus for 8 days. Tubules of nucleoprotein are irregularly distributed throughout the nucleus (thin arrow). A large aggregate of such tubules, many of them intertwined to form coiled structures (heavy arrow), is in the center of the illustration (× 60,000). (From Howatson, 1971.)

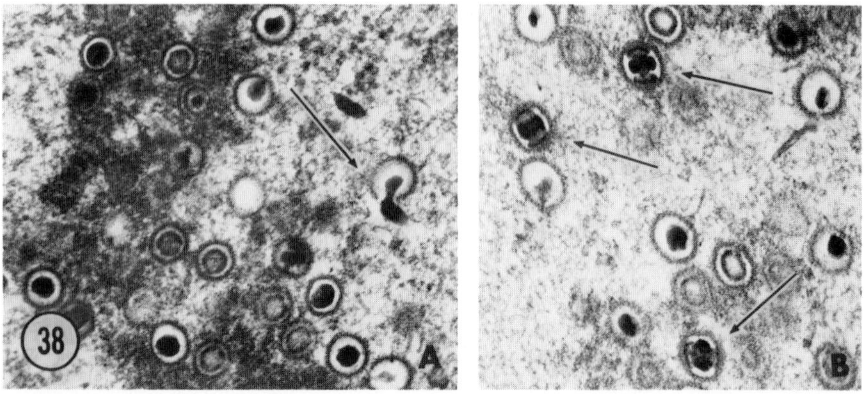

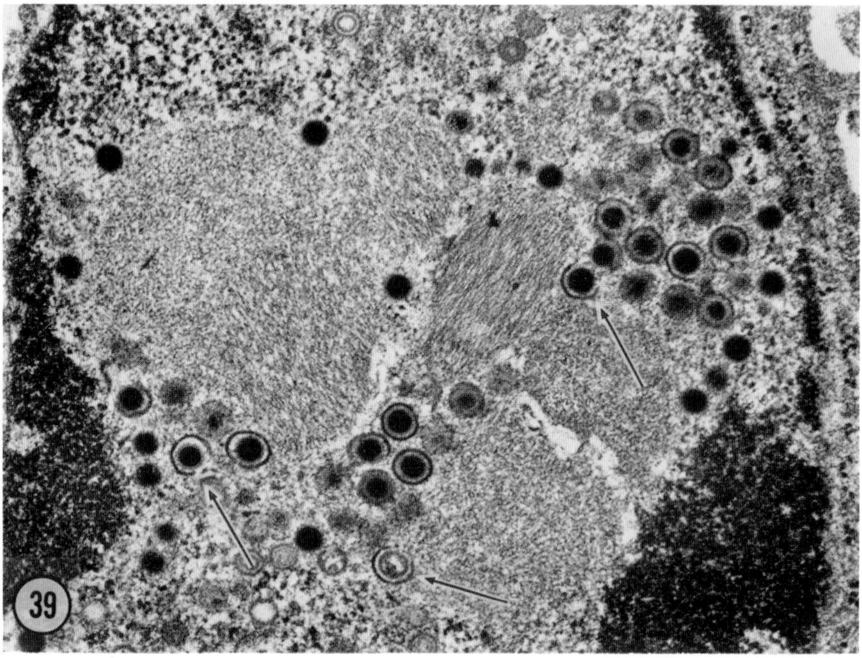

Fig. 38 Cells infected *in vitro* with a human cytomegalovirus isolated from a Kaposi sarcoma. (A) Three different types of immature virions are present (single and double shells, and those with an electron-dense core), in addition to a tadpolelike structure (arrow). (B) Cores of toroid composition (arrows) (× 64,000).

Fig. 39 African green monkey kidney cell infected with *Herpesvirus saimiri*. Chromatin is condensed adjacent to the nuclear envelope. The electron-dense cores of the virions are outstanding. Masses of fibrils cut at different angles occupy large parts of the nucleus. The arrows indicate virions at different stages of envelopment (× 35,000). (From Heine *et al.*, 1971.)

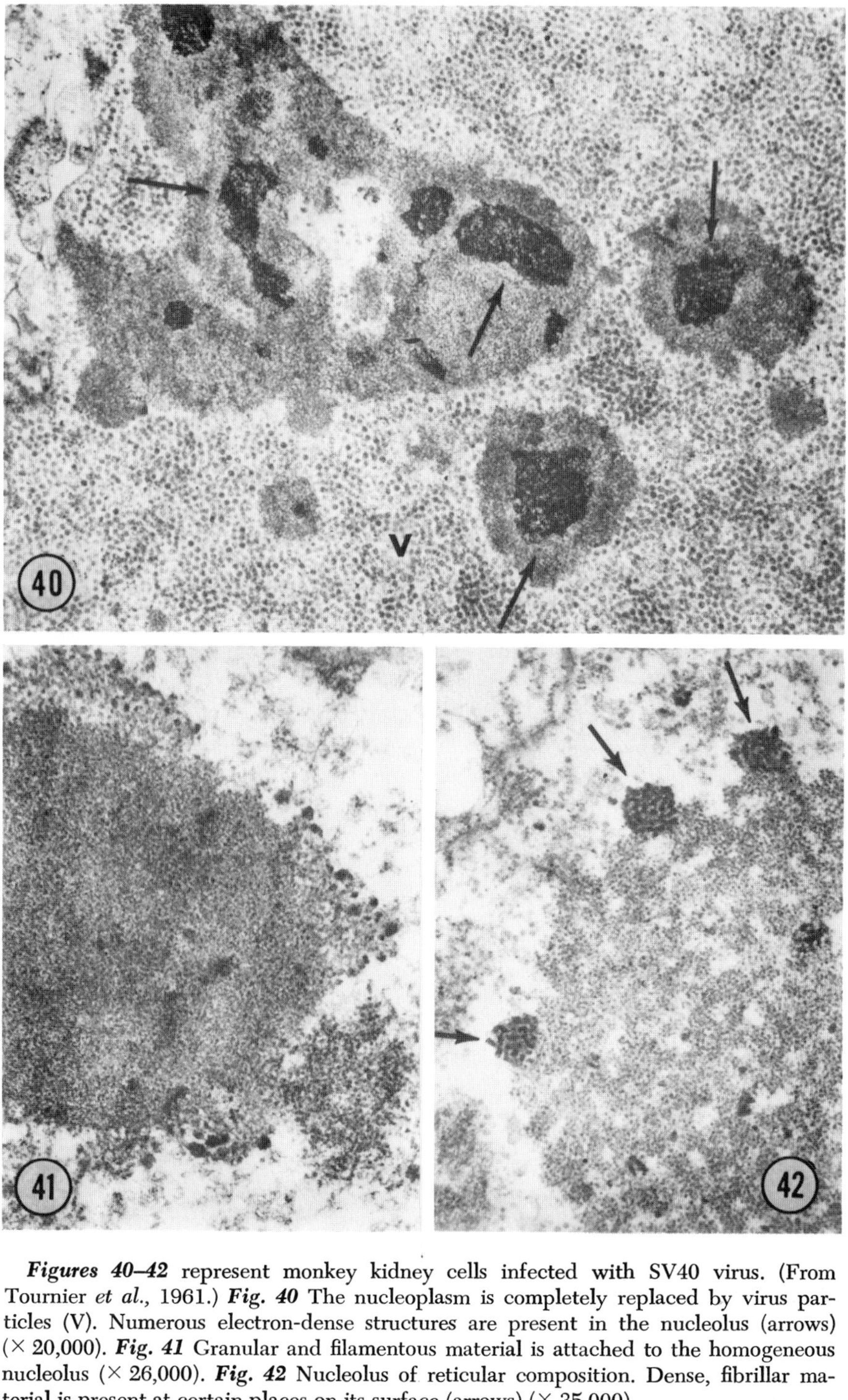

Figures 40–42 represent monkey kidney cells infected with SV40 virus. (From Tournier *et al.*, 1961.) ***Fig. 40*** The nucleoplasm is completely replaced by virus particles (V). Numerous electron-dense structures are present in the nucleolus (arrows) (× 20,000). ***Fig. 41*** Granular and filamentous material is attached to the homogeneous nucleolus (× 26,000). ***Fig. 42*** Nucleolus of reticular composition. Dense, fibrillar material is present at certain places on its surface (arrows) (× 35,000).

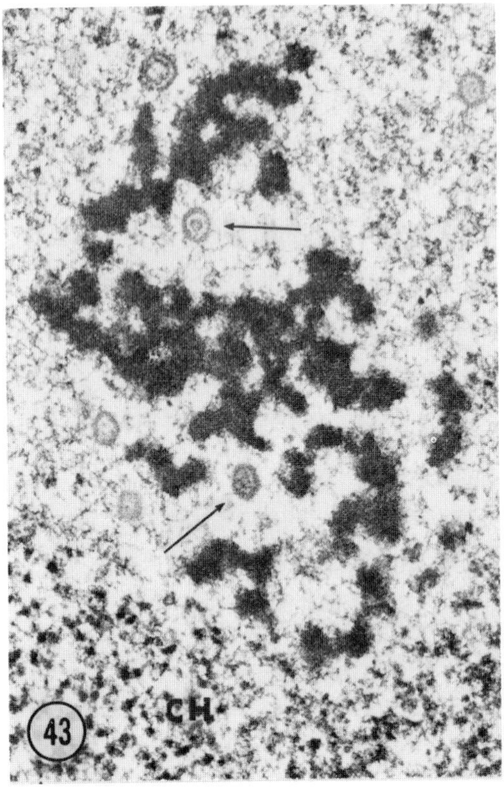

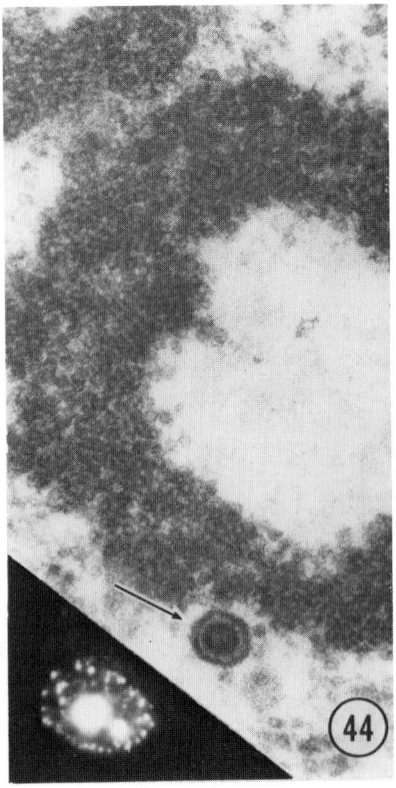

Fig. 43 W1–38 cell infected with herpesvirus simplex I. The nucleolus is fragmented, and its granular component is missing. Incomplete virions are designated by arrows. CH, condensed chromatin (× 35,000).

Fig. 44 Human embryo lung cell infected with adenoassociated virus and herpesvirus simplex, 36 hr after infection. Virus capsids have replaced the nucleolus. A herpes-type virion is present (arrow) (approx. × 70,000). Inset: Immunofluorescent staining of a nucleus with adenoassociated virus-1 conjugated antibody. The nucleolus is intensely stained (× 650). (From Henry *et al.*, 1972.)

Fig. 45 KB cell infected with adenovirus, type 5. The nucleus contains a large, paracrystalline inclusion body and numerous virions distributed at random (× 26,000). (From Marusyk *et al.*, 1972.)

Fig. 46 HeLa cell infected with adenovirus, type 2. Higher magnification of the crystalline inclusion illustrates its fine structure. (A) Cross section shows hollow, circular profiles in hexagonal arrays (approx. × 80,000). (B) Longitudinal section exhibits parallel arrangement of microtubules (approx. × 100,000). (From Henry *et al.*, 1971.)

Fig. 47 Vero cell infected with adenovirus, type 2. (A) A cytoplasmic and a nuclear inclusion body are present. Hematoxylin–eosin (× 1900). (B) Similar preparation but stained with fluorescein isothiocyanate-labeled adenovirus 2 antibody. The cytoplasmic inclusion and circumscribed areas in the nucleus exhibit positive fluorescence (× 1800). (From Henry *et al.*, 1971.)

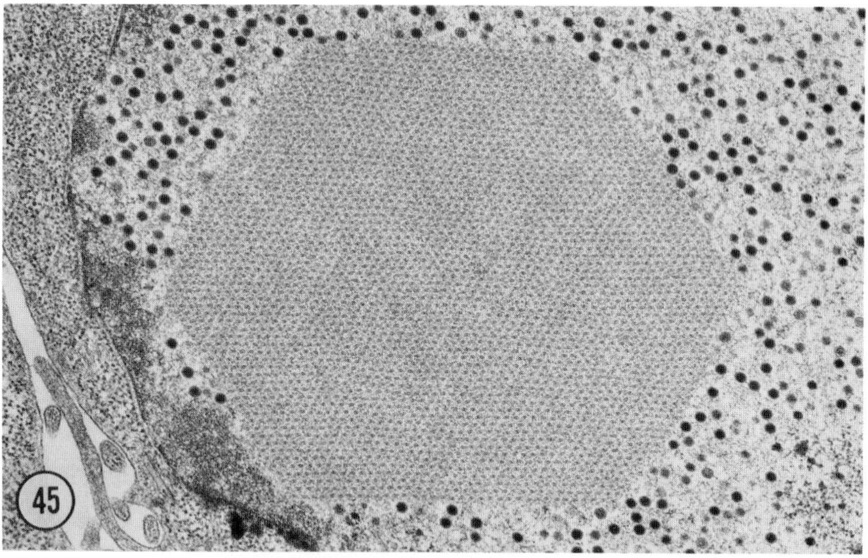
45

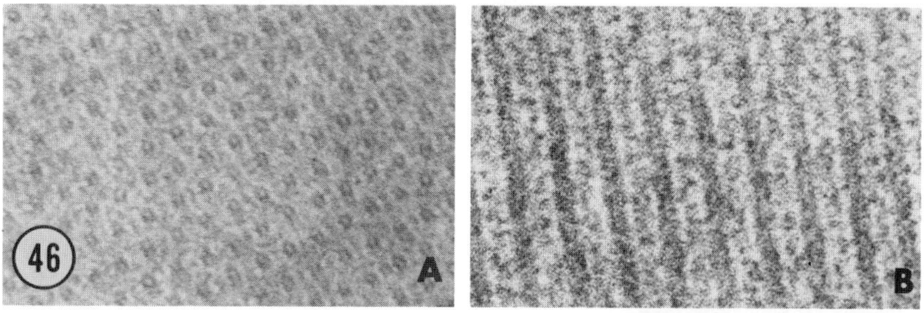
46
A
B

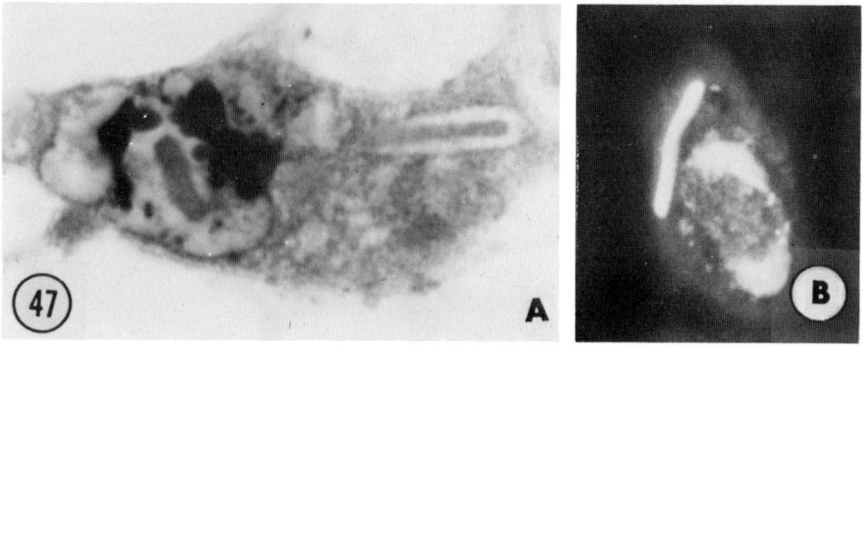
47
A
B

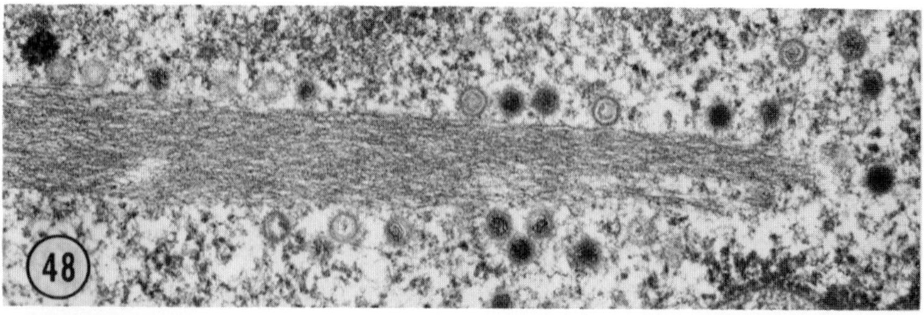
48

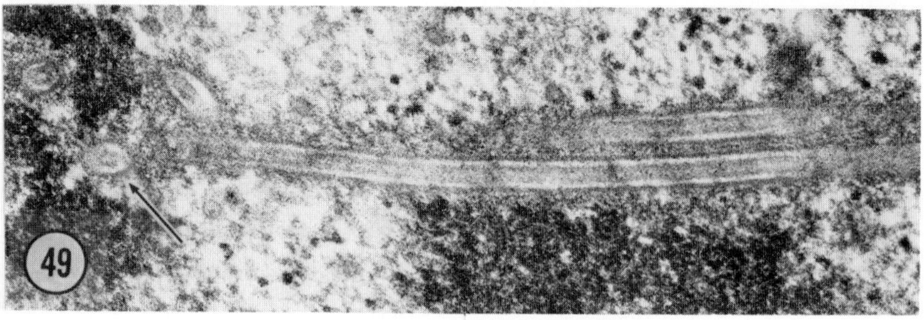
49

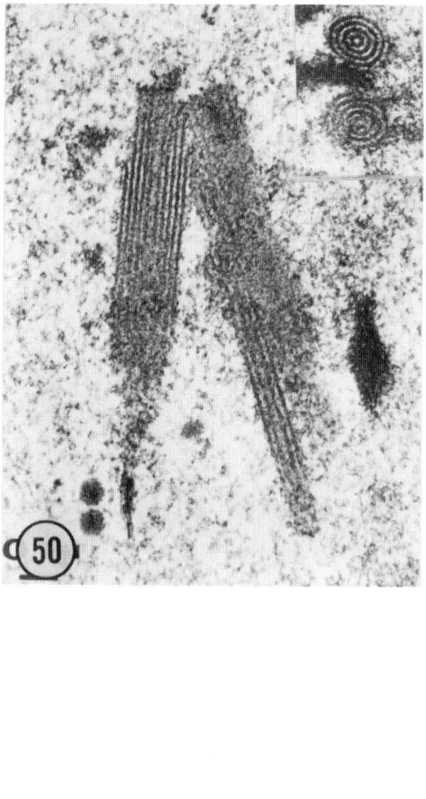
50

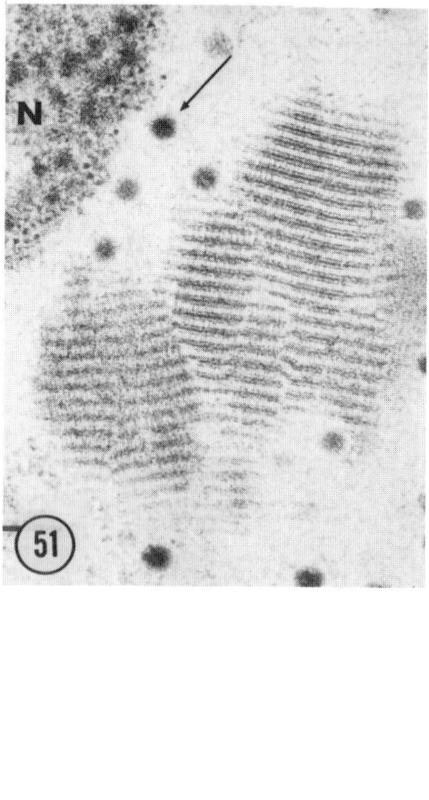
N
51

2. ENVELOPED VIRIONS

The egress of herpes-type virions from the nucleus into the cytoplasm is paralleled by their conversion to mature forms due to the addition of one or more envelopes. This process may take place at any cellular membrane (Becker *et al.*, 1965; Heine and Hinze, 1972; McCracken and Clarke, 1971; Siminoff and Menefee, 1966; Stackpole, 1969), but apparently the inner nuclear membrane is a preferred site (Darlington and Moss, 1968; McCracken and Clarke, 1971; Nii *et al.*, 1968; Stackpole, 1969). The immature particle, the nucleocapsid, is enveloped by budding through this membrane, which may undergo morphological changes at the same time (Fig. 56). Budding of herpes-type virus was also observed at the plasma membrane (Fig. 57), at numerous membranes in the cytoplasm (Fig. 58), at the rough endoplasmic reticulum, and at membranes of the Golgi zone (McCracken and Clarke, 1971). The mature particles are released either by passing to the cell surface in cytoplasmic vacuoles (Darlington and Moss, 1968; Heine and Dalton, 1972), or by using cytoplasmic channels for the egress (Jasty and Chang, 1972), a process termed "reverse phagocytosis" (Nii *et al.*, 1968).

A unique formation of the enveloping virus membrane was observed in cells infected with *Herpesvirus saimiri*. As illustrated in Fig. 39, viral envelopes may be formed *de novo* inside the nucleus, often in close association with bundles of fibers (Heine *et al.*, 1971). In this case, mature virions accumulate in the nuclei and are released by lysis of the cells.

Another interesting variation of maturation and viral egress is exemplified in cells infested with the herpes-type virus of frog renal adenocarcinoma. This virus passes through different developmental stages: The virions are first enveloped at the inner nuclear membrane, but are immediately unenveloped when crossing the outer membrane. Within the

---

Fig. 48 Owl monkey kidney cell infected with *Herpesvirus saimiri*. Bundles of thin fibers are common in the infected nuclei. Immature virions are in close association with the fibers (× 55,000). (From Heine *et al.*, 1971.)

Fig. 49 Rabbit kidney cell infected with *Herpesvirus sylvilagus*. Long tubules are present in the nucleus. Their diameter is similar to that of viral nucleocapsids. In cross sections (arrow) the tubules resemble immature virions lacking the DNA core (× 35,000). (From Heine and Hinze, 1972.)

Fig. 50 Part of the nucleus of a KB cell infected with adenovirus, type 12. Transverse and cross sections through cylindrical bodies of unknown origin (× 54,000). (From Martinez-Palomo, 1968.)

Fig. 51 KB cell infected with adenovirus, type 12. Bundles of fibers with regular periodicity are in the nucleoplasm. Virions are designated by an arrow and N represents part of a spotted nucleolus (× 60,000). (From Martinez-Palomo *et al.*, 1967.)

cytoplasm they acquire a fibrillar coat, possibly consisting of virus-specific antibody, and are enveloped by budding into vesicles. From there they are released into the extracellular space (Stackpole, 1969). In addition, the release of mature virions was observed infrequently at the plasma membrane (Fig. 57).

Some of the oncogenic herpesviruses which may be closely related to the cytomegaloviruses (Lee *et al.*, 1969) and herpesviruses of group B exhibit a considerable difference in the appearance of the mature virions depending on the site where the envelopment takes place (Ahmed and Schidlovsky, 1968; Heine *et al.*, 1971; Heine and Dalton, 1972; Heine and Hinze, 1972; Nazerian, 1970; Nazerian *et al.*, 1971). Virions acquiring their envelope at the nuclear membrane or *de novo* inside the nuclei are very uniform, whereas those virions that obtain their envelope at cytoplasmic membranes are extremely pleomorphic and characterized by the presence of an electron-dense matrix of as yet unknown composition (Cook and Stevens, 1968; Craighead *et al.*, 1972; McCracken and Clarke, 1971; McGavran and Smith, 1965; Stackpole, 1969), varying considerably in amount under the viral envelope. Only the latter virions have been found in the extracellular spaces (Heine and Dalton, 1972; Heine and Hinze, 1972; Heine *et al.*, 1972). It was suggested that the two morphologically distinct viral entities may exhibit different biological activities (Ablashi *et al.*, 1972; Heine and Dalton, 1972).

Another form of viral maturation and egress has been observed in cells infected with bacilliform plant viruses (Chen and Shikata, 1971; Chiu *et al.*, 1970; McLeod, 1968; Wolanski and Chambers, 1971). Here also an extension of the inner nuclear membrane constitutes the outer viral envelope (Fig. 59); but, unlike the illustration referred to in the preceding paragraph for herpes-type viruses, morphologically defined immature forms are not recognizable in the adjacent nucleoplasm. Rather, small portions of the viroplasm which replaces large areas of the nucleoplasm (Fig. 30) assemble in the cores of the newly formed virions during the budding process. After release from the nuclear membrane, mature virions accumulate in large vacuoles limited by the inner and outer nuclear membrane and are transported from there into the extracellular space.

D. *Cytoplasmic Changes*

Alterations of the cytoplasm are usually minute during early stages of viral infection, but may be intensive during late phases. Two alterations are especially spectacular. One is expressed by an extensive vacuolization of the cytoplasm, i.e., in SV40-infected cells (Gaylord and Hsiung, 1961; Granboulan *et al.*, 1963) and less impressively, in cells containing the

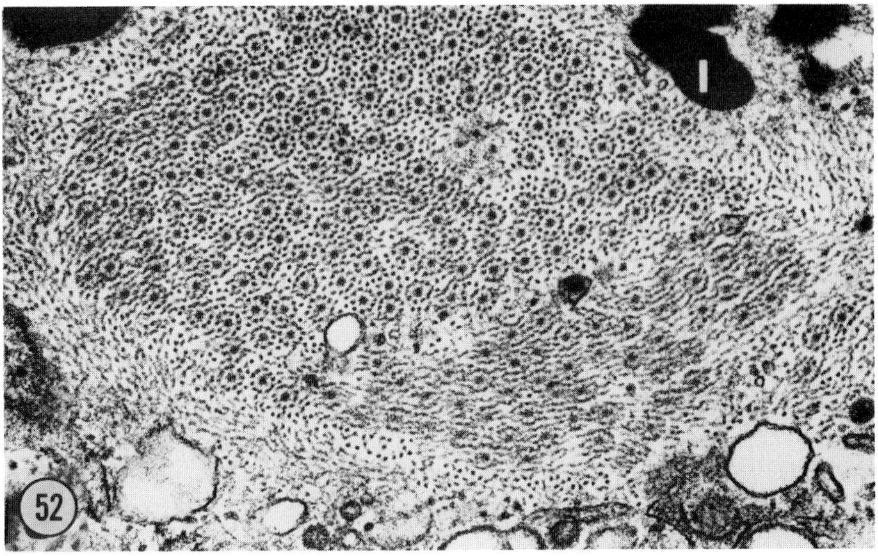

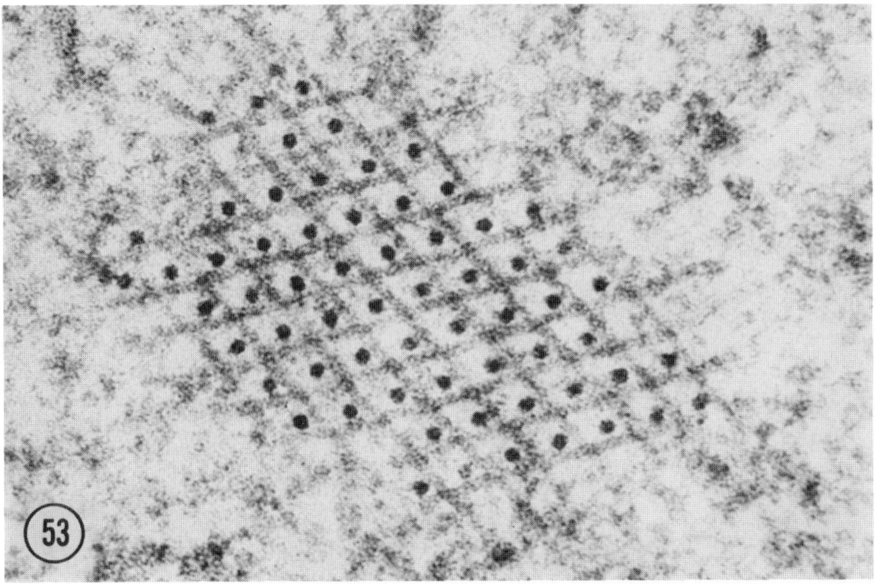

Fig. 52 Crystalloid inclusion in the cytoplasm of a dog kidney cell four days after infection with canine hepatitis virus. Cylindrical bodies are of uniform size, composed of a circular envelope of numerous fibers and a dense axial body; l-lipid body. × 45,000. (From Matsui and Bernhard, 1967.)

Fig. 53 Mouse cell infected with cytomegalovirus. Crystalloid structure in the nucleus. Probably, these crystalloids develop from structures as illustrated in Fig. 52. × 100,000. (Courtesy of Dr. A. F. Howatson, unpublished observation.)

virus of human demyelinating disease (ZuRhein, 1969). In both cases, nuclear alterations appear in advance of cytoplasmic modifications.

The other alteration is exemplified by the appearance of numerous electron-dense, cytoplasmic inclusion bodies late during infection in cells containing a cytomegalovirus (Ahmed and Schidlovsky, 1968; Becker *et al.*, 1965; Berezesky *et al.*, 1971; Kazama and Schornstein, 1972; Luse and Smith, 1958; McGavran and Smith, 1965; Nazerian and Witter, 1970). Only recently, attention was called to the presence of two kinds of inclusion bodies, one of lysosomal nature, the other most probably consisting of antigenic material synthesized by the infected cells (Craighead *et al.*, 1972).

Other aspects of cytoplasmic changes due to viral infection are variations in the appearance of the mitochondria. Giant forms have been seen after infection with adenovirus, type 12 (Bernhard and Tournier, 1966; Bernhard *et al.*, 1965), especially dense mitochondria have been seen in adenovirus 2-infected cells (Philips and Raskas, 1972), and "beaded" mitochondria are common in cells containing the genome of the Epstein-Barr virus (Dalton and Manaker, 1967; Epstein *et al.*, 1965). Further devi-

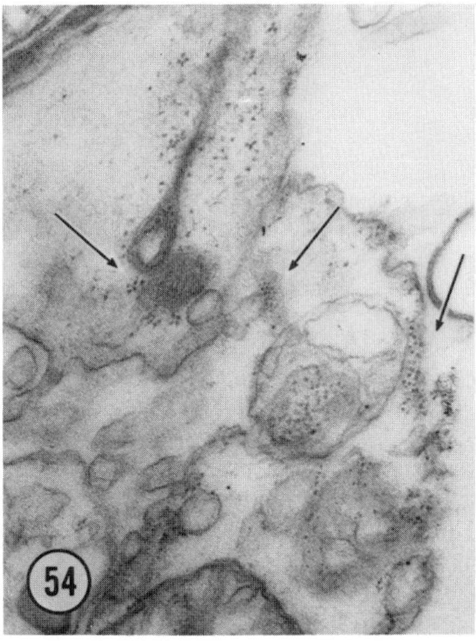

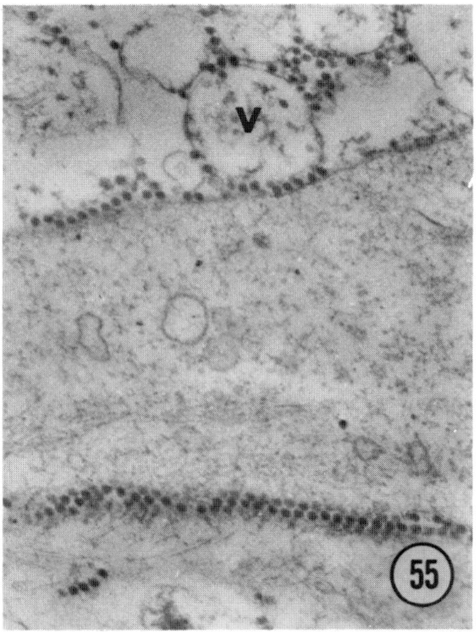

Fig. 54 Monkey kidney cell 48 hours after simultaneous infection with adenovirus and adenosatellite virus. Adenoassociated virus is distributed throughout the cytoplasm (arrows). × 52,800. (From Mayor *et al.*, 1967.)

Fig. 55 African green monkey kidney cell infected with SV40 virus. Virions are attached to the plasma membrane and surround a vacuole (V) in the extracellular space. × 50,400. (Mayor *et al.*, 1962.)

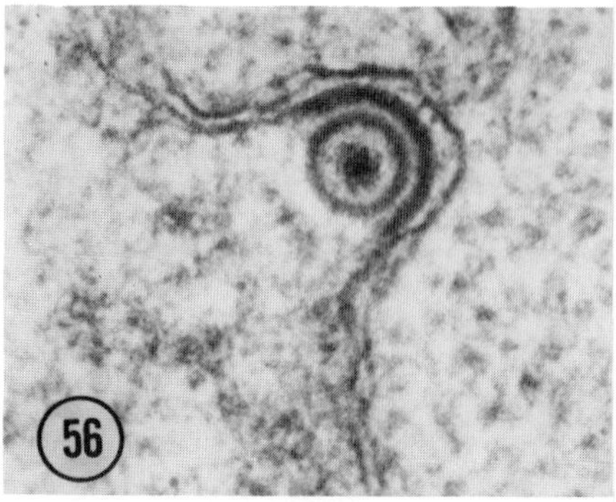

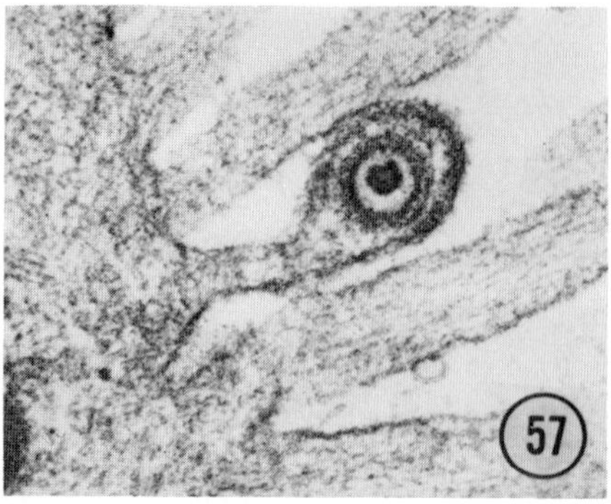

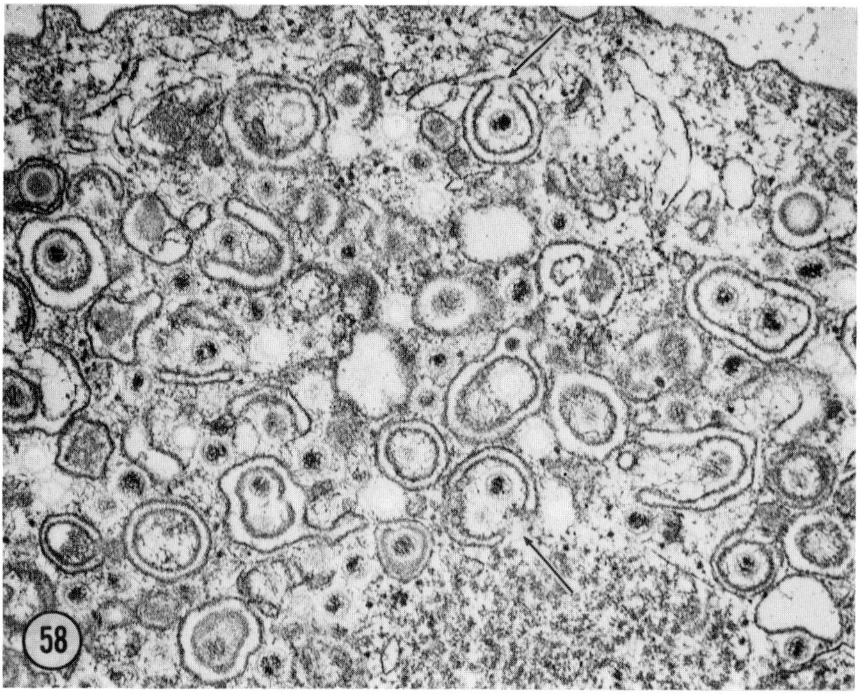

Fig. 56 Herpes-type virus (Epstein-Barr virus) budding at the nuclear membrane. The inner nuclear membrane is thickened adjacent to the nucleocapsid (approx. × 80,000). (Courtesy of G. Schidlovsky, unpublished observation.)

Fig. 57 Herpes-type virus of frog renal adenocarcinoma. Budding at the plasma membrane. The nucleocapsid is surrounded by a fuzzy coat. The plasma membrane surrounding the nucleocapsid is thicker than the one in adjacent areas (approx. × 65,000). (Courtesy of G. Schidlovsky, unpublished observation.)

Fig. 58 Rabbit kidney cell infected with herpesvirus simplex, type I. Numerous virions acquire their outer envelope by budding through cytoplasmic membranes (arrows) (× 40,000). (From Heine and Dalton, 1972.)

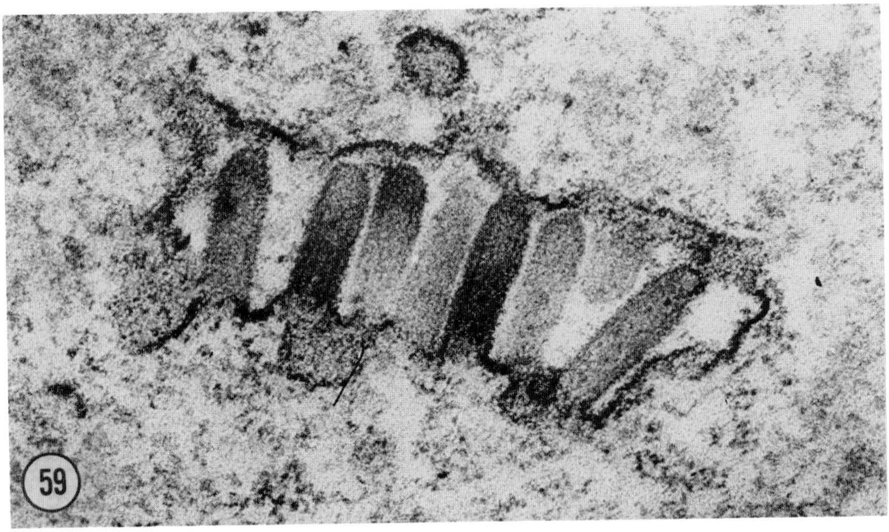

Fig. 59 Cell of *Nicotiana rustica* L. infected with potato yellow dwarf virus. Virus is assembled at the inner nuclear membrane and released into a nuclear vacuole. Although numerous virions are formed at the nuclear membrane the interior of the nucleus is completely void of immature particles (approx. × 100,000). (Courtesy of Dr. R. McLeod, unpublished observation.)

ations from the normal are manifested by an intense growth of nuclear membranes (Bedoya *et al.*, 1968; Bierwolf *et al.*, 1968; Heine and Dalton, 1972; Nii *et al.*, 1968), by the presence of numerous fibers (Lunger, 1964; Stackpole, 1969), large numbers of polysomes (Heine and Dalton, 1972), elaborate tubuloreticular inclusions (Heine and Dalton, 1972; Schaff *et al.*, 1972), and the formation of phagocytic vacuoles often containing numerous virus particles.

IV. Concluding Remarks

It is evident that intranuclear viruses comprise a wide variety of viruses belonging to different families. An effort has been made to elucidate not only differences in the appearance of the virions, but, in addition, to give the reader an introduction to the numerous morphological changes which are often the result of viral interaction with the cell. An immense amount of knowledge is presently available in this area of research. Therefore, the examples discussed can give only a limited impression of the wide variety of the intriguing interplay between viruses and cells. The interested reader is advised to consult the literature for further detailed information. Limitation in space was the cause for omitting one virus group completely: the insect viruses giving rise to nuclear polyhedrosis. These viruses have been discussed in detail in recent articles by Smith (1967, 1971).

Knowledgeable colleagues may excuse not more extensive literature citations but, it appears to be nearly impossible to give proper acknowledgment to all the publications relevant to the material discussed.

ACKNOWLEDGMENTS

This chapter would not have been completed without the help of my colleagues who most generously contributed many excellent illustrations. Herewith the author expresses her sincere gratitude to them. She is also indebted to Dr. A. J. Dalton who made available the preprint of "Ultrastructure of Animal Viruses and Bacteriophages: An Atlas." (A. J. Dalton and F. Haguenau, eds.), to be published by Academic Press. Likewise she acknowledges the excellent help of Mrs. M. Cottler Fox in editing the manuscript.

REFERENCES

Ablashi, D. V., Armstrong, G. R., and Blackham, E. A. (1972). *Amer. J. Vet. Res.* **33,** 1689.

Ahmed, M., and Schidlovsky, G. (1968). *J. Virol.* **2,** 1443.

Al-Lami, F., Ledinko, N., and Toolan, H. (1969). *J. Gen. Virol.* **5,** 485.

Almeida, J. D., Howatson, A. F., and Williams, M. G. (1962). *J. Invest. Dermatol.* **38,** 337.

Atchison, R. W., Castro, B. C., Hammon, W. McD. (1966). *Virology* **29,** 353.

Barbanti-Brodano, G., Swetly, P., and Koprowski, H. (1970). *J. Virol.* **6,** 78.

Beard, J. W. (1968). *In* "Zinsser Microbiology" (D. T. Smith, N. F. Conant, and H. P. Willet, eds.), pp. 883–955. Appleton, New York.

Becker, P., Melnick, J. L., and Mayor, H. D. (1965). *Exp. Mol. Pathol.* **4,** 11.

Bedoya, V., Rabson, A. S., and Grimley, P. M. (1968). *J. Nat. Cancer Inst.* **41,** 635.

Berezesky, J. K., Grimley, P. M., Tyrrell, S. A., and Rabson, A. S. (1971). *Exp. Mol. Pathol.* **14,** 337.

Bernhard, W., and Granboulan, N. (1968). *In* "The Nucleus" (A. J. Dalton and F. Haguenau, eds.), pp. 81–149. Academic Press, New York.

Bernhard, W., and Tournier, P. (1966). *Int. J. Cancer* **1,** 61.

Bernhard, W, Fébvre, H. L., and Cramer, R. (1959) *C. R. Acad. Sci. Paris* **249,** 483.

Bernhard, W., Vasquez, C., and Tournier, P. (1962). *J. Microsc.* **1,** 343.

Bernhard, W., Kasten, F. H., and Chany, C. (1963). *C. R. Acad. Sci. Paris* **257,** 1566.

Bernhard, W., Tournier, P., and Lorans, G. (1965). *C. R. Acad. Sci. Paris* **261,** 2137.

Bierwolf, D., Graffi, A., and Baumbach, L. (1961). *Acta. Morphol. Acad. Sci. Hung.* **X,** 357.

Bierwolf, D., Schramm, T., Bender, E., Cunderlik, V., and Graffi, A. (1968). *Arch. Geschwulstforsch.* **32,** 35.

Boulanger, P. A., Torpier, G., and Biserte, G. (1970). *J. Gen. Virol.* **6,** 329.

Chai, L. S. (1971). *62nd Ann. Meeting Amer. Ass. Cancer Res., Chicago, Illinois* **12,** 18.

Chandra, S., and Toolan, H. W. (1961). *J. Nat. Cancer Inst.* **27,** 1405.

Chardonnet, Y., and Dales, S. (1970a). *Virology* **40,** 462.

Chardonnet, Y., and Dales, S. (1970b). *Virology* **40,** 478.

Chardonnet, Y., and Dales, S. (1972). *Virology* **48,** 342.

Chen, M-J., and Shikata, E. (1971). *Virology* **46,** 786.
Chiu, R. J., Liu, H. Y., McLeod, R., Black, L. M. (1970). *Virology* **40,** 387.
Cook, M. L., and Stevens, J. G. (1968). *J. Virol.* **2,** 1458.
Craighead, J. E., Kanich, R. E., and Almeida, J. D. (1972). *J. Virol.* **4,** 766.
Crawford, L. V., and Crawford, E. M. (1963). *Virology* **21,** 258.
Crawford, L. V., and Follett, E. A. C. (1969). *J. Gen. Virol.* **4,** 37.
Dales, S. (1962). *J. Cell Biol.* **13,** 303.
Dales, S. (1965). *Progr. Med. Virol.* **7,** 1.
Dalton, A. J., and Manaker, R. A. (1967). *In* "Carcinogenesis. A Broad Critique," Univ. of Texas, M. D. Anderson Hosp. and Tumor Inst. at Houston, pp. 59–90. Williams and Wilkins, Baltimore, Maryland.
Dalton, A. J., Kilham, L., and Zeigel, R. F. (1963). *Virology* **20,** 391.
Darlington, R. W., and Moss, L. H., III (1968). *J. Virol.* **2,** 48.
Epstein, M. A., Henle, G., Achong, B. G., and Barr, Y. M. (1965). *J. Exp. Med.* **121,** 761.
Esau, K., and Hoefert, L. L. (1972a). *J. Ultrastruct. Res.* **40,** 556.
Esau, K., and Hoefert, L. L. (1972b). *Virology* **48,** 724.
Farley, C. A., Banfield, W. G., Kasnic, G., Jr., and Foster, W. S. (1972). *Science* **178,** 759.
Fawcett, D. W. (1956). *J. Biophys. Biochem. Cytol.* **2,** 725.
Fazekas de St. Groth, S. (1948). *Nature* (*London*) **162,** 294.
Fenner, R. (1968). "The Biology of Animal Viruses." Academic Press, New York.
Finch, J. T., and Klug, A. (1965). *J. Mol. Biol.* **13,** 1.
Furlong, D., Swift, H., and Roizman, B. (1972). *J. Virol.* **10,** 1071.
Gaylord, W. H., and Hsiung, G. D. (1961). *J. Exp. Med.* **114,** 987.
Ginsberg, H., and Dingle, J. (1965). *In* "Viral and Rickettsial Infections of Man" (F. Horsfall and J. Tamm, eds.), 4th ed., pp. 860–891. Lipincott, Philadelphia, Pennsylvania.
Graffi, A., Schramm, T., Graffi, I., Bierwolf, D., and Bender, E. (1968). *J. Nat. Cancer Inst.* **40,** 867.
Graffi, I., Bierwolf, D., Schramm, T., Bender, E., and Graffi, A. (1972). *Arch. Geschwulstforsch.* **40,** 191.
Granboulan, N., and Tournier, P. (1965). *Ann. Inst. Pasteur* **109,** 837.
Granboulan, N., Tournier, P., Wicker, R., and Bernhard, W. (1963). *J. Cell Biol.* **17,** 423.
Gross, L. (1970). "Oncogenic Viruses," 2nd ed. Pergamon Press, Oxford.
Heine, U., and Dalton, A. J. (1972). *In* "Molecular Studies in Viral Neoplasia," Univ. of Texas at Houston, M. D. Anderson Hospital and Tumor Inst. *Annu. Symp. Fundamental Cancer Res. 25th,* Williams and Wilkins, Baltimore, Maryland (in press).
Heine, U., and Hinze, H. C. (1972). *Cancer Res.* **32,** 1340.
Heine, U., Ablashi, D. V., and Armstrong, G. R. (1971). *Cancer Res.* **31,** 1019.
Heine, U., Kondratick, J., Ablashi, D. V., Armstrong, G. R., and Dalton, A. J. (1971). *Cancer Res.* **31,** 542.
Henry, C. J., Slifkin, M., Merkow, L. P., and Pardo, M. (1971). *Virology* **44,** 215.
Henry, C. J., Merkow, L. P., Pardo, M., and McCabe, C. (1972). *Virology* **49,** 618.
Horne, R. W. (1973). *In* "Ultrastructure of Animal Viruses and Bacteriophages: An Atlas" (A. J. Dalton and F. Haguenau, eds.). Academic Press, New York.
Horne, R. W., and Wildy, P. (1963). *Advan. Virus Res.* **10,** 101.
Horne, R. W., Brenner, S., Waterson, A. P., and Wildy, P. (1959). *J. Mol. Biol.* **1,** 84.

Howatson, A. F. (1971). *In* "McGraw-Hill Yearbook of Science and Technology," pp. 263–265. McGraw-Hill, New York.
Howatson, A. F., and Almeida, J. D. (1960a). *J. Biophys. Biochem. Cytol.* **8,** 828.
Howatson, A. F., and Almeida, J. D. (1960b). *J. Biophys. Biochem. Cytol.* **7,** 753.
Howatson, A. F., Nagai, M., and ZuRhein, G. M. (1965). *Can. Med. Ass. J.* **93,** 379.
Hummeler, K., Tomassini, N., and Sokol, F. (1970). *J. Virol.* **6,** 87.
Jasty, V., and Chang, P. W. (1972). *J. Ultrastruct. Res.* **38,** 433.
Karasaki, S. (1966). *J. Ultrastruct. Res.* **16,** 109.
Kazama, F. Y., and Schornstein, K. L. (1972). *Science* **177,** 696.
Klug, A. (1965). *J. Mol. Biol.* **11,** 424.
Klug, A., and Finch, J. T. (1965). *J. Mol. Biol.* **11,** 403.
Lee, L. F., Roizman, B., Spear, P. G., Kieff, E. D., Burmester, B. R., and Nazerian, K. (1969). *Proc. Nat. Acad. Sci. U.S.* **64,** 952.
Lunger, P. D. (1964). *Virology* **24,** 138.
Luse, S. A., and Smith, M. G. (1958). *J. Exp. Med.* **107,** 623.
Maramorosch, K., and Kurstak, E. eds. (1971). "Comparative Virology." Academic Press, New York.
Martinez-Palomo, A. (1968). *Pathol. Microbiol.* **31,** 147.
Martinez-Palomo, A., and Granboulan, N. (1967). *J. Virol.* **1,** 1010.
Martinez-Palomo, A., LeBuis, J., and Bernhard, W. (1967). *J. Virol.* **1,** 817.
Marusyk, R., Norrby, E., and Marusyk, H. (1972). *J. Gen. Virol.* **14,** 261.
Matsui, K., and Bernhard, W. (1967). *Ann. Inst. Pasteur* **112,** 773.
Mayor, H. D. (1973). *In* "Ultrastructure of Animal Viruses and Bacteriophages: An Atlas" (A. J. Dalton and F. Haguenau, eds.). Academic Press, New York.
Mayor, H. D., and Jamison, R. M. (1966). *Progr. Med. Virol.* **8,** 183.
Mayor, H. D., and Jordan, L. E. (1966). *Exp. Mol. Pathol.* **5,** 580.
Mayor, H. D., and Melnick, J. L. (1966). *Nature (London)* **210,** 331.
Mayor, H. D., Stinebaugh, S. E., Jamison, R. M., Jordan, L. E., and Melnick, J. L. (1962). *Exp. Mol. Pathol.* **1,** 397.
Mayor, H. D., Jamison, R. M., and Jordan, L. E. (1963). *Virology* **19,** 359.
Mayor, H. D., Ito, M., Jordan, L. E., and Melnick, J. L. (1967). *J. Nat. Cancer Inst.* **38,** 805.
McCracken, R. M., and Clarke, J. K. (1971). *Arch. Ges. Virusforsch.* **34,** 189.
McGavran, M. H., and Smith, M. G. (1965). *Exp. Mol. Pathol.* **4,** 1.
McLeod, R. (1968). *Virology* **34,** 771.
McLeod, R., Black, L.M., and Moyer, F. H. (1966). *Virology* **29,** 540.
Melnick, J. L. (1962). *Science* **135,** 1128.
Miyamoto, K., and Morgan, C. (1971). *J. Virol.* **8,** 910.
Morgan, C., Godman, G. C., Rose, H. M., Howe, C., and Huang, J. S. (1957). *J. Biophys. Biochem. Cytol.* **3,** 505.
Morgan, C., Rosenkranz, H. S., and Meduis, B. (1969). *J. Virol.* **4,** 777.
Nakai, T., Shand, F. L., and Howatson, A. F. (1968). *Proc. Electron Microsc. Soc. Amer.* (C. J. Arceneaux, ed.), 27th Annu. Meeting, pp. 206–207. Claitor's Publ. Baton Rouge, Louisana.
Nakai, T., Shand, F. L., and Howatson, A. F. (1969). *Virology* **38,** 50.
Nazerian, K. (1970). *In* "Microscopie Electronique" (P. Favard, ed.), Vol. 3, p. 939. Soc. Fr. Microsc. Electron., Paris, France.
Nazerian, K., and Witter, R. L. (1970). *J. Virol.* **5,** 388.
Nazerian, K., Lee, L. F., Witter, R. L., and Burmester, B. R. (1971). *Virology* **43,** 442.
Nii, S., Morgan, C., and Rose, H. M. (1968). *J. Virol.* **2,** 517.

Norrby, E. (1969). *J. Gen. Virol.* **5,** 221.
Norrby, E. (1971). *In* "Comparative Virology" (K. Maramorosch and E. Kurstak, eds.), pp. 105–134. Academic Press, New York.
Philipson, L. (1963). *Progr. Med. Virol.* **5,** 43.
Philipson, L. (1967). *J. Virol.* **1,** 868.
Phillips, D. M., and Raskas, H. J. (1972). *Virology* **48,** 156.
Pope, J. H., and Rowe, W. P. (1964). *J. Exp. Med.* **120,** 577.
Rivers, T. M., and Horsfall, F. L., Jr. (eds.) (1959). "Viral and Rickettsial Infections of Man," 3rd ed. Lipincott, Philadelphia, Pennsylvania.
Roizman, B., and Spear, P. G. (1973). *In* "Ultrastructure of Animal Viruses and Bacteriophages: An Atlas" (A. J. Dalton and F. Haguenau, eds.). Academic Press, New York.
Rouse, H. C., and Schlesinger, R. W. (1972). *Virology* **48,** 463.
Schaff, Z., Heine, U., and Dalton, A. J. (1972). *Cancer Res.* **32,** 2696.
Schlesinger, R. W. (1969). *Advan. Virus Res.* **14,** 1.
Scott, T. F. McN. (1959). *In* "Viral and Rickettsial Infections of Man" (T. M. Rivers and F. L. Horsfall, Jr., eds.), pp. 762–765. Lippincott, Philadelphia, Pennsylvania.
Shikata, E., and Maramorosch, K. (1966). *Virology* **30,** 439.
Siegel, G., Hallauer, C., Novak, A., and Kronauer, G. (1971). *Arch. Ges. Virusforsch.* **35,** 91.
Siminoff, P., and Menefee, M. G. (1966). *Exp. Cell Res.* **44,** 241.
Sirtori, C., and Bosisio-Bestetti, M. (1967). *Cancer Res.* **27,** 367.
Smith, D. T., Conant, N. F., and Willet, H. P. (eds.) (1968). "Zinsser Microbiology," 14th ed. Appleton, New York.
Smith, K. M. (1967). "Insect Virology." Academic Press, New York.
Smith, K. M. (1971). *In* "Comparative Virology" (K. Maramorosch and E. Kurstak, eds.), pp. 479–505. Academic Press, New York.
Stackpole, C. W. (1969). *J. Virol.* **4,** 75.
Stackpole, C. W., and Mizell, M. (1968). *Virology* **36,** 63.
Stone, R. S., Shope, R. E., and Moore, D. H. (1959). *J. Exp. Med.* **110,** 543.
Sussenbach, J. S. (1967). *Virology* **33,** 567.
Torikai, K., Ito, M., and Mayor, H. D. (1970). *J. Virol.* **6,** 363.
Torpier, G., and Petitprez, A. (1968). *J. Microsc. Paris* **7,** 411.
Tournier, P., Granboulan, N., and Bernhard, W. (1961). *C. R. Acad. Sci. Paris* **253,** 2283.
Valentine, R. C., and Perreira, H. G. (1965). *J. Mol. Biol.* **13,** 13.
Wildy, P. (1971). *In* "Monographs in Virology," Vol. 5. Karger, New York.
Wildy, P., Russell, W. C., and Horne, R. W. (1960a). *Virology* **12,** 204.
Wildy, P., Stoker, M. G. P., McPherson, J., and Horne, R. W. (1960b). *Virology* **11,** 444.
Wilner, B. I. (1964). "A Classification of the Major Groups of Human and Other Animal Viruses," 4th ed. Burgess Publ., Minneapolis, Minn.
Wolanski, B. S., and Chambers, T. C. (1971). *Virology* **44,** 582.
Zeigel, R. F., and Clark, H. F. (1972). *Infect. Immun.* **5,** 570.
ZuRhein, G. M. (1969). *Progr. Med. Virol.* **II,** 185.
ZuRhein, G. M., and Chou, S. M. (1965). *Science* **148,** 1477.

Author Index

Numbers in italics refer to the pages on which the complete references are listed.

B

C

E

H

M

S

Y

Z

Subject Index

D

I

O

P

A 4
B 5
C 6
D 7
E 8
F 9
G 0
H 1
I 2
J 3